Thick Coal Seam Underground Mining

Jiachen Wang · Weijie Wei

Thick Coal Seam Underground Mining

Science Press
Beijing

Springer

Jiachen Wang
School of Energy and Mining Engineering
China University of Mining
and Technology-Beijing
Beijing, China

Weijie Wei
School of Energy and Mining Engineering
China University of Mining
and Technology-Beijing
Beijing, China

ISBN 978-981-96-3873-4 ISBN 978-981-96-3874-1 (eBook)
https://doi.org/10.1007/978-981-96-3874-1

Jointly published with Science Press
The print edition is not for sale in China mainland. Customers from China mainland please order the print book from: Science Press.

This Springer imprint is published by the registered company Springer Nature Singapore Pte Ltd.
The registered company address is: 152 Beach Road, #21-01/04 Gateway East, Singapore 189721, Singapore

Preface

Over the past 40 years, China has made significant progress in thick coal seam underground mining technology (with thickness greater than 3.5 m), forming a technical system centered around top coal caving and large mining height. In terms of mining theory and technology, coal machinery manufacturing, strata control technology, disaster prevention and mitigation technology, technical and economic indicators, as well as levels of intelligence, these technologies have reached a world-leading level. A comprehensive summary of China's achievements in thick coal seam mining technology is of great significance for promoting the high-quality development of thick coal seam mining and enhancing the level of new productive forces.

Thick coal seams are the main coal seams for high-yield and high-efficient mining in China, accounting for over 50% of both production and reserves. Over the past 20 years, the production from mines and panels extracting thick coal seams has rapidly increased, with single well coal production reaching 28 million tons per year and panel production reaching 16 million tons per year. The thickness of coal seams mined in full seam mining has also gradually increased; for instance, the thickness of coal seams mined in a single operation in a longwall top coal caving panel has reached 20 m, while large mining height mining has achieved a thickness of 10 m. The rapid increase in mine and panel production and the increase in coal seam thickness in full seam mining are primarily attributed to innovations in mining concepts and theories, advancements in mining technology and equipment, and strong market demand for coal products. Social demand is the primary driving force behind technological innovation. Although coal's share in China's primary energy consumption has slightly decreased in recent years due to global environmental concerns and domestic "dual carbon" goals and the impact of new energy sources (55.3% in 2023), the total demand for coal has not only remained stable but has increased. In 2023, domestic coal production reached 4.71 billion tons, with coal imports at 474 million tons and exports at 4.47 million tons, leading to a domestic coal consumption of 5.18 billion tons, a historical high. This clearly demonstrates the important position of coal in China's energy landscape and its irreplaceable role in supporting national economic development. In the future, as breakthroughs in clean coal utilization technology and increased coal consumption for coal chemical industries occur, China's demand for coal is expected to remain high. The high-yield and high-efficient mining technology for thick coal seams is fundamental to support China's coal production. Without the leapfrog development of thick coal seam mining technology, it would be challenging to ensure today's coal supply in China and achieve safe, green, and intelligent development in the coal industry.

This book is authored based on the achievements of the research team over nearly 30 years in thick coal seam mining technology, as well as some typical cases of thick coal seam mining, with the aim of introducing China's technology and achievements in thick coal seam mining to readers and peers both domestically and internationally, as well as promoting advanced theories and technology in this field. The book is divided into eight chapters, with Chaps. 1–3, and 6 written by Jiachen Wang, and Chaps. 4, 5, 7, and 8 written by Weijie Wei. During the writing process, special thanks to Prof. Yang Li for his translation of the draft of Chaps. 1–4, and support and assistance were also received from Prof. Shengli Yang, Prof. Weidong Pan, Prof. Xianlei Xu, Assoc. Prof. Jinwang Zhang, Assoc. Prof. Zhaohui Wang, Lecturer Lianghui Li, as well as Dr. Yuesong Tang, Dr. Liu Yang, Dr. Tao Li, Dr. Meng Li, Dr. Hao Yue, Dr. Shanxi Wu, Dr. Xin Zhang, Dr. Jialong Li, Dr. Yunxi Liu, Dr. Tianrui Fan, Dr. Zhifeng Wang, Dr. Zheng Li, Dr. Boyuan Cheng, Dr. Bochao An, Dr. Haoyu Hu, Dr. Chaoshan Zhang, Dr. Shixiong Song, Dr. Yongsheng Chen, Dr. Ruihao Zhai, Master's students Jiqing He, Shenyi Zhang, Huashun Xie, Han Zhang, Minghui Yu, Yuhang Song, Xinlin Li, Zewei Wang, Yubing Wang, Yaochen Wang, Xiujuan Li, Yanjun Song, and Dongxin Hou, among others. The research process for this book was supported by the National Key R&D Program Project (2023YFC2907501), the National Natural Science Foundation Key Project (51934008), the National Natural Science Foundation General Project (52374106) and the Young Scientists Fund Project (52204163), the China Postdoctoral Science Foundation (Grant no. 2024T171006), and the Fundamental Research Funds for the Central Universities (2024ZKPYNY04). I would like to express our gratitude for this support.

Beijing, China
May 2024

Jiachen Wang

Brief Introduction

This book systematically introduces the main technology and latest advancements of China's thick coal seam underground mining technology. It covers the mine development methods, the top coal caving technology and drawing law, the measurement of top coal recovery and technology to improve the ratio, the large mining height technology and coal wall stability control, large cross-section roadway support technology, slicing mining roadway layout and mining technology, intelligent mining technology and advancements, theories and techniques for controlling surrounding rock in thick coal seam, as well as some typical cases of top coal caving and large mining height.

This book can serve as a teaching reference for graduate students in mining engineering and senior undergraduate students at higher education institutions. It is also suitable for researchers, engineering technicians, designers, and relevant technology management personnel engaged in coal mining.

Contents

Chapter 1
Thick Coal Seam Mining in China

1.1 Overview of Coal Mining

1.1.1 Characteristics of Coal Resources

China is the world's largest energy consumer and coal producer, and its annual energy consumption accounts for about 23% of the world's total energy consumption, reaching 5.41 billion tons of standard coal in 2022. China's coal production was 4.56 billion tons in 2022, accounting for 51.8% of the world's coal production, of which 82% came from underground mining, and 290 million tons of coal was imported. Figure 1.1 shows the changes of China's annual coal production and consumption from 2001 to 2022. In general, the main users of coal consumption in China are the power industry, accounting for 52%. Secondary users are the steel industry, chemical industry, civil and other industries, accounting for 17%, 7% and 11% respectively. Among them, the steel industry primarily uses coking coal, while other industries mainly use thermal coal. With the development of renewable energy industries such as wind and solar energy, the proportion of coal used for power generation is expected to decrease, however, the demand for coal in the chemical industry may increase.

Currently, coal occupies a dominant position in China's energy structure, a situation unlikely to change in the short term. In 2022, coal consumption accounted for 56.2% of primary energy consumption. Although this proportion will gradually decline, it is expected to remain above 40% for a considerable period. Due to China's huge economic aggregate, the GDP in 2023 was 17.89 trillion dollars, and maintain an annual growth ratio of more than 5%, which will continue to increase the demand for energy. Despite China's goals of carbon peaking by 2030 and carbon neutrality by 2060, coal consumption is expected to remain around 4–5 billion tons annually over the next decade. In 2023, coal production reached 4.71 billion tons, coal imports were 474 million tons, and the total consumption of coal exceeded 5 billion tons, reaching the highest level in history.

J. Wang and W. Wei, *Thick Coal Seam Underground Mining*,
https://doi.org/10.1007/978-981-96-3874-1_1

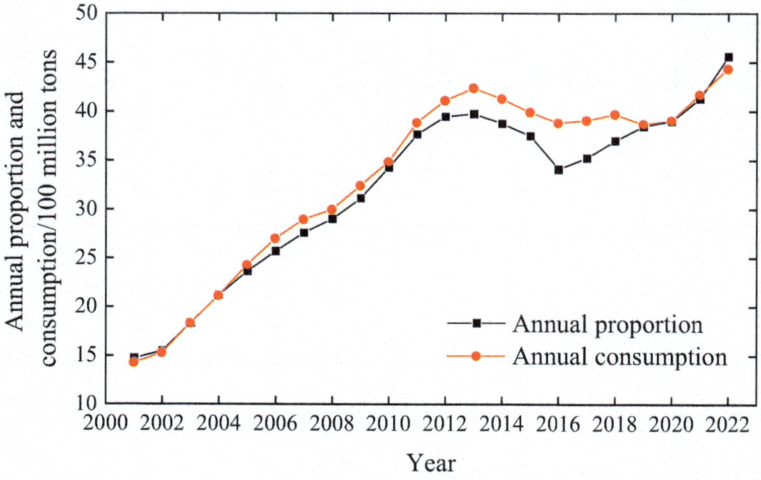

Fig. 1.1 Changes of China's annual coal production and consumption in 2001–2022

China's coal resources are abundant but unevenly distributed. Most coal seams are deep, with more than 2000 m being common, making extraction challenging. With recoverable reserves estimated at 58.2 trillion tons [1], China ranks third globally, after the United States and Russia. However, globally, per capita recoverable coal reserves are only half the global average. Despite this, coal remains the most abundant fossil energy resource in China, with relatively stable supply compared to oil and natural gas. Given the sufficient geological reserves, coal will continue to be a fundamental energy source for a long time.

Coal resources are inversely related to economic development and water resources distribution. For example, the northwest regions, though economically less developed, have abundant coal reserves. However, due to lower economic development and water scarcity, the development and utilization of coal resources in these regions are constrained. The constraints in economic development and water resource distribution have led to certain difficulties, increasing the costs of coal conversion and transportation. In recent years, China has implemented the Western Development Strategy and the Belt and Road Initiative, which will drive the development and utilization of coal in the central and western regions. Coal development has also shifted to the west. Figure 1.2 compares coal reserves, coal production, and GDP among major coal-producing provinces in China in 2022.

Due to energy constraints, environmental issues, and carbon reduction targets, the focus will be on efficient, clean, high-tech coal extraction and utilization in the future. This will be the basic approach to coal development and utilization in the next 20 years, supported by advancements in scientific mining technologies and cleaner production methods.

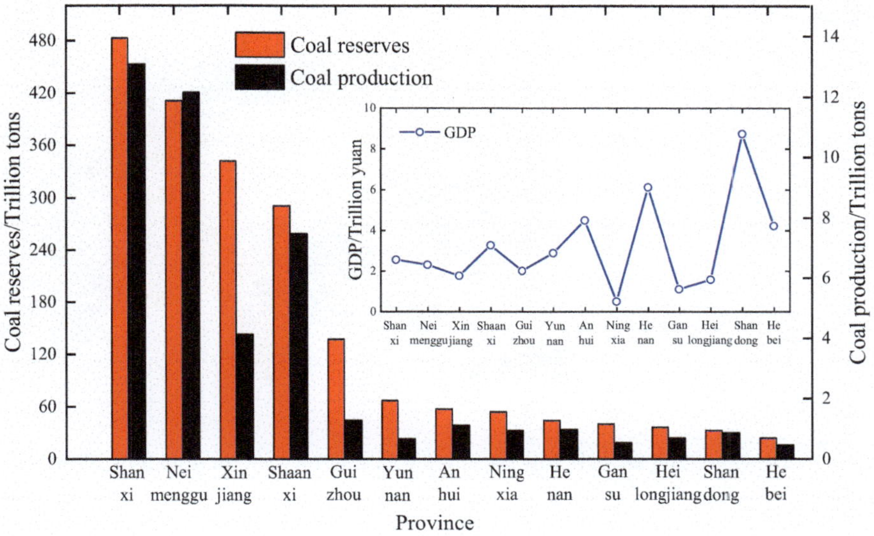

Fig. 1.2 Coal reserves, production, and GDP in major coal-producing provinces

1.1.2 History of Coal Mining

China is one of the earliest countries to discover, extract, and utilize coal in the world. The earliest records of coal can be traced back to the period between 200 and 100 BC in the ancient text "The Classic of Mountains and Rivers", which describes a place 300 miles southwest with Women Mountain whose south side is covered with red copper while north side is stone coal [2]. The term "stone coal" in the text refers to coal, mainly used at the time for dyeing and decoration.

With the demand for traditional technology, China ushered in its first peak in coal usage. Since the melting point of iron (1535 °C) is much higher than that of copper (1083 °C), coal was used extensively in smelting processes in the Han Dynasty (202 BC–220 AD). The site of an ancient smelting furnace in Zhengzhou, Henan Province, is currently the largest in the world, with the highest and longest usage of coal. The site measures 400 m in length and 300 m in width, covering an area of 120,000 m^2, making it the largest coal-fired smelting site of its era, as shown in Fig. 1.3.

During the Song Dynasty (960–1127 AD), China experienced its second peak in coal usage. At this time, coal had officially entered large-scale mining operations and was utilized as an industrial energy source for activities such as iron smelting and porcelain firing. Coal mining industries had become quite advanced in regions such as Shanxi, Hebei, Shaanxi, and Henan. In 1959, the discovery of a Song Dynasty coal mine site in Zhongxin Coal Mine, Hebi City, Henan Province, marked the earliest and relatively well-preserved coal pit ruins found in China. These pits were circular shafts with a diameter of 2.5 m and a depth of 46 m, situated centrally within their respective fields. They were directly excavated into the coal seams, with four coal

Fig. 1.3 Ancient coal-fired smelting site in Zhengzhou, Henan Province

mining drifts and ten panels below the shaft, totaling over 500 m in length, alongside a drainage well.

Italian traveler Marco Polo (1254–1324 AD) documented the use of coal in Yuan Dynasty of China in his book "The Travels of Marco Polo", stating, "Throughout the whole country of China, there is found a certain black stone (coal), which they dig out of the mountains and use it for fuel. It is found in veins, like in mines underground, and when lighted, it burns like charcoal and makes a great heat. And this coal never burns out, and is exceedingly cheap." [3].

The earliest scientific description of coal mining techniques in China can be found in "Tiangong Kaiwu" (The Exploitation of the Works of Nature) by Song Yingxing, written in 1637 AD. It mentions, "Those who have experienced coal extraction, after a long time, can discern its presence from the color of the soil, and then begin excavation. When digging about fifteen meters deep, coal is obtained. At first sight of coal, the toxic fumes are severe. Some use large bamboo poles, sharpened at the end, inserted into the coal. The poisonous smoke comes out through the bamboo, and people collect coal from below. Some descend into a single well, where coal is widely spread vertically and horizontally, and extract it as needed. Support beams are placed above to prevent collapse." [4] (Fig. 1.4).

The Keelung Coal Mine in Taiwan, which was built and put into operation in 1878, utilized a 40-horsepower steam boiler, winches, and drainage machines, marking the first use of mechanization in coal mining in China. The main shaft had a depth of 89.92 m and a diameter of 3.81 m, with a coal seam thickness of 1.06 m and a daily coal production of 200 tons. Coal mining in Taiwan began before the year 1600, focusing on thin coal seams. The last coal mine was closed in 2000, as shown in Fig. 1.5. The coal production peaked in 1964 at 5.027 million tons, the highest in history [5].

In June 1878, China established the Kaiping Mining Bureau in Tangshan, Hebei Province. The Tangshan Coal Mine was built and put into operation in 1881, with a shaft depth of over 200 m. By 1898, the mine produced 730,000 tons of coal and employed more than 3000 Chinese and foreign workers, as shown in Fig. 1.6. Mechanized operations were used for hoisting, ventilation, and drainage. A coal

Fig. 1.4 Coal mining schematic

Fig. 1.5 Closed coal mine in Taiwan

transport railway was laid from Tangshan to Xugezhuang, covering a total length of 9.3 km. Herbert Clark Hoover, the 31st President of the United States, worked at the Tangshan Coal Mine between 1899 and 1900. The Tangshan Coal Mine is still in operation today, with an annual coal production of 2.4 million tons. The Keelung Coal Mine and the Tangshan Coal Mine were the first in China to adopt mechanized operations, with machinery purchased from the United Kingdom.

In 1949, China's coal production was only 32 million tons, primarily using the room-and-pillar mining method. Since 1953, China has vigorously promoted the

Fig. 1.6 Tangshan mine of Kailuan Group

longwall mining method, which accounted for 92% of the total coal production by 1960. Currently, almost all underground coal mines in China use the longwall mining method. After 1960, China began developing mechanized coal mining technology. In 1964, the first mechanized coal mining panel equipped with a single-drum shearer was introduced, although individual props were still used for support. After 1970, China started to widely promote mechanized coal mining technology and developed comprehensive mechanized coal mining technology. In 1970, the first fully mechanized coal mining panel was equipped at the Meiyukou Mine of the Datong Mining Bureau. In 1974 and 1977, China imported 43 sets and 100 sets of comprehensive mechanized coal mining equipment from the UK, Germany, and Poland, respectively, while also producing 500 sets domestically [6], significantly advancing comprehensive mechanized coal mining technology in China.

In April 1984, China's first longwall top coal caving (LTCC) panel underwent industrial testing at the Puhe Coal Mine of the Shenyang Mining Bureau, marking the rapid development of top coal caving technology for thick coal seams in China. The large mining height technology originated from the introduction of German large mining height supports such as the G320-20/37 and G320-23/45 models in 1978, successfully trialing 3.3–4.3-m-thick coal seams, with an average monthly coal production of 70,819 tons at that time [7]. Currently, the height of large mining height supports can reach up to 10 m, the highest in the world, as shown in Fig. 1.7.

Based on different coal seam thicknesses and dip angles, coal mining in China can currently be divided into the following techniques:

(1) Thick Seam Mining: For thick seams (coal seam thickness being more than 3.5 m), the primary techniques are top coal caving and large mining height extraction. The slicing mining method, which was widely used before 1990, is now almost obsolete.
(2) Medium-Thick Seam Mining: Medium-thick seams are defined as coal seams with a thickness of 1.3–3.5 m. These seams are generally mined using fully

Fig. 1.7 Mining panel with a 10-m cutting height

mechanized methods that extract the entire seam thickness in one go. This type of coal seam mining is relatively straightforward.

(3) Thin Seam Mining: Thin seams are those with a coal seam thickness of less than 1.3 m. Apart from a few panels that use coal plows, thin seam mining predominantly employs low-profile, small-drum shearers.

(4) Longwall Mining for Large Dip Angle Seam: Longwall mining along the seam strike is the basic method of coal mining in China. However, for seams with a large dip angle (greater than 35°), the equipment on the mining panel is prone to toppling and sliding, making the mining process challenging. Currently, China has developed longwall mining technology with large dip angle seam through innovative panel layout, equipment development, and mining techniques. Comprehensive mechanized mining has been achieved for seams with dip angle up to 70°, and LTCC has been successfully implemented for seams with dip angle up to 60°.

(5) Horizontal Sublevel Top Coal Caving (STLL) for Steep Thick Seams: When the dip angle exceeds 50° and the seam thickness is greater than 15 m, horizontal sublevel top coal caving is usually employed. The typical sublevel height ranges from 20 to 25 m, and research is now underway to develop mining techniques for sublevel heights of 35 m.

(6) Intelligent Mining: In the past decade, significant progress has been made in intelligent mining based on electro-hydraulic control systems. Currently, there are 1400 intelligent panels in China, with 730 mines having intelligent panel, accounting for 59.5% of the total production capacity. The main control projects for intelligent panels include intelligent coal cutting by shearers, automatic support movement and height adjustment, and automatic advancement and alignment of scraper conveyors. Many panels still require remote control and manual intervention either in the roadway or on the surface, especially when geological conditions change, indicating that it is not yet fully intelligent. Nonetheless, this represents a significant step forward from traditional manual-controlled mechanized mining towards intelligent mining. In recent years, substantial investment and interest from coal enterprises in intelligent mining have greatly promoted its development.

1.1.3 Major Coal Producing Enterprises

Coal is a typical resource-based industry with the dual attributes of providing both energy and industrial raw materials. It is the most important foundational energy source supporting the rapid development of the national economy and serves as a crucial industrial raw material for industries such as steel and chemicals. According to the "National Mineral Resources Planning 2016–2020" released by the state in 2016, 14 key coal bases were identified for focused development: Shendong, North Shanxi, Central Shanxi, East Shanxi, East Mongolia (Northeast), Yunnan-Guizhou, Henan, West Shandong, Huainan-Huaibei, Huanglong, Central Hebei, Ningdong, North Shaanxi, and Xinjiang. These coal bases are essential pillars for securing China's coal supply.

Statistics show that by the end of 2022, there were 4618 coal enterprises above designated size nationwide, with a total of 4342 coal mines. Of these, 3985 were underground mines and 357 were open-pit mines, with underground mines accounting for over 90% of the total. There were more than 1100 large coal mines (with a production capacity of over 1.2 million tons per year), accounting for about 85% of the total production, and more than 3700 mines with a capacity of over 300,000 tons per year. In September 2023, during the China Top 500 Enterprises Summit Forum, the China Enterprise Confederation and the China Entrepreneur Association jointly released the "Top 500 Chinese Enterprises" list. Among them, 22 coal enterprises were listed.

As shown in Fig. 1.8, according to statistics from the China National Coal Association, the top 10 coal enterprises produced a total of 2.4 billion tons of raw coal in 2023, an increase of 84.12 million tons year-on-year, accounting for 51.3% of the raw coal production of enterprises above designated size, including China Energy Investment, Jinneng Holding Group Co., Ltd, Shandong Energy Group Co., Ltd, China National Coal Group Corporation, Shaanxi Coal and Chemical Industry Group Co., Ltd., Shanxi Coking Coal Group Co., Ltd., China Huaneng Group Co., Ltd., Lu'an Chemical Group Co., Ltd., State Power Investment Co., Ltd., and Huaihe Energy (Group) Co., Ltd.. They are the major coal producers in China.

1.2 Mine Development Methods

China's coal seam conditions are complex, leading to a variety of roadway layouts for mine development. Common methods include vertical shaft development, inclined shaft development, horizontal adit development, and combined development. In recent years, with the use of trackless rubber-tired vehicles and the increase in mine production capacity, inclined shaft development has become popular. This method is particularly common in western regions such as Inner Mongolia and Shaanxi, where seam conditions are relatively simple and seam dip angle are small. However, early

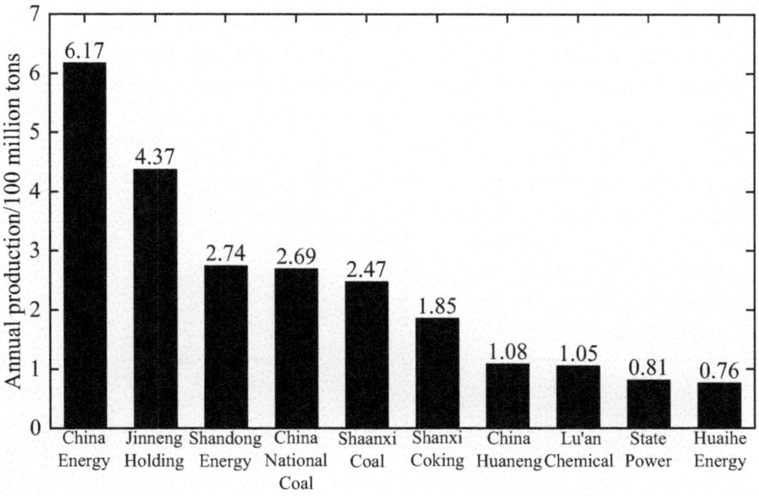

Fig. 1.8 Major coal producers in China in 2023

designed mines and those in eastern regions such as Shandong, Anhui, Hebei, and Henan, where coal seams are buried deeper, generally use vertical shaft development.

1.2.1 Vertical Shaft Development

Vertical shaft development involves using vertical shafts that directly connect to the surface as the main and secondary shafts for mine development. This method is typically employed in mining areas with flat terrain, deep coal seam burial, and thick overburden layers. Vertical shaft development is highly adaptable and is not constrained by the seam dip angle, thickness, depth, gas content, or hydrogeological conditions. It has significant advantages for mining deeply buried coal seams [8]. A typical vertical shaft development layout is shown in Fig. 1.9.

1.2.2 Inclined Shaft Development

Inclined shaft development utilizes inclined shafts that connect directly to the surface as the main and secondary shafts for mine development. This method is widely used in western mining areas with simple conditions. The construction of inclined shafts is relatively straightforward and quick. Inclined shafts equipped with belt conveyors can achieve continuous coal transportation to the surface, offering high transport capacity and efficiency. Figure 1.10 illustrates a schematic of inclined shaft development.

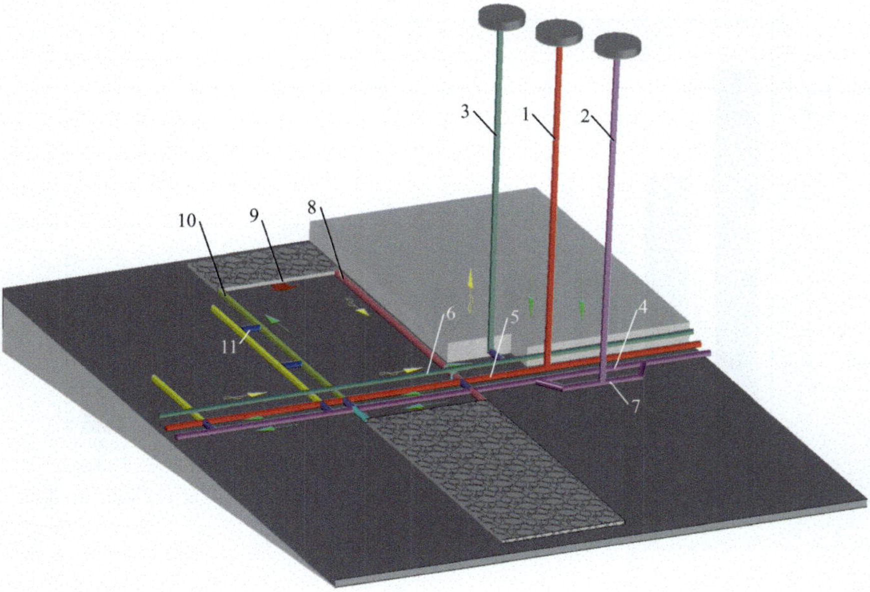

Fig. 1.9 Schematic diagram of vertical shaft development (1-main shaft; 2-secondary shaft; 3-return air shaft; 4-auxiliary transportation roadway; 5-transportation roadway; 6-retum air roadway; 7-shat station; 8-section return air inclined roadway; 9-panel; 10-section transportation inclined roadway; 11-crossheading)

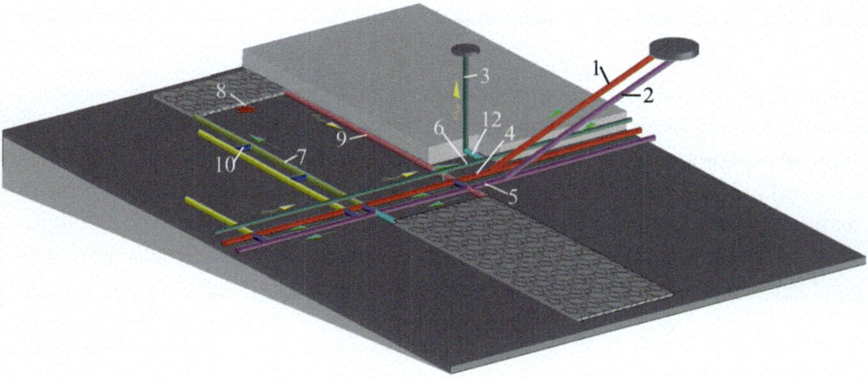

Fig. 1.10 Schematic diagram of inclined shaft development (1-main inclined shaft: 2-secondary inclined shaft; 3-return air shaft; 4-transportation roadway; 5-auxiliary transportation roadway; 6-retum air roadway; 7-section transportation inclined roadway; 8-panel; 9-section return air inclined roadway; 10-crossheading)

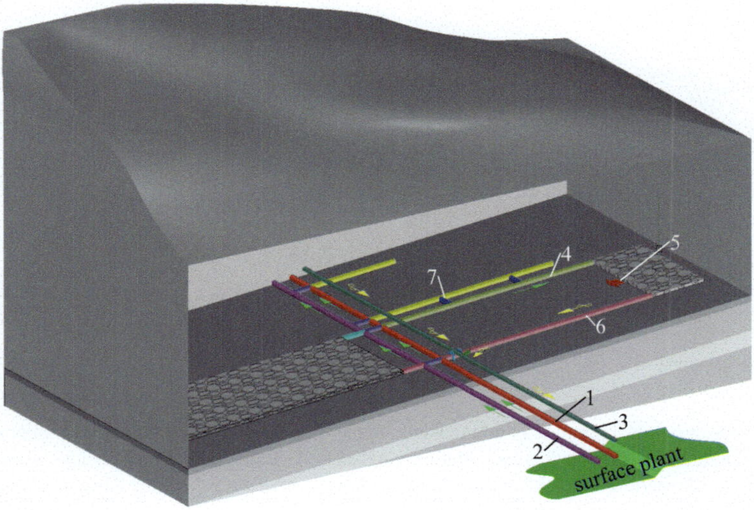

Fig. 1.11 Schematic diagram of adit development (1-main adit; 2- secondary adit; 3-return air adit; 4-section transportation inclined roadway; 5-panel; 6-section return air inclined roadway; 7-crossheading)

1.2.3 Horizontal Adit Development

Horizontal adit development utilizes horizontal shafts that connect directly to the surface as the main and secondary shafts for mine development. This is the simplest and most advantageous development method, characterized by simple construction techniques and equipment, fast construction speed, short mine development periods, fewer transportation stages and equipment, a simple system, and low costs. It is widely used in mining areas with mountainous and hilly terrain. A typical schematic of horizontal adit development is shown in Fig. 1.11.

1.2.4 Combined Development

Combined development involves using any two or more of the following development methods: vertical shaft, inclined shaft, and horizontal adit. The essence of combined development is to combine different shaft forms based on the specific conditions of coal seam extraction in a mine, optimizing their advantages. Depending on the combination of shafts, combined development can be divided into three basic types: inclined shaft-vertical shaft, horizontal adit-inclined shaft, and horizontal adit-vertical shaft. It can also include various arrangements such as single-level, multi-level, uphill, and downhill. A typical schematic of combined development is shown in Fig. 1.12.

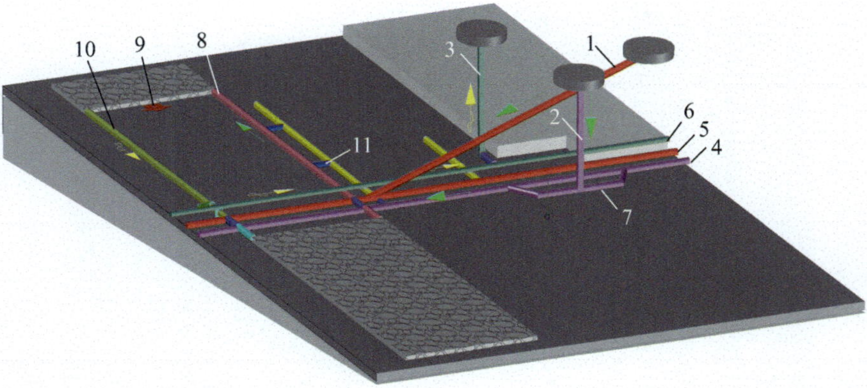

Fig. 1.12 Schematic diagram of combined development (1-main inclined shaft; 2-secondary shaft; 3-return air shaft; 4-auxiliary transportation roadway; 5- transportation roadway; 6-retum air roadway; 7-shat station; 8-section transportation inclined roadway; 9-panel; 10-section return air inclined roadway; 11-crossheading)

1.3 Three Mining Techniques for Thick Coal Seam

Within the current coal reserves and production, the reserves and production of underground thick coal seams (with a thickness greater than 3.5 m) both account for about 50%. Thick coal seams are the main seams for achieving high-yield and high-efficiency mining, with an annual production of thick coal seams approaching 2 billion tons from underground mining. Due to their considerable thickness, various methods can be chosen for their extraction. With the improvement of the coal market and the urgent need for high-yield and high-efficiency mining, top coal caving technology and large mining height technology, which have high production and efficiency, have been rapidly developed and widely applied. However, coal mining is closely related to specific geological conditions and mining conditions, thus the mining of thick coal seams should adopt appropriate mining methods based on coal seam conditions and technical conditions. Furthermore, with the increase in seam thickness and mining intensity, many urgent new issues need to be addressed, such as improving coal resource recovery ratio, strata control, and prevention and control of gas and rock burst. Currently, the main technologies used for mining thick coal seams in China are top coal caving, large mining height, and slicing mining, with slicing mining being rarely used.

1.3.1 Top Coal Caving

Top coal caving, also known as fully mechanized top coal caving mining, involves arranging a longwall panel with a height of 2–7 m (mostly 2–3 m) at the bottom of a thick coal seam. The coal is cut by a shearer and transported out of the panel by an armored face conveyor (AFC) in the front of the panel. Furthermore, using the mining-induced stress or assisted by methods such as pre-loosening blasting, the top coal is broken into loose bodies above the support, which are then drawn through the drawing opening (DO) at the tail of the support. The drawn top coal is transported out of the panel by an AFC at the rear of the panel, achieving coal removal from both the front and rear of the panel with AFCs, which improves mining efficiency and production.

The technological characteristics of top coal caving determine that this method has the advantages of low roadway driving rate, low investment, low mining cost, large production, and high efficiency. However, it also requires high demand in seam hardness and degree of fracture development. It requires the top coal to be broken into suitable block sizes for discharge. Moreover, due to its high production, large mining height, and high absolute gas emission from the panel, a certain amount of residual coal is left in the goaf, which brings some difficulties to gas management, goaf fire prevention, and ground settlement prevention and control work.

In 1982, China began researching and experimenting with top coal caving technology. In April 1984, the first gently inclined LTCC panel conducted underground industrial trials at Puhe Coal Mine, Shenyang Mining Bureau, however, the experiment failed due to unreasonable support design and goaf fires. In 1986, successful trials of STLL in steeply inclined extra-thick coal seams were conducted at the Second Mine of Yaojie Mining Bureau, Gansu. After 1987, top coal caving technology began to be tested in gently inclined thick coal seams. By the end of 1990, there were already 32 LTCC panels across the country. At that time, the thickness of the coal seam in LTCC panel was mostly between 5–8 m, with a seam dip angle of less than 20°, support working resistance around 4,000 KN, and a cutting height of 2–2.5 m. The coal seam thickness of STCC panel was between 15–30 m, with a seam dip angle of over 45°.

With the rapid development of LTCC, the production of the panel has rapidly increased. It increased from 1 million tons per year in 1990 to 3 million tons per year in 1995. Furthermore, the experimental LTCC technology was accelerated in soft coal seams (Zhengzhou), high gas coal seams (Yangquan), and large dip angle coal seams (more than 30°). In 1993, the Ministry of Coal Industry established the Top Coal Mining Technology Center at China University of Mining and Technology (Beijing). In 2014, the center was renamed the Coal Industry Engineering Research Center of Top coal Caving Mining by the China National Coal Association, further deepening and expanding the related research fields. From 1996 to 2005, LTCC technology developed rapidly, proving its advantages of low cost, high efficiency, and large output. Therefore, various coal enterprises took the lead in applying and actively developing LTCC technology. After 2006, LTCC technology gradually entered a mature stage, and bold innovations were carried out. Successful developments included LTCC technology with large mining height (i.e., cutting coal height being more than 3.5 m), STCC in steeply inclined thick coal seam, and along-the-strike LTCC technology in steeply inclined thick coal seam. Two-column hydraulic supports were introduced, and drawing supports developed towards high resistance direction. The working resistance of the four-column drawing supports at Huangyuchuan Coal Mine of Shenhua Group reached 20,000 KN. Currently, LTCC technology has become the main mining technology for thick coal seams in China, with approximately 1 billion tons of coal coming from LTCC panels each year.

In November 2004, Yanzhou Coal Mining Group registered Yan Coal Australia Limited in Australia. The company acquired the Austar Coal Mine in the southern part of New South Wales, Australia, on December 24, 2004. In October 2006, the company provided the first set of LTCC equipment, which was put into use at the Austar Coal Mine, establishing Australia's first LTCC panel. In October 2013, the company provided a set of two-column drawing supports and complete equipment, which were put into use at the North Goonyella Mine of the BHP Billiton company in Queensland. Besides Australia, countries such as India, Turkey, Russia, and Vietnam are also applying LTCC technology and conducting some fundamental research [9]. Some of the equipment for LTCC panels is also provided by China.

1.3.2 Large Mining Height

Large mining height refers to the extraction of coal with a single-cut height exceeding 3.5 m. According to the Chinese standard MT550-1996 "Technical conditions of large mining height support", hydraulic supports with a maximum cutting height greater than or equal to 3800 mm, used for full-seam mining, are referred to as large mining height hydraulic supports, and the corresponding extraction panel is called a large mining height panel. When the cutting height in LTCC panel exceeds 3.5 m, it is commonly referred to as large mining height LTCC technology.

Since the 1980s, China has gradually developed a series of products adapted to the geological conditions of Chinese coal mines based on the introduction of equipment from countries such as Germany. These products have undergone industrial trials and actual production, achieving success. So far, the large mining height technology has been applied in multiple mining areas and has achieved gratifying results, such as in the Shendong mining area, Jincheng mining area, Xingtai mining area, Datong mining area, Yulin mining area, and others. With the advancement of mining and related technologies, large mining height methods will be further promoted and applied.

Large mining height is characterized by tall supports, powerful coal-cutting machines, strong AFC, and corresponding large-scale roadway and auxiliary equipment. It requires large initial investment and imposes significant size requirements on shaft design, underground roadways, and chambers. However, it features high production, high efficiency, suitability for centralized production, and simple underground layout.

Large mining height technology has been widely recognized and rapidly developed in the past 20 years. In the early stages, due to the constraints of equipment such as supports and shearer, coupled with the high investment in large mining height panels, the promotion of this technology encountered certain difficulties. In recent years, with the acceleration of relevant technology solutions and the localization of related equipment, coupled with the improvement in economic benefits of coal mining enterprises, large mining height technology has developed rapidly. In September 2018, the Shendong mining area's Shangwan coal mine adopted the ZY26000/40/88D support, with a maximum supporting height of 8.8 m, a support center distance of 2.4 m, and a rated working resistance of 26,000 kN. The average thickness of the coal seam was 9.16 m, with a dip angle of 1°–3°. The panel was 299.2 m long, with a driving length of 5254.8 m and an annual output of 16 million tons, which improved mining efficiency and coal recovery ratio. In November 2023, Shaanxi Coal Group's Caojiatan Mining Company pioneered the successful mining of a 10-m extra-high large mining height panel, marking a significant breakthrough in China's domestic coal mining equipment and mining technology. The complete set of equipment for this fully mechanized mining can cut coal every 40 min, with a production capacity of 3500 tons, increasing the annual output of the coal mine from the previous 13 million tons to 20 million tons, and increasing the coal recovery ratio by more than 10%.

1.3.3 Slicing Mining

Before the 1980s, thick coal seams were mainly mined using slicing mining, which involved dividing the thick coal seam into several layers of 2.0–3.0 m each and then extracting them layer by layer from top to bottom. Some mines also employed bottom-to-top slicing mining.

In the case of top-to-bottom slicing mining, after the upper layer is mined, the next layer is mined beneath the collapsed roof of the upper layer. To ensure the safety of the lower layer's extraction, an artificial roof must be laid or a regenerated roof must be formed in the upper layer. Currently, metal mesh is often used between layers as an artificial roof for the next layer's mining. Mining of the lower layer is carried out under the protection of this artificial roof, known as downward slicing mining. In some mining areas, upward filling mining is used for ground protection, or in the case of extra-thick coal seams prone to spontaneous combustion, and it is also called upward slicing mining.

Slicing mining has the advantage of relatively mature technology and has been a long-term mining method applied in China. It requires less equipment investment, has a small single-cut height, relatively mature gas control technology, and can achieve slow subsidence of overlying strata and the surface. However, slicing mining also has some disadvantages, such as high roadway driving ratio, low output, high mining cost, difficulty in supporting lower roadways, large loss of coal pillars in sections, repeated disturbance of goaf, and susceptibility to goaf spontaneous combustion. Due to these shortcomings of slicing mining, China began researching and applying LTCC technology and large mining height technology for thick coal seam around the mid-1980s. Currently, these technologies have basically replaced slicing mining technology.

References

1. Peng S (2019) The strategy of coal resource power. Science Press, Beijing
2. Ni T, Qian F (2006). The classic of mountains and rivers. Chongqing Press, Chongqing
3. Polo M (2009) The travels of Marco Polo. China Book Press, Beijing
4. Song Y (2022) Tiangong Kaiwu. Phoenix Press, Nanjing
5. Lai K, Liu Y, Xie J (1992) Coal mines in Taiwan. Hiking Culture Co., Ltd.
6. Hu S, Liu X, Cheng Y (2010) Technical revolution in coal mining history: 40 years development of fully mechanized coal mining in China. J China Coal Soc 35(11):1769–1771
7. Wang J (2009) Thick coal seam mining theory and technology. Metallurgical Industry Press, Beijing
8. Du J, Meng X (2009) Mining science. China University of Mining and Technology Press, Xuzhou
9. Tien DL (2017) Longwall top coal caving mechanism and cavability assessment. The University of New South Wales

Chapter 2
Top Coal Caving

LTCC technology is an effective method to mine thick coal seams, realizing one-time mining of thick coal seams at full height, which is a revolutionary technology to mine thick coal seams, solving the problems existing in the slicing mining technology, such as stress concentration caused by the coal pillar left behind in the upper layer, the easy to fire in the goaf and difficulty in the roadway support at the lower layer, the large amount of relative emission of gas in the first layer, the low output and the high cost. Compared with large mining height technology, LTCC has the advantages of less investment, low cost, low energy consumption, low emission, and adaptable to large changes in coal seam thickness. China started to research and test LTCC technology in 1982, and has achieved great success and major scientific and technological innovations in the past 40 years. LTCC has become the main method of extracting thick coal seams in China, and it is also the landmark technology of China in the world coal mining industry. LTCC is a comprehensive technology, which mainly includes panel layout, coal mining system, top coal fracturing, top coal drawing technology, surrounding rock control, and coal mining equipment.

2.1 Top Coal Caving Technology

2.1.1 Layout Methods of LTCC Panel

2.1.1.1 One-Time Mining Full-Thickness LTCC

One-time mining full-thickness LTCC refers to the method where the entire seam thickness is extracted in one operation through LTCC. This is the main and the most fundamental layout method for LTCC panel. In the lower part of the thick coal seam, a fully mechanized mining panel is arranged along the seam floor. The coal above the panel (commonly referred to as top coal) is crushed under the influence

J. Wang and W. Wei, *Thick Coal Seam Underground Mining*,
https://doi.org/10.1007/978-981-96-3874-1_2

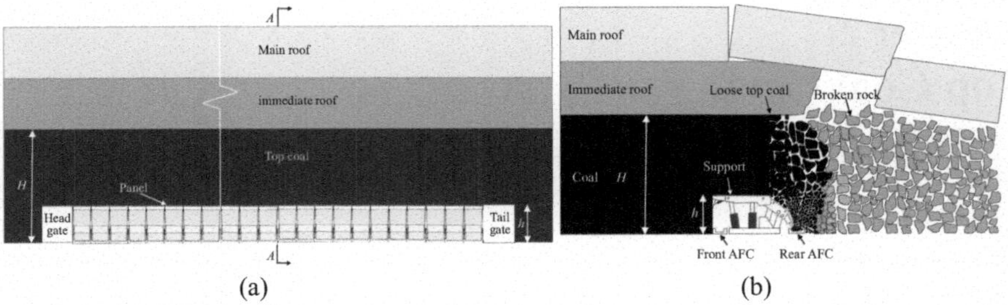

Fig. 2.1 One-time mining full-thickness LTCC: **a** the layout direction; **b** the advancing direction

of mine pressure and drawn through DO at the tail beam of support onto the rear AFC. This coal, along with the coal cut from the coal wall and transported on the front AFC, is carried out of the panel (Fig. 2.1). To improve the top coal recovery, prevent excessive gas outbursts, and avoid the impact of roof collapse in the goaf, the "Coal Mine Safety Regulations" [1] stipulate that the drawing height of the top coal in LTCC panel should not exceed three times the cutting height. That is, $H - h \leq 3h$, where H is the coal seam thickness, and h is the cutting height.

The layout of one-time mining full-thickness LTCC is relatively simple, with fewer roadways and effective top coal size due to mining-induced stress. However, when the coal seam thickness is excessive, the top coal recovery decreases [2]. This method is typically suitable for coal seams with a thickness ranging from 5 to 20 m. The cutting height is generally 2–3 m. For thicker coal seams, to enhance the top coal recovery or increase the ventilation cross-section and address high gas emission issues, the cutting height has been increasing in suitable mines at recent years, such as 4-m cutting height at the Wangzhuang Coal Mine of Lu'an Group, and 5-m cutting height at the Tashan Coal Mine of Datong Coal Mine Group. At the Jinjitan Coal Mine of Yankuang Group, located in Yulin City, Shaanxi Province, due to the high hardness of the coal seam (coefficient of hardness $f = 2.8$), the caving characteristics of the top coal are poor. To improve the recovery ratio, the ZFY21000/35.5/7 two-leg drawing support was employed in a near-horizontal coal seam with a thickness of 8.0–12.5 m, achieving a cutting height up to 6.5 m. The panel has an inclined length of 300 m and an extractable length of 5,093 m. Production commenced at the beginning of 2019, with an average daily output of 52,500 tons.

To distinguish it from conventional LTCC panels, a LTCC panel with a cutting height greater than 3.5 m is typically referred to as a large mining height LTCC panel. Currently, large mining height LTCC are also widely utilized.

2.1.1.2 Pre-mining Top Layer LTCC

When the coal seam is thick, and the gas content is high, even with the risk of outbursts, or when the roof is hard, the pre-mining top layer LTCC can be employed. The layout of the panel involves initially arranging a conventional comprehensive

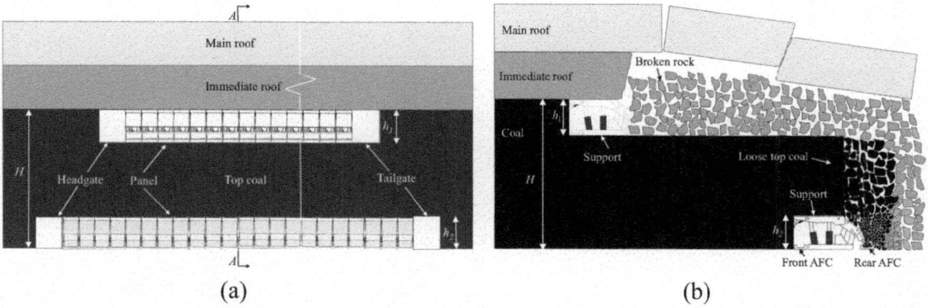

Fig. 2.2 Pre-mining top layer LTCC: **a** the layout direction; **b** the advancing direction

mechanized mining panel along the roof within the upper part of the coal seam, where a metal mesh is usually laid during extraction. After the extraction of the top layer comprehensive mechanized mining panel is completed, another LTCC panel is arranged along the floor of the coal seam to extract the lower coal. Generally, the extraction of LTCC panel should lag the top layer comprehensive mechanized mining panel by more than half a year, as shown in Fig. 2.2.

In high-gas outburst thick coal seams, pre-mining top layer LTCC can release a large amount of gas, eliminating the danger of gas outbursts in the lower part of the coal seam, which is conducive to ensuring the safety of the lower LTCC panel. For example, the Zhuxianzhuang Coal Mine and Luling Coal Mine of Huaibei mining area have both adopted the pre-mining top layer LTCC in high-gas outburst coal seams. The average thickness of the coal seam is 10.11 m, with a hardness coefficient ($f < 0.3$), belonging to an extra-soft coal seam, and the gas content is 16 m^3/t, with a risk of outbursts. The cutting height of the pre-mining top layer is 2.5 m [3]. For coal seams with hard roofs, pre-mining top layer can cause the roof to break and fracture, reducing the impact of the hard roof during the mining of the lower LTCC panel, which is beneficial for safe mining. When the top layer is mined and the metal mesh is laid, it is necessary to cut the metal mesh at DO during the mining of the lower LTCC panel. The metal mesh can prevent the gangue from the goaf entering DO, which is beneficial for improving the top coal recovery and reducing the gangue mixed ratio. After pre-mining top layer, it will weaken the crushing effect of the mine pressure on the lower coal seam. When the coal seam is hard, it will affect the crushing effect of the top coal on the lower LTCC panel, thus affecting the top coal recovery. Therefore, this method is generally used for coal seams with lower hardness.

2.1.1.3 Slicing LTCC

Slicing LTCC is generally used for the extraction of extra-thick coal seams with a dip angle of less than 45°. It involves dividing the extra-thick coal seam into several layers of 10–20 m parallel to the floor of the coal seam, with a LTCC panel arranged at

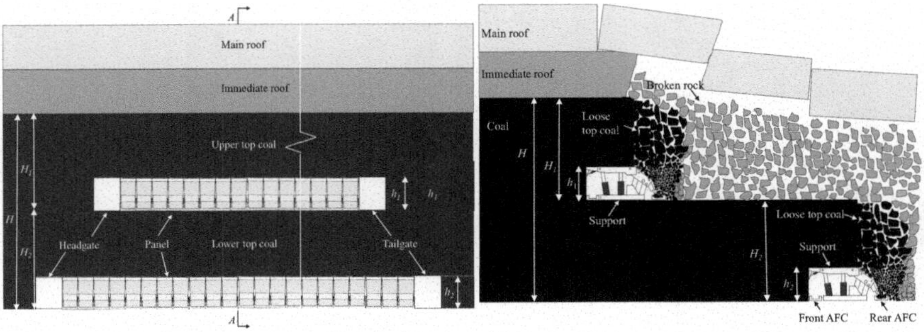

(a) (b)

Fig. 2.3 Slicing LTCC: **a** the layout direction; **b** the advancing direction

each layer and mining the whole coal seam from top to bottom, as shown in Fig. 2.3. For example, at the Laohutai Coal Mine of the Fushun mining bureau, the seam thickness being mined is about 60 m, with a dip angle of 15°–32°, generally divided into 4 layers with each layer being about 15 m thick. To improve the recovery ratio of the lower layer and reduce the gangue mixed ratio, a metal mesh is laid during the mining of the first layer.

2.1.1.4 Horizontal Sublevel Top Coal Caving

For steeply inclined thick coal seams, when the seam thickness is greater than 15 m, the sublevel top coal caving (STCC) can be used, as shown in Fig. 2.4. STCC is similar to the slicing LTCC, however, the panel of STCC is arranged along the horizontal thickness direction of the steeply inclined thick coal seam, and the length of the panel is relatively short, generally only the horizontal thickness of the coal seam. The roadways of the panel are arranged in the coal seam next to the roof and floor and are parallel to the direction of the coal seam. The sublevel height is generally between 20–30 m. When the sublevel height is too large, the top coal recovery will decrease. When the sublevel height is too low, the mining efficiency will decrease.

"Coal Mine Safety Regulations" stipulate that for STCC panel, the drawing height of top coal should not exceed eight times the cutting height of the panel. The Wudong Coal Mine of the CHN Energy applies the STCC. In the mine's northern area, the coal seam thickness is about 40 m, with a dip angle of 43°–51°. The sublevel height is 25 m, the cutting height is 3 m, and the panel has an annual output of 3 million tons.

Due to the small length of the STCC panel and the large thickness of the top coal, the fracturing effect of the top coal is poor, especially near the triangular area close to the roof and floor. Therefore, improving the fracturing effect of the top coal and increasing the top coal recovery are the main directions for the technology's breakthrough. Blasting and hydraulic fracturing are the main methods to improve the fracturing effect of the top coal.

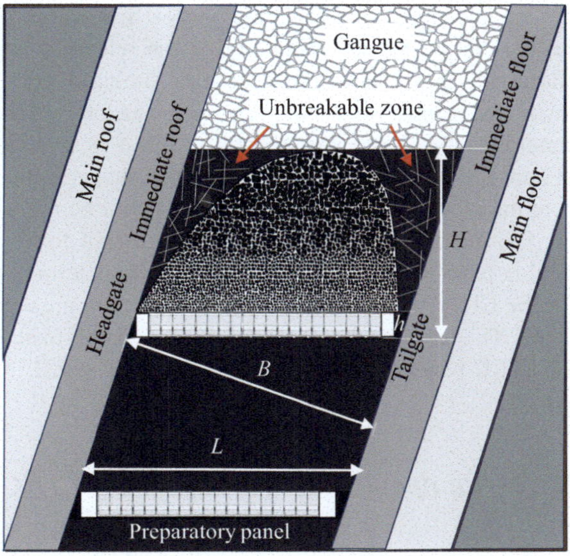

Fig. 2.4 Horizontal sublevel top coal caving

2.1.1.5 Pre-mining Middle Layer LTCC

The pre-mining middle layer LTCC introduced in 2015 is a novel top coal caving mining method [4], primarily designed for hard coal layers and high-gas extra-thick coal seams. The principle of this technique is depicted in Fig. 2.5. Initially, a comprehensive mechanized mining panel is arranged in the middle-lower part of the extra-thick coal seam for extraction. After the completion of the comprehensive mechanized mining, a LTCC panel is set up in the reserved bottom coal below the comprehensive mechanized mining panel. This panel is responsible for mining the reserved bottom coal and releasing the top coal above the comprehensive mechanized mining panel.

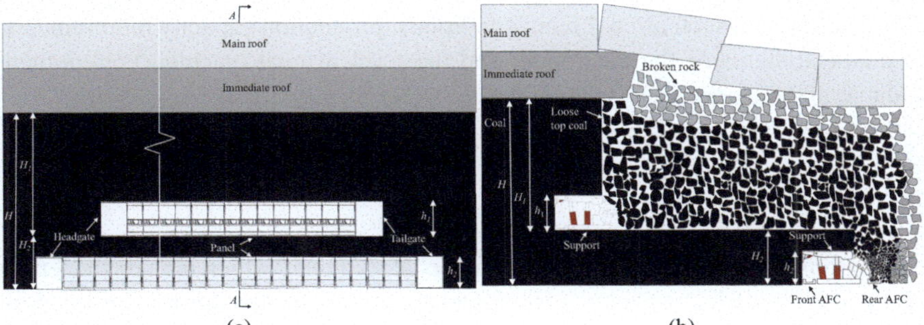

Fig. 2.5 Technology of partial layer caving of top coal in pre-mining: **a** the layout direction; **b** the advancing direction

The advantage of this technology is that for high-gas or hard extra-thick coal seams, the pre-mining middle layer can release a large amount of coal seam gas, addressing the issue of gas exceeding limits in subsequent LTCC panel. For hard thick coal seams, after comprehensive mechanized mining, the top coal above the panel becomes loose and broken, which is beneficial for the subsequent recovery of hard top coal. The height of the comprehensive mechanized mining panel can be determined based on actual conditions, with being generally suitable between 3–5 m. The thickness of the reserved bottom coal for the comprehensive mechanized mining panel should be 5–8 m, specifically determined by the actual height needed for the LTCC panel, which is generally 2–3 m greater than the height of the LTCC panel. This ensures that there are 2–3 m of unbroken top coal above the support of the LTCC panel, preventing leakage from the panel.

2.1.2 Equipment and Parameters of LTCC Panel

2.1.2.1 Equipment Arrangement of LTCC Panel

LTCC is similar to the traditional longwall comprehensive mechanized mining methods, fundamentally utilizes equipment such as shearer, hydraulic support and AFC, collectively referred to as "three machines' of the panel. In addition, within the roadway, there are also transloader, crushers, belt conveyors, pumping stations, and auxiliary transportation equipment for personnel and materials on the panel [5], as shown in Fig. 2.6.

The production capacity of the panel necessitates different capabilities and dimensions for the shearer, the front AFC and rear AFC. For high-capacity LTCC panels, it is essential to equip with powerful three machines. Generally, the thickness of the top coal exceeds the cutting height, and the volume of drawing top coal surpasses the volume of cutting bottom coal. Therefore, the capacity of the rear AFC must be greater than that of the front, or at least not less than the front. The specific selection and matching of equipment for the panel should be determined by a comprehensive analysis of factors such as coal seam conditions, production capacity requirements, the financial status of the enterprise, and the level of coal machinery equipment available in the market.

2.1.2.2 Panel Parameters

Panel parameters refers to the main parameters of the panel, which include the length of the panel, the advancing length of the panel, the cutting height, the drawing interval, the cutting-drawing ratio, and the cutting depth of shearer.

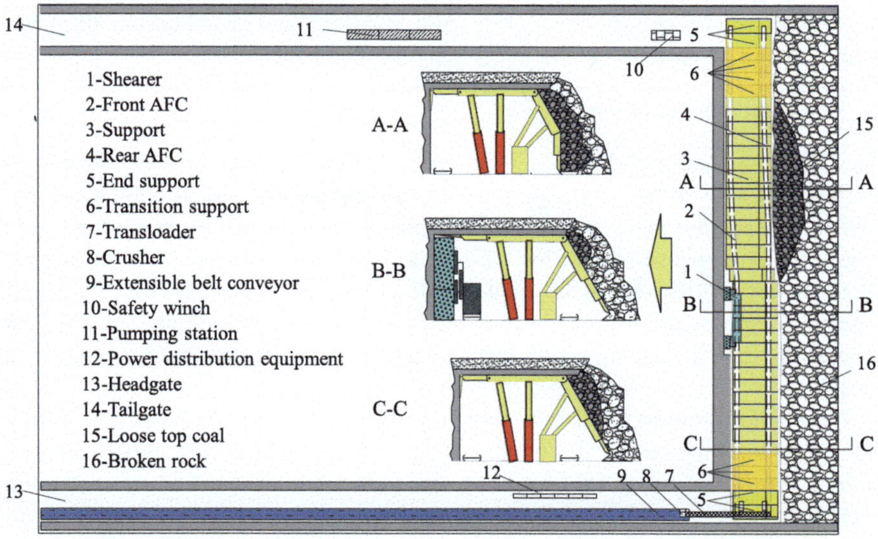

Fig. 2.6 Equipment layout of LTCC panel

1. Length of the panel

The length of the panel is one of the key parameters for LTCC panels. For panels with simple geological conditions, the length is primarily determined based on the reasonable daily advancing distance and the required daily production. Shorter panels can speed up the advancing of the panel, which is beneficial for preventing spontaneous combustion of coal in the goaf. However, shorter panels result in a higher proportion of auxiliary operation time at both ends of the panel, such as the time for cutting at the end, maintenance of roadways, and relocation of belt conveyors and transloader, leading to lower production per advancing meter. Additionally, due to the lower top coal recovery at both ends of the panel, when the panel is short, it occupies a larger proportion of the panel, which also affects the overall recovery rate. Therefore, while ensuring the advancing ratio of the panel, it is advantageous to increase the length of the panel as much as possible to improve the recovery ratio, reduce the loss of coal pillars, and increase the production and mining efficiency of the panel. However, the length of the panel is influenced by factors such as the capacity of the AFC, the required advancing velocity of the panel, ventilation and geological conditions. In high-gas mines, the ventilation capacity of the panel is an important factor limiting its length. Currently, the length of LTCC panels generally ranges from 150 to 300 m, with a few exceeding 300 m, and some high-production and high-efficiency LTCC panels reaching over 350 m. When the dip angle of the panel is large, shorter panels are used, even less than 100 m, which is beneficial for the management of supports and other equipment.

The actual length of a panel must consider the specific geological conditions. Factors such as the distribution of geological structures, coal seam thickness and strength, parting, rock properties of the roof and floor, and gas emission all play a crucial role in determining the practical length of the panel.

2. Advancing length of the panel

The advancing length of the panel is primarily influenced by geological conditions, the laying length of belt conveyors, the major overhaul period of equipment, the costs of installation and relocation of equipment, as well as the maintenance and transportation expenses of section roadways. Generally, increasing the advancing length of the panel can reduce the frequency of relocating the panel, extend production time, and decrease the loss of top coal during initial and final mining. However, an excessively long advancing length can lead to difficulties in maintaining the retreat roadways.

The main difference in the advancing direction between LTCC panels and fully mechanized mining panels lies in the issue of top coal loss during initial and final mining. Generally, during initial mining, due to low mining-induced stress, the top coal is not prone to fall or to the large size top coal, which makes it difficult to release. Most LTCC panels do not release top coal for 8–12 m at the initial mining (Fig. 2.7a), resulting in this portion of top coal being lost in the goaf. The final mining typically requires laying mesh on the roof or climbing to the roof (Fig. 2.7b, c), and the top coal in 12–15-m length cannot be drawn, thus leading to a loss of top coal during initial and final mining. Additionally, for coal seams prone to spontaneous combustion, the mining time of the panel should not be too long to prevent fires. The advancing length of LTCC panels should ideally be 2000–3000 m, and it can reach 4000–5000 m in favorable conditions. Generally, it should not be less than 2000 m if not limited by geological conditions.

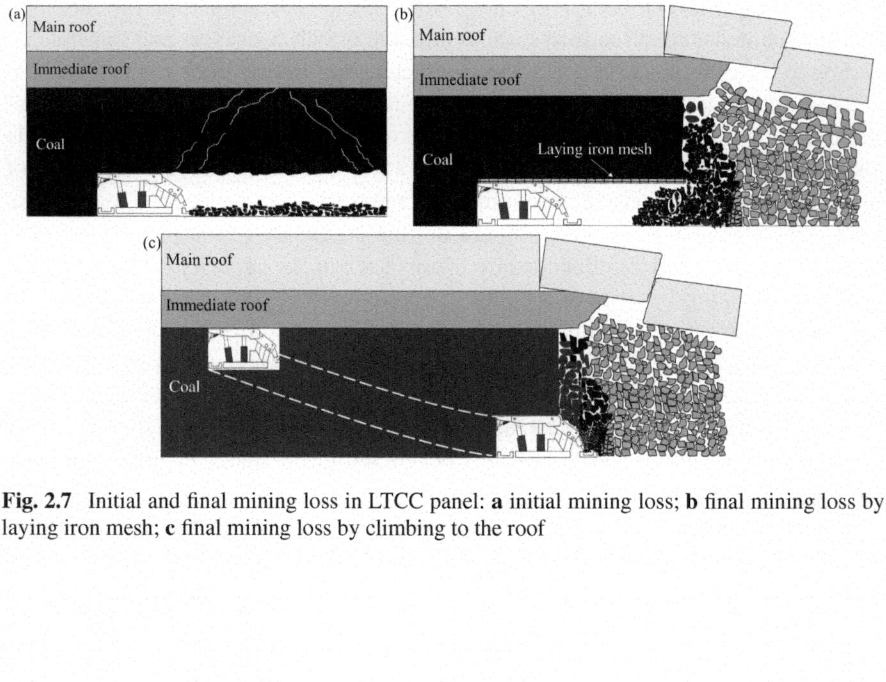

Fig. 2.7 Initial and final mining loss in LTCC panel: **a** initial mining loss; **b** final mining loss by laying iron mesh; **c** final mining loss by climbing to the roof

The advancing length is also affected by the laying length of the belt conveyor. In recent years, domestically produced telescopic belt conveyors have made significant progress, with multi-drive conveyors reaching lengths of up to 6000 m and annual transportation mass exceeding 20 million tons. The major overhaul period of the panel equipment is also a factor that affects the advancing length. The progress in shearer has been substantial in recent years, fully meeting the needs of extra-large panels. The determination of the advancing length should also consider the impact on power supply, ventilation, and auxiliary transportation. Excessively long advancing lengths can cause difficulties in power supply, ventilation, and auxiliary transportation.

The determination of the actual advancing length is greatly influenced by geological conditions, such as the distribution of faults and coal seams. In mines with complex geological conditions, the advancing length is often determined by the boundaries of geological structures. For mines with simple geological conditions, the advancing length is typically determined by comprehensively considering factors such as the laying length of the conveyor belt, roadway maintenance, power supply, ventilation, and auxiliary transportation.

3. Cutting Height

The coal output from a LTCC panel consists of two parts: the coal cut by the shearer and the coal drawn from the top coal. Generally, the top coal recovery ranges between 75–90%, which is lower than the recovery ratio of coal cut by the shearer. Therefore, in hard coal seams where the breaking and caving of top coal are not effective, it is advisable to increase the cutting height as much as possible to improve the overall recovery ratio of the panel.

The determination of the cutting height primarily considers factors such as coal wall stability, ventilation requirements, inclination of the panel, recovery ratio of the panel, reasonable operating space, and the cutting-drawing ratio. In nearly horizontal thick coal seams, the cutting height for LTCC panels is generally 2.0–3.0 m. However, to improve the recovery ratio of the panel, the cutting height can be increased, such as the Jinjitan Coal Mine of Yanzhou Mining Group, where the designed cutting height reaches 7 m in the 8.0–12.5 m thick nearly horizontal coal seam. For soft coal seams, top coal is easily released and coal wall stability is poor, a lower cutting height is generally used, not exceeding 3 m. When the panel has a significant inclination, a lower cutting height is also adopted. For hard coal seams with good coal wall stability but lower top coal recovery, a greater cutting height is used, typically between 3–5 m. To increase the ventilation cross-section of high-gas thick coal seams and improve the ventilation conditions of the panel, a larger cutting height is generally employed.

For gently inclined and inclined thick coal seams, the cutting height should not be less than one-third of the top coal thickness. For horizontal STCC panel in the steeply inclined coal seam, the cutting height should not be less than one-eighth of the top coal thickness. Some coal mines increase the cutting height to meet the cutting-drawing ratio requirements of the "Coal Mine Safety Regulations". For example, the Tashan and Tongxin Coal Mines within a 20-m-thick coal seam, the cutting height was increased to 5 m to meet the cutting-drawing ratio requirement of 1:3, resulting in great mining outcomes.

The greater the cutting height, the taller the support needed, and the power and cutting height of the shearer must also increase accordingly, leading to higher equipment investment. Especially when the support height reaches a certain level, the weight and cost of the support will significantly increase. Therefore, when determining the cutting height, in addition to considering the seam thickness, the dip angle, the recovery ratio of the panel, and the ventilation cross-section, the one-time investment cost of the equipment should also be considered.

4. Drawing interval

The drawing interval refers to the distance the panel advances between two caving process. An optimal drawing interval should result in the highest top coal recovery ratio and the lowest gangue mixed ratio. Therefore, it is related to factors such as the size and quality of the coal and gangue, resistance to movement, direction of movement, degree of gangue mixing, and distance to DO. It also correlates with the strength and size of the coal and rock, thickness of the top coal, drawing angle, height of the gangue pile, and the angle of repose. In practice, the drawing interval is a multiple of the shearer's cutting depth, typically following "one-cutting and one-caving", "two-cutting and one-caving" and "three-cutting and one-caving" pattern. For drum shearer machines, the drawing interval varies between 0.6–1.8 m. For large mining height LTCC panels, the shearer's cutting depth can reach 0.8–1.2 m, resulting in a drawing interval variation between 0.8–2.4 m. Due to the small range of variation and complex influencing factors, it is challenging to determine an exact value through calculation using a equation. Generally, a qualitative analysis based on actual conditions is used to determine it. For hard coal, one-cutting and one-caving with a smaller drawing interval is suitable. For soft coal, broken top coal is small with less movement resistance, a larger drawing interval can be used. The block size of the falling top coal is generally smaller at the bottom and larger at the top. The lower part of the top coal loosens fully, while the upper part lags. Therefore, for thicker top coal, the caving time and width should be increased to allow the upper top coal to loosen and fall sufficiently, making a larger drawing interval preferable. For thinner top coal layers, due to the large space for the lower part to fall, it is very prone to mixing with gangue, thus the number of natural falls should be reduced to allow timely release before mixing, making a smaller drawing interval preferable for caving.

2.1.2.3 Coal Mining Process

The coal mining process at the retreat mining panel refers to the sum of spatial and temporal relationships between various operations that must be carried out on the retreat mining panel, as well as the mechanical equipment and various types of work required to complete these operations. The space of the retreat mining panel generally includes the entire operation area from the headgate, tailgate, the end of the coal wall, the coal wall, and the goaf. Some panels extend this space to include the hydraulic pump station, mobile substation or loading point of the mining area. Spatial changes at the panel refer to changes in the position of the coal wall and goaf. In terms of time, a coal mining cycle can vary from several hours to more than 24 h. Different coal mining methods require different mining operations and different spatiotemporal coordination relationships. In addition to the cutting coal, moving support, and advancing AFCs operations common to ordinary longwall comprehensive mechanized mining, the LTCC panel also adds a top coal caving operation. Generally, more than half of the coal volume at the panel comes from top coal caving. Therefore, reasonably arranging the relationship between coal cutting and coal caving in terms of time and space becomes a fundamental issue that must be addressed in the production process of the LTCC panel. As shown in Fig. 2.6, the general sequence of operations at the panel is: coal cutting by the shearer, moving supports following the machine, advancing the front AFC, then opening DO to release coal, and finally pulling the rear AFC. After all operations at the panel are completed, a complete mining cycle is finished. A-A cross-section in the figure shows the equipment position during top coal caving, with the rear AFC in the last position and the front AFC already advanced. B–B cross-section shows the equipment position before moving supports after coal cutting by the shearer, which is the state of maximum control distance. C–C cross-section shows the equipment position at the minimum control distance, with at least one cutting depth between the front end of the support base and the front AFC, and the rear AFC close to the rear of the support base.

The haulage equipment in a LTCC panel is similar to that in conventional longwall mining, consisting of bridge transloader and extendable belt conveyors. The difference lies in the requirement for a crusher to be installed on the bridge transloader, especially for coal seams with greater hardness and undeveloped fractures. This is because the rear AFC transports a large amount of large coal blocks, and it would cause the belt conveyor to misalign and block the coal bunker without secondary crushing, preventing the panel from reaching its full production capacity.

In a LTCC panel, two AFCs are arranged in the front and rear, the two AFCs are positioned side by side on the horizontal loading section of the bridge transloader. As the panel advances, the movement of the bridge transloader is more frequent than that is in conventional comprehensive mechanized mining panels, requiring almost every shift. Depending on the coordination of coal cutting and coal caving processes, there are four main methods for top coal caving: sequential top coal caving with the shearer, segmented top coal caving with the shearer, cross-operation of coal cutting and caving, and independent coal cutting and coal caving.

1. Sequential top coal caving with the shearer

The shearer cuts coal in the front, and the sequential top coal caving begins behind 15 m. The process involves: cutting coal, advancing the front AFC, moving the support, caving top coal, and pulling the rear AFC. To control the top coal in time and prevent leakage in front of the support, the process can also be: cutting coal, moving the support, advancing the front AFC, caving top coal, and pulling the rear AFC. This method simplifies labor organization and is widely adopted. To prevent the caving from affecting the cutting speed of the shearer, a multi-person and multi-support simultaneous caving method can be employed.

2. Segmented top coal caving with the shearer

The mining process involves a technique where top coal is released in three sections (upper, middle, and lower) behind the shearer, with each section operated by a separate group. The workflow includes cutting coal, moving supports, advancing the front AFC, retracting the rear AFC, and caving top coal. This method allows for the caving operation to be independent of other processes, with segmented and grouped top coal caving. However, each section must complete its top coal caving before the next cutting cycle begins. A potential drawback is the simultaneous caving in multiple sections, which can overload the rear AFC and cause shutdowns. The set-up room position can be adjusted based on the condition of the top coal at the panel, allowing for two-cutting and one-caving or three-cutting and one-caving, depending on the characteristics and thickness of the top coal. One disadvantage of this technique is that the caving operation on the downwind side can be affected by dust from the upwind side.

3. Cross-operation of coal cutting and caving

The mining process is divided into upper and lower sections of the panel, with the upper part cutting coal and the lower part caving top coal; then the lower part cuts coal while the upper part releases top coal, operating in a cross manner. This technique generally involves initiating the cut in the middle of the panel, often uses two-cutting and one-caving method. This is particularly advantageous in longwall panels.

4. Independent coals cutting and coal caving

The above three techniques allow for simultaneous cutting and caving operations on long panels to enhance mining efficiency. For shorter panels or when mining steeply inclined coal seams in segments, a single operation of cutting without caving or caving without cutting is generally adopted. The process includes: cutting coal, moving supports, advancing the front AFC, retracting the rear AFC, and caving top coal. This method often employs the two-cutting and one-caving or the three-cutting and one-caving approach, where the cutting entry method and position are not affected by the caving process.

2.2 Top Coal Fracturing Mechanism

The study of the fracturing mechanism of top coal under mining pressure is a unique research aspect and is also the fundamental work that determines whether a thick coal seam is suitable for LTCC. Top coal that is easily fractured and has a smaller broken size is suitable for mining using LTCC. The fracturing of top coal is related to the strength of the coal body, the distribution of fractures, the partings, and mining pressure. It also has a certain relationship with the resistance of the support and the mining process. Among these factors, the distribution and development of coal body fractures often play a major role in the fracturing of top coal, especially for hard coal seams [6].

2.2.1 Top Coal Destruction Process

2.2.1.1 Top Coal Destruction Zoning Model

In the LTCC panel, the top coal (micro-unit in Fig. 2.8a) undergoes a transition from an initially intact state to a post-support collapse state. During this process, the vertical stress on the top coal is continuously loaded. Based on the relationship between the post-mining vertical stress and its initial value, the post-mining vertical stress distribution can be divided into the original stress area, stress-concentrated area, and stress-relaxed area. After the coal is mined out, the overlying strata in the goaf lose support and gradually sink and move, causing the vertical stress in the top coal in front of the panel to start increasing at Point A and reach a peak at Point B, where the vertical stress in the top coal increases to its limit bearing capacity and fails. As the overlying strata continue to sink, the vertical stress continues to load, the micro-fractures in the top coal increase, the bearing capacity decreases, and the vertical stress begins to decrease. In the upper top coal above the support, the vertical stress is reduced to a residual level, which is provided by the immediate roof, fractured main roof rock blocks, and the following rock layers above the main roof rock blocks. The lower top coal is directly subjected to the repeated support of the support, bearing a cyclic load with a maximum value of support strength and a minimum value of its own weight. Under the action of the load, fractures in the top coal body continue to develop and expand, and the fracture development process experienced by different types of top coal is completely different. Hard top coal undergoes two evolutionary stages: local densification, local expansion, and overall densification, overall expansion; medium-hard top coal mainly densifies fractures before the peak of vertical stress and expands after the peak; soft top coal has a strong local characteristic in the densification and expansion of mining-induced fractures.

Previous studies have considered the fracturing under support pressure to be the decisive factor for the broken size of top coal, neglecting the impact of horizontal stress changes on top coal destruction. In the actual mining process, the excavation causes the horizontal constraints near the panel side along the direction of the seam to weaken or even disappear, resulting in a gradual decrease in horizontal stress along the direction of the seam in the top coal, while the horizontal stress along the inclined direction is less affected by mining activities, as shown in Fig. 2.8a. Overall, the stress path experienced by the upper top coal is a process of vertical stress loading and horizontal stress unloading; then vertical stress unchanged and horizontal stress reverse loading, while the lower top coal experiences a stress path of vertical stress loading, horizontal stress unloading; then vertical stress cyclic loading and unloading, and horizontal stress remains unchanged. Based on the different stages of destruction mechanisms of the upper and lower top coal, it can be determined that the upper top

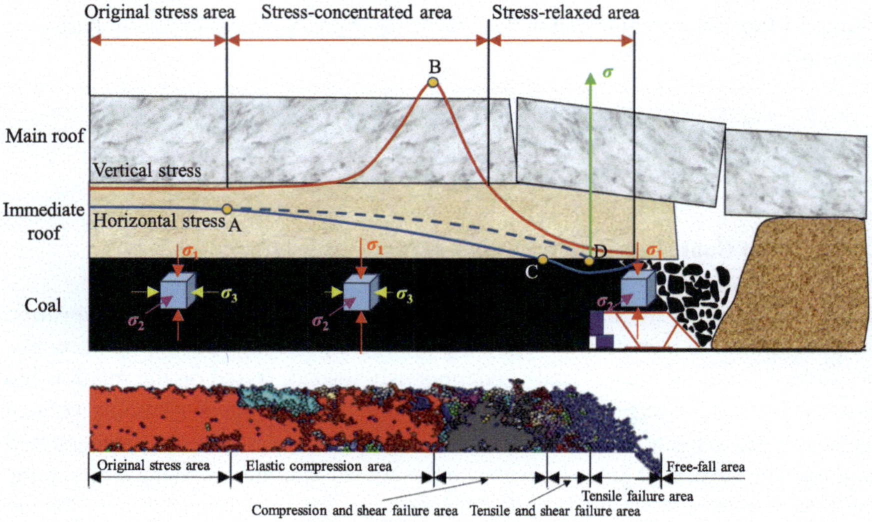

Fig. 2.8 Stress state of top coal and the upper top coal destruction zoning model

coal has gone through the original stress area, elastic compression area, compression and shear failure area, tensile and shear failure area, tensile failure area, and free-fall area, as seen in Fig. 2.8b. The lower top coal, on the other hand, has gone through the original stress area, elastic compression area, compression and shear failure area, cyclic damage area, and free-fall area. The destruction mechanisms of the upper and lower top coal in the Section AC in front of the panel are the same, however, after Point C, the destruction mechanisms of the upper and lower top coal are no longer consistent. The upper top coal is significantly affected by the rotation of the roof, while the lower top coal is significantly affected by the repeated effect of the support.

2.2.1.2 Measured Characteristics of Top Coal Deformation

To characterize the deformation and failure of the top coal, inclined boreholes are drilled between hydraulic supports, and a YTJ20 type strata behavior recorder is used to observe the deformation and failure process of the top coal. The deformation and failure process of the top coal is influenced by the deformation state of the contact surface between the top coal and the immediate roof. The shear slip failure of the contact surface reflects the extent of top coal damage, and the difficulty of top coal failure is directly proportional to the strength parameters of the shear surface. Therefore, observations were made on both the contact surface between the top coal and the immediate roof and the deformation of the top coal itself, with results shown in Fig. 2.9. Points A, B, and C represent the deformation and failure of the contact surface, while points D and E represent the deformation and failure of the top coal itself. In front of the panel at point A, the contact surface remains intact; at this time, the bending deformation of the roof is low, and the normal stress and strength parameters of the contact surface are sufficient to resist and transmit the horizontal shear stress exerted by the roof on the top coal. There is no dislocation slip on the contact surface, maintaining an intact state. Above the coal wall at Point B, obvious cracks appear on the contact surface; at this time, the contact surface is damaged and cannot withstand tensile stress, reducing its ability to transfer horizontal shear stress. Small vertical fractures develop in the top coal below the damaged contact surface, indicating that the top coal has been damaged at this location. Above the roof control area at Point C, the lower surface of the top coal becomes a stress boundary condition supported by the supports. During the advancing of the panel, the supports repeatedly lift and drop the columns, causing the lower surface of the top coal in the roof control area to periodically lose support, becoming a free boundary condition. The top coal in the roof control area freely bends and sinks, ultimately leading to complete destruction of the contact surface and significant dislocation slip. The diameter of the drill bit used for this borehole is 36 mm, and by comparing the slip amount with the borehole diameter, it can be determined that the maximum slip amount between the top coal and the immediate roof can reach 10 mm.

Corresponding to the deformation state of the contact surface, at Point D in front of the panel, no significant macroscopic fractures appear in the top coal, and the borehole only exhibits minor shrinkage deformation with a small amount of coal dust falling from the borehole walls. This is because the contact surface at this location

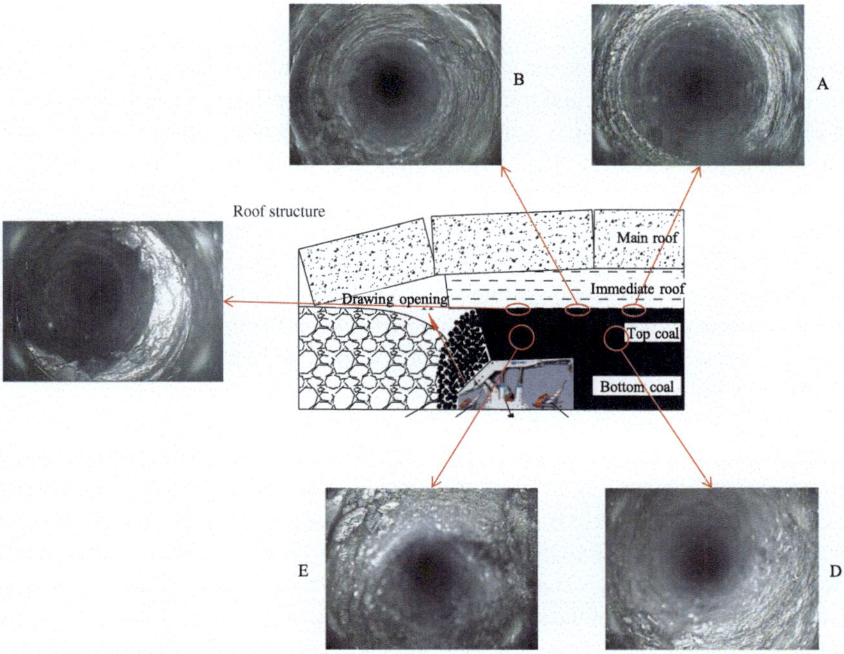

Fig. 2.9 Characteristics of top coal deformation and failure

remains intact, capable of transmitting high levels of horizontal shear stress. Under the confinement of the immediate roof, the initial horizontal stress in the top coal cannot be effectively released, and the top coal remains in a triaxial stress state with high confining stress. Experiments have shown that under high confining stress, the triaxial compressive strength of the coal is substantial. In this stress environment, the compressive strength of the top coal is sufficient to resist the high level of support stress, and the top coal maintains a relatively intact state. At Point E within the roof control area, the top coal is severely damaged, with frequent occurrences of borehole collapse during measurement, and the top coal is fully fractured, resulting in smaller coal blocks being cut. This is because the contact surface within this area is destroyed, reducing or even eliminating the confining effect on the top coal. The top coal has undergone multiple cycles of support column lifting and dropping, during which the initial horizontal stress in the top coal is fully released, and even a certain level of tensile stress appears in the top coal under the cantilever effect. At this point, the top coal is in a uniaxial compressive, or vertical compressive and horizontal unidirectional tensile stress environment. Experiments indicate that under this stress state, the compressive strength of the coal is very low. Therefore, within the roof control area, the top coal undergoes significant deformation and a higher degree of damage under the stress of the roof.

2.2.2 Stress Field in the Top Coal

During the mining process, the stresses in the surrounding rocks of the panel and in the top coal will change, and the typical stress distribution in the top coal is shown in Fig. 2.8a. The following is a simulation of the stress field variation in the top coal using numerical methods. The basic condition of simulation is to assume that the maximum principal stress is 10 MPa in the vertical direction of the original ground stress, the minimum principal stress is 3 MPa in the horizontal y-axis direction (the direction of the panel), and the intermediate principal stress is 6 MPa in the horizontal x-axis direction (the direction of the panel advancing). A total of five schemes were simulated and analyzed, in which the angle between the direction of the minimum principal stress and the panel was 0°, 30°, 45°, 60°, and 90°, which were labeled as N0, N30, N45, N60, and N90, respectively.

Figure 2.10 shows the distribution of maximum and minimum principal stresses in the top coal when the angle is different. The maximum principal stress starts to increase slowly at 80 m in front of the panel under the influence of mining, then increases sharply at 20 m in front of the panel and reaches the peak point at 7 m in front of the panel with the peak coefficient of 2.5, and then the top coal starts to be damaged, the bearing capacity decreases, and the maximum principal stress decreases rapidly to the residual strength of the top coal. The influence of different advancing programs of the panel on the maximum principal stress is small. The minimum principal stress in the top coal has a similar change law with the maximum principal stress, after the peak, the excavation unloading effect starts to play a dominant role, the minimum

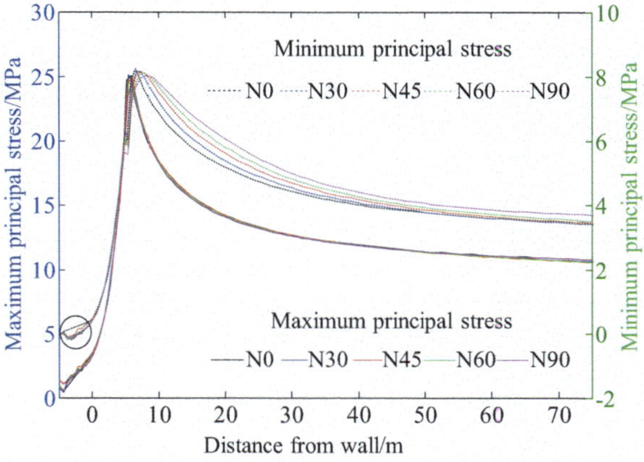

Fig. 2.10 Distribution of principal stress in top coal

principal stress in the top coal starts to decrease, and the stress environment of the top coal tends to be poor. Near the coal wall of the panel, the minimum principal stress in the top coal decreases to the level of 0. After that, under the influence of the cantilever effect of the top coal in the roof control area, the minimum principal stress in the top coal starts to load in the reverse direction, which transforms into the tensile stress, and ultimately prompts the top coal to fall behind the shelf. The distribution of the minimum principal stress is greatly influenced by the direction of panel advancing.

After the panel advances the same distance, the distribution of principal stresses in the top coal of the five scenarios is shown in Fig. 2.11. After the coal seam is mined, the stresses in the goaf decrease rapidly, and the overburden gravity in this range transfers to the solid coal in the surrounding area. The coal body at the proximity side enters the plastic failure area, the bearing capacity is reduced, and the high stress concentration area is transferred to the deep coal body, and the maximum principal stress peak of the top coal appears in the middle of the panel unlike the end distribution form of the peak of the abutment pressure of the coal body in front of the panel. The minimum principal stress also has an obvious concentration phenomenon, which is different from the usually considered continuous reduction process of horizontal stress in the coal seam under the action of excavation unloading, indicating that the horizontal stress is subject to the double influence of excavation unloading and overburden subsidence movement. Compared with the maximum principal stress, the minimum principal stress changes relatively slowly. Comparing the stress distributions in the five models, the maximum and minimum principal stresses in the coal seam do not change much with the change of advancing direction, and the peak coefficients of the maximum and minimum principal stresses are both about 2.5, which indicates that the change of advancing direction does not have a significant effect on the size of the stress path in the top coal. Combined with the characteristics that the location of the peak of the minimum principal stress in the top coal exceeds the peak of the maximum principal stress, the stress path experienced by the top coal is that the maximum and minimum principal stresses are loaded at the same time at first, and the external load and its own bearing capacity of the coal body increase at the same time, however, the external load is always smaller than the critical load that the coal body can withstand. The minimum principal stress reaches the peak, and it is gradually reduced by unloading under the action of the excavation, and the coal body's bearing capacity decreases, and the maximum principal stress quickly reaches the peak, and the coal wall reaches the limit equilibrium state. With the unloading caused by excavation and the continuation of the phenomenon of roof settlement, the top coal undergoes damage, the strength decreases, and the level of the maximum and minimum principal stresses in the coal body shows a decreasing trend at the same time, the number of the fractures of the top coal is constantly increasing, and the broken size decreases.

2.2.3 *Coal Mechanical Property*

2.2.3.1 Fracture Distribution of the Coal Body

1. Fracture survey

The coal body contains many fractures, which have a great influence on the overall mechanical characteristics of the coal body, especially for the higher strength (uniaxial compressive strength (UCS) being greater than 20 MPa), the fractures have a more obvious influence. It affects both the mechanical characteristics and the crushing degree of the coal body. Usually, we can use the Scan Line method suggested by the International Society for Rock Mechanics to investigate the distribution of fractures on the surface of the coal body, see Fig. 2.12, and use the compass, tape measure, steel tape measure on the coal wall of the roadway or the panel to measure the fracture surface intersecting with the scan line in terms of its production shape, distance D, half-trace length L.

Through many measurements, it is possible to obtain the statistical distribution of the production shape, distance, half-trace length on the surface of the coal body, and give the distribution type and parameters. The following are the results of the fracture investigation and statistical analysis of the coal body in the LTCC Panel 8914 of Xinzhouyao Coal Mine, Datong Coal Mining Group. Because the coal body of the panel is strong and the fractures are not developed, two blasting roadways are dug in the top coal, which are called Roadway 1 and Roadway 2, used for construction of blastholes and pre-blasting of the top coal, see Fig. 2.13. The fracture investigation of the coal body was carried out in Roadway 1, Roadway 2 and the panel, and the statistical results are shown in Tables 2.1, 2.2 and 2.3.

The distribution of fractures in the coal body appears to be disordered on the surface, however, in fact their distribution has a certain statistical law. The fracture in the coal body usually has 2 or 3 groups of dominant fracture groups, and the production of dominant fracture groups is often related to the cause of coal and geological movement, see Fig. 2.14. The fracture distance and half-trace length obey

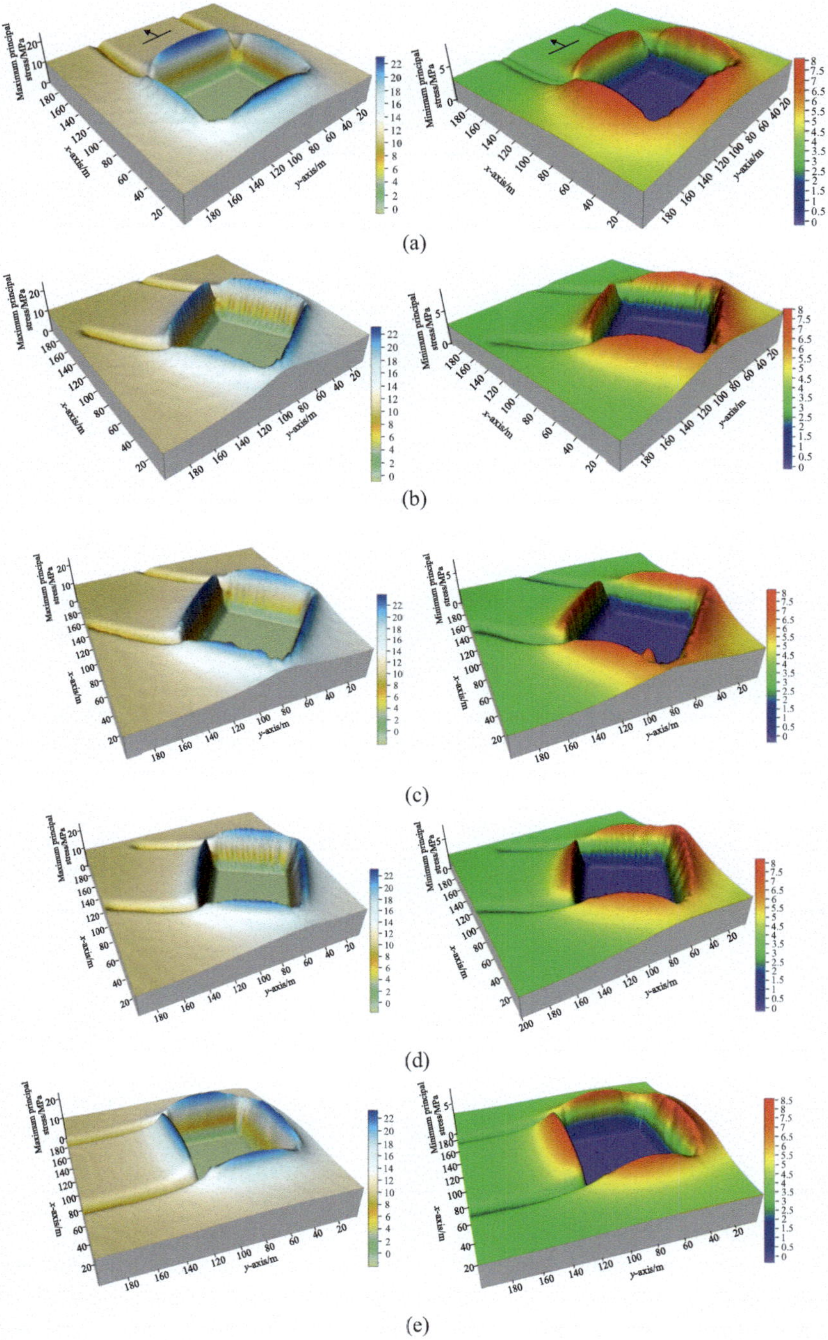

Fig. 2.11 Distribution of principal stress magnitude: **a** Scheme N0; **b** Scheme N30; **c** Scheme N45; **d** Scheme N60; **e** Scheme N90

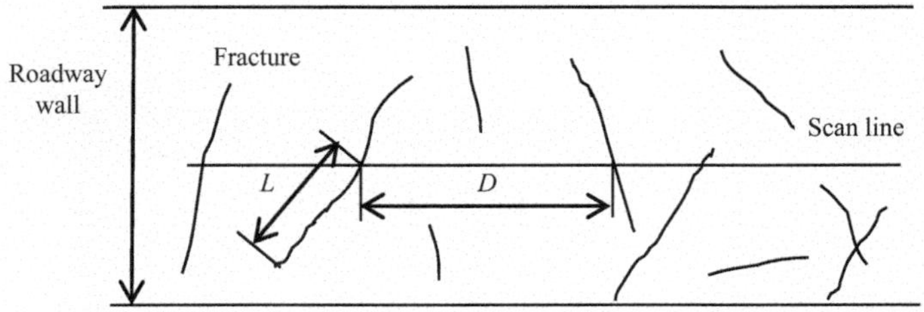

Fig. 2.12 Fracture distribution survey by Scan Line method

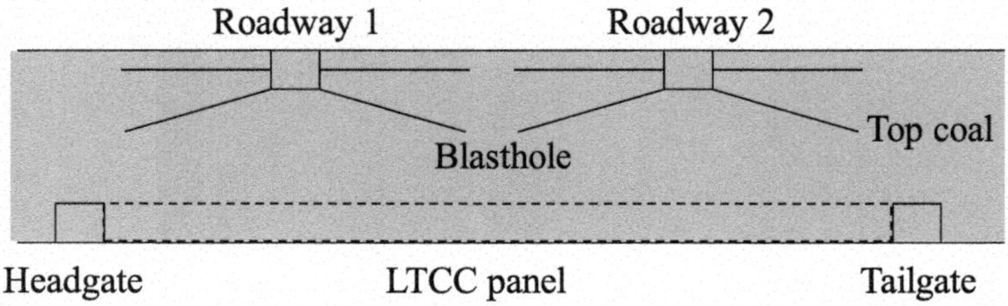

Fig. 2.13 Roadway layout of LTCC Panel 8914

Table 2.1 Characteristic parameters of *D*

	Average value (cm)	Standard deviation (cm)	Variation coefficient	Density (article/m)	Distribution function
Roadway 1	65.85	53.48	0.8523	1.5188	Negative exponential
Roadway 2	62.5714	49.615	0.7979	1.598	Negative exponential
Panel	108.6	61.333	0.5649	0.9028	NA

Table 2.2 Characteristic parameters of *L*

	Average value (cm)	Standard deviation (cm)	Variation coefficient	Distribution function
Roadway 1	57.826	31.38	0.5427	Lognormal
Roadway 2	76.9139	40.2116	0.5228	Lognormal
Panel	39.2273	20.3421	0.5187	NA

a certain distribution in the statistical law. The measured distance of the coal body fractures in Panel 8914 obeys the negative exponential distribution (Fig. 2.15), the half-trace length of the fractures obeys the lognormal distribution (Fig. 2.16), and the other researches on the distribution of the rock fractures have the similar statistical results, however, some differences are existing in the specific distribution function [7, 8].

Table 2.3 Parameter distribution of dominant fracture groups

Fracture zone	Group I	Group II	Group III
Average value of dip direction	40°	170°	350°
Distribution of dip direction	Normal	Normal	Normal
Distribution of dip angle	Exponential	Exponential	Exponential
Distribution of *D*	Negative exponential	Negative exponential	Negative exponential
Distribution of *L*	Lognormal	Lognormal	Lognormal

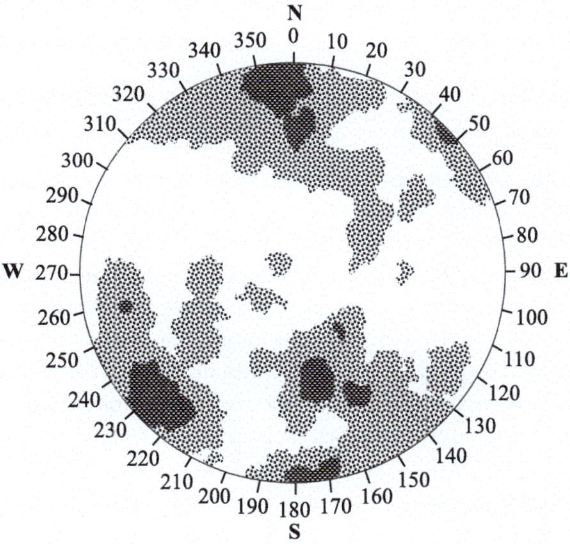

Fig. 2.14 Distribution of fracture poles on the surface of coal body

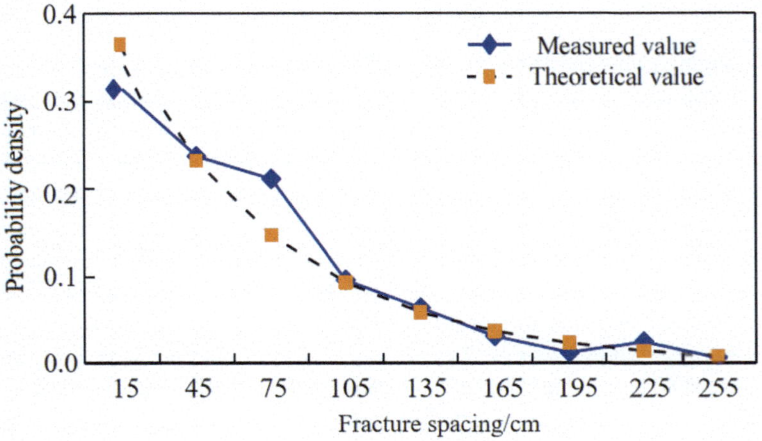

Fig. 2.15 Measured and theoretical fitting curves for the fracture distance *D* in Roadway 1

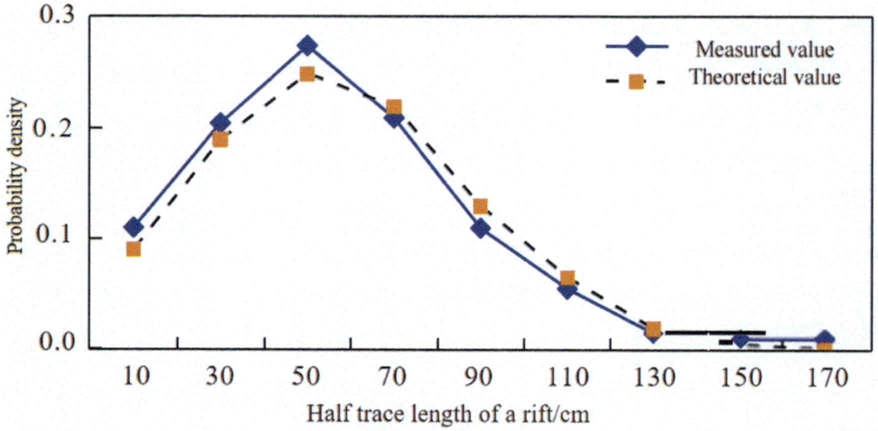

Fig. 2.16 Measured and theoretical fitting curves for the half-trace length L of in Roadway 1

In addition to the traditional manual on-site measurements, the non-contact measurement methods for surface fracture counting with the help of the multi-source photography principle has been developed in the recent years. For example, the ShapeMetriX3D system developed by Austrian Startup Company can realize the 3D non-contact measurement of geometric parameters of coal and rock body fractures. The ShapeMetrix3D system mainly consists of imaging system, software analysis system and sighting ruler. A calibrated digital camera is used to image the designated area at the two positions of the rock body, and then the acquired photos are imported into the analysis software package, and through a series of techniques (datum calibration, pixel matching, and image deformation deviation correction, etc.), 3D geometric images are synthesized to realize the reconstruction of a true 3D model of the rock body surface.

2. 3D simulation of coal body fractures

The investigated and statistically analyzed fractures are only the fractures on the surface of the coal body, or the intersection line between the fracture surface and the exposed surface inside the coal body, which cannot reflect the distribution of fractures inside the coal body yet. With the Poisson disk model, the 3D distribution of fractures inside the coal body can be constructed with the help of the statistical distribution of fractures on the surface of the coal body and Monte Carlo simulation. The Beacher disc model considers that the fracture surface inside the coal body is several discs, and the fractures observed on the surface of the coal body are the intersection lines between the fracture surfaces of these discs and the surface of the coal body, as shown in Fig. 2.17. In the actual simulation, it is assumed that the fracture surfaces inside the coal body are a few spatially distributed discs, and each fracture disc can be represented by the center coordinates, the diameter and the shape of the production. Spatially the coordinates of the disk center position obey the Poisson Distribution, the dip and dip angle are calculated according to the

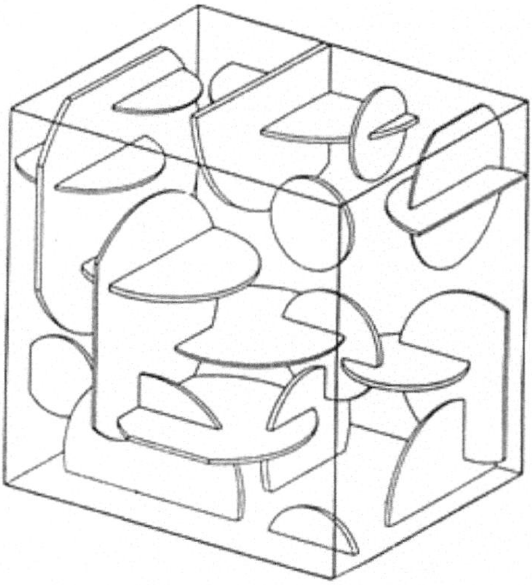

Fig. 2.17 Beacher disc model

fitted distribution of the measured sample data in the field, and the disk diameters of the dominant fracture group are derived according to the fitted distribution of the measured data of the half-trace length in the field, and then stochastic simulation is carried out using the Monte Carlo method.

The distribution of fracture disc center and half-trace length is independent of the distribution of fracture dip and dip angle, and thus the effect of dip and dip angle distribution need not be considered when simulating the half-trace length and disc center position. Statistical results show that the distribution of the fracture disc center position is uniformly distributed in the statistical region, and the half-trace length obeys a lognormal distribution. The simulation results are shown in Fig. 2.18 [3].

2.2.3.2 Physical Characteristics of Fractured Coal

1. Stress–strain relationship

In fact, it is difficult to obtain real coal samples containing various fracture distributions, thus the similarity theory is applied to conduct stress–strain relationship experiments using synthetic briquette coal containing fractures. Cubic briquette coal with each side length of 100 mm, are made by combining coal powder, cement, and water according to a certain ratio. In Fig. 2.19, uniaxial compression experiments were carried out on different fracture angles, fracture number, and fracture lengths, respectively [9].

Figure 2.20 shows the comparison of the experimental results between the briquette coal and the real coal sample, where σ_a is the axial stress, ε_a is the axial strain, ε_r is the radial strain, and ε_v is the volumetric strain. During the initial loading stage (*OA*), the coal sample undergo an obvious initial crack closure process, and then the coal sample enters the linear elastic stage (*AB*). When the axial stress reaches

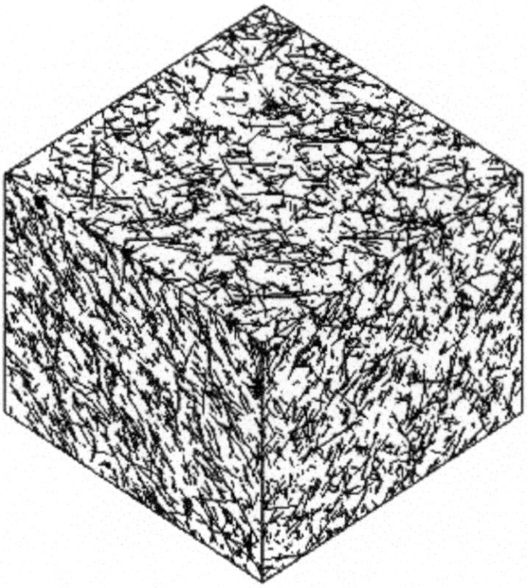

Fig. 2.18 3D network of fractures

the initial yield strength of the coal sample, the elastic deformation stage ends, and the coal sample enters the strain-hardening stage (*BC*), which produces irrecoverable plastic deformation in the coal sample, and the radial strain increment is larger than axial strain increment, and the compressive strength of the coal sample at this stage shows a nonlinear increase with the increase of plastic deformation, and the shape of the stress–strain curve is up-convex. After the axial stress reaches the peak strength, the deformation of the coal sample enters the strain softening stage (*CD*), in which the tiny cracks produced by plastic deformation in the coal sample rapidly expand and penetrate, compared with the initial volume before loading, the coal sample transitions from compression to expansion near the peak point, and with the accumulation of plastic deformation and development of microscopic cracks, macroscopic cracks appear on the surface of the sample, the load carrying capacity continues to decrease, and ultimately, only the residual stress. Compared with the full stress–strain curve of the real coal samples (Fig. 2.20b), the macroscopic stress–strain curve of the

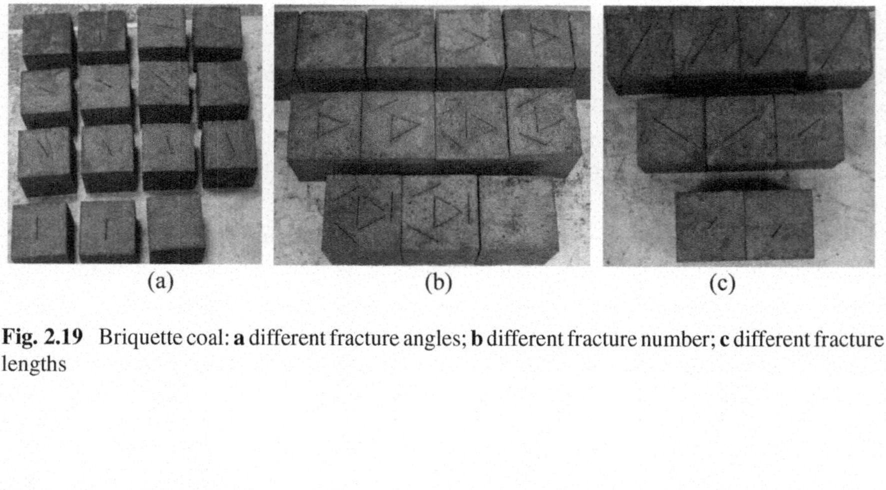

(a) (b) (c)

Fig. 2.19 Briquette coal: **a** different fracture angles; **b** different fracture number; **c** different fracture lengths

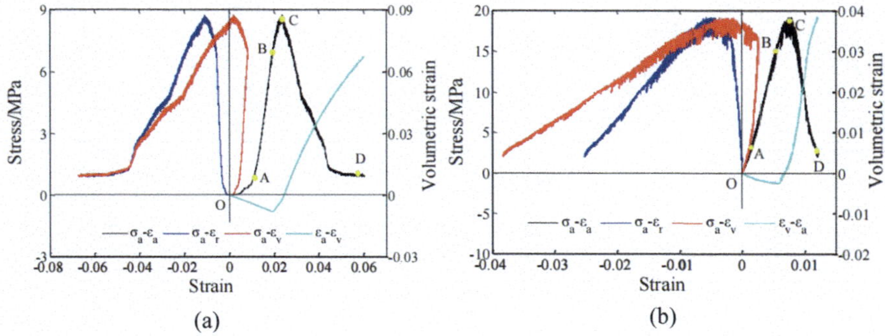

Fig. 2.20 Comparison of stress–strain curves: **a** briquette coal; **b** real coal sample

synthetic briquette coal also has the initial fracture compaction stage, linear elasticity stage, strain-hardening stage and strain-softening stage, and the briquette coal show obvious plastic flow behavior and shear expansion effect after reaching the initial yield, and the deformation of the briquette coal is transformed from compression to expansion near the peak, and the growth ratio of volumetric shear expansion deformation decreases with the increase of the degree of plastic deformation in the strain-softening stage, and the briquette coal can reflect the macroscopic deformation and damage characteristics of the real coal body to a high degree.

The results of uniaxial compression experiments on intact briquette coal, as well as coal specimens with different fracture angles (angle between the fracture and the horizontal plane, θ, °), different fracture lengths (l, cm), and different fracture number (n) are shown in Fig. 2.21. In Fig. 2.21, the above factors all influence the stress–strain relationship, in which an increase in the number and length of fractures significantly reduces the strength and elastic modulus of the briquette coal.

2. Strength characteristics
(1) Shear strength of briquette coal

The results of the shear experiments of intact briquette coals are shown in Fig. 2.22, and the shear strength increases with the increase of normal stress. From the theory of rock mechanics, there is the following relationship between the shear strength of coal and rock and normal stress:

$$\tau = C + \sigma_n \tan \varphi \tag{2.1}$$

where C is the cohesive force of the coal body, MPa; φ is the angle of internal friction of the coal body, °; τ and σ are the shear strength of the coal body and the positive stress acting on the shear surface, MPa, respectively. Fitting the experimental data in Fig. 2.22, the cohesive force and internal friction angle of the complete briquette coal are 4.5 MPa and 32°, respectively. The red line in Fig. 2.22 is the fitting curve, and the correlation coefficient between the experimental data and the fitting result reaches 0.91, which is a great match.

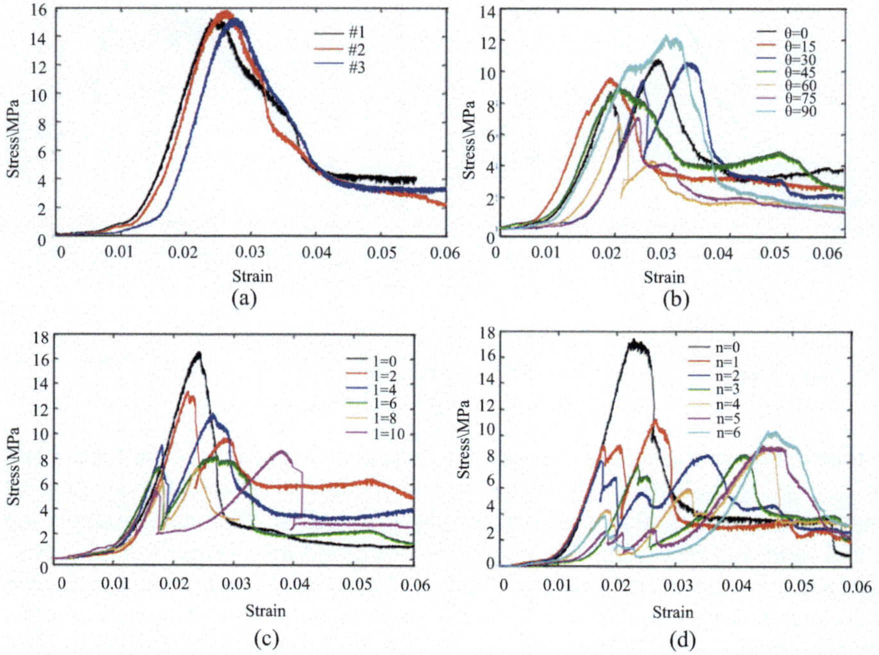

Fig. 2.21 Stress–strain curve **a** intact sample; **b** change in fracture angle; **c** change in fracture length; **d** change in fracture number

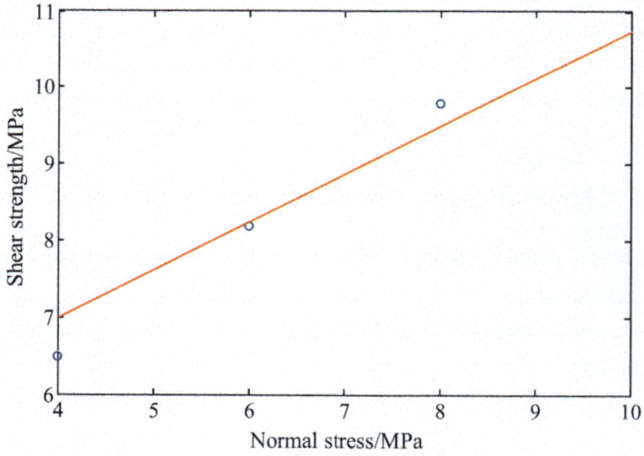

Fig. 2.22 Shear strength of intact briquette coals

(2) The effect of fractures on the elastic modulus of a coal body

As shown in Figs. 2.21b, c, the impact of fracture angle and length on the elastic modulus of coal body is relatively small. Therefore, the effects of these two factors on the elastic modulus are temporarily not taken into consideration. The impact of the fracture number on the elastic modulus of coal body is shown in Fig. 2.23. The

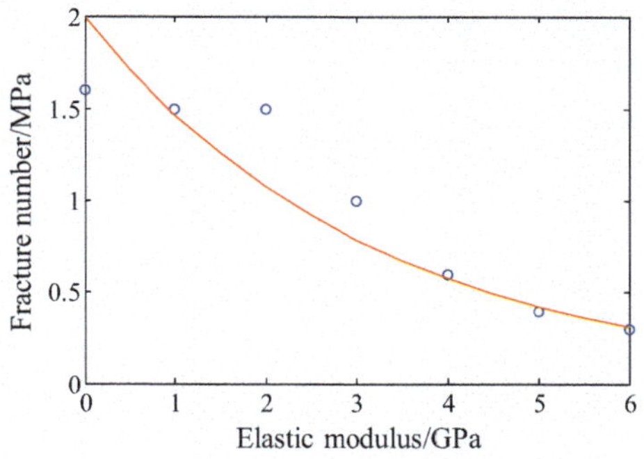

Fig. 2.23 Effect of fracture number on elastic modulus

elastic modulus decreases continuously with the increase of the fracture number in the elastic modulus. The fitting of the experimental data shows that the change trend of the elastic modulus of the coal body with the fracture number can be expressed by Eq. (2.2):

$$E = E_i \exp(-\alpha n) \tag{2.2}$$

where E_i represents the elastic modulus of intact briquette coals, GPa; α is the fitting constant; and n stands for the fracture number. Through data fitting of Fig. 2.23, the values of E_i and α are obtained as 2 GPa and 0.31, respectively. The fitting curve derived from Eq. (2.2) is indicated by the red line in the figure, which aligns well with the experimental data, achieving a correlation coefficient of 0.92.

(3) Effect of fractures on UCS

The effects of fracture angle (Fig. 2.24), fracture length, and fracture number on UCS of coal body are shown in Figs. 2.25, 2.26 and 2.27. The relationship between UCS and fracture angle is not monotonic, however, exhibits a trend of first decreasing and then increasing. The fitting analysis of the data in Fig. 2.25 indicates that the influence of fracture angle on UCS can be expressed by Eq. (2.3):

$$\sigma_c = \sigma_{c0}\left(1 - \frac{\beta_1}{\sqrt{2\pi}\beta_2}\exp\left(-\frac{(\theta-\beta_3)^2}{2\beta_2^2}\right)\right) \tag{2.3}$$

where σ_{c0} represents UCS of the coal sample when the fracture angle is 0, MPa; β_1, β_2, and β_3 are the fitting constants, among $\beta_3 = \pi/4 + \varphi/2$. Through data fitting of Fig. 2.25, the values of β_1, β_2, and β_3 are obtained as 16, 15, and 64, respectively. Substituting these parameters into Eq. (2.3) yields the fitting curve, as shown by the red line in Fig. 2.25. The fitting results exhibit the same trend as the experimental data, demonstrating a high degree of agreement, with a correlation coefficient of 0.88.

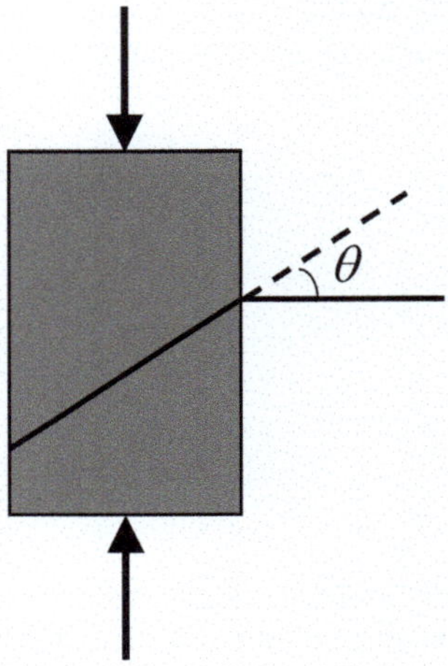

Fig. 2.24 Schematic diagram of fracture angle

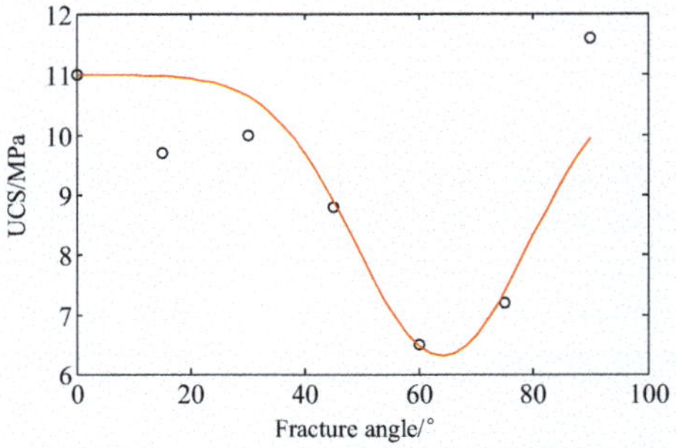

Fig. 2.25 Relationship between fracture angle and UCS

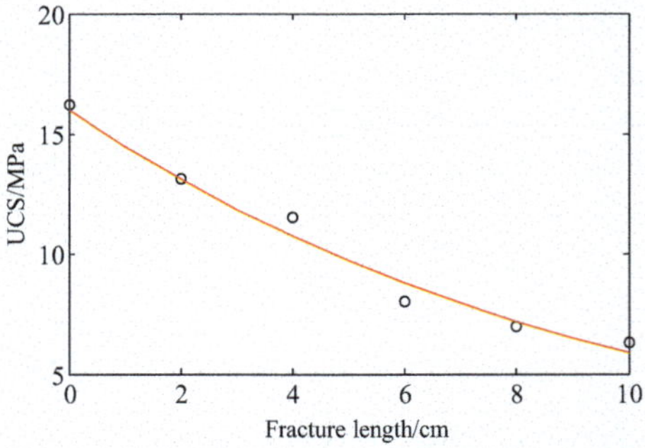

Fig. 2.26 The relationship between fracture length and UCS

The influence of fracture length on UCS of coal body is shown in Fig. 2.26. As the fracture length increases, UCS of the coal body continuously decreases, however, the ratio of decrease diminishes with the increase in fracture length. Through data fitting, the relationship between UCS of the coal body and the fracture length can be expressed by Eq. (2.4):

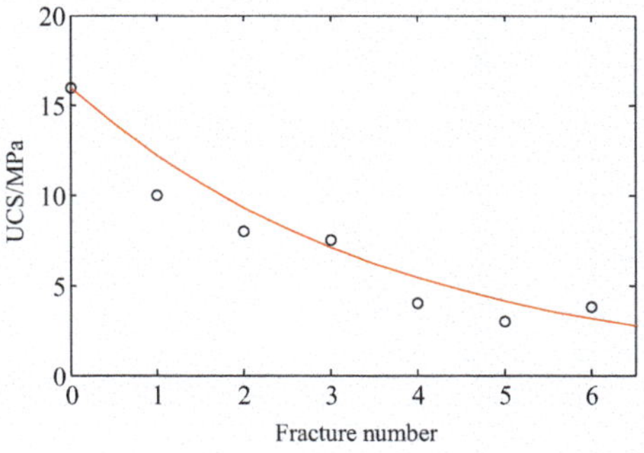

Fig. 2.27 Relationship between the fracture number and UCS

$$\sigma_c = \sigma_{ci} \exp(\eta l) \quad (2.4)$$

Where σ_{ci} represents UCS of intact coal, MPa; η is a fitting constant. The fitting results of the experimental data indicate that σ_{ci} and η are 16 MPa and 0.1, respectively. By substituting the above parameters into Eq. (2.4), the fitting curve is obtained as shown by the red line in Fig. 2.26. The fitting results show good consistency with the experimental data, with a correlation coefficient of 0.99.

The relationship between the fracture number in the briquette coal and UCS is shown in Fig. 2.27. The variation trend of UCS with the fracture number is similar to that with the fracture length, showing a negative correlation. The influence of the fracture number on UCS is more significant. When the fracture number increases from 0 to 6, UCS of the briquette coal decreases from 16 to 4 MPa.

By fitting the experimental data in Fig. 2.27, the relationship between the fracture number and UCS of the coal body can be expressed by Eq. (2.5):

$$\sigma_c = \sigma_{ci}\exp(-\kappa\, n) \quad (2.5)$$

where κ represents the fitting constant. The fitting result of the data in Fig. 2.27 indicates that its value is 0.27. Substituting this value into Eq. (2.5), the fitting curve is obtained as shown by the red line in Fig. 2.27. The fitting results show a high degree of agreement with the experimental data, and the correlation between the two sets of data can reach 0.96.

3. Effect of different loading paths on the strength of coal samples

Figure 2.28 shows the axial loading experiment of coal samples under different confining pressures.

Figure 2.29 depicts the experiment where the confining pressure is first increased to 7.5 and 15 MPa, followed by unloading of the confining pressure while the axial loading continues. Compared with the results of conventional triaxial experiments, under the condition of the same initial confining pressure, in the unloading confining pressure experiment, due to the gradual reduction of the confining pressure, the peak strength of the coal sample decreases, the residual strength of the coal sample after the peak is small, and it loses the load-bearing capacity, and under the condition of

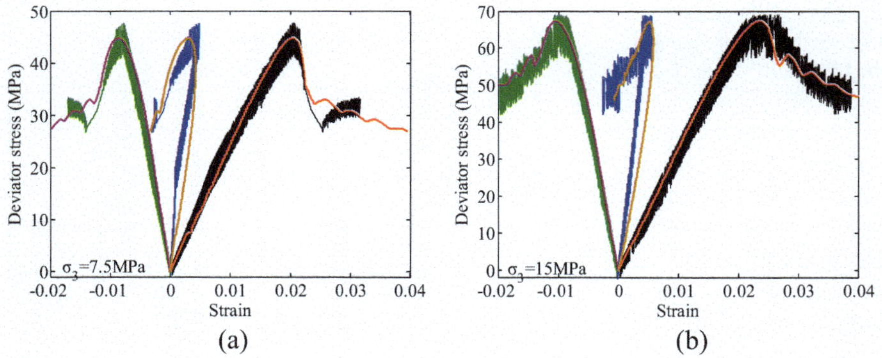

Fig. 2.28 Axial loading experiment with fixed confining pressure: **a** confining pressure 7.5 MPa; **b** confining pressure 15 MPa

Fig. 2.29 Experiment of unloading confining pressure and adding axial pressure

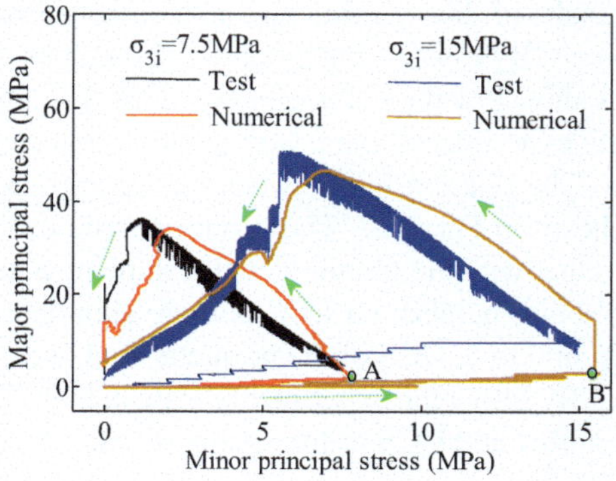

unloading the confining pressure, the coal sample is easy to be destroyed. The top coal above the support for LTCC is then in a state where it is loaded in the vertical direction and unloaded in the advancing direction.

2.2.4 Theoretical Analysis of Top Coal Damage Process

According to the results of the previous study, the top coal has small deformation and low damage degree in front of the panel, and large deformation and high damage degree behind the panel. Combined with the shear sliding of the contact surface between the top coal and the immediate roof, it can be inferred that the top coal shows completely different deformation and damage characteristics in the two stages of the front and back of the coal wall, and this phenomenon is caused by the change of the boundary conditions of the top coal's upper and lower surfaces. Therefore, the top coal damage mechanism should be studied by dividing the top coal into two stages: the front of the coal wall and the above the roof control area with the coal wall as the limit.

2.2.4.1 Damage Mechanism in the Front of the Coal Wall

1. Determination of Vertical/Horizontal Stress

The excavation of the panel causes the horizontal constraint on the coal seam to disappear, and the horizontal principal stress distributed in the coal seam gradually decreases from the initial value (original stress). This phenomenon is known as the unloading effect caused by excavation. As the distance from the panel decreases, the unloading effect caused by excavation in the coal body gradually increases. After the coal body ahead of the panel is affected by mining activities, the principal stress in the coal body gradually decreases and reaches zero at the panel. According to previous research, the change in horizontal principal stress caused by excavation from the initial mining influence point to the panel approximately follows the following negative exponential relationship:

$$\sigma_h = \sigma_{hi}(1 - \exp(-\alpha x)) \tag{2.6}$$

where σ_{hi} represents the initial horizontal principal stress, MPa, α is a fitting constant that controls the range of stress relief and the gradient of stress change, and this parameter is influenced by mining height, panel length, and panel advancing speed; x represents the distance from any point in the coal mass in front of the panel to the panel, taking the position of the coal wall in the panel as the origin of the coordinate, m. In addition to causing lateral unloading effects, excavation also induces subsidence movement of the overlying rock. Before being affected by mining, the overlying rock is in an initial equilibrium state. After being influenced by mining, part of the coal seam is recovered, forming a goaf. The overlying strata in this area lose support and start to subside. The gravity of these strata is transferred to the unrecovered part of the coal seam, increasing the vertical stress in this section of the coal body, thus generating support pressure ahead of the panel (see Fig. 2.8). Contrary to the trend of horizontal stress reduction due to excavation unloading, vertical stress increases from the initial mining influence Point A. The increase in vertical stress can be fitted by the following equation:

$$\sigma_v = (K - 1)\gamma H \exp((1 - x)/\beta) + \gamma H \tag{2.7}$$

where β is the fitting constant, K the peak coefficient of vertical stress, l is the distance from the peak vertical stress point to the panel.

Numerical simulations and field measurements show that the maximum value of the vertical stress peak coefficient during underground mining can reach 4. Therefore, the value of K ranges between 1 and 4:

$$\sigma_{v\max} = K\gamma H, K = 1 - 4 \tag{2.8}$$

At the peak stress point, the top coal enters a state of ultimate equilibrium, and due to plastic deformation and the development of microfractures, damage begins to appear in the top coal. Assuming no softening of the top coal, within the range from the peak vertical stress position to the panel, the top coal remains in a state of ultimate equilibrium, satisfying the M-C yield criterion and deformation consistency conditions:

$$\sigma_v = N_\phi \sigma_h + 2\sqrt{N_\phi} C \tag{2.9}$$

where $N_\varphi = (H\sin\varphi)/(1 - \sin\varphi)$, φ is the internal friction angle of coal. From Eq. (2.9), due to the continuous decrease in horizontal stress, the vertical stress decreases from the peak value, eventually dropping to UCS of the top coal at the panel. By substituting Eqs. (2.6) and (2.8) into Eq. (2.9), the horizontal distance from the peak vertical stress point *B* to the panel can be obtained:

$$1 = -\frac{1}{\alpha}\ln\left(1 - \frac{1}{\sigma_{hi}N_\varphi}\left(K\sigma_{vi} - 2\sqrt{N_\varphi}C\right)\right) \tag{2.10}$$

Substituting Eq. (2.10) into Eq. (2.6) gives the horizontal stress value corresponding to the peak vertical stress position:

$$\sigma_{h\max} = \sigma_{hi}\left(1 - \exp\left(\ln\left(1 - \frac{1}{\sigma_{hi}N_\varphi}\left(K\sigma_{vi} - 2\sqrt{N_\varphi}C\right)\right)\right)\right) \tag{2.11}$$

Combining Eqs. (2.6)–(2.11), the characteristics of vertical and horizontal principal stress changes in the top coal can be preliminarily obtained, forming the basis for understanding the damage mechanism of the top coal ahead of the panel.

2. The risk coefficient of top coal failure

After obtaining the characteristics of the principal stress distribution in the top coal, to facilitate the analysis of the risk distribution of the top coal failure, we first define the risk coefficient of top coal failure. The relative positional relationship between the Mohr–Coulomb strength curve and the Mohr circle is shown in Fig. 2.30. The risk coefficient of top coal failure (f_d) is defined as the difference between the radius of the Mohr stress circle and the vertical distance from the center of the circle to the strength curve. According to the geometric relationship, the risk coefficient of failure f_d is related to the principal stresses σ_1, σ_3, and the strength parameters of the top coal (c, φ) as follows:

$$f_d = \frac{1}{2}(\sigma_1 - \sigma_3) - \frac{1}{2}(\sigma_3 - \sigma_1)\sin\varphi - c\cos\varphi \tag{2.12}$$

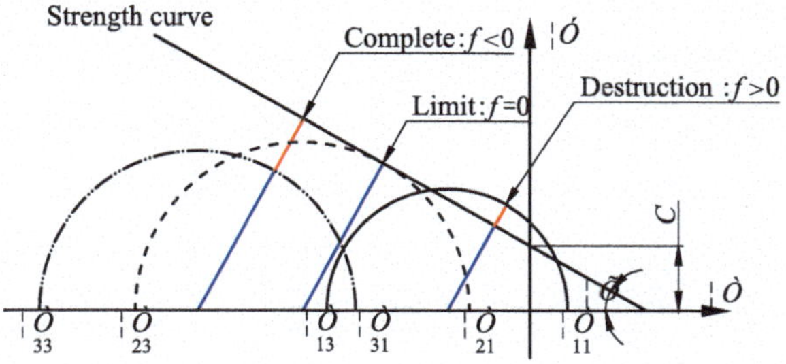

Fig. 2.30 Top coal damage hazard factor

Equation (2.12) essentially represents the expression of the principal stress under the M-C strength criterion. If the Mohr's stress circle determined by the stress state of the top coal is separated from the strength curve, the top coal is in an elastic state, where $f_d < 0$; if the Mohr's stress circle is tangent to the strength curve, the top coal is in a limit equilibrium state, where $f_d = 0$; if the Mohr's stress circle intersects the strength curve, the top coal enters a failure state, where $f_d > 0$. Therefore, the risk of top coal failure increases as f_d increases, and the two are positively proportional.

3. Analysis of the top coal damage process

Assuming that the initial vertical and horizontal principal stresses in a coal seam are 12 MPa and 8 MPa, respectively, with a peak support pressure coefficient K of 2, and the coal's cohesive strength and internal friction angle are 2 MPa and 30°, respectively. The unknown constants α and β in Eqs. (2.6)–(2.11) are taken as 0.1 and 20, respectively. Substituting these parameters into Eqs. (2.7) and (2.10) yields the distribution curves of vertical and horizontal principal stresses in front of the coal wall, as shown in Fig. 2.31. Under the influence of mining, the stress distribution in the top coal begins to change at 45 m (Point A) in front of the panel. The vertical stress increases under the transfer of overburden load, however, the change in horizontal stress is not significant. Its value begins to decrease under the effect of excavation unloading at about 20 m from the panel, indicating that the influence of excavation unloading on horizontal stress lags the influence of overburden subsidence on vertical stress. The peak vertical stress occurs at 13 m in front of the panel (Point B), reaching 18 MPa. From Point A to Point B, the top coal failure risk coefficient gradually increases, however, in the AA' segment, only the vertical stress changes, and the effect of excavation unloading is not significant, resulting in a small growth ratio of the top coal failure risk coefficient. In the A'B segment, the top coal is simultaneously affected by overburden subsidence and excavation unloading, and the growth ratio of the failure risk coefficient increases significantly. Comparing the slopes of the failure risk coefficient curves in the AA′ segment and the A′B segment, it can be judged that the role of excavation unloading (horizontal stress unloading) in promoting top coal failure is significantly greater than that of overburden settlement (vertical stress loading). At Point B, the top coal failure risk coefficient increases to 0, indicating that the top coal enters a limit equilibrium state and begins to damage. Subsequently, as the horizontal stress continues to decrease under the effect of excavation unloading, the vertical stress also shows a decreasing trend after the peak point.

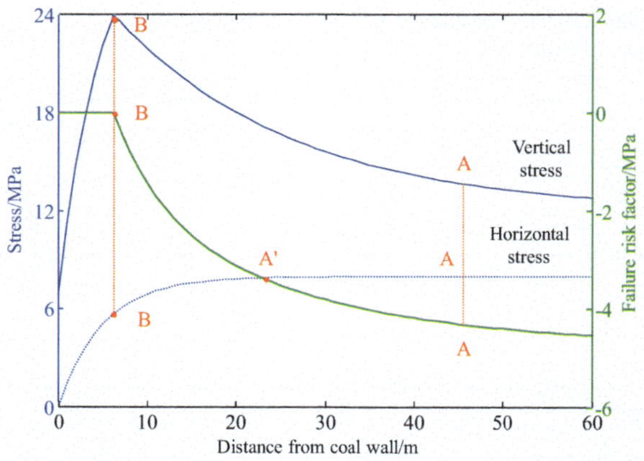

Fig. 2.31 Characteristics of vertical and horizontal stress changes in top coal

As shown in Fig. 2.31, the distribution of vertical and horizontal stresses in the top coal essentially represents the stress path experienced by any point in the top coal during the transition from the original rock stress state to the coal wall position. In front of the peak point, the top coal is in the elastic deformation stage. Although the top coal failure risk coefficient continues to increase during this stage, the top coal will not fail, and its load-bearing capacity remains within the triaxial ultimate bearing capacity of the top coal. According to the theory of elasticity and plasticity, the degree of top coal damage is only related to the stress history experienced during the subsequent yield stage. Therefore, the stress path in this stage (AB) will not affect the damage and failure of the top coal.

2.2.4.2 Damage Mechanism Above the Roof Control Area

In the roof control area, the main factors causing continuous damage to the upper and lower top coal differ. The upper top coal undergoes tensile failure due to horizontal tensile stress induced by roof rotation, while the lower top coal experiences shear or tensile failure due to repeated support by the supports. Therefore, their damage mechanisms differ significantly.

1. Upper top coal

Without considering the effect of support loads, the upper top coal in the roof control area can be regarded as a cantilever beam under uniform load. The horizontal stress distribution in the top coal under this condition is shown in Fig. 2.32. The horizontal coordinate origin is at the panel, and the area behind the panel is negative. As shown in Fig. 2.32, horizontal tensile stress occurs in the upper top coal, while compressive stress occurs in the lower top coal. The maximum tensile stress is located at the upper surface of the fixed section (panel). According to the theory of material mechanics, the horizontal tensile stress in the top coal increases with the increase in roof load and the decrease in top coal thickness. Therefore, the top coal recovery can be significantly

Fig. 2.32 Horizontal stress distribution in the top coal of the fixed support beam

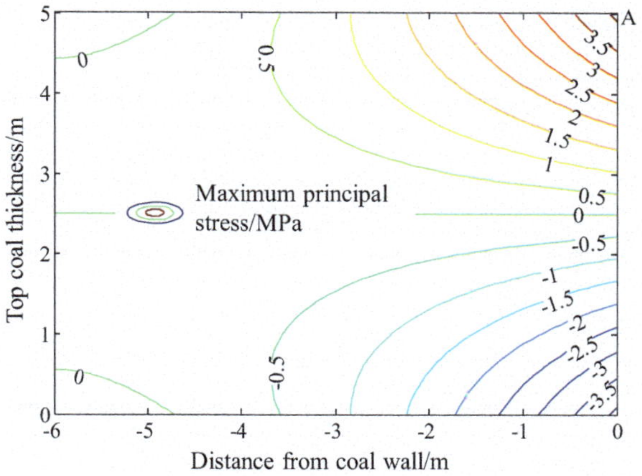

improved by adopting LTCC for medium-hard roof conditions that fracture and fall promptly and coal seams of suitable thickness (<12 m).

2. Lower top coal

Within the roof control range, the repeated support action on the top coal can be considered as cyclic loading. Experiments show that cyclic loading and unloading during the post-peak stage can effectively reduce the residual compressive strength of the coal body. As shown in Fig. 2.33, unloading at Point *A* after the peak, the residual peak strength of the coal body is less than the axial stress at the time of unloading during reloading, indicating a reduction in residual strength.

Laboratory experiments indicate that the relationship between the damage variable and permeability of the coal body and the number of cycles during cyclic loading and unloading tests is as shown in Fig. 2.34. Permeability reflects the development degree of fractures in the coal body to a certain extent. Under the condition of constant maximum cyclic load, both the damage variable and permeability of the coal body increase with the number of cycles, reaching a stable state after ten cycles.

Fig. 2.33 Stress–strain curve of coal body under cyclic loading and unloading [10]

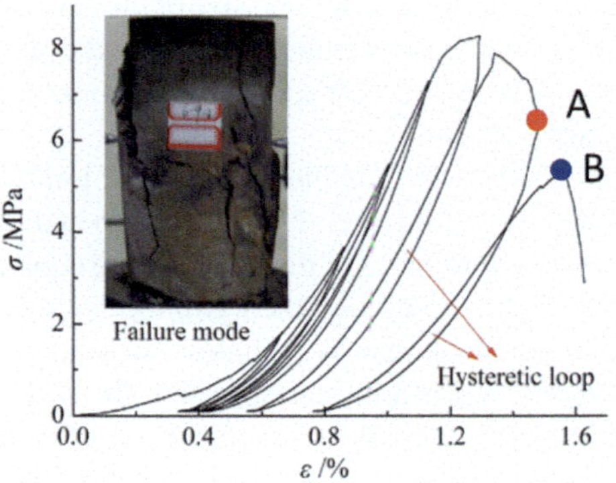

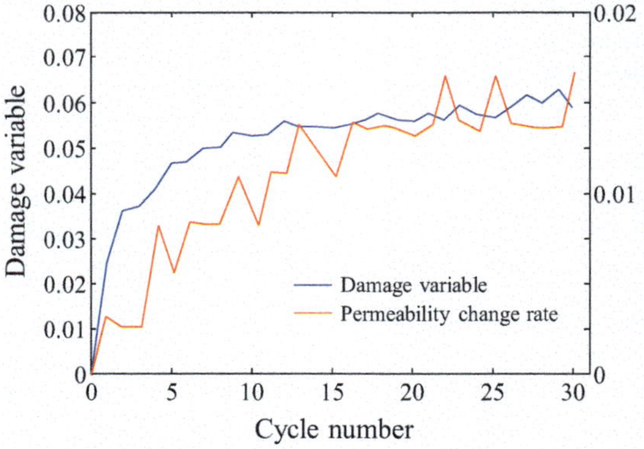

Fig. 2.34 Relationship between damage variables and permeability with number of cycles [11]

As inferred from Fig. 2.34, with the maximum support load of the supports remaining constant, reducing the shearer's cutting depth can increase the number of times the top coal undergoes repeated support within the roof control area. That is, enhancing the load cycles experienced by the top coal can effectively enhancing the degree of top coal damage, the recovery and extraction efficiency of the top coal can ultimately be improved.

2.2.5 Top Coal Size Distribution

2.2.5.1 Prediction of Top Coal Size

For coal seams with hard coal bodies (UCS > 30 MPa) and undeveloped fractures, one of the main challenges in applying LTCC technology is the difficulty in caving the top coal, resulting in large sizes and challenges in extraction. Therefore, studying the size of the broken top coal is particularly important.

The natural fracturing in hard coal bodies, like in other rocks, is characterized by the spatial distribution of fracture surfaces and their developmental extent, often determining the shape and size of the resulting broken blocks. Thus, investigating and analyzing these fracture surfaces, and establishing a 3D polyhedral block prediction model based on their spatial distribution, form the foundation for studying coal body fragmentation.

Firstly, based on the principles of joint network generation and the observed and fitted data from Sect. 2.2.3.1, the dip and dip angle of the joint surface, disc radius, center coordinates of the disc, and the equation of the joint surface are derived. Then, the simple homology theory in topology is used to establish a 3D block prediction model, which mainly includes: obtaining the edges of all polyhedra from the equations of each joint surface, and all edges have two directions; starting from the bidirectional chain, obtain the 1D unidirectional closed chain, which is the polygon of all the closed surfaces that make up the polyhedron; Through simple homology

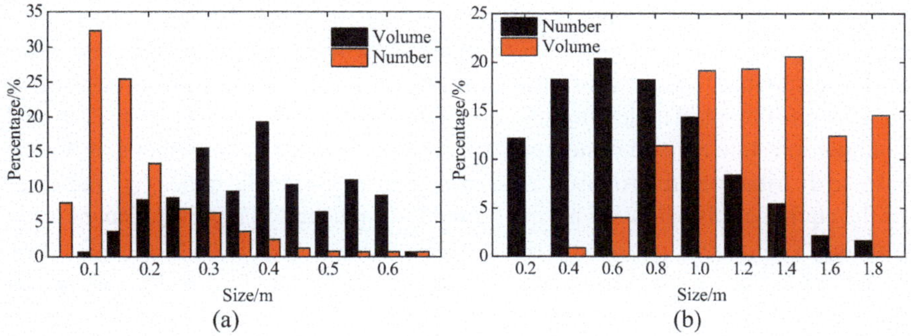

Fig. 2.35 Distribution of top coal block size: **a** initial block size; **b** top coal size after pre-blasting

operation, the unidirectional 2D closed chain is obtained from the unidirectional closed chain, which is the closed surface of all polyhedra. In the data structure of the computational program, spatial coordinate values and function Equations are used to represent the vertices and joint surfaces of polyhedra, to determine the maximum line size and polyhedral volume of each polyhedron. Finally, the maximum unidirectional size of the polyhedral block is used to represent the block size of each block, and the block size distribution of the top coal according to the original fracture can be obtained [12].

Taking the statistics of coal body fractures in Panel 8914 panel of Xinzhouyao Coal Mine as an example for actual calculation, the block size distribution of the top coal according to the original fracture is shown in Fig. 2.35a. According to the maximum opening size of DO of 1.2 m (ZFS6000-22/35 type drawing support), and calculated with a utilization coefficient of 0.9 based on the size of DO, the top coal with a block size of 1.08 m can be drawn. Therefore, the top coal recovery is less than 49.93%, which cannot meet the requirement of coal recovery ratio being 80% in the panel, and manual auxiliary crushing is required (Fig. 2.13). The distribution of top coal size after auxiliary blasting is shown in Fig. 2.35b. The weighted average block size of the top coal is 0.38 m, the top coal recovery is 75.7%, and the recovery ratio of the panel is 84.67%, which meets the requirements of LTCC under these conditions.

2.2.5.2 Measurement Method of Top Coal Size Distribution

1. Determination of measurement method

The bottom coal of the lower part of the LTCC panel is cut by the shearer and transported out of the panel by the front AFC. However, the top coal is gradually crushed into loose size under the combined effect of mining stress, repeated support, and lateral stress relief, etc., and accumulated above and behind the tail beam of the support, as shown in Fig. 2.36a. When DO is opened (Fig. 2.36b), the granular top coal flows into the rear AFC, then enters the belt conveyor through the transfer machine and crusher, and is finally transported out of the panel. As shown in Fig. 2.37, considering the complete procedure of top coal drawing and transportation out of the panel, the suitable locations for measuring the top coal size are mainly on the belt conveyor and the space under the tail beam of the support. Combined with the top coal accumulation state at the measurement time, three measurement methods of the top coal size distribution are proposed, as shown in Table 2.4. In the table, "√" indicates that the equipment is operating normally, and "×" indicates that the equipment is not operating. The specific procedures of the three measurement methods are as follows.

Method 1: First, a suitable measuring support is selected for coal drawing operation, while the rear AFC is operating normally and the crusher and front AFC do not operate temporarily. After the granular top coal reaches the belt conveyor through the AFC and transfer machine, the belt conveyor is closed. The surveyors select a suitable measurement position and length on the belt conveyor, which is the position (1) in Fig. 2.37. The surveyors first measure the size and weight of the large top coal, then sample the small top coal and measure these in a safe area nearby. With this

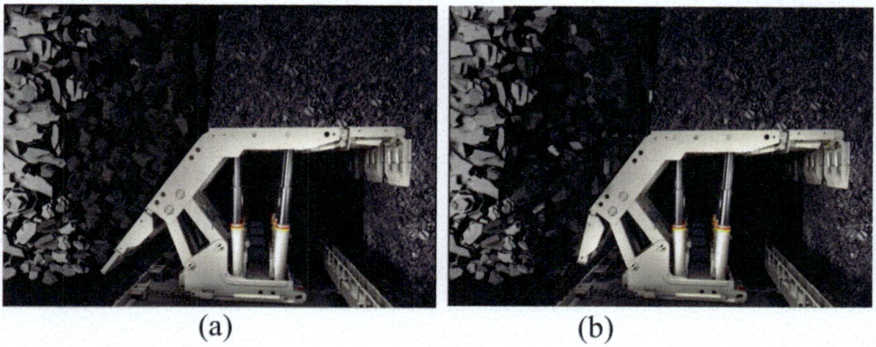

Fig. 2.36 Status of top coal caving support: **a** DO closed; **b** DO opened

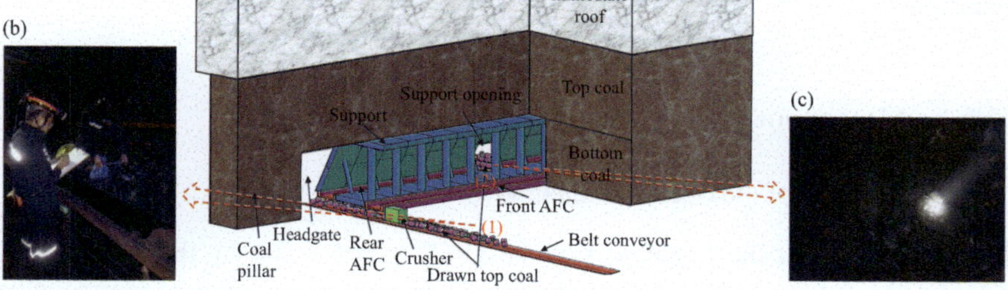

Fig. 2.37 The measurement methods of the top coal size distribution: **a** schematic diagram of the LTCC panel; **b** measurement in the headgate; **c** measurement below the shield beam

Table 2.4 Equipment operation and top coal stacking under the three methods

	Name	Rear AFC	Transfer machine	Crusher	Belt conveyor	Front AFC	Top coal status
Method 1	Dispersed measurement method on belt	√	√	×	√	×	Dispersion
Method 2	Accumulated measurement method on AFC	×	×	×	×	×	Accumulation
Method 3	Dispersed measurement method on AFC	√	√	×	√	×	Dispersion

method, the top coal is evenly distributed and dispersed on the belt, and hence, this method is termed as the dispersed measurement method on the belt.

Method 2: First, the rear AFC, front AFC, transfer machine and crusher of the panel are shut down. When the coal is fully drawn from the selected support, the surveyors enter the lower space of the support tail beam, i.e., the position (2) in Fig. 2.37, through the gap between two supports. When less top coal is drawn, the size and weight of all drawn top coal can be measured. However, when a large quantity of top coal is drawn, the size and weight of the measured top coal are selected on the surface of the drawn top coal to the maximum possible extent. In this method, the top coal does not move with the rear AFC, and accumulates under DO in an accumulated form. Hence, this method is termed as the accumulated measurement method on the conveyor.

Method 3: First, the front AFC and crusher are shut down and the rest of the equipment operate normally. When coal drawing from the selected support is finished, the surveyors enter the lower space of the support tail beam, i.e., the position (2) in Fig. 2.37, through the gap between two supports, and select a suitable measurement position and length to measure the size and weight of the drawn top coal. In this method, the top coal moves with the rear AFC, and the top coal in the measurement area is dispersed. Hence, this method is termed as the dispersed measurement method on the conveyor.

To reduce the error of the measurement system to the maximum possible extent, 3–5 supports should be evenly selected in the panel to measure the size distribution of the drawn top coal, if the conditions allow.

Figure 2.38 shows the comparative analysis of the three methods in terms of the measurement environment, operation difficulty, applicable conditions and safety when measuring the top coal size distribution. It is evident that Method 1 has the advantages of a large measurement space, great air quality, unrestricted measurement time and high safety factor; however, secondary crushing during transportation to the rear AFC and transfer machine maybe produce, and it is also not suitable for

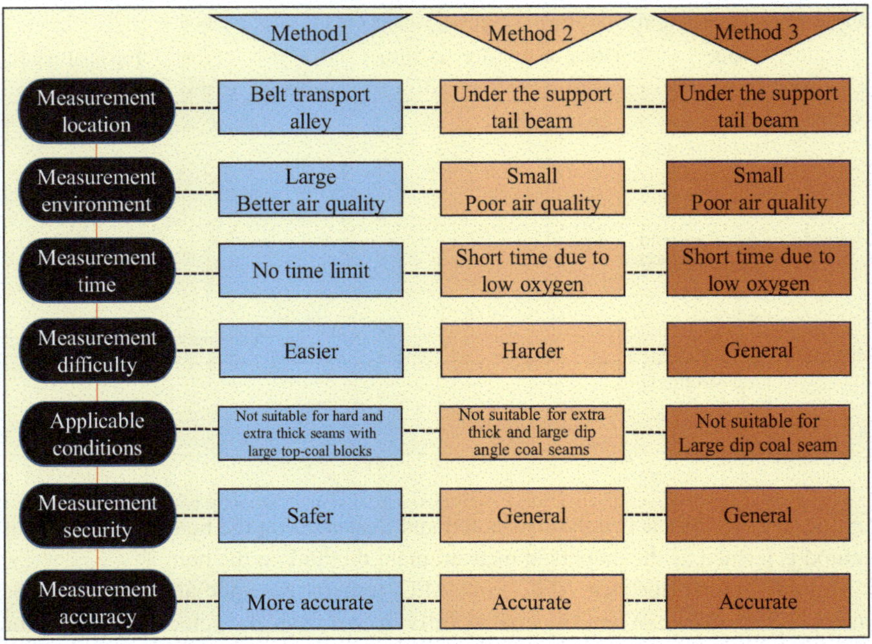

Fig. 2.38 Comparison of the advantages and disadvantages of the three methods

the hard and extra-thick seam with large top coal blocks. Method 2 is faster and more accurate for the measurement of thin top coal seams, and there is no secondary crushing in the subsequent transportation procedure. However, there are problems such as small measurement space, poor air quality, lack of general safety and thick top coal accumulation, which makes measurement difficult. Method 3 combines the advantages of Method 1 and Method 2. There is no possibility of re-crushing after the granular top coal flows out of DO, and the top coal is dispersed and displayed on the rear AFC, making sampling less difficult and more accurate.

All three methods can be used to measure the top coal size distribution, and each has its advantages and disadvantages. However, it is important to note that all three methods measure the size of the drawn top coal, which has been extruded and crushed through the coal drawing procedure, rather than the initial crushed size accumulated above the support tail beam before drawing. Considering the measurement accuracy and safety, Method 3 and Method 1 are preferred in this study, followed by Method 2, for the field measurement of the top coal size distribution in the panel. When Method 3 and Method 1 are selected, it requires a longer time and more effort to measure all the drawn top coal, because they are dispersed and widely distributed on the AFC or belt conveyor. Therefore, it is necessary to further determine the appropriate measurement parameters, such as the measurement position and measurement range on the belt or scraper, which can speed up the progress of measurement, reduce the measurement system error and ensure the field measurement accuracy. Therefore, based on the geological conditions of a certain LTCC panel, a PFC^{3D} numerical calculation model is established to analyze the top coal particle transport and drawing law, which provides a basis for selecting the best measurement area for the top coal size distribution.

2. Determination of measurement parameters

In Fig. 2.39, the initial model of the top coal drawing test is established based on the geological conditions of a certain LTCC panel. In Fig. 2.39a, it is the shape of top coal drawing body in the direction of panel layout. The top coal with different drawing heights is marked with different colours, and the granular top coal particles are gradually drawn in an approximate shape of cutting ellipsoid. With the increase of drawing height, the top coal drawing bodies gradually develop from the stocky shape to the tall and thin shape, referred as to the eccentricity gradually increases. As shown in Fig. 2.39b, it is the distribution of top coal particles on the rear AFC with different coal drawing heights, at 2.0, 2.5, 3.0, 3.5, 4.0, and 4.5 m. The drawing height is 2.0 m, referred as to the top coal particles at 2.0 m reach the AFC for the first time. Similarly, the drawing height is 4.5 m, that is, the top coal particles at 4.5 m exit DO for the first time. In the figure, 1/2/3 is the number of the support, and the red box is the position of the support, whose length is the length of DO. Furthermore, due to the increasing distribution length of drawn top coal, to clearly display top coal particles of different colours, some particles are omitted from the distribution of particles at the drawing height of 2.5–4.5 m.

Figure 2.40 shows the accumulated distribution length and volume of the drawn top coal with different drawing height. With an increase in the coal drawing height, the accumulated distribution length and volume of the drawn top coal both increase approximately in the form of a quadratic function, and the correlation coefficient is as high as 0.99. This indicates that with increase in the coal drawing height, the accumulated distribution length increases and the volume growth ratio of the drawn top coal becomes faster. This implies that more top coal particles are drawn under the interval of a particular coal drawing height.

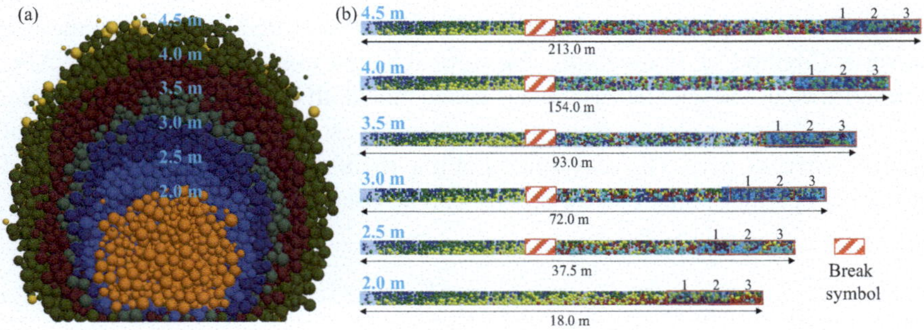

Fig. 2.39 The drawing body shape and distribution of drawn top coal on the rear AFC: **a** top coal drawing body with different drawing heights; **b** top coal particle distribution characteristics

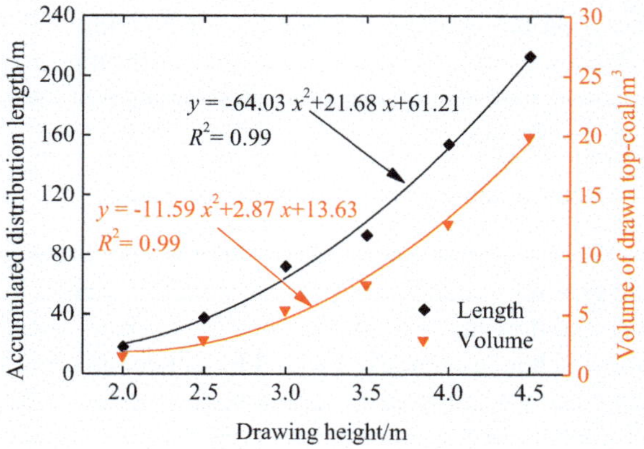

Fig. 2.40 The accumulated distribution length and volume of the top coal with drawing height

In summary, with an increase in the coal drawing time, the drawn top coal gradually expands outward. Therefore, the size distribution of the drawn top coal on the last position of the AFC, closer to the DO, can basically represent the top coal particle size distribution law at different layers above the supports. If Method 1 is used for measurement on the belt conveyor, the top coal sampled at the last position on the belt conveyor, near the part of the mining area, is selected for measurement. If Method 2 is used to measure the top coal size, the coal on the accumulated top coal body surface should be selected as far as possible.

According to the distribution law of the drawn top coal particles on the AFC, the top coal size distribution within the range of 0.5, 1.0, 1.5, 2.0 m, and all drawn on the AFC, were counted separately, and their top coal size distribution curves are shown in Fig. 2.41. It is evident that when the statistical length is 0.5 m, the measured top coal size distribution curve diverges more and is less accurate compared with the overall top coal size distribution curve. When the statistical length is more than or equal to 1.0 m, the measured top coal size distribution curve fits well with the overall top coal size distribution curve, and can represent the top coal size distribution in the area.

Figure 2.42 shows a schematic diagram of the variation relationship between the measured length L_1 on the AFC (Method 3) and the measured length L_2 on the belt conveyor (Method 1). In the figure, the vertical height from the surface of the AFC to the surface of the transfer machine is h, and the width of the central trough of the AFC is w. Moreover, the traction speed of the AFC is v_1, and the traction speed of the belt conveyor is v_2. The acceleration of gravity is g. The measured length L_2 of the top coal particles on the belt conveyor can be obtained as

$$L_2 = w + v_2\sqrt{\frac{2h}{g}} + \frac{v_2}{v_1}L_1 \tag{2.13}$$

As an example, with $w = 0.8$ m, $v_2 = 1$ m/s, $h = 0.1$ m, $g = 10$ m/s^2, $L_1 = 1$ m, and $v_1 = 1$ m/s, the length L_2 is obtained as 1.94 m by substituting into Eq. (2.13).

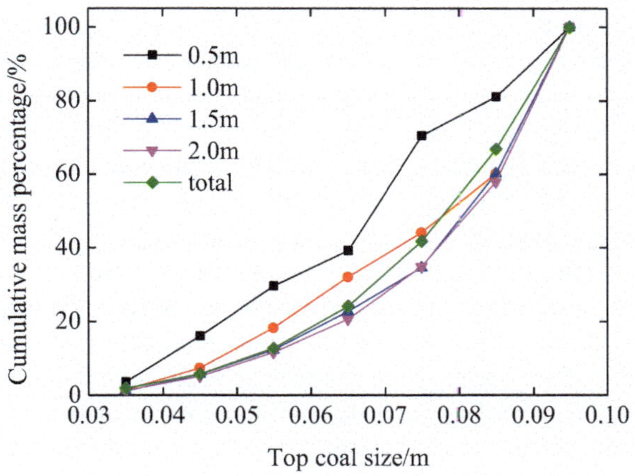

Fig. 2.41 The size distribution of the top coal under different measurement lengths

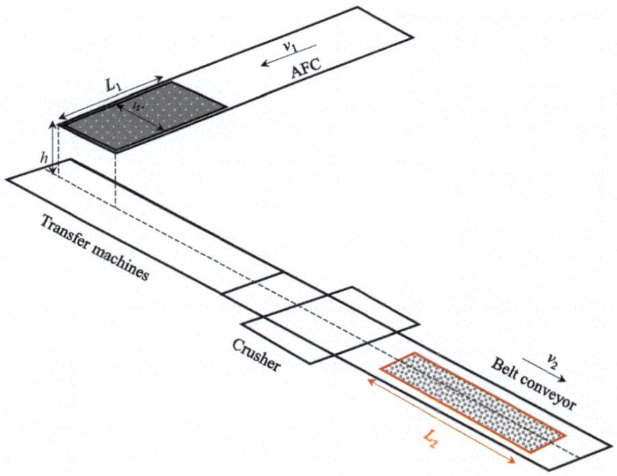

Fig. 2.42 The relationship between L_1 and L_2

Therefore, the measurement length should be in the range of 1.0–1.5 m when using Method 3 to measure the top coal size distribution in the field panel. When using Method 1, the required measurement length should be calculated by Eq. (2.13) according to the actual site conditions to obtain a more accurate top coal size distribution curve.

3. Top coal size Distribution Measurement Results

Based on previously proposed measurement methods and measurement parameters, we selected four LTCC panels with representative geological conditions in Shanxi and Henan provinces to carry out size measurements. After measurement of the top coal size distribution in each panel, the mean value and standard deviation of the top coal size in each panel were calculated separately, and the results are shown in Figs. 2.43 and 2.44. With an increase in the coal seam thickness, the average values of the top coal sizes of the four panels show a trend of gradual increase. Among them, the average size in the 15081 panel is the smallest at 7.3 mm, which means that the top coal in this panel is sufficiently broken. However, the phenomenon that serious coal leakage from the headgate and gap between the supports often occurs, which is not conducive to the safety management of the panel. The average sizes in panels 8103b and 8103r are in the middle, while the average size in panel 8404 is the largest at 353.4 mm. Owing to the larger size of the drawn top coal, it is necessary to pay attention to the rear AFC chain frequently. If a failure occurs during drawing or transportation of coal and rocks, timely maintenance is required.

In addition, with an increase in the average top coal size, the standard deviation of the size first increases and then gradually stabilizes. The standard deviation of the 15081 panel is the smallest at 11.26 mm, while the standard deviations of the 8103r and 8404 panels are basically stable at approximately 160 mm. As the average top coal size of each panel varies considerably, the standard deviation alone cannot accurately determine the degree of dispersion of size in each panel. Hence, according to probability theory, the concept of coefficient of variation *CV* is introduced, as shown in Eq. (2.14).

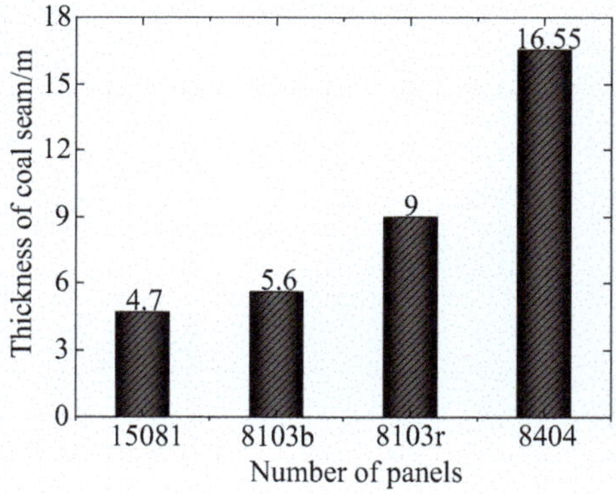

Fig. 2.43 Thickness of coal seam

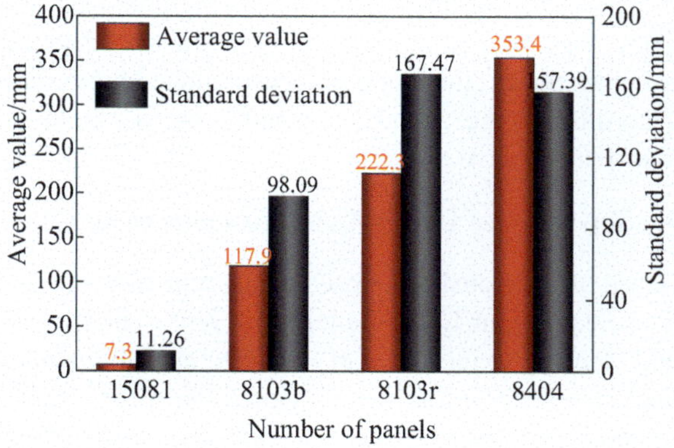

Fig. 2.44 Average value and standard deviation

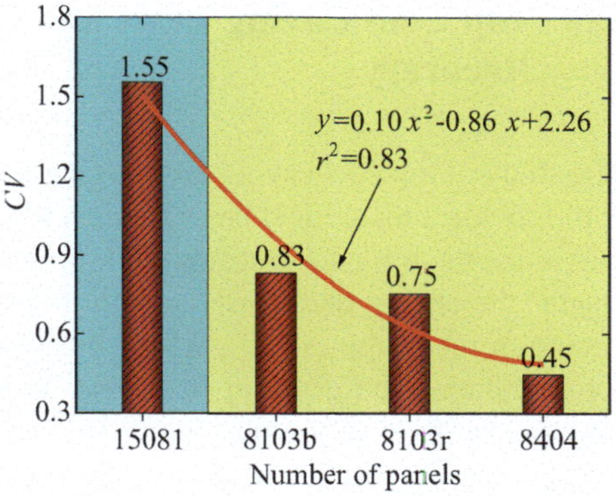

Fig. 2.45 Coefficient of variation of top coal size in the four panels

$$CV = \frac{\delta_t}{d_{av}} \tag{2.14}$$

where, δ_t is the standard deviation of the top coal size and d_{av} is the average size.

CV of each panel are shown in Fig. 2.45. For the four panels, *CV* shows a trend of gradual decrease with an increase in d_{av} or seam thickness. Combined with Fig. 2.44, although d_{av} and δ_t of the 15081 panel are small, the maximum *CV* is 1.55, which indicates large dispersion of the top coal size. d_{av} and δ_t in panel 8404 are larger, while *CV* is minimum at 0.45, which indicates that the top coal size is relatively concentrated and larger. The possible reason is that with the advance of the panel, the vertical abutment pressure on the top coal gradually increases, while the horizontal principal stress gradually decreases, and fractures in the top coal body begin to expand, thus breaking the top coal into loose blocks.

Therefore, the thickness of the top coal is a key factor that affects the top coal size distribution in the panel. If the top coal thickness is larger, d_{av} and δ_t are also relatively larger. The coal quality, fracture development and parting are also important factors affecting the degree of top coal crushing. Under the condition of nearly equal thickness of the top coal, if the coal is hard, and less-fracture with parting, d_{av} is larger and more concentrated. If the coal is soft, more-fracture without parting, d_{av} is smaller, but more dispersed. In addition, the fracturing degree of top coal at different positions of inclined panel is slightly different. In general, the fracturing degree of top coal in the middle is better than that in the lower part and that in the lower part is better than that in the upper part. The measurement of top coal size distribution in each panel lays a foundation for the subsequent research of the effect of top coal size distribution on the law of top coal caving.

2.3 Top Coal Caving Laws and Observation of Top Coal Recovery

The study of top coal caving laws is one of the core directions in LTCC. Its purpose is to provide a theoretical basis for developing precise caving techniques, aiming to maximize the top coal recovery and minimize the gangue mixed ratio. In the early stages, the caving theory was mainly borrowed from the ellipsoidal theory of ore caving in metal mines [13], which theoretically provided mathematical equations for calculating the drawing body and the top coal boundary. This has played an important role in promoting the theoretical research of drawing laws and guiding the engineering practice of top coal caving. With the deepening of research, theories such as the stochastic medium theory [14], the Bergmark-roos model [15], and the kinematic model [16] have gradually been introduced into the study of top coal caving laws, analyzing the flow characteristics of top coal from various perspectives. Based on previous and existing researches, the author's research team has successively proposed the theory of granular medium flow [17, 18], BBR research system [19], etc. In addition to paying attention to the study of drawing body, attention is paid to the change of top coal boundary [20, 21], the effect of support on top coal caving [22, 23], the drawing laws along the direction of the panel [24, 25], the influence of dip angle and top coal size distribution [26–30].

There are substantial differences between top coal caving in coal mines and ore caving in metal mines. One distinct feature lies in the existence of the hydraulic support. In addition, the mining process involves a cyclical process of support movement and top coal caving, which renders the top coal caving pattern unique. After proposing the granular medium flow theory and the BBR research system, Wang Jiachen conducted in-depth and systematic research work respectively on the 3D equation of the drawing body and top coal boundary, the prediction model of top coal recovery, and the relationship between top coal recovery and gangue mixed ratio, achieving systematic research results in the caving theory.

2.3.1 Theory of Granular Medium Flow

The theory of granular medium flow was first proposed by Wang Jiachen in 2001 and later refined in relevant articles published in the Journal of China Coal Society in 2002 and 2004 [17, 18]. According to the theory, during the advancing of the panel, the top coal and the immediate roof have already been completely broken into loose bodies, thus their movement and release conform to the laws of granular flow. In the composite granular media composed of loose top coal and loose roof, DO of the support becomes the free boundary for the flow of media and the release of the interaction stress between the media particles. The loose media above and behind the support will gradually move towards DO along the path of minimum resistance, forming a movement field similar to traction flow within the granular media. This

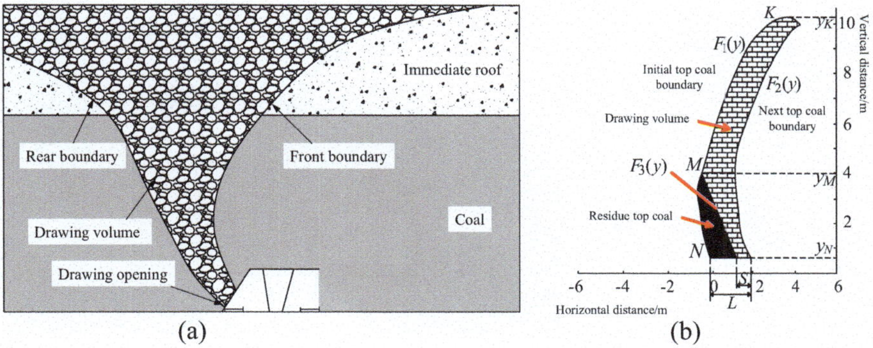

Fig. 2.46 **a** The theoretical model of granular medium flow; **b** the calculation of top coal caving amount

process of top coal flow and release is referred to as the theoretical model of granular medium flow for top coal motion, as shown in Fig. 2.46a.

The theory of granular medium flow proposes a theoretical calculation method for top coal drawing volume, which is defined as the volume enclosed by the initial top coal boundary before coal caving and the next top coal boundary after coal caving, minus the top coal remaining in the goaf. This represents the theoretical drawing volume for that round of coal caving, as shown in Eq. (2.15) and Fig. 2.46b.

$$Q = \int_{y_H}^{y_K} F_1(y)dy - \int_{y_N}^{y_k} F_2(y)dy - \left(\int_{y_N}^{y_M} F_1(y)dy - \int_{y_N}^{y_M} F_3(y)dy \right) \tag{2.15}$$

2.3.2 *BBR Caving Theory ("Four Elements" Caving Theory)*

BBR caving theory, also known as the "four-element" caving theory, refers to a comprehensive study of four key elements in the process of top coal caving, including the top coal boundary, the drawing body of top coal, the top coal recovery, and the gangue mixed ratio, as well as their interrelationships. This theory was proposed by the author in 2015 based on the concept of granular medium flow theory for top coal caving, serving as a specific research support work [19], as shown in Fig. 2.47.

The fundamental academic ideology of the BBR research system is to conduct a unified study of four interrelated spatio-temporal elements: the initial and final top coal boundary shape, the development process and shape of the drawing body, the volume of drawing top coal, and the volume of rock mixed, within each coal caving cycle. This approach aims to establish a comprehensive research system that accurately reflects the actual coal caving process, scientifically elucidating the four

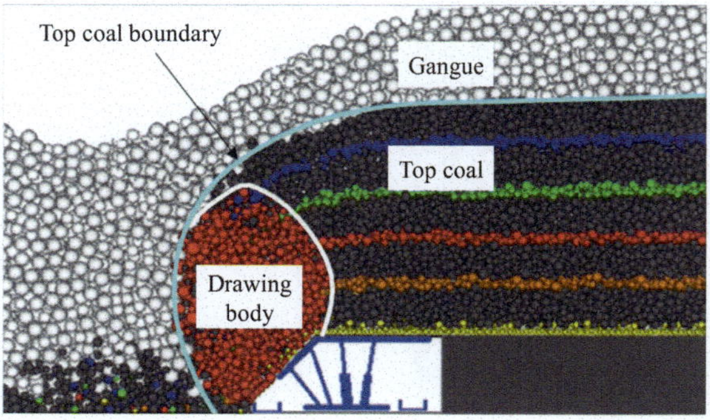

Fig. 2.47 Schematic diagram of BBR research system

elements and their mutual relationships. The goal is to provide scientific guidance for improving the top coal recovery ratio and reducing the gangue mixed ratio.

2.3.3 Shape of the Drawing Body

The drawing body refers to the restoration of the drawn top coal to its original spatial form within the original top coal body. It is an inverse, invisible, but able to reflect the boundary shape of a certain support's caving top coal range and its spatial relationship with the top coal boundary (Fig. 2.48). By studying the shape characteristics of the drawing body, we can understand the top coal caved and coal loss, thereby providing theoretical guidance for developing a precise coal caving technology with high resource recovery ratio.

2.3.3.1 Drawing Body in the Advancing Direction

In the advancing direction of the panel, the shield beam of the support has a significant impact on the evolution of the drawing body. As shown in Fig. 2.49, the presence of the support results in significantly different boundary effects at both ends of DO. The interaction between the loose top coal and gangue is in the goaf side. On the other side, the interaction between the loose top coal and the shield beam is in the support side. The mechanisms of contact and friction coefficients between these two interactions are different, leading to asymmetric characteristics in the shape development of the drawing body on the goaf side and the support side. The basic principle is that the smooth caving shield increases the critical motion angle of the granular particles, affecting the initial position of the particles on the right side of the drawing body and thus altering the shape of the right boundary of the drawing body. Based on the consideration of the influence of the support on the shape of the drawing body in top coal caving, the Bergmark-Ross model [31] (original B-R model, as shown in Fig. 2.49a) in granular medium mechanics is modified. The modified B-R model can

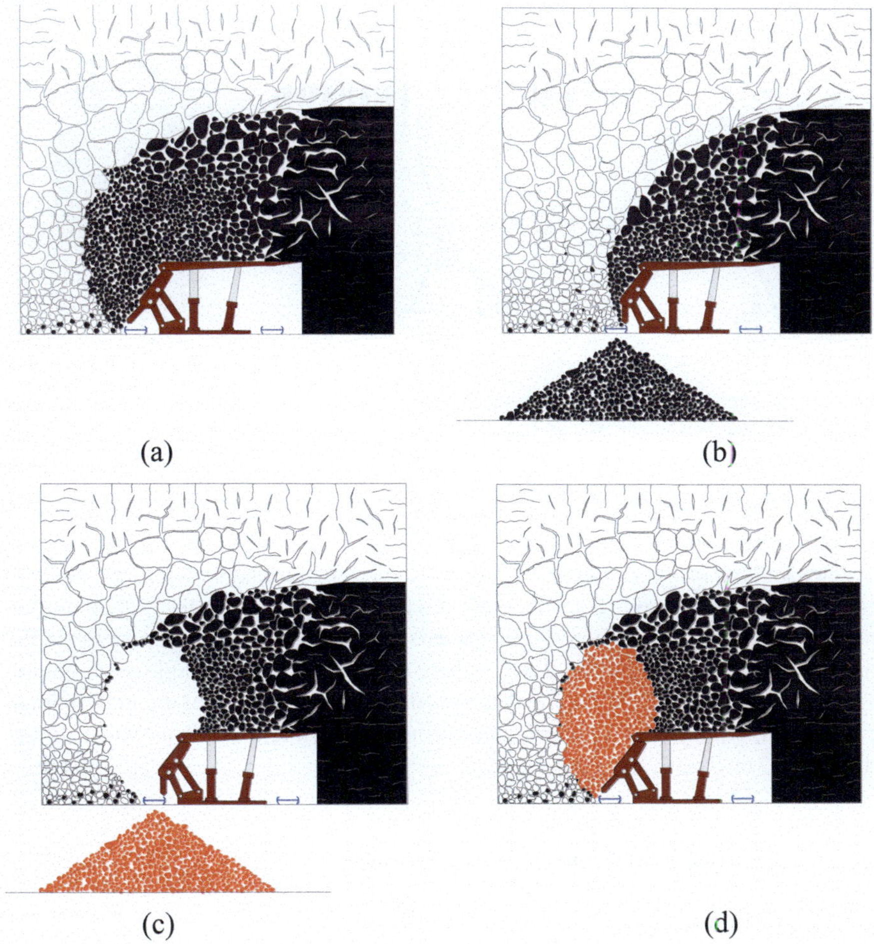

Fig. 2.48 The drawing body and top coal boundary: **a** top coal boundary before caving process; **b** top coal boundary after caving process; **c** drawing top coal; **d** the drawing top coal being reversed to its original position (drawing body)

describe the shape characteristics of the drawing body in top coal caving mining [15], and its theoretical shape and theoretical calculation Equation are shown in Fig. 2.49b and Eq. (2.16).

$$\begin{cases} r_{\max} = r_D + K(\frac{gt_2^2}{2})(1 - \cos\theta_G) \\ r_0(\theta, r_{\max}) = (r_{\max} - r_D)\frac{(\cos\theta - \cos\theta_G)}{1 - \cos\theta_G} + r_D \end{cases} \tag{2.16}$$

where $r_{\max}$ represents the distance from the farthest starting point to the origin of the polar coordinate system, m; r_{D} represents the length of the tail beam of the support,

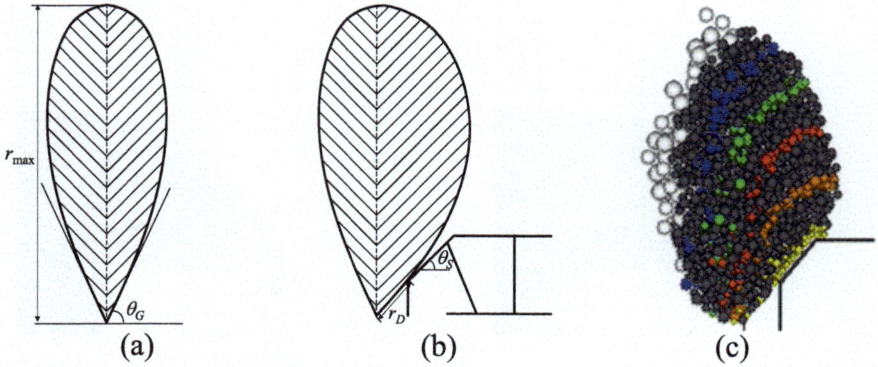

Fig. 2.49 Theoretical and numerical shape of the top coal caving considering the influence of support: **a** original B-R model; **b** modified B-R model; **c** numerical result

m; g represents the gravitational acceleration, m/s^2; t represents the caving time, s; θ represents the motion angle (i.e., the angle between the line connecting the top coal particles and the origin of the polar coordinate system and the vertical line, positive for clockwise and negative for counterclockwise), °; θ_G represents the critical motion angle, $\theta_G = 45° - \varphi_0/2$ when $\theta < 0°$ (where φ_0 is the internal friction angle of the particles); $\theta_G = 90° - \theta_S$ when $\theta > 0°$ (where θ_S is the inclination angle of the caving shield of the support), °; K_m represents the correction coefficient of the gravitational acceleration, which can be determined through experiments, and its value differs when $\theta < 0°$ or $\theta > 0°$.

2.3.3.2 Drawing Body in the Inclined Direction

1. Theoretical equation

In the inclined direction of the panel, the adjacent support tail beams and the inclination of the panel directly affect the shape characteristics of the drawing body. When DOs are inclined on the inclined panel, the boundary conditions on both sides of DOs are different. The lower end of DO can be regarded as an infinite boundary condition, while the upper end is an inclined boundary condition. Since the frictional contact between particles and the support at the upper end of DO is less than the frictional contact between particles, the inclined coal seam facilitates the caving of coal at the upper end. Figure 2.50 shows the calculation model of the drawing body in an inclined coal seam based on the modified B-R model. When the particles are within the range of the upper head DA_1 of the drawing body, if the particles continue to move along a straight line, they will collide with the adjacent support tail beam. Therefore, if the particles follow the principle of flowing out of the DO along the direction of minimum resistance path, the movement path of the particles within the range of DA_1 must change under the constraint of the support tail beam. Furthermore, since the friction coefficient between the support tail beam and coal particles is less

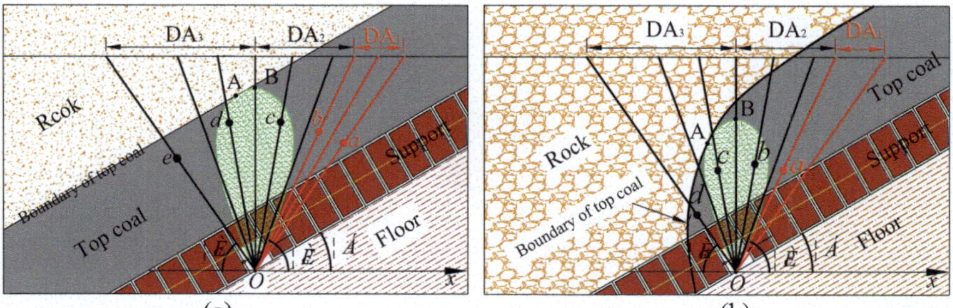

Fig. 2.50 Calculation model for the drawing body in a inclined coal seam: **a** initial drawing stage; **b** normal drawing stage

than the friction between coal particles, the top coal particles at the upper end of the DO are quickly drawn. When the particles are within the range of DA_2 and DA_3, their movement path can be considered as not significantly affected by the support tail beam, and there is no significant difference compared to the horizontal condition. The particles flow towards the trajectory center *O* point along a straight line and flow out of DO.

Using the established calculation model of the drawing body along the dip direction of coal seam, the theoretical equation of the drawing body when the dip angle of the panel is greater than 0° is derived, as shown in Eq. (2.17).

$$\begin{cases} l_1 : \rho(\theta_{O'}) \frac{\rho(\alpha)}{(\sin\alpha - f_1)} (\sin\theta_{O'} - \frac{M \sin\alpha - \sin\theta_{G'}}{M-1}), \alpha \leq \theta_{o'} \leq \theta_{G'} \\ l_{2-3} : \rho(\theta) = \rho(\theta_A) \frac{\sin\theta - \sin\theta_G}{\sin\theta_A - \sin\theta_G}, \theta_{G'} \leq \theta \leq 180^\circ - \theta_G \end{cases} \quad (2.17)$$

where α represents the dip angle of the coal seam, °; $\theta_{G'}$ represents the angle when the particle just passes through the upper boundary of DO along a straight line, °; θ_A represents the polar angle at Point *A*, the tangent point, °; $\rho(\theta_A)$ represents the polar radius at Point *A*, m; f_1 and *M* are parameters of the equation.

2. Shape-difference and volume-symmetry characteristic

As can be seen from Fig. 2.51 and Eq. (2.17), the variation in particle velocity field and path in the inclined LTCC panel leads to the variation in the boundary shape of the drawing body, with the shape rapidly developing in the upper direction of the panel after the variation. Using numerical simulation and physical experiments to investigate the top coal caving process under different coal seam dip angles, it was found that the drawing body in the inclined direction exhibits the characteristic of "shape-difference and volume-symmetry" [15]. With the vertical line passing through the center of DO as a reference, there are differences in the shape of the drawing body on both sides of the central vertical line, however, the volumes on both sides are basically equal. As shown in Fig. 2.51, as the dip angle of the panel increases, the position of the inflection point on the right side of the drawing body rises higher, and the drawing body on the right side is increasingly influenced by boundary conditions,

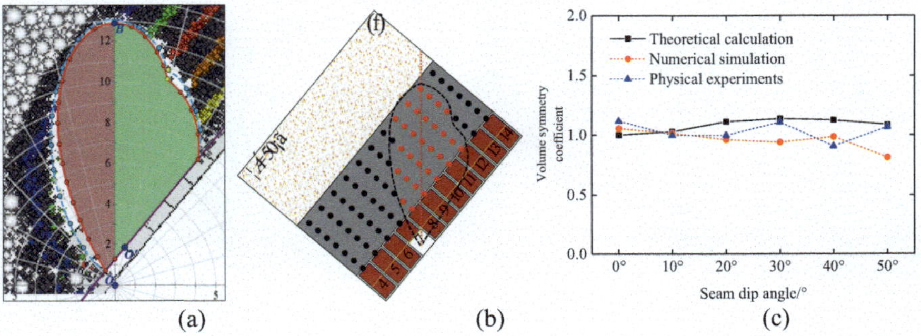

Fig. 2.51 Shape-difference and volume-symmetry characteristic of the drawing body and physical verification: **a** numerical result; **b** physical test result; **c** volume symmetry coefficient

resulting in a more pronounced difference in the shape of the drawing body on both sides of the central vertical line.

To further validate the shape-difference and volume-symmetry characteristic of the drawing body, the volumes of the drawing body on both sides of the central vertical line in the theoretical calculations, numerical simulations, and physical experiments were statistically analyzed, and the volume symmetry coefficient (i.e., the ratio of the volume on the upper side to the volume on the lower side of the drawing body) was calculated. It was found that the volumes of the drawing bodies on both sides were basically equal, as shown in Fig. 2.51c. In other words, the influence of the dip angle of the panel on the drawing body is mainly reflected in its shape, rather than its volume.

2.3.3.3 Drawing Body Near the Roof and Floor Sides in STCC

For STCC panel in steeply inclined thick coal seams, due to its short panel and large thickness of top coal, the influence of the initial drawing body affected by the inclined boundaries near the roof and floor sides on the top coal recovery is greater than that in traditional LTCC. As shown in Fig. 2.52, based on the BBR theory, the theoretical calculation model of the corresponding drawing body has been established by considering the influence of the inclined boundaries on the roof and floor sides on the drawing body [32].

In the theoretical model of drawing body affected by boundary conditions, a polar coordinate system is established with the axis of the drawing body ON' as the polar axis and counterclockwise rotation as the positive direction. The axis of the drawing body is the straight line where the maximum motion path of the granular particles in the drawing body region exists. Based on the degree of influence of the inclined boundary on the motion of the granular top coal, the drawing body region is divided into four parts. By calculating and combining them separately, the theoretical boundary equation of the drawing body affected by the inclined boundary is obtained.

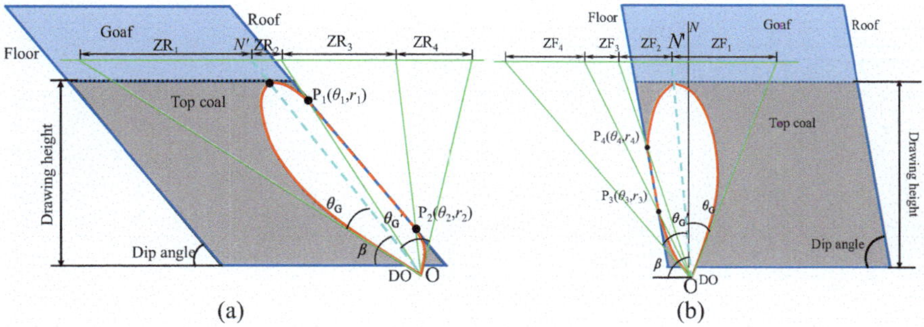

Fig. 2.52 Theoretical model of drawing body affected by inclined boundary: **a** near roof side; **b** near floor side

Taking the near roof side as an example, the boundary equation of the drawing body affected by the inclined boundary is shown in Eq. (2.18).

$$r_R(\theta) = \begin{cases} r_{D\beta} + \left(r_{\max\beta} - r_{D\beta}\right)\frac{\cos\theta - \cos\theta'_G}{1-\cos\theta'_G}, & \theta \in [\theta_{G'}, \theta_2] \cup [\theta_1, 0] \\ \frac{(2D+L)\sin\alpha}{\sin(\theta-\alpha-\beta)} - \frac{2D\sin(\beta-\theta_{G'})\sin(\beta-\theta_G+\alpha)}{\sin(\theta_G-\theta_{G'})\sin(\theta-\alpha-\beta)}, & \theta \in [\theta_2, \theta_1] \\ r_{D\beta} + \left(r_{\max\beta} - r_{D\beta}\right)\frac{\cos\theta - \cos\theta_G}{1-\cos\theta_G}, & \theta \in [0, \theta_G] \end{cases} \tag{2.18}$$

where $r_{\max\beta}$ represents the distance from the farthest initial starting point to the origin of the polar coordinate system when affected by the boundary, m; $r_{D\beta}$ represents the distance from DO boundary to the origin of the polar coordinate system, m; β represents the axis inclination angle (the maximum particle motion path is the axis of the drawing body, and the angle between the axis and the vertical normal is the axis inclination angle), °; D represents half of the width of DO, m; L represents the length of the no drawing section (i.e., the horizontal distance from DO boundary to the nearest boundary), m; $\theta_{G'}$ represents the critical motion angle affected by the boundary inclination angle, which has a linear relationship with the coal seam inclination angle and can be determined by experiments, °; θ_1 and θ_2 represent the polar angles corresponding to the intersection points P_1 and P_2 between the drawing body boundary and the inclined boundary on the roof side in Fig. 2.52a.

2.3.3.4 Drawing Body at the End Area

When the top coal above the roadway support and transition support becomes loose blocks under the action of mining-induced stress and auxiliary fracturing, the drawing operation of roadway support and transition support begins. As shown in Fig. 2.53, when top coal particles (such as Particles *a*, *b*, *c*) are in the DA_1 region ($\theta_G \le \theta \le \theta_L$), they can still move to the center point of DO along a straight line, where θ_L is the angle when a particle move in a straight line just past the upper boundary point (Point *B*) of the roadway. However, when top coal particles (such as Particle *a*) are in the DA_2 region ($\theta_L \le \theta \le \pi - \theta_G$), the broken top coal inside the roadway can

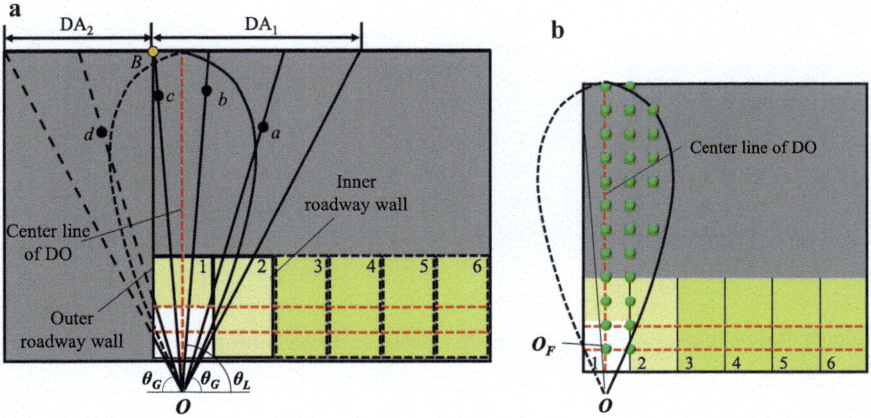

Fig. 2.53 Drawing body shape of roadway support in the layout direction: **a** theoretical model of drawing body at the end area; **b** initial drawing body shape at the end area in the drawing experiment

normally flow out of DO, while the complete top coal outside the outer roadway wall cannot be drawn. Considering that the friction between top coal particles and roadway wall is much smaller than that between top coal particles, the flow speed of top coal particles near the outer roadway wall is relatively fast, which makes gangue particles flow out of DO in advance and then causes incomplete development of the drawing body on the right side of the center line of DO. Figure 2.53b shows the distribution of the drawn marked particles in the layout direction in the drawing experiment. As a result of the test device boundary (simulating the outer roadway wall) limitation, the drawing body on the left side is cut by roadway, and that on the right side is in the shape of the upper part over-development and the lower part defect, however, the drawing body is in line with the B-R model overall [33].

Considering the constraints of roadway sidewalls and drawing support, the boundary equation of drawing body under the condition of roadway support at the end of the panel is shown in Eq. (2.19).

$$\rho(\theta) = \begin{cases} \left(H_C - \frac{L_{OLP}}{2} + \frac{L_{OWP}}{2\tan\theta_G}\right)\frac{\sin\theta - \sin\theta_G}{1-\sin\theta_G}, \theta_G \le \theta \le \theta_L \\ -\frac{L_{OWP}}{2\cos\theta}, \quad \theta_L \le \theta \le \pi - \theta_G \end{cases} \tag{2.19}$$

where H_c represents the thickness of coal seam, m; L_{OLP} stands for the projection length of DO, m; L_{OWP} stands for the projection width of DO in the layout direction of the panel, m.

2.3.3.5 Drawing Body Under the Effect of Size Distribution

The top coal drawing from the LTCC panel is composed of coal blocks of different sizes, and the particle size distribution characteristics of the drawing top coal have a significant impact on the drawing theory. To quantitatively describe the influence of the particle size distribution of top coal on the equation and shape of drawing body, drawing experiments under different average sizes of top coal (the average size is within the range that can be smoothly drawn) were conducted. The results show that under the same length of DO, the volume of the drawing body increases approximately linearly with the increase of the average size of the top coal. According to the relationship between the volume of the drawing body and θ_G [31], combined with the experimental data, an approximate linear fitting relationship between θ_G and the average size of the top coal is discovered. Based on the variation law of θ_G, the boundary equation of the drawing body under the condition of horizontal coal seam considering the influence of the average size of the top coal can be obtained, as shown in Eq. (2.20).

$$\rho(\theta) = H_f \frac{\sin\theta - \sin(ad_{av} + b)}{1 - \sin(ad_{av} + b)} \tag{2.20}$$

where H_f represents the drawing height, m; d_{av} stands for the average size of the top coal, m; a and b are equation parameters that are related to the average size of the top coal and the length of DO.

As shown in Fig. 2.54, the theoretical shapes of the drawing body under different average sizes of top coal and lengths of DOs are plotted. Under the same coal seam thickness, when the length of DO is fixed, the volume of the drawing body increases as the average size of the top coal increases, and the maximum width of the drawing body also becomes larger. Similarly, when the average size of the top coal is fixed, as the length of DO increases, the volume of the drawing body increases with a decreasing growth ratio, indicating that although increasing the length of DO can increase the volume of the drawing body, the width of DO will limit its continuous increase [34].

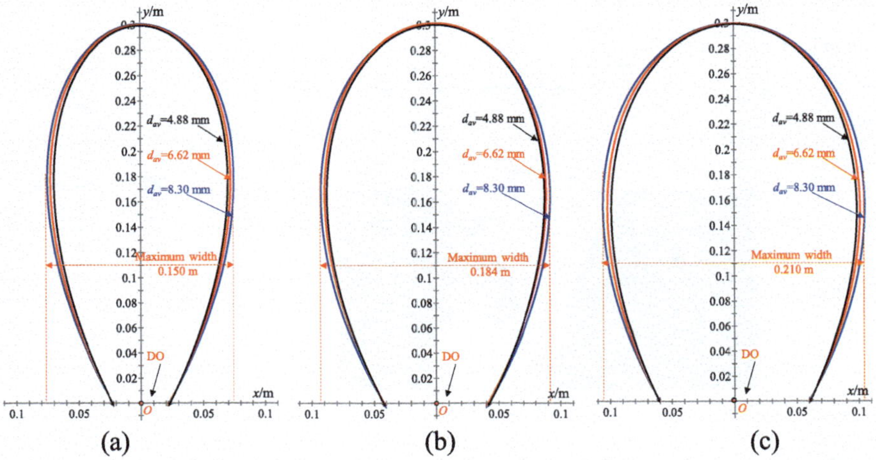

Fig. 2.54 Theoretical shapes of the drawing body under different drawing experiments: **a** DO = 50 mm; **b** DO = 100 mm; **c** DO = 150 mm

2.3.4 *Equation and Shape of the Top Coal Boundary*

2.3.4.1 Top Coal Boundary During the Initial Drawing Stage

The top coal boundary refers to the actual surface between top coal and gangue, and its shape characteristics can directly affect the next round of top coal caving process, playing a constraining role in the development and evolution of the top coal drawing body. As shown in Fig. 2.55, to deeply analyze the shape characteristics of the top coal boundary, a force analysis was conducted on any particle on the cross-section of the top coal boundary, and a mechanical model of the shape of the top coal boundary was established. It is assumed that the particle on the surface of the top coal boundary is subjected to a load of $\gamma H_g \varDelta S$ from the upper broken gangue pillar, a horizontal lateral pressure $F_s = K_h \gamma H_g \varDelta S$, a friction force f along the tangent direction of the top coal boundary, and a supporting force F_N acting on the particle. Here, $\varDelta S$ represents the area of the load applied to that point, and K_h is the lateral pressure coefficient. Based on the force balance relationship of the particle, the 3D equation of the top coal boundary at the initial gangue exposure is obtained as shown in Eq. (2.21). This equation quantifies the relationship between the length (y) of each cross-section of the top coal boundary and the distance (H_g) from the floor, and its shape characteristics are shown in Surface 1 in Fig. 2.56 [20].

$$y^2 + z^2 = k^2 \left[-\frac{(1+\mu^2)}{m} \ln(mH_g + \mu) + \mu H_g + \frac{(1+\mu^2)}{m} \ln\left(m\left(H_\gamma + H_c\right) + \mu\right) - \mu\left(H_\gamma + H_c\right) \right]^2 \tag{2.21}$$

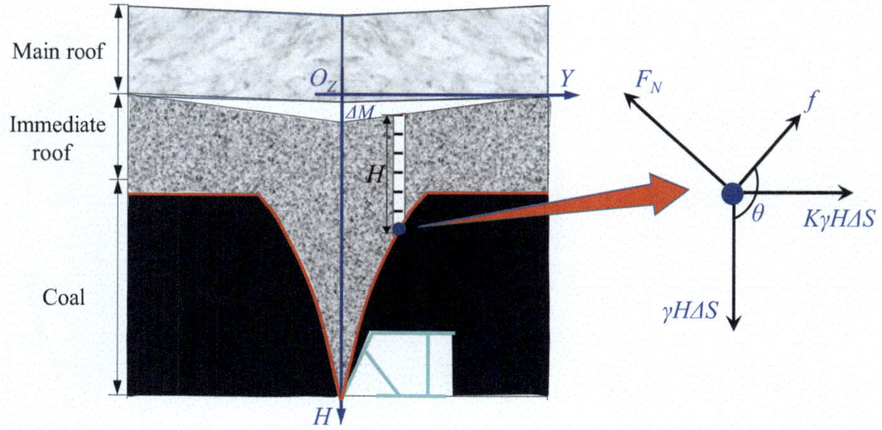

Fig. 2.55 Force model of particles on the surface of the initial top coal boundary

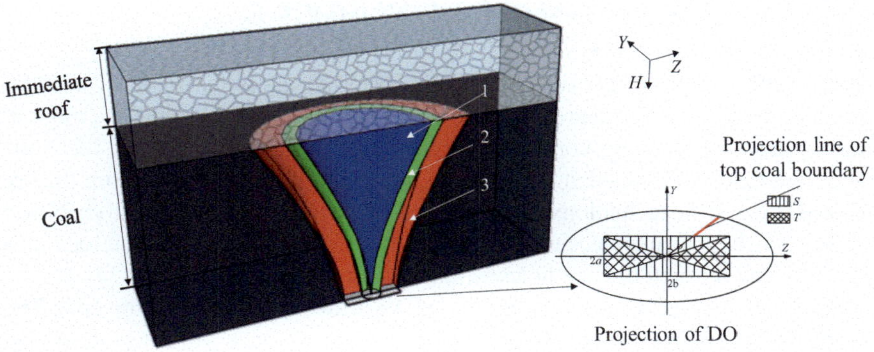

Fig. 2.56 Development process of top coal boundary during excessive drawing stage

where μ represents the friction coefficient between particles; m represents the lateral pressure transfer coefficient; H_g represents the height of the overburden rock column above any point on the top coal boundary, m; H_r represents the thickness of the gangue, m; H_c represents the thickness of the coal seam, m; and k represents the correction coefficient of the equation.

2.3.4.2 Top Coal Boundary During Excessive Drawing Stage

As shown in Fig. 2.56, in actual LTCC panels, to extract as much coal resource as possible, caving operators often perform excessive drawing operations. During the process of excessive drawing, the shape of the top coal boundary gradually evolves from a "quasi-logarithmic curve funnel shape" (Surface 1) to a "rectangular bottom and elliptical top funnel shape" (Surface 3). The equation of the 3D top coal boundary under excessive drawing conditions is shown in Eq. (2.22). The top coal boundary after excessive drawing has a more significant constraint effect on the subsequent coal drawing process. Therefore, while increasing the top coal recovery through excessive drawing, the impact on the subsequent coal drawing process and the amount of gangue discharge should also be considered. This is an important research content for intelligent drawing control.

$$\begin{cases} y|_{\varphi_p \in S} = ky + (1-k)\left|\frac{a}{\cos\varphi_p}\right|, \ \varphi_p \in S \\ y|_{\varphi_p \in T} = ky + (1-k)\left|\frac{b}{\sin\varphi_p}\right|, \ \varphi_p \in T \end{cases} \tag{2.22}$$

where a represents the half-length of the short side of the projection of DO (the advancing direction of the panel), m; b represents the half-length of the long side of the projection of DO (the layout direction of the panel), m; S and T represent the angular range of the projection line of the top coal boundary, °.

2.3.5 *Field Measurement of Top Coal Recovery*

Improving the top coal recovery has always been an important topic of concern for coal enterprises and researchers. Among them, accurately grasping the motion, drawing law, and recovery conditions of top coal at different strata and locations is the foundation for enhancing the top coal recovery. The top coal motion tracker invented using RFID technology can effectively solve the above-mentioned problems [35, 36]. The basic principle is to place top coal tracking marks at different locations in the top coal through drilling holes with a certain density beforehand [37]. As the top coal is released, some tracking marks will be drawn along with the top coal. The receiving base stations installed on the belt conveyor in advance will detect the drawn top coal tracking marks and their installation locations in the top coal, thus accurately deducing the drawing conditions of top coal at various locations in the top coal, so as to adopt targeted coal drawing techniques to improve the top coal recovery. This technology has been promoted and applied in more than 20 LTCC panels, including Tashan Coal Mine in Datong, Luling Coal Mine in Huaibei, Wangzhuang Coal Mine in Lu'an, and Xinyang Coal Mine in Fenxi.

2.3.5.1 Development of Testing Device

1. The first-generation top coal motion tracker

As shown in Fig. 2.57, to accurately measure the top coal recovery ratio in the LTCC panel, the research team independently developed the first-generation top coal motion tracker (Patent No. ZL200910080005.9) [35]. This tracker mainly consists of two parts: underground and aboveground. The underground part includes: mark, receiving base station, parameter setter, data collectors, and stabilized power supply. The aboveground part includes: USB wireless receiver, computer, analysis software, and data storage.

2. The third-generation top coal motion tracker

Figure 2.58 shows the third-generation top coal motion tracker used to measure top coal recovery. The system mainly includes signal receiver, power supply, display screen, mark, auxiliary mounting device.

Each component is characterized as follows:

(1) The signal receiver is mainly used to enter the number information to the mark and receive the specific signal from the mark, with a convenient operation method, large signal reception range and high sensitivity.
(2) The power supply is used to power the signal receiver and display screen, and is powered by a portable 7.4v lithium-ion battery that can last up to 12 days, which fully meets the time for measuring the recovery ratio at the panel.
(3) Configured with a visual display screen, when the signal receiver receives the mark signal, it can record its drawing order and drawing time, which is directly displayed on the display screen.

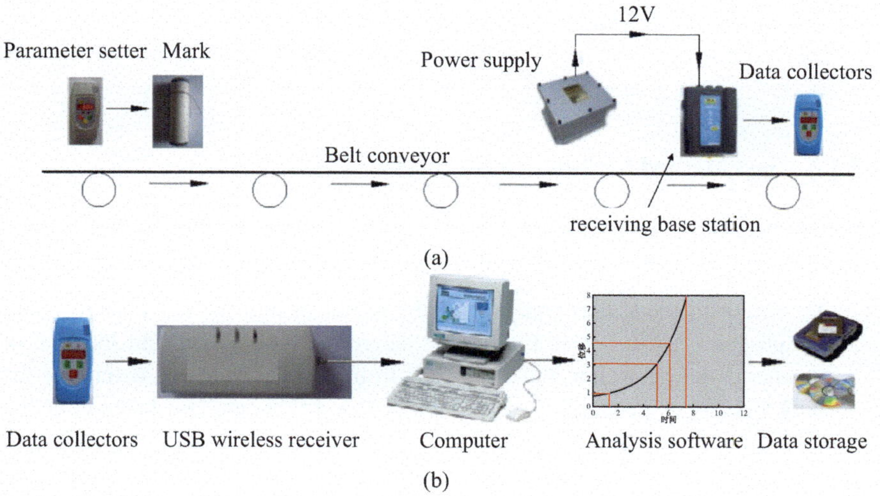

Fig. 2.57 Composition of the first-generation top coal motion tracker: **a** underground part; **b** aboveground part

Fig. 2.58 The third-generation top coal motion tracker

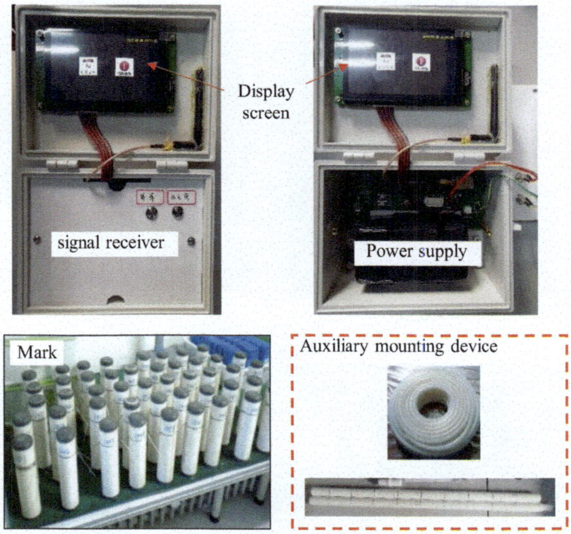

(4) The mark's simple internal connections and easily replaceable batteries provide enough power to keep the mark working all day long for up to 30 days, which is enough time to meet your measurement needs.
(5) The auxiliary mounting device includes two kinds of white hard push rod and PVC hose, which are used for on-site auxiliary installation of marks. It is more

convenient to use white hard push rod when the top coal is thin. If it is extra-thick coal seam, PVC hose can greatly accelerate the installation speed of the mark.

2.3.5.2 Test Method for Top Coal Recovery

When the measuring panel advances normally, the geological structure-free zone is selected for on-site top coal recovery measurement, and the basic steps are as follows.

(1) According to the actual situation of the panel, design the top coal recovery test program, as shown in Fig. 2.59a, determine the number and location of punched holes, as well as the distance l_1 from the coal wall, and determine the mark spacing l_2, generally l_1 is taken to be 0.5–1.0 m, and l_2 is taken to be 0.5 m [37].

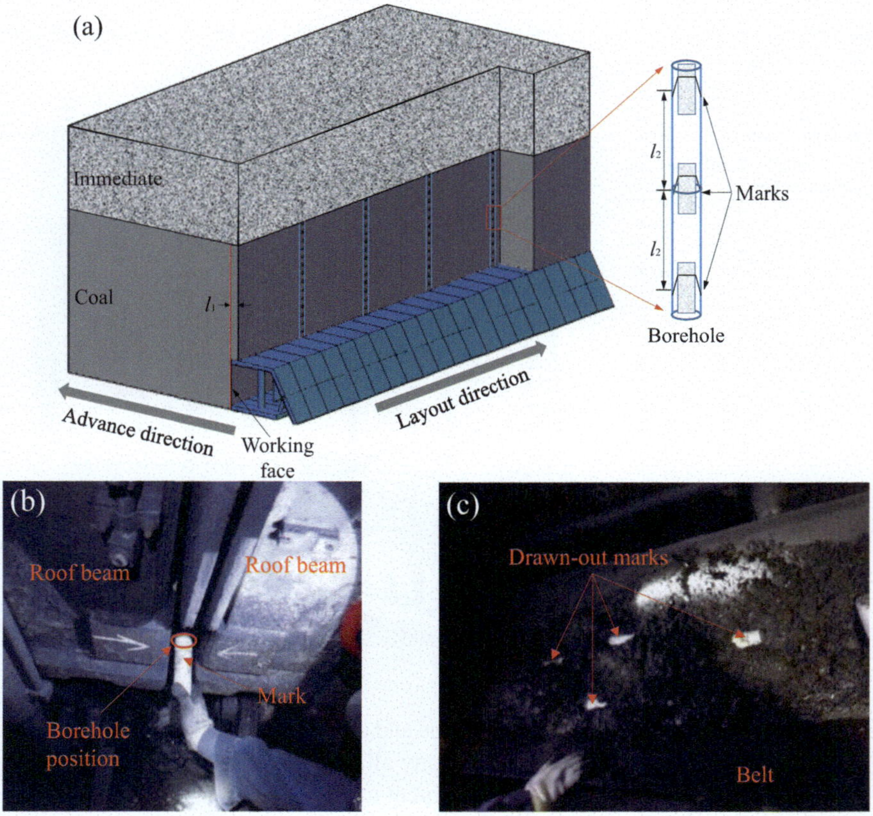

Fig. 2.59 Main measurement steps of top coal recovery: **a** measurement scheme; **b** drilling boreholes and installing marks; **c** drawn-out marks on the belt conveyor

(2) Prior to installation in the field, check that the components of the measurement device are functioning correctly and make sure that the signal receiver and mark batteries are fully charged.
(3) In Fig. 2.59b, when the panel is the maintenance shift, according to the design scheme, the drilling rig is used to drill into the top coal in the LTCC panel (according to the feasibility of the site, the drilling position can be adjusted appropriately), and then the pre-numbered mark is installed in the corresponding position of the top coal with the push rod or PVC hose. Considering the length of the maintenance shift and the workload of the personnel, it is generally appropriate to install 30–50 marks.
(4) Signal receivers are installed above or to the side of the belt at a position 50–100 m from the panel in the headgate. As the panel advances, the marks installed in the top coal are drawn along with the bulk top coal from DO, transported out of the panel by the rear AFC and then transported to the ground via the transfer machine and belt conveyor. As shown in Fig. 2.59c, during the transportation process, the mark sends out a specific message every second to be received by the signal receiver installed in the headgate. And then the signal receiver will record the number of the mark and the time of reception, and show it on the display screen, which is convenient for the surveyor to always observe.
(5) After the panel passes through the mark installation area, that is, the signal receiver no longer receives the mark signal, the signal receiver is removed and brought back to the ground, and the top coal recovery of the panel is calculated and analyzed.

2.3.5.3 Analysis of Top Coal Recovery Test Results

Since the development of RFID technology-based top coal motion tracker, field measurements have been carried out in many different conditions of the LTCC panel, and the measurement results can reflect the change of top coal recovery in different locations and layers of the panel, and the application effect is good, and some of the existing measurement results are shown in Table 2.5. Overall, when the thickness of the seam is smaller (≤ 6.5 m), the top coal recovery is relatively higher, at about 80%. With the increase of seam thickness, the top coal recovery basically shows the trend of increasing first and then decreasing. Possible reason is that, without considering the difference of fracture distribution and coal quality of each panel, with the increase of top coal thickness, the top coal broken bulkiness should show a gradual increasing trend, which makes the recovery ratio increase first and then decrease.

To further compare and analyze the recovery situation of different layers of top coal in each panel, after sorting the panels according to coal thickness, the panels were unified to the median thickness of 9 m (Ruilong Coal Mine) i.e., the different LTCC panels were converted according to the proportion of coal thickness, and the recovery ratio of different layers of top coal in each panel was finally plotted in a diagram, and the results are shown in Fig. 2.60. The recovery ratio of middle and lower top coal in the panel of different coal mines is high, with an average of about

Table 2.5 Measurement of top coal recovery of different LTCC panels (sorted by seam thickness)

Serial number	Coal mine	Panel	Sema conditions	Installation mark/pc	Received mark/pc	Top coal recovery/%
A	Jiulishan	15081	Seam thickness 4.7 m Cutting height 2.5 m Drawing height 2.2 m Dip angle 11.5°	42	36	85.7
B	Beixinyao	8103	Seam thickness 5.6 m Cutting height 3.0 m Drawing height 2.6 m Dip angle 22°	31	23	74.2
C	Xinliu	241103	Seam thickness 6.2 m Cutting height 2.5 m Drawing height 3.7 m Dip angle 3°	24	20	83.3
D	Dayuan	1203	Seam thickness 6.5 m Cutting height 2.5 m Drawing height 4.0 m Dip angle 45°	37	29	78.4
E	Wangzhuang	4331	Seam thickness 7.18 m Cutting height 3.0 m Drawing height 4.18 m Drawing interval 0.8 m	23	16	69.9

(continued)

Table 2.5 (continued)

Serial number	Coal mine	Panel	Sema conditions	Installation mark/pc	Received mark/pc	Top coal recovery/%
F	Ruilong	8103	Seam thickness 9.0 m Cutting height 3.0 m Drawing height 6.0 m Drawing interval 0.8 m Dip angle 22°	42	33	78.6
G	Xinjulong	1302N	Seam thickness 9.4 m Cutting height 4.0 m Drawing height 5.4 m Dip angle 3°	25	22	88.0
H	Baojishan	703	Seam thickness 11.0 m Cutting height 2.5 m Drawing height 8.5 m Drawing interval 1.2 m Dip angle 45°	73	50	68.5
I	Luling	II927	Cutting 9#coal, caving 8#coal 9#coal seam thickness 2.5 m Parting thickness 3.0 m 8#coal sema thickness 6.25 m Dip angle 12°	54	37	68.5

(continued)

Table 2.5 (continued)

Serial number	Coal mine	Panel	Sema conditions	Installation mark/pc	Received mark/pc	Top coal recovery/%
G	Tashan	8105	Seam thickness 14.5 m Cutting height 4.2 m Drawing height 10.3 m Drawing interval 0.8 m	48	28	58.3
K	Jinzhuang	8404	Seam thickness 16.55 m Cutting height 3.9 m Drawing height 12.65 m Drawing interval 0.8 m Dip angle 6°	50	33	66.0

70%, while the recovery ratio of upper top coal is relatively low, with an average of about 53%. This is due to two reasons, one is the flow characteristics of the loose top coal block itself, which can be seen from the shape of the top coal drawing body, the volume of middle top coal release is higher, while the volume of upper top coal release is lower. The other is that due to the relatively large broken block size of the upper top coal, especially under the condition of the extra-thick coal seam, it is easy to form an arch structure in the caving process, which reduces the volume of upper top coal release and ultimately makes it far less than the middle and lower top coals.

The actual on-site measurement results of the top coal recovery provide reliable data for verifying the results of the indoor similar simulation and numerical calculation and provide a new method for revealing the drawing law in depth.

2.3.6 Top Coal Recovery and Gangue Mixed Ratio

2.3.6.1 Relationship Between Top Coal Recovery and Gangue Mixed Ratio

Improving the top coal recovery and reducing the gangue ratio is an important research content of LTCC, however, the two are also a pair of contradictions. In the early caving stage, pure top coal can be released, and the drawing body is completely

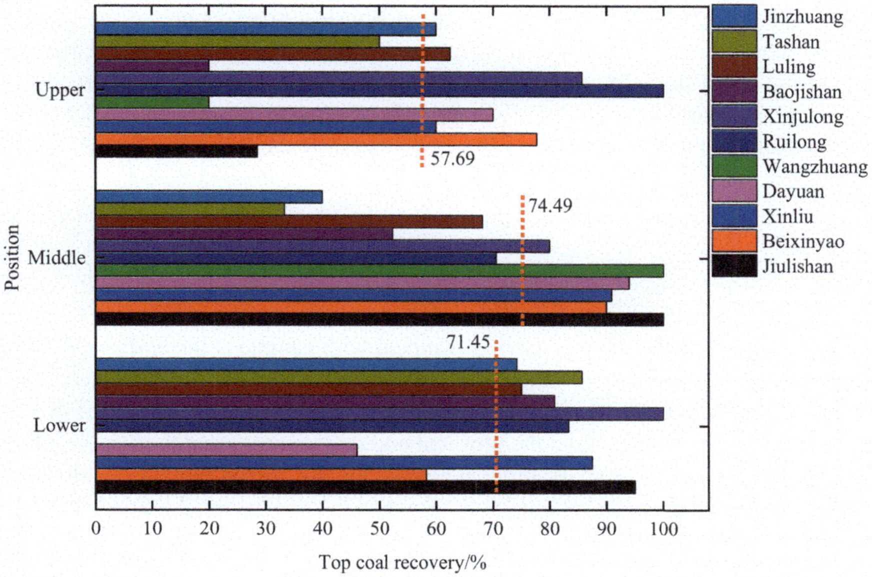

Fig. 2.60 Changes in top coal recovery of different seams in various coal mines

composed of top coal. With the top coal continue caving, the volume of the drawing body increases, the broken immediate roof rock (gangue) will enter the drawing body, forming a mixed gangue. At this time a part of the top coal is still not released, and thereafter the more top coal released, the more rock mixed into the amount will be more, the higher the gangue mixed ratio. Figure 2.61a shows the numerical simulation data of the relationship between top coal recovery and gangue mixed ratio, and Fig. 2.61b is the comparative results of filed measurements and physical tests of the top coal in 4#coal and 9#coal of Pingshuo mine area.

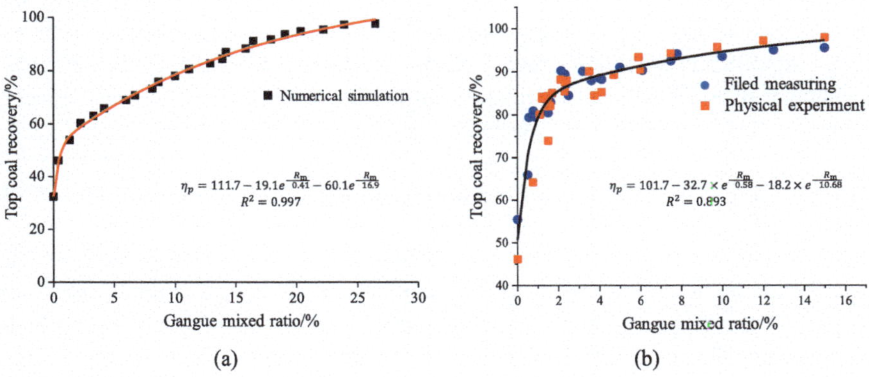

Fig. 2.61 Relationship between top coal recovery and gangue mixed ratio: **a** numerical simulation **b** field measuring and physical experiments

In Fig. 2.61, no matter numerical simulation, physical experiment, field measuring data, with the increase of gangue mixed ratio in the caving process, the total recovery ratio in the panel firstly increases rapidly and then grows slowly, and the recovery ratio increases with the increase of gangue mixed ratio is a compound exponential function, and its basic relationship can be expressed by the following Eq. (2.23):

$$\eta_{\mathrm{p}} = \mathrm{a} + \mathrm{b}_1 \times \mathrm{e}^{\frac{R_M}{C_1}} + \mathrm{b}_2 \times \mathrm{e}^{\frac{R_M}{C_2}} \tag{2.23}$$

where: η_p is the recovery rate, %, R_m is the gangue mixed ratio, %; a, b_1, b_2, c_1, c_2 are the equation coefficients.

Therefore, when the top coal caving is mixed with gangue at the early stage. The top coal recovery increases significantly with the increase of gangue mixed ratio. When the gangue mixed ratio increases to a certain value, at about 10%, the top coal recovery increases slowly. Take the physical and on-site measured data in Pingshuo mine area in Fig. 2.61b as an example, when the gangue mixed ratio is 2%, the total recovery ratio of the panel is about 75–80%; when the gangue mixed ratio is 4%, the total recovery ratio of the panel is about 85–88%; when the gangue mixed ratio is more than 10%, the overall change of the total recovery ratio of the panel is small, and it is not suitable to continue to carry on the excessive caving process. Therefore, for different coal seam conditions and mining technology, to determine the reasonable gangue mixed ratio threshold in the caving process is important research content to improve the top coal recovery and mining efficiency.

2.3.6.2 Top Coal Recovery Prediction Model

Accurate on-site measurement of top coal recovery is crucial to the study of top coal drawing law, for this reason, the research team has developed three generations of top coal motion tracking instrument, which is used to monitor the top coal recovery in the panel and trace back the location of the top coal drawn, providing the equipment basis for the accurate top coal caving process in the panel. In addition to the actual measurement of top coal recovery, the prediction research of top coal recovery is also an important content in the theory of top coal caving. By studying the influence mechanism of top coal size distribution on the drawing law, a top coal recovery prediction model based on size distribution is proposed [28], as shown in Eq. (2.24).

$$\eta_p = a_1(d_r - d_{rm})^2 + k \tag{2.24}$$

where d_r is the relative size of top coal, which is related to the average value of top coal size and the length of DO; d_{rm} is the relative size of top coal when the top coal recovery is maximum; a_1 and k are parameters of the equation, and $a_1 < 0$, which can be obtained by analyzing the statistics of test data. Figure 2.62 shows the change trend of top coal recovery with relative size of top coal. When the relative size of top coal is larger ($d_r \geq 0.079$), the top coal recovery measured on-site and the numerical

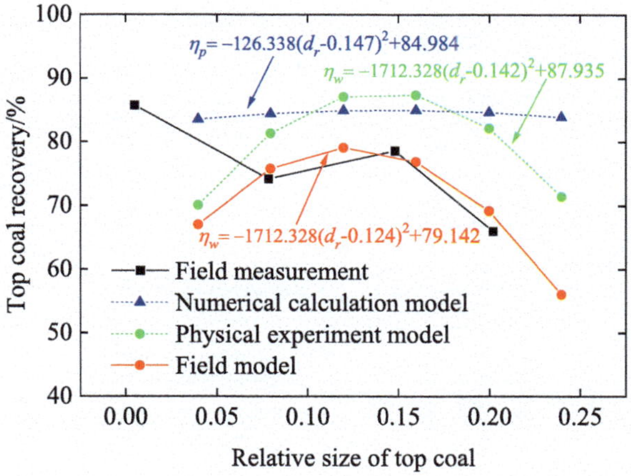

Fig. 2.62 The variation of top coal recovery with relative size of top coal

calculation and physical experiment results all show the change law of increasing and then decreasing. To predict the top coal recovery more accurately, the simulation of top coal recovery prediction obtained from the test was corrected. and the modified recovery ratio prediction model can fit the top coal recovery of each panel better. If the relative size of the top coal is measured on-site, the prediction model can be used to predict the top coal recovery of the panel.

2.4 Caving Technology with High Top Coal Recovery

The caving technology is the core of improving the top coal recovery. Through long-term theoretical research and field industrial tests, the top coal loss characteristics under different seam conditions and caving technology have been summarized. Targeted optimization measures of caving technology have been formulated to form a precise caving system under complex conditions, as shown in Fig. 2.63, to further improve the recovery ratio of coal resources in LTCC.

2.4.1 Caving Technology in Nearly Horizontal Extra-Thick Coal Seams

2.4.1.1 Precise Control Multi-round Caving Technology

For LTCC panels with extra-thick top coal, adopting a multi-round caving technology helps ensure a uniform sinking of the top coal boundary, preventing the early intrusion of gangue and achieving a higher top coal recovery. However, it is difficult to precisely control the time of multi-rounds caving when done manually. When

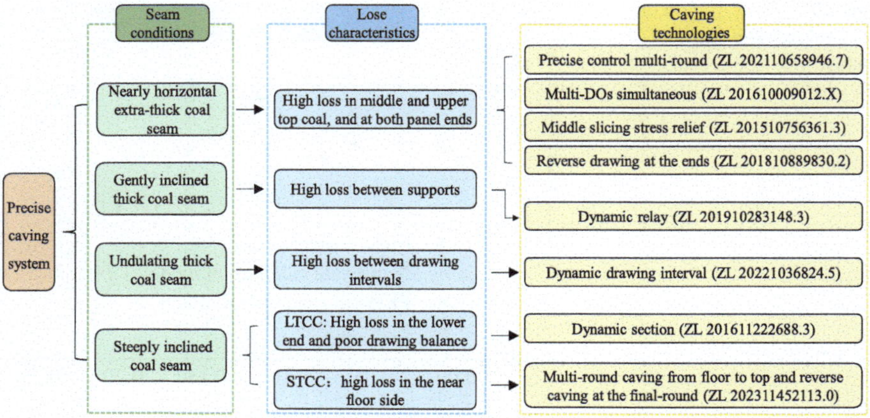

Fig. 2.63 Precise caving system (ZL 20… is the Chinese invention patent number)

a dirt band within the top coal, coal miners often mistake the drawn dirt band for roof rock and prematurely close the DOs, leading to significant losses of top coal. Addressing the above issues, the author's team has developed a top coal drawing time measurement system that can accurately record the drawing time of top coal from different layers. Combining this with a gangue mixed ratio detection technique using image recognition, a precise control multi-round caving technology [38] has been developed, as shown in Fig. 2.64.

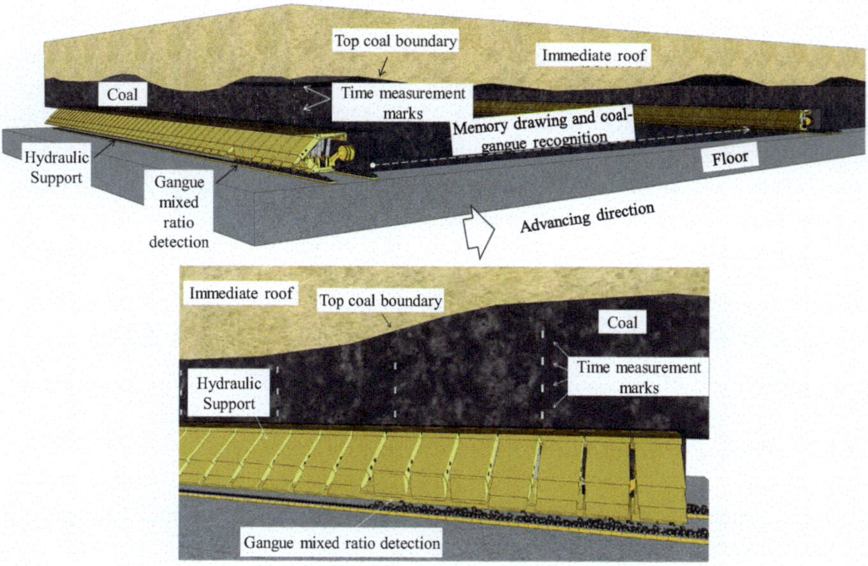

Fig. 2.64 Precise control multi-round caving technology

The precise control multi-round caving technology primarily includes the determination of the multi-round caving technology, measurement of multi-round drawing time parameters, and multi-round memory drawing and parameter correction.

(1) Determination of the multi-round caving technology. Based on the geological conditions of the panel and the physical characteristics of coal and gangue, numerical simulations and physical experiments are used to study the recovery ratio and gangue mixed ratio of different multi-round caving technology, determining the drawing rounds, drawing intervals, and other parameters, as well as the layout of top coal motion time measurement marks.
(2) Measurement of multi-round drawing time parameters. According to the determined drawing rounds and mark layout, several layers of top coal motion time measurement marks are arranged in the top coal of the panel from bottom to top until the top coal boundary. When a mark from a certain layer fall onto the AFC, it serves as the end mark for a round of drawing. The time from the start of movement to the fall of each layer of marks onto the AFC is automatically recorded as the drawing time for that round, which serves as the time parameter for subsequent automatic multi-round drawing.
(3) Multi-round memory drawing and parameter correction. During the advancing of the panel, memory drawing is conducted based on the established multi-round drawing technology and time parameters. Furthermore, the coal-gangue recognition of the drawn top coal is performed using gangue mixed ratio detection technology. The decision to close DOs and the correction of drawing parameters are made based on the recognition results. When significant changes occur in the thickness of the top coal or geological conditions, or when it is determined through gangue mixed ratio detection results that the current process parameters are no longer applicable, the top coal motion time measurement marks are rearranged to determine the process parameters for a new cycle of multi-round drawing.

2.4.1.2 Multi-drawing Openings Simultaneous Caving Technology

For nearly horizontal extra-thick coal seams, the adoption of traditional single-DO caving technology tends to result in many top coal boundaries and an increased probability of gangue mixing. Therefore, from the perspective of improving recovery ratio and reducing gangue mixed ratio, the multi-DOs simultaneous caving technology can be employed, as shown in Fig. 2.65. This technology refers to simultaneously opening two or more DOs of the support along the layout direction of the panel for drawing operations. These DOs can be adjacent or spaced apart by a certain distance. The multi-DOs simultaneous caving technology was applied in early drawing practices, however, due to its complexity compared to single-DO drawing, higher requirements on top drawing control and the capacity of the rear AFC, as well as a lack of theoretical understanding of the top coal drawing law in multi-DOs simultaneous caving, its early adoption was limited. In recent years, with the rapid upgrading of coal mining equipment and the advancement of intelligent mine construction, the cost of high-capacity AFCs has decreased, and the popularity of electro-hydraulic controls has made it easier to control multiple hydraulic supports simultaneously. As a result, this multi-DOs simultaneous caving technology has gradually begun to be adopted on site [29, 39]. Figure 2.66 shows the top coal caving situations under various DOs simultaneous caving. When using LTCC in nearly horizontal coal seams, adopting a multi-DOs simultaneous caving technology can reduce the number of top coal boundaries and losses between supports, thereby reducing the probability of gangue mixing and improving the top coal recovery. For extra-thick coal seams, the improvement effect is even more significant.

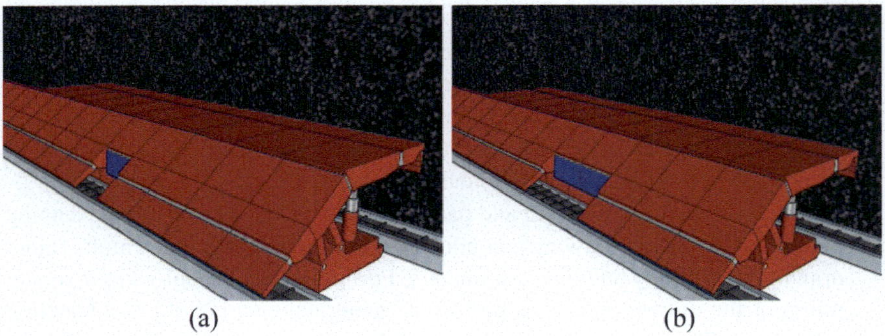

Fig. 2.65 Multi-DOs simultaneous caving technology: **a** ordinary single-DO; **b** multi-DOs

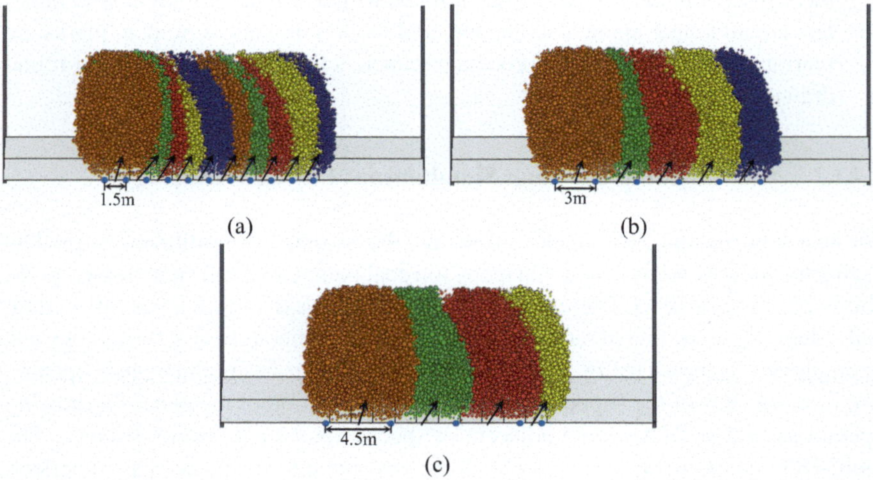

Fig. 2.66 The influence of DO number on the drawing body of each support: **a** single DO; **b** two Dos; **c** three DOs

2.4.1.3 Middle Slicing Stress Relief Caving Technology

For extra-thick coal seams with a thickness of over 20 m, which are nearly horizontal or gently inclined, the middle slicing stress relief caving technology can be adopted to achieve efficient and high recovery ratio mining. The basic process is illustrated in Fig. 2.67a. The extra-thick coal seam is divided into three layers: *A*, *B*, and *C*. The stress relief panel is arranged in Layer *B*, and fully mechanized mining is used for extraction. During the extraction of this layer, the coal in Layer *C* first undergoes mining-induced loading, then unloading after the panel passes through, collapsing and piling up on top of Layer *A*. Subsequently, under the effect of overburden subsidence, it begins to bear load, entering the stress recovery process (the above coal fracture is considered as primary fracture). Finally, a LTCC panel is arranged in layer *A*. Shearer is used to extract the coal in Layer *A*, while the top coal Layer *C*, which has already collapsed and fragmented above Layer A, is extracted through top coal caving (the coal fracture in this stage is considered as secondary fracture). This mining technology is suitable for hard extra-thick coal seams or extra-thick coal seams with high gas content. The mining of Layer *B* can achieve pre-fracturing of hard coal or effectively release coal seam gas. The top coal undergoes three processes: unloading mining influence, stress recovery and reloading, and top coal caving. Under the influence of multiple disturbances, the top coal is highly damaged, enabling one-time extraction of extra-thick coal seams of 20 m or more, greatly improving resource mining efficiency.

Figure 2.67b compares the broken size of top coal between the stress relief panel and the LTCC panel. During the secondary top coal caving mining stage, the cumulative volume of top coal blocks is approximately 119 m^3, and the cumulative volume below 108 m^3 are the most concentrated. Compared to after stress relief, the block ratio increased by 3.9%, and the number of blocks increased by 3110. After secondary

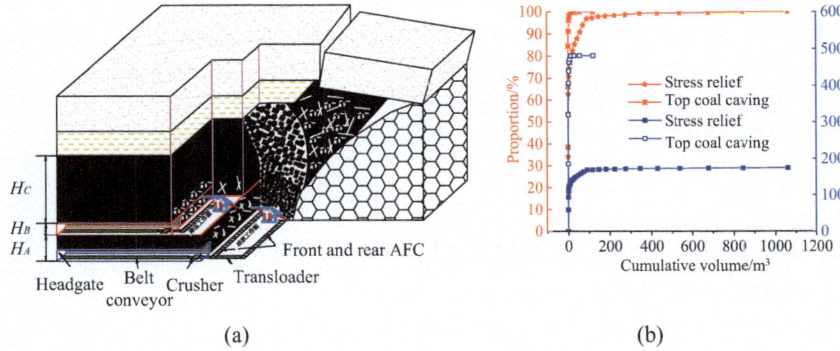

Fig. 2.67 Middle slicing stress relief caving technology: **a** schematic diagram of caving technology; **b** comparison of top coal broken effect

fracturing, the volume of top coal blocks significantly decreased, and the number of smaller-volume blocks significantly increased, indicating a significant secondary fracturing effect [4].

2.4.1.4 Reverse Drawing at the Ends

The coal loss near the ends of the LTCC panel is an essential part of the mining area's coal loss. For extra-thick coal seams, due to the large thickness of the top coal, the coal loss at the ends also increases. To reduce the coal loss at the ends of extra-thick coal seams, the reverse drawing at the ends is proposed based on the characteristics of the top coal boundary at the upper and lower ends. This process involves setting up a reverse drawing section near the ends of the panel, where the drawing direction is opposite to that of the normal drawing section. By adjusting the top coal boundary shape through the reverse technology, the top coal recovery at the lower end is improved [39].

The reverse drawing at the ends is shown in Fig. 2.68a, where all the drawing supports in the panel, except for the no drawing sections at the upper and lower ends, are divided into a normal drawing section and a reverse drawing section. The normal drawing section refers to the section where the drawing direction is the same as the cutting direction of the shearer (downward drawing in Fig. 2.68); the reverse drawing section refers to the section where the drawing direction is opposite to the cutting direction of the shearer (upward drawing in Fig. 2.68). The reverse section is adjacent to the no drawing section at the lower end, and its length is approximately 3–5 times the width of the support. In field production, if two-way cutting is adopted, with two cuts in one round trip, and drawing occurs in both upward and downward directions, the reverse section will alternately appear at the upper and lower ends. After adopting the reverse drawing technology at the ends, the top coal boundary shape at the ends is more conducive to reducing coal loss, as shown in Fig. 2.68b. According to the difference between the reverse and normal top coal boundaries at the ends, the volume of top coal recovered at the ends using the reverse technology is approximately V_1–V_2, and this drawing technology is more effective in reducing coal loss in extra-thick coal seams.

2.4.2 Dynamic Relay Caving Technology for Gently Inclined Thick Seam

When adopting the multi-drawing openings simultaneous caving technology for gently inclined coal seams, the interaction between adjacent two drawing bodies during their development is not significant, and the coal loss between supports mainly comes from the upper top coal. Therefore, the dynamic relay caving technology has been developed to improve the top coal recovery in gently inclined thick

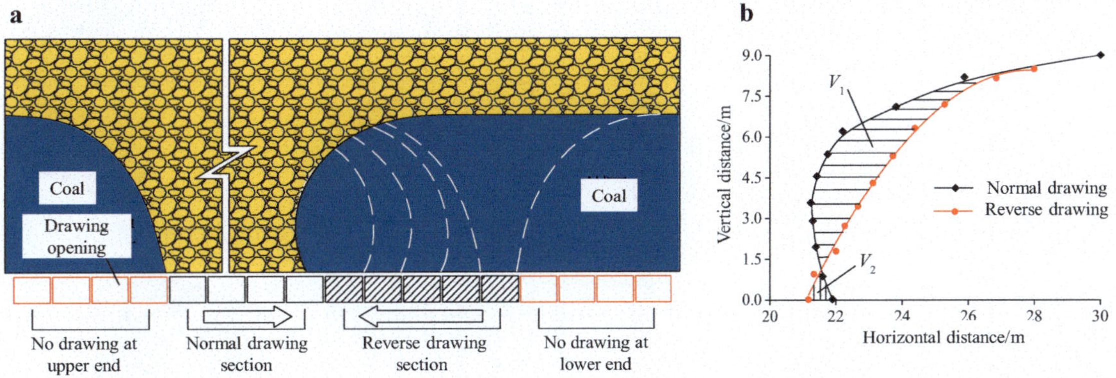

Fig. 2.68 Reverse drawing at the ends in LTCC panels: **a** schematic diagram of reverse drawing at the ends; **b** difference of top coal boundaries between reverse and normal drawing

coal seams [24]. The schematic diagram of this technology is shown in Fig. 2.69, and the basic flow is as follows:

(1) First, open the DO 1 to release top coal. When approximately half of the top coal has been drawn from DO 1, open DO 2 to release top coal simultaneously with DO 1. During the drawing process, gangue gradually invades the broken top coal, forming the top coal boundary.
(2) As shown in Fig. 2.69a, when gangue is first encountered during the drawing process of DOs 1 and 2 (at this time, top coal above DO 2 has released approximately half of the coal), close DO 1 and simultaneously open the DO 3. This allows DOs 2 and 3 to release top coal simultaneously. During this period, the left top coal boundary is relatively steep but remains largely unchanged in subsequent drawing, while the right top coal boundary continues to develop, exhibiting a relatively gentle shape and extending towards the upper end.
(3) When gangue is encountered again, close DO 2 and simultaneously open the DO 4 to release coal (at this time, top coal above DO 3 has released approximately half of the coal, as shown in Fig. 2.69b). This ensures that there are always two supports releasing top coal simultaneously in the panel, however, a gap in the drawing progress between the two supports. This process is repeated throughout the entire panel. This drawing process can effectively improve the top coal recovery and the stability of supports in gently inclined thick coal seams.

Figure 2.70 shows that the dynamic relay caving technology has the highest top coal recovery, which is increased by about 8.44% compared with the single-DO sequential coal caving technology. Meanwhile, the drawing time under this technology is greatly shortened, saving about 46% of the time, indicating that this caving technology can accelerate the advance speed of the panel, increase the annual production of the panel, and comprehensively reflect the advantages of the high top coal recovery of the single-DO caving technology and the short drawing time of the two-DOs caving technology under such panel conditions.

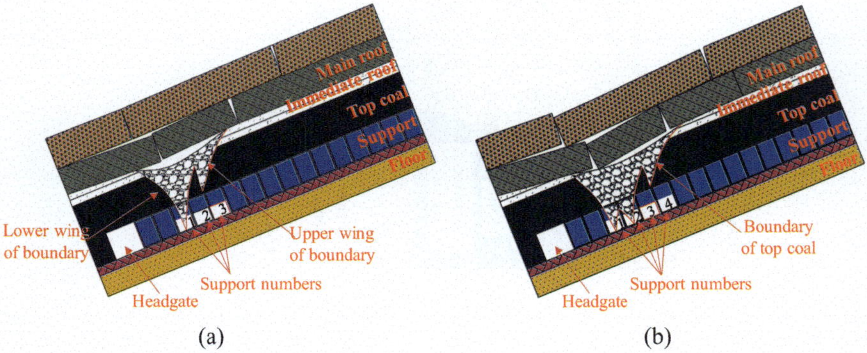

Fig. 2.69 Dynamic relay caving technology: **a** top coal drawing from DOs 1 and 2, **b** top coal drawing from DOs 2 and 3

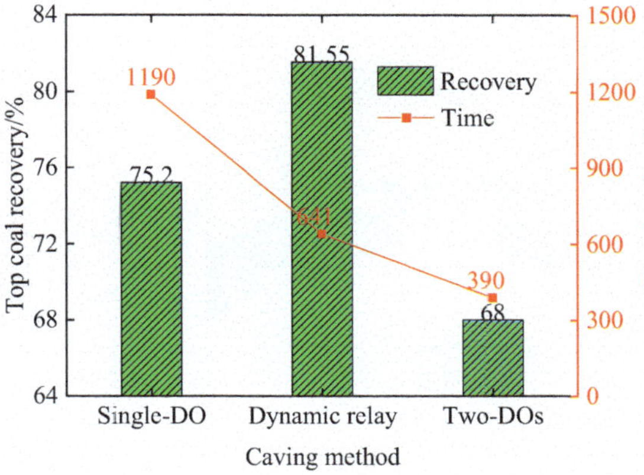

Fig. 2.70 Top coal recovery under different caving technologies

2.4.3 Dynamic Drawing Interval for Undulating Thick Coal Seam

When thick coal seams are affected by folds, local undulations may occur. When using LTCC to mine such coal seams, local overhand, underhand, or alternating overhand-underhand mining conditions may be encountered, leading to issues such as poor drawing balance and low top coal recovery. Through research on top coal drawing law, it has been found that as the overhand mining angle increases, the top coal above the support will be prematurely released, and the coal loss left in the goaf also increases. In such cases, the coal drawing interval should be appropriately reduced to minimize coal loss. Conversely, as the underhand mining angle increases, it will cause premature mixing of gangue in the goaf, reducing the overall amount of top coal drawing in this round and increasing the gangue mixed ratio. In this case, the drawing interval should be appropriately increased to reduce the mixing of gangue during underhand mining. Therefore, for LTCC of undulating thick coal seams with alternating overhand-underhand mining conditions, the dynamic drawing interval technology can be adopted. This involves using a dynamic and variable drawing interval at different coal seams with different inclinations in the transition section to improve top coal recovery, effectively reducing the top coal that is prone to entering the goaf during overhand mining and the gangue that is prone to being prematurely released during underhand mining [40].

Figure 2.71a is a schematic diagram of the dynamic drawing interval technology, defining the overhand/underhand mining angle corresponding to different drawing interval switches as the critical angle β' for dynamic drawing interval. Taking the example of overhand-underhand transition top coal caving mining, in the overhand mining section, if the overhand mining angle $\beta \geq \beta'$, a small drawing interval is used; if $\beta < \beta'$, a medium drawing interval is used. In the horizontal mining stage, a medium drawing interval is adopted. In the underhand mining section, if the underhand mining angle $\beta < \beta'$, a medium drawing interval is used; if $\beta \geq \beta'$, a large drawing interval

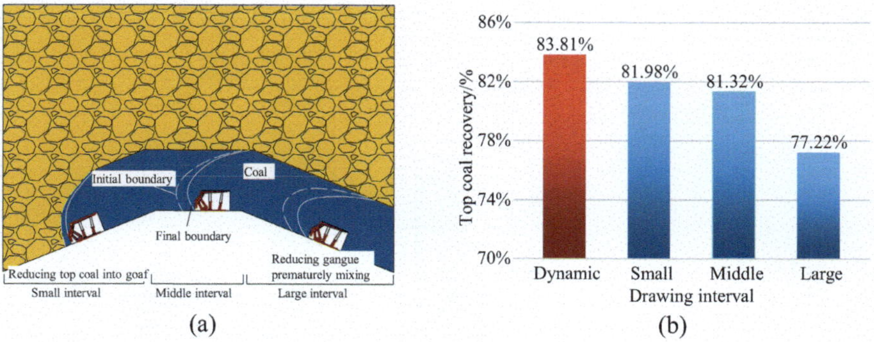

Fig. 2.71 **a** Dynamic drawing interval technology; **b** comparison of top coal recovery

is adopted. The critical angle β' for dynamic drawing interval can be determined by the following Eq. (2.25):

$$\beta' = 45^\circ - \frac{\alpha}{2} \tag{2.25}$$

where α refers to the dip angle of the shield beam, (°); β' refers to the critical angle of dynamic interval, (°). Verified by numerical simulation, compared with the traditional fixed drawing interval, the average top coal recovery using the dynamic drawing interval technology increased by about 3.71%, which can effectively reduce the waste of coal resources, as shown in Fig. 2.71b.

2.4.4 Caving Technology for Steeply Inclined Thick Coal Seams

2.4.4.1 Dynamic Section Caving Technology in LTCC

During the mining process of LTCC in steeply inclined coal seams, coal losses are mainly concentrated at the lower end of the panel. Additionally, due to the large dip angle of the coal seam, the traditional bottom-up caving technology can lead to the downward flow of top coal above the support when top coal caving at the lower part of the panel, reducing the force between the upper support, top coal, and roof, and weakening the stability of the support and the drawing balance. Therefore, based on the movement and drawing law of top coal in steeply inclined thick coal seams, the downward dynamic segmentation and upward drawing within segments technology for LTCC in steeply inclined thick coal seams has been developed, referred to as the dynamic section caving technology [22].

The schematic diagram of the dynamic section caving technology is shown in Fig. 2.72a. Its basic technological process includes: the shearer cuts coal from top

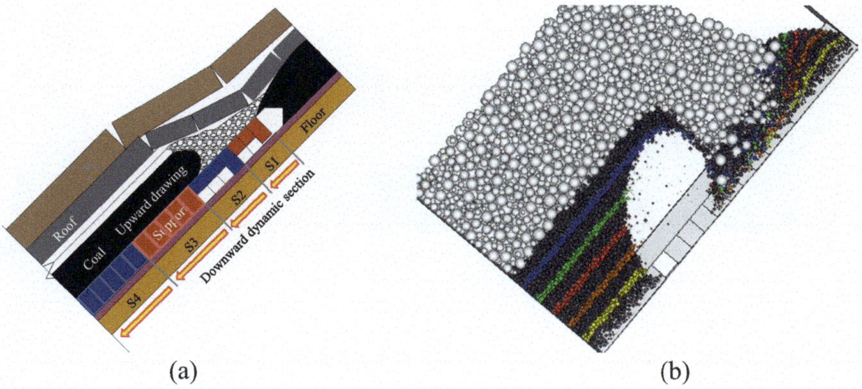

Fig. 2.72 **a** dynamic section caving technology; **b** the drawing body in the first section

to bottom; the support is moved from top to bottom; dynamic section is performed from top to bottom, and top coal is released from bottom to top within each section; the front AFC and rear AFC are pushed forward and pulled backward from bottom to top. The dynamic section from top to bottom refers to determining the number of supports in each section based on the top coal boundary and the shape of the top coal drawing body (Fig. 2.72b). The dynamic section caving technology leverages the advantages of LTCC in steeply inclined thick coal seams. Using this technology under the conditions of steeply inclined coal seams can achieve balanced coal caving on the panel, improve the anti-skid performance of equipment, enhance the safety of panel operations, and improve top coal recovery and resource utilization efficiency.

2.4.4.2 Multi-round Caving with Reverse Caving at the Final Round for STCC

Under the conditions of STCC, adopting a larger sublevel height can reduce roadway excavation, improve mining efficiency, and lower mining costs. However, increasing the sublevel height means that the motion path of the top coal becomes longer, increasing the probability of mixing between the top coal and gangue during motion. During drawing process, the gangue from adjacent supports can easily intrude into the flowing coal stream on the side, causing premature closure of DO, thus increasing the gangue mixed ratio or residual coal volume. Furthermore, due to the shorter horizontal sublevel panel, the coal loss on the roof and floor sides accounts for a larger proportion. In addition to improving the top coal recovery on the floor side by drawing top coal from the floor to the roof, the multi-round caving process also helps the top coal boundary to sink relatively evenly and slowly, ensuring uniform force on the support and avoiding stress concentration, which can reduce the gangue mixed ratio and increase the top coal recovery in the middle of the panel.

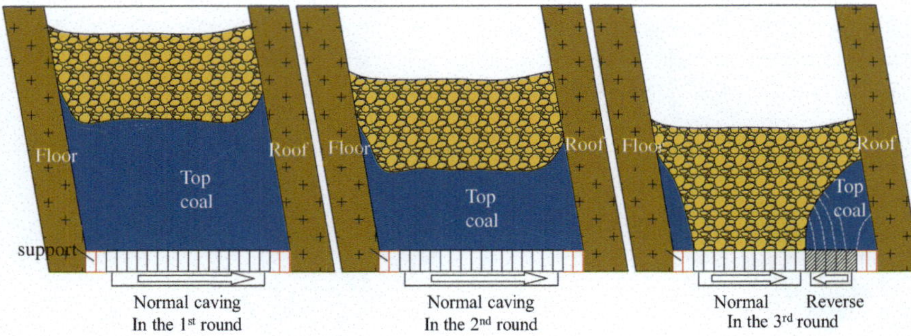

Fig. 2.73 Multi-round caving with reverse caving at the final round for STCC

For coal loss on the roof side of the horizontal sublevel, based on top coal drawing from the floor to the roof, the "multi-round normal caving and reverse caving in final round" technology can be adopted to specifically improve the top coal recovery on the roof side of the horizontal sublevel. Taking three-round normal caving and reverse caving in final round as an example, its technological process is shown in Fig. 2.73 [32].

When using this technology, two rounds of normal caving from the floor to the roof are first performed, as shown in Fig. 2.73a, b; then, in the final round, i.e., the third round of top coal caving, the panel is divided into normal and reverse caving sections. The normal caving section releases top coal from the floor to the roof, while the reverse caving section releases top coal from the roof to the floor (the length of the reverse caving is generally 3–5 support widths), as shown in Fig. 2.73c. This caving technology allows the top coal boundary to sink slowly, reduces the mixing of gangue, and minimizes the amount of residual coal near the roof side, effectively improving the top coal recovery.

2.4.5 Equipment and Technology for Top Coal Caving at the Ends

The crushing and caving of top coal in the end areas (transition support and roadway support areas) of LTCC have also been a research focus in recent years. As shown in Fig. 2.74, the research team took the lead in designing special supports and supporting transportation equipment for coal extraction at the ends (roadway coal extraction) in the Chuancaogeda Coal Mine and conducted underground tests [33]. Figure 2.74b shows the physical diagram of the modified caving support, where the shield beam positions of the two supports in the roadway have been modified, and the DO location has been redesigned. It is equipped with a coal breaking mechanism, allowing the crushed top coal to flow out of DO through the gap in the iron mesh, realizing the top coal caving at the end area. Figure 2.74c shows the modified transloader mechanism, and Fig. 2.74d shows the spatial position relationship between the rear AFC and the extended transloader. By extending the transloader to the lower end of DO of the roadway support, the crushed top coal can be transported out of the panel. After field use, the special equipment for top coal caving at the ends has achieved significant results.

In the "three-front and four-rear" transition support area of the LTCC panel, due to the layout of the front and rear AFCs' heads and tails, the caving support lacks caving space, resulting in only a small amount of top coal or no top coal being drawn, causing significant waste of coal resources. To address this issue, the team has also initially developed caving supports for the transition area (Fig. 2.75). This involves installing a skylight on the shield beam and a DO insert device that can crush and release top coal. The drawn top coal can be directly transported out of the panel by the rear AFC, achieving top coal caving in the transition area.

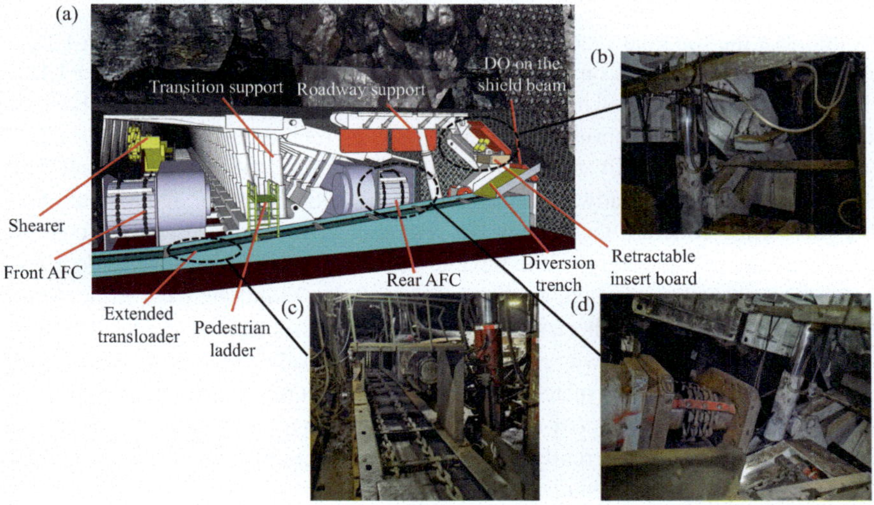

Fig. 2.74 Special support and matched transportation equipment: **a** drawing equipment layout at the end area; **b** modified caving support; **c** extended transloader; **d** spatial layout between the rear AFC and the extended transloader

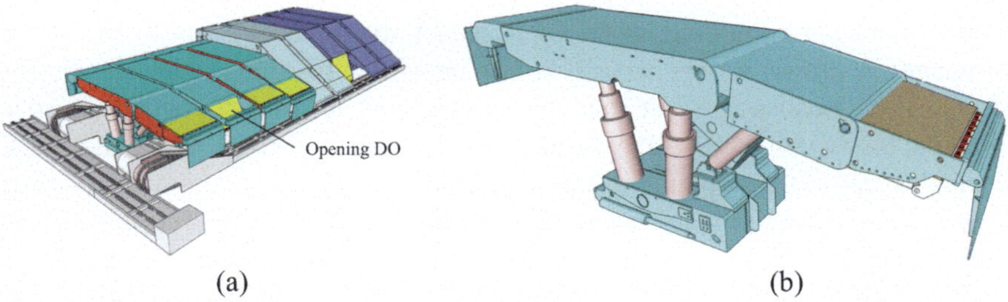

Fig. 2.75 Special caving support for the transition area

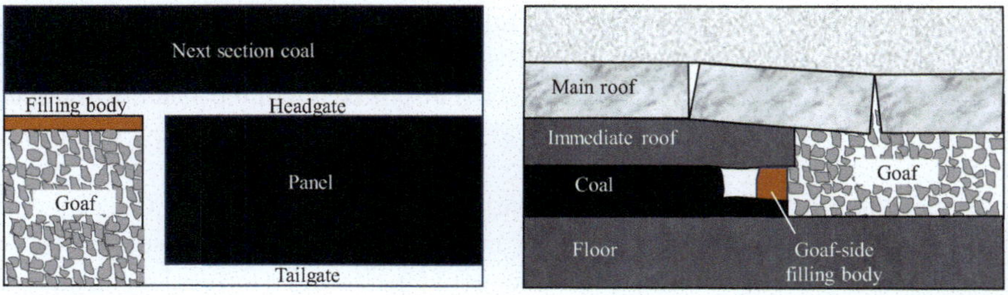

Fig. 2.76 Goaf-side roadway retaining technology for LTCC panel

2.4.6 *Other Technologies*

2.4.6.1 Goaf-Side Roadway Retaining Technology for LTCC Panel

To improve the top coal recovery and reduce roadway excavation costs in LTCC, research and experiments on the goaf-side roadway retaining technology for LTCC panels have been conducted in recent years. Due to the low-strength top coal above the roadway, the implementation of goaf-side roadway retaining technology for LTCC panels is more difficult. To address this issue, Changcun Coal Mine of Lu'an Group excavated the haulage roadway of the LTCC panel along the roof and adopted flexible membrane concrete technology for retaining goaf-side roadway. To increase the overall support capacity of the flexible membrane concrete wall, the floor coal under the wall was reinforced, as shown in Fig. 2.76. Xinjulong Coal Mine, Shandong, located in a nearly horizontal LTCC panel with a mining depth of 810 m, constructed goaf-side roadway retaining, which is a roadway along the floor, using gangue cementitious filling material combined with bolt reinforcement. Xiegou Coal Mine, Shanxi, constructed goaf-side roadway retaining along the floor through roof cutting and stress relief. Currently, goaf-side roadway retaining technology for LTCC panels has been widely recognized and promoted [41].

2.4.6.2 Increasing Production Technology at the Initial and Final Mining for LTCC

In the initial stage of mining on the LTCC panel, the top coal is not fully crushed and cannot be drawn or only partially drawn, resulting in initial mining losses. The stress distribution characteristics and the effect of coal and rock strength can lead to a top coal loss width of 10–20 m during the initial mining period. When the LTCC panel advances to the stopping line, to ensure the safety of the withdrawal roadway support, it is usually necessary to leave top coal or climb ahead of time during the final mining stage to ensure the surrounding rock stable, resulting in final mining losses. Initial and final mining losses are significant components, with a loss ratio ranging from 0.7 to 10%, averaging approximately 2%. To reduce the initial and final mining losses in LTCC panels, the Panel 403 at the Pubaijianzhuang Coal Mine of Shaanxi Coal Industry has been designed with a strike length of 3580 m. The extended panel advancing length has effectively lowered the loss ratio of top coal during the initial and final mining stages. Mines such as Daliuta Coal Mine of China Energy, Yankuang Group in Shandong, Qianqiu Coal Mine of Yima Gropu, Wangzhuang Coal Mine of Lu'an Mining Group, and Pingshuo No. 2 Mine have made significant contributions in panel design and on-site management. During the initial mining stage, measures such as deep-hole blasting to release the top coal and excavating auxiliary roadways to alter the boundary conditions of the top coal are commonly employed, especially for top coal caving under "two hard" conditions (hard coal and hard roof). During the final mining stage, three technologies are typically used: climbing to find the roof for closure, directly closing along the floor, and pre-excavating a roof roadway for closure. For "three soft" coal seams, additional measures such as laying artificial supports are required to strengthen the support system.

2.5 Development of LTCC

After 40 years of exploration, innovation and development, China's LTCC has made world-renowned achievements, from the early blasting mining LTCC, general mining LTCC, lightweight support LTCC, fully mechanized mining LTCC, to today's comprehensive promotion of LTCC, and the formation of a variety of conditions of the LTCC. In terms of caving technology, it has developed from high-level drawing and middle-level drawing at the initial stage of applying LTCC to the low-level drawing method commonly used at present. In the last 20 years, coal mining enterprises have widely adopted LTCC in thick coal seams, and the support tends to be heavy and large, and the rated working resistance of the support gradually increases. At present, the maximum height of caving support reaches 7 m, and the rated working resistance is up to 21,000 kN. The type of support is four-pillar in the early stage, and the four-pillar type and two-pillar type is concurrent nowadays. Toward the coal seam thickness of up to 20 m in LTCC, and the maximum dip angle of the seam can be 60°, the mining height of the panel can be from 2 to 7 m, and the maximum production of the panel can be up to 15 million t/a. For the steeply inclined seams with the thickness more than 20 m, the technology of STCC has been developed, and the production of the panel can be up to 4 million t/a. The depth of the mined area can be from 100 m to 1000 m, and the UCS of the seams can be up to 35 MPa. China has shown great technical advantages and wide scope of application of LTCC, which was exported to Australia in 2004, put into use in Ausdal Coal Mine in 2006, and exported to Turkey and other European countries in 2022, truly realizing the magnificent transformation from technology introduction to technology export.

2.5.1 Early Development of LTCC

The idea of LTCC originated from the high drop mining technique for thick seams. This technique was first developed in the nineteenth century during the manual mining of coal, which was also one of the main methods for mining thick coal seams in China before the 1950s. Kailuan and Lu'an mining district after the 1950s had experimented with wood and metal friction pillars to mine the lower layer and recover the middle layer of LTCC technology. The panel using a AFC to transport coal in the early, and later the panel used two AFC, to transport the blasting coal of the coal wall and the drawn top coal respectively [2], which is the earliest blasting mining LTCC. This technology was later developed into the monolithic hydraulic pillar LTCC, suspension support LTCC, net-type support LTCC, etc. Due to the poor stability of the monolithic pillar, low support resistance, the blasting mining LTCC has been gradually eliminated in China by the beginning of the twenty-first century.

At the beginning of the twentieth century, Europe began to use room and silo type LTCC in some complex geological conditions of coal seams. Along with the development of longwall mining method, slicing mining method for thick seams appeared in the 1930s, such as the upward filling mining method which was popular in Europe at that time, and the inclined slicing and horizontal slicing mining method which was popularized in China in the1950s, etc. Although these methods have higher coal extraction ratio and better safety, the process is complicated and the cost is high. For this reason, France and Yugoslavia began to use the long-wall or short-wall top coal caving method with monolithic pillars and roof beams.

In 1954 the UK equipped the world's first fully mechanized longwall panel, which greatly increased productivity and safety. As a result, from the mid-late 1950s to the 1960s, the fully mechanized mining technology in Europe developed rapidly and gradually took the leading position, and began to study and test the fully mechanized LTCC for thick coal seams. 1957, the Former Soviet Union firstly used the KTy two-pillar shield hydraulic support to mine thick coal seams with an inclination angle of 5°–18° and a thickness of 9–12 m. The panel was firstly mined the top coal and laid the bottom mesh, then the bottom seams were mined and the middle seams were released through the skylight on the roof beam of the support. This is the earliest pre-mining top layer LTCC in the world, and this method of Former Soviet Union lasted until the end of the twentieth century, and in recent years, individual panels in Russia still use the LTCC. 1964, France successfully tested the one-time mining full-thickness LTCC technology in the Darcy D mine of the Blanchy mine in the south-central coal field, which is also the one-time mining full-thickness LTCC in the world. The panel is equipped with two AFCs in the front and rear realizing the parallel operation of cutting and caving. The coal thickness was 6–8.5 m, the average monthly production was 49,600 t, the recovery ratio of the panel was 90%, and the manpower efficiency was 33.6 t/worker, which was the best technical and economic index achieved in the early stage of LTCC.

Although the early stage of LTCC in Europe did not play its technical potential of high yield and high efficiency, however, the mining of unstable thick coal seam, corner coal and coal pillar is still widely applied. After the 1970s, Eastern European countries such as the former Soviet Union, Yugoslavia, Poland, and Hungary followed the technical route of a single AFC on the panel. Based on the KTy shield support of the former Soviet Union, DO was changed from the front of the top beam to the rear of the top beam, and a hydraulic prop was designed to control the opening and closing of DO. Among them, the VHP-type caving support in Hungary is the most representative, as shown in Fig. 2.77. The support was introduced to China in 1987.

France, Germany, Britain, Spain, and other countries follow the technical route of double AFCs on the panel, and continue to develop on the basis of banana-type supports in France, forming a series of median and low-level caving supports represented by 400 t shield supports of British Dorset Company, as shown in Fig. 2.78.

The main types of foreign caving support are shown in Table 2.6. Due to the depletion of coal resources in thick coal seams in most European countries, and the coal industry has shrunk, and LTCC technology in Europe has not been well

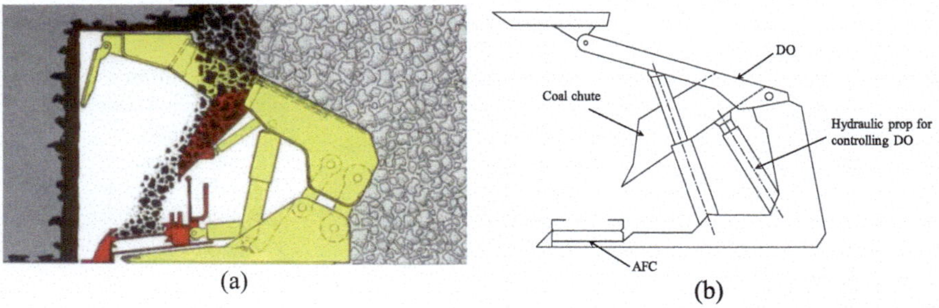

Fig. 2.77 Caving support with single AFC: **a** KTy type shield caving support of former Soviet Union; **b** Hungarian VHP caving support

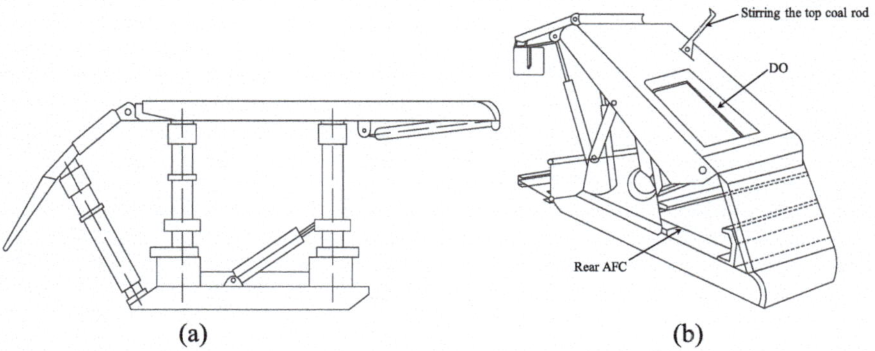

Fig. 2.78 Caving support with two AFCs: **a** France banana-type caving support; **b** Britain 400 t shield caving support

developed, and the technical advantages have not been fully utilized. By the end of the twentieth century, except for Russia and other individual countries, LTCC in Europe gradually withdrew. In recent years, China's great achievements in coal mining have enabled individual countries in Europe, such as Turkey, to start research and application of LTCC technology [42].

It has been 40 years since China began to study and introduce the LTCC technology in 1982. The development and stage division of LTCC technology are expounded

Table 2.6 List of main types of foreign top coal caving supports [2]

Nationality	Type	Schematic diagram	Characteristics and caving method	Year
Former Soviet Union	KTy	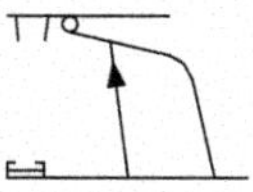	Single AFC, skylight in roof beam, high-level caving method	1957
France	BANANA	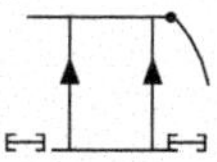	Two AFC, single twisted door type, low-level caving method	1963
US		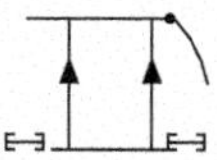	Two AFC, insert plate type, low-level caving method	1977
France	MB170	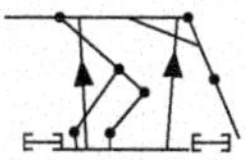	Two AFC, four-bar insertion plate, low-level caving method	1978
France	FBS-4-340	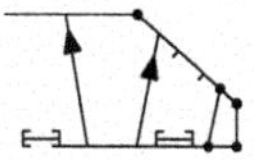	Two AFC, skylight in shield beam, middle-level caving method	1980
Germany	1000 kN-19/28	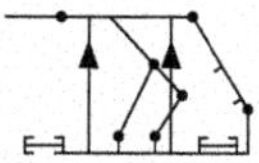	Two AFC, four-bar linkage, skylight in shield beam, middle-level caving method	1982
Hungary	VHP730	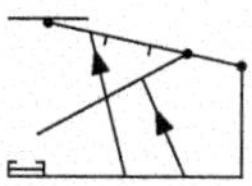	Single AFC, skylight in shield beam, high-level caving method	1982
UK	4L-4000 kN-19/28	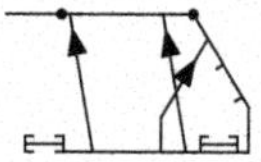	Two AFC, skylight in shield beam, middle-level caving method	1983

in the articles of different periods in the past [43–45]. To facilitate memory, the 40-year development of China 's LTCC technology is divided into two stages, namely, the exploration and application stage in the first 20 years, and the technological innovation and export stage in the latter 20 years.

2.5.2 *Exploration and Application Stage*

2.5.2.1 The First Caving Mining Industrial Test

The first industrial test of LTCC technology in China was carried out in the second section of the north third mining area of Puhe Coal Mine of Shenyang Mining Bureau in June 1984, with a panel length of 65 m. The coal seam is 12–14 m, the dip angle is 5°–14°, the coal type is lignite, the coal quality is medium hard, the joints are developed, the roof and floor are soft, and the coal seam is easy to spontaneous combustion. The support is FY4000-14/28 low-level caving support, which is designed by the cooperation of Beijing Mining Institute of China Coal Research Institute and Shenyang Coal Research Institute. From June to December 1984, the experiment lasted for seven months, with an average monthly yield of 13,100 t. Due to the unreasonable force and serious damage of the four-bar linkage of the support, the slow advance speed of the panel and the spontaneous combustion of the goaf, the test was forced to be interrupted. Later, the support was changed to FYA4000-16/26 skylight type middle-level caving support. In June 1986, the second test was carried out on the right wing of the second section of the third mining area in the north. The panel was 53 m long and the maximum monthly coal production was 20,100 t [44]. In general, the industrial test of the first LTCC in Puhe Coal Mine is not good, however, this LTCC test has accumulated valuable experience and lessons for the later research and development of LTCC technology in China. First, it is particularly important to select a reasonable support type and structural parameters; the second is to ensure the advancing speed of the panel and realize the regular cutting and caving operation cycle; the third is to train technical personnel, management personnel and workers.

2.5.2.2 Horizontal Sublevel Caving Mining in Steeply Inclined Thick Coal Seam

After the first industrial test of LTCC in gently inclined thick coal seam did not achieve ideal results, the horizontal sublevel top coal caving mining (STCC) test in some steeply inclined thick coal seams was successful and achieved good results.

1. STCC in Mine 2 of Yaojie Mining Bureau

In April 1986, the industrial test of horizontal STCC in steeply inclined thick coal seam was carried out in Well 4 of Yaojie Mine 2. The coal seam is 20 m thick, the coal seam dip angle is 55°, the coal quality is medium hard ($f = 1.2$), the panel is 660 m long, and the panel is 20 m width. The main equipment of the panel is: total 15 of FY2800-14/28 shield type low-level caving support, MLS31-170 double drums type shearer, SGWD-180PB type front AFC, SGW-40T type rear AFC. The support at the ends is DZ25 type single hydraulic prop with self-made long hinged top beam (3.6 m) double-row step support [43, 46].

Tests were carried out on different sublevel heights, and it was considered that under the condition of coal seam, the cutting height was 2.5 m and the sublevel height was 10–12 m. The two-cutting and one-caving method is adopted, and the footage of each knife is 0.5–0.6 m. Furthermore, the top coal recovery and the composition and gangue mixed ratio of the lost top coal were preliminarily studied. The average monthly coal production of the panel is 19,000 tons, the average efficiency of the panel is 124,400 tons per worker, the mining ratio of the panel is 85.9%, and the gangue mixed ratio is 15–20%. The project won the second prize of the National Science and Technology Progress Award in 1990 [45].

2. STCC in Meihekou Coal Mine

There are 7 pairs of inclined shafts in Meihekou Mine of Liaoyuan Mining Bureau in Jilin Province, mainly mining No.12 coal seam ($f = 0.9$–1.4). The coal seams of the third and fourth wells located at the west and east ends of the coalfield are thicker, 20–70 m, and the dip angle is more than 45°. In November 1986, STCC test was carried out in four wells. The panel is 25–38 m long, the height of the sublevel is 15 m, and the YFY2000-16/26 light high-level caving support is adopted. The average monthly output of the panel is 25,200 t, and the recovery ratio of the panel is 85%. In October 1987, the test was carried out in the third well. The panel is 74 m long, the mining height is 17.5 m, and the FYC4000-16/28 sky window type middle-level caving support is adopted. The average monthly output of the panel is 40,000 tons, and the recovery ratio of the panel is 87.1%. Compared with the previous oblique slicing mining with metal mesh false roof, the monthly production of the third and fourth wells increased by 3.4 times and 2 times respectively, the recovery ratio of the mining area increased by 3–5%, the excavation ratio of the mining area decreased by 80% and 58% respectively, and the direct cost of the panel decreased by 35% and 25% respectively [44]. After 1988, STCC was applied in the third and fourth wells of Meihekou Mine. The annual output of the panel exceeded 600,000 tons for more than 10 years, which was the highest record in the same kind of coal seam in China. The STCC showed obvious technical and economic advantages.

In the mining of the third well of Meihekou Mine, through the study of geological conditions, overburden lithology, 'two zones' height prediction, a large amount of clay infiltrated into the sand layer to block cracks, limited thickness mining, uniform caving and other measures, the overlying water-bearing sand layer was successfully passed, which accumulated experience for the mining of STCC under water body. The two roadways of the panel are arranged in the roof and floor of the coal seam, which avoids the loss of the roadway coal pillar. The measures of high-pressure water injection and pre-blasting top coal were adopted for the phenomenon of large block stuck DO in local coal hardening, and the height of lower sublevel was reduced from 15 m to 10–12 m, which achieved great results. Furthermore, according to the change of coal seam thickness, the length of panel is adjusted by adding and subtracting supports to reduce coal loss. The strict multi-round one-cutting and one-caving method is implemented to ensure that the recovery ratio is above 85%.

3. STCC in Urumqi Liudaowan Coal Mine

The dip angle of coal seam in Liudaowan Coal Mine is 64°–71°, the thickness of coal seam is 33–50 m, the coal is soft and the endogenous fracture is developed. The horizontal sublevel height is 10 m, and the cutting height is 2.5 m. The MS-950 short fuselage shearer (fuselage length 3 m) produced by SDS company in France, FYS3000-19/28 skylight type middle-level caving support, 2 × 18/400 type front AFC matched with shearer, and domestic SGW-40 T rear AFC are adopted in the panel.

From November 1987 to January 1988, more than 20,000 tons of coal were produced per month for three consecutive months, and the recovery ratio of the panel was 87.04%. The hydraulic support, shearer and AFC have great adaptability, and the designed support has good adaptability. The imported short-body shearer and the matching front AFC have good use effect. However, there are also some problems in the test, such as the serious leakage in front of the support, which affects the mining efficiency; the size of DO is small, and it is easy to plug.

After exploration and innovation, the STCC technology in steeply inclined thick coal seam has achieved initial success, and the recovery ratio of panel is more than 85%. It solves the problems of high cost, poor safety and low efficiency in the past, such as slicing mining and high caving mining in roadway. The reasonable sublevel height is 10–15 m, and the reasonable mining and caving parameters are obtained, which accumulates experience for the later LTCC in gently inclined thick coal seam.

2.5.2.3 LTCC Test of Imported Support

In May 1987, a set of 80 VHP-732 shield type skylight high-level caving supports (Fig. 2.77b) was imported from Hungary in No.1 Coal Mine of Pingdingshan Mining Bureau in Henan Province. On September 15, 1987, the panel began production. The total thickness of the coal seam is 7.3 m, the dip angle of the coal seam is 7°–9°, the buried depth is 464 m, the hardness of the coal seam is 1.5–1.8. The panel is 87 m long, the advancing length is 489 m, the two roadways are excavated along the middle of the coal seam, and U29 four-section retractable arch support [44].

The MXA-300/3.5 double drums shearer is used in the panel, and SGZ-730/320 front AFC is arranged. The top coal flows into the front AFC through the caving skylight. Three-cutting and one-caving method was employed, the caving interval of 2 m. During normal mining, the monthly output of the panel reaches 42,000 t, the recovery ratio of the panel is 84.6%, and the economic benefit is good. However, in the mining test, it is also encountered that the dust is large when the coal is put in the high skylight, and the dust control effect of coal body water injection in the tailgate is limited. The air leakage from the two roadways to the goaf, the residual coal in the goaf is prone to spontaneous combustion.

2.5.2.4 High Yield and Efficiency LTCC Technology in Gently Inclined Thick Seam

After the practice of Shenyang, Yaojie, Liaoyuan, Urumqi, Pingdingshan and other bureaus, especially the successful industrial test of top coal caving mining in steeply inclined thick coal seam, the experience of gas, spontaneous combustion and dust prevention and control has been accumulated. Yangquan, Lu'an and Yanzhou mining bureaus have carried out the bold practice of LTCC in gently inclined thick coal seam. The support type has also changed from high-level and middle-level to low-level, which has truly realized the high yield and high efficiency of top coal caving mining.

1. Technical breakthroughs in LTCC technology at Yangquan Mining Bureau

The Yangquan Mining Bureau initiated its exploration of LTCC in Mine No. 1, targeting the No. 15 coal seam. This seam has an average thickness of 6.5 m, a dip angle of less than 5°, medium-hard coal ($f = 2$–3). In December 1988, an experimental LTCC operation was conducted at the Panel 8605 using ZFS4400 middle-level caving supports. The panel length was 42 m, achieving a maximum monthly production of 32,900 tons with a recovery ratio of 81.52%. This trial demonstrated that LTCC is feasible in medium-hard coal seams, enabling parallel operations of bottom coal mining and top coal caving, effectively doubling the productivity of the panel. It revealed the high potential for achieving high yields and efficiency through LTCC, resolving issues associated with traditional methods. These issues included low production (400,000 tons annually), difficulties in roof management due to stratified roof conditions, short spontaneous combustion periods, high gas emissions, and poor economic performance. Following this success, the bureau expanded the use of ZFS4400 middle-level caving supports across four mines. By 1990, with the panel lengths ranging from 114 to 180 m, annual production reached 1 million tons, achieving high yield and efficiency goals.

In July 1996, Mine No. 1 began low-level caving tests in the No. 15 coal seam, employing ZFS4800-17/28B four-link low-level caving supports, alongside transitional and end supports. The shearer used was the AM-500, and front and rear AFC were SGZ-764/220 models. The panel featured advanced equipment with high power, strength, weight, throughput, and flow rates. The panel measured 180 m in length and 1242 m in strike, with an average monthly production of 148,600 tons. The one-cutting and one caving method achieved a recovery of 87.74% and a face efficiency of 51.78 tons per worker. The low-level caving eliminated coal loss in the arch zone, resulting in higher top coal recovery. Compared with middle-level caving operations in 1994 and 1995, recovery improved by 6.7% and 5.26%, respectively. Low-level caving demonstrated superior technical advantages, with extended face lengths serving as the foundation for increased efficiency.

All mines under the Yangquan Mining Bureau are high-gas mines. Gas from the coal seam and adjacent layers was effectively managed through a comprehensive system involving strike high-drainage roadways, inclined high-drainage roadways, and large-diameter boreholes. The successful application of LTCC in the No. 15 coal seam addressed the issues of low production and poor efficiency associated with traditional mining methods in Yangquan. It significantly contributed to the reform of mining methods and enhanced extraction efficiency. Moreover, it provided valuable insights and direction for the nationwide adoption of LTCC in thick coal seams, paving the way for high-yield, efficient coal mining practices.

2. High-yield and efficient LTCC at Lu'an Mining Bureau

The mines under the jurisdiction of the Lu'an Mining Bureau primarily extract coal from Seam No. 3 in the eastern part of the Qinshui Coalfield. The coal seam is approximately 7 m thick with a dip angle of 3°–7°. By 1993, all mines under the Lu'an Mining Bureau had adopted LTCC technology, playing a significant role in advancing this method within China.

In March 1989, Wangzhuang Mine conducted a test of high-level top coal caving technology with a skylight-type caving system at the Panel 4309. The support system used ZFD4000-17/23 high-level caving supports, while the AFC was an SGZC-730/320 model. Due to the hardness of the coal seam ($f = 1.5$–2.5) and its underdeveloped fissures, water injection into the top coal via boreholes in the panel roadways effectively softened the top coal. The panel had a length of 125 m, a cutting height of 2.8 m, and a caving height of 4.06 m. The two-cutting and one-caving method was employed with an interval caving technique, resulting in a monthly production of 92,000 tons and a top coal recovery of 78.97%.

In June 1991, low-level top coal caving trials were conducted at the Wuyang Mine using ZZPF4000-1.7/3.5 supports. The coal seam was 6.42 m thick, with a cutting height of 2.6 m and a panel length of 165 m. The one-cutting and one-caving sequential interval multi-round caving method was implemented, achieving parallel mining and caving operations. The panel reached a peak monthly production of 200,000 tons and an average monthly output of 120,000 tons, with a recovery of 91.1%, demonstrating the high production and efficiency potential of top coal caving.

In July 1999, the Wangzhuang Mine carried out trials on an extra-long LTCC panel spanning 270 m. The power of the shearer and the front and rear AFCs was appropriately increased. Additionally, roadway support technology was improved, introducing small coal pillar and bolt-mesh support techniques. The panel achieved an annual production of 4 million tons, with a recovery exceeding 85%.

To investigate coal loss in LTCC panels, the Wangzhuang Mine conducted a study in 1994. A detection roadway was driven from the original tailgate coal pillar into the goaf area of the 4318 low-level top coal caving panel to directly observe residual coal loss. The detection roadway had a height of 2 m. The panel was 171 m long, with a coal seam thickness of 6.8 m and a mechanized mining height of 3 m. As shown in Fig. 2.79, residual coal distribution was observed in the goaf. Near the edge of the tailgate coal pillar, coal remnants were detected up to 6 m into the goaf, with

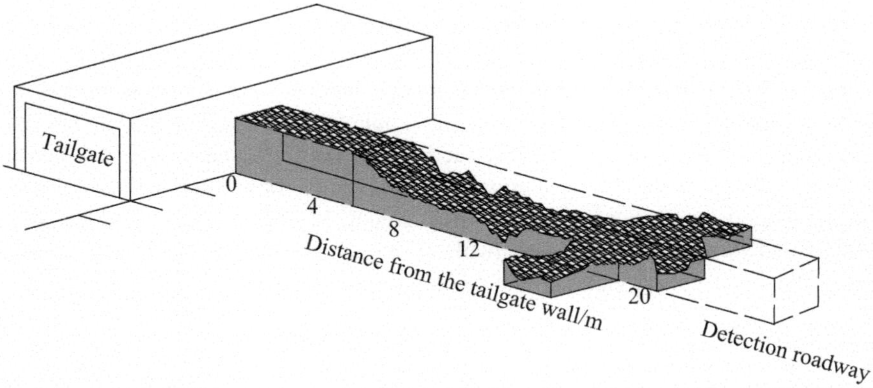

Fig. 2.79 Distribution of residual coal observed in detection roadway of the goaf [47]

visible waste rock beyond this point, indicating coal losses at the panel ends. Within the goaf, residual coal distribution became more uniform. The average thickness of residual coal in the goaf was 0.94 m, resulting in a calculated recovery of 86.18% for the panel [47]. This study significantly contributed to understanding coal loss distribution in top coal caving goafs and provided insights into improving recovery rates in such panels.

3. High-yield and efficient LTCC at Yanzhou Mining Bureau

LTCC at the Yanzhou Mining Bureau began in 1992 at the Xinglongzhuang Mine, where the average coal seam thickness was 7.25 m, the cutting height was 2.8 m, and the panel length was 160 m. The mine adopted ZFS5200-1.7/3.5 type caving supports, ZFGT5200-1.8/3.2 transitional supports, AM-500 shearers, SGBW-764/264 front AFC, and SGZD-764/320 rear AFC. The panel achieved an average monthly output of 100,000 tons. Previously, Yanzhou Mining Bureau primarily employed the slicing mining method for thick coal seams, which posed significant challenges for roof management in the lower layers. The introduction of LTCC technology resolved these issues and reduced the mining cost by 10.47 yuan per ton compared to slicing mining. In March 1993, Baodian Mine implemented LTCC, with a coal seam thickness of 8.5 m and a panel length of 125 m. The equipment configuration was nearly identical to that of Xinglongzhuang Mine. The panel achieved an average monthly coal output of 171,700 tons and a working efficiency of 74.8 tons per worker, demonstrating significant economic benefits. Given the advantages of LTCC in terms of investment, cost, output, and efficiency, the Yanzhou Mining Bureau actively promoted this technology. The Bureau also conducted systematic experiments on various mining and caving methods, such as two-cutting and one-caving, one-cutting and one-caving, single-round sequential caving, double-round sequential caving, single-round interval caving, and double-round interval caving. Consequently, both the output and efficiency of top coal caving panels improved rapidly. In 1994, the LTCC panel of the first mechanized mining team at Xinglongzhuang Mine achieved an output of 2.3 million tons, while the second mechanized mining team at Baodian Mine achieved 2.2 million tons. By 1995, the Bureau had implemented LTCC at 17 panels across mines such as Xinglongzhuang, Baodian, Nantun, and Dongtan. The first mechanized mining team at Xinglongzhuang Mine produced 3 million tons, and the second mechanized mining team at Baodian Mine produced 2.46 million tons.

In 1999, Xinglongzhuang Mine adopted ZF6200-16/35 high-resistance caving supports to mine an 8.2 m thick coal seam with a dip angle of 4°–12°. The panel length was 162 m, employing the one-cutting and one-caving single-round sequential caving method with parallel mining and caving operations. The panel achieved an average monthly output of 350,000 tons, with a recovery of 85.17% and an average production efficiency of 177.98 tons per worker.

The Dongtan Mine featured coal seams with thicknesses ranging from 5.6 to 6.5 m and dip angles of 3°–8°. LTCC was introduced in 1994, and by 1997, the second mechanized mining team produced 4.1 million tons of raw coal with a recovery ratio of 85.93%. Through measures such as increasing panel length and advancement, optimizing equipment configurations, and upgrading auxiliary production systems, the second mechanized mining team achieved 5.01 million tons in 1998 and 5.51 million tons in 2001, further demonstrating the potential of top coal caving technology.

The Yanzhou Mining Bureau, benefiting from strong mechanized mining foundations and favorable conditions such as moderate coal seam thickness, inclination, and hardness, developed a distinctive high-efficiency LTCC technology. The Bureau summarized valuable experiences in optimizing panel layouts, roadway support, production processes, end support improvements, equipment maintenance, system capacity enhancements, workforce training, and strict panel management. In 1998, the Bureau's achievements in LTCC were recognized with the National Science and Technology Progress Award (First Prize) for the "Comprehensive research and development of coal production technology in the Yanzhou Mining Area".

Yangquan, Lu'an, and Yanzhou were among the earliest mining bureaus in China to adopt LTCC technology. Their shared characteristics include nearly horizontal or gently inclined coal seams with thicknesses of 6–10 m and moderate conditions of coal seam hardness and gas content, making them well-suited for LTCC. The development progressed through stages of high-level, middle-level, and ultimately low-level top coal caving, achieving high production and efficiency while promoting the application and innovation of this technology across China.

2.5.2.5 Exploration of LTCC Technology Under Complex Conditions

1. LTCC in the "Three soft" thick coal seam

(1) LTCC in "three soft" unstable thick coal seams at Zhengzhou Mining Bureau

The coal seams in the Zhengzhou mining area are loose, with a coal hardness coefficient (f) ranging from 0.1 to 0.9, and both the roof and floor strata have low strength. The coal seam thickness varies significantly from 0 to 25.26 m, with an average thickness of 7.5 m. The production of panels using a slicing mining method is low, and material consumption is high, with the annual output generally hovering around 250,000 tons. In September 1988, Zhengzhou Mining Bureau conducted an experimental LTCC trial in the No. 2-1 coal seam at Micun Mine. The coal thickness ranged from 1.19 to 20.87 m, with an average thickness of 7.8 m, and the coal seam hardness coefficient was between 0.3 and 0.8. Both the roof and floor strata consisted of muddy shale, typical of "three soft" unstable thick coal seams. The initial trial failed due to the inadequate design of the support structure under these "three soft" conditions. In 1991, the ZFS4400-19/28 support system was modified, changing the telescoping beam from a glove-type to an inward telescoping type, adding telescoping cylinders, and increasing the stroke of the telescoping beams. The support force at the end face was improved by increasing the diameter of the rear column cylinders and modifying the four-bar linkage structure to a single hinge structure. Additionally, the space for the rear AFC was expanded. These modifications led to a successful trial, with the annual output of the panel reaching 600,000 tons. By 1994, the annual output at Micun and Chaohua mines reached one million tons, and by 2000, the monthly production reached 150,000 to 170,000 tons. In 1994, Micun Mine successfully conducted a downward LTCC trial in the "three soft" unstable thick coal seam with a coal seam dip of 19°–21° [48], with a panel length of 72 m and a monthly output of 53,200 tons.

(2) LTCC in "three soft" thick coal seams at Xuzhou Mining Bureau

LTCC at Xuzhou Mining Bureau began in 1993 at the Sanhejian Coal Mine, where the No. 7 coal seam was mined using this method. The average coal thickness was 9 m, with an average monthly production of 83,000 tons and a mining recovery of 82%. In 1994, a slicing mining method was implemented in the East No. 1 mining area, achieving an annual output of one million tons. In 1997, the light-duty support system was applied at Hanqiao Mine, and in 1998, it was used at Quantai Mine, with an average monthly production of 50,000 to 80,000 tons.

Xiaokang Coal Mine of Tiefa Mining Bureau also exhibits typical "three soft" coal seams, with a coal thickness of 7–8 m. LTCC technology was applied there in 1992, with a panel length of 150 m and an average monthly production of 82,000 tons.

The main challenges in LTCC in "three soft" thick coal seams are preventing top coal leakage at the face and between the supports, preventing coal wall spalling, controlling dust during the caving process, and supporting soft rock roadways.

2. LTCC in hard thick coal seams

(1) LTCC in "two hard" thick coal seams at Datong Mining Bureau

The Jurassic coal seams in Datong are hard, with a coal hardness coefficient ranging from 2.9 to 4.4. The seams have little development of fractures and are difficult to break. The roof is also hard and difficult to cave, often referred to as "two hard" coal seams. In 1991, Datong Mining Bureau began conducting LTCC trials in the Meiyukou Coal using a high-level caving method. They developed the ZFD5600, ZFD4500, and ZFD4400 types of caving supports. However, due to the high cost of mesh installation and the complexity of the slicing mining process, the monthly production of the panel was only around 60,000 tons.

In 1992, Xinzhouyao Mine adopted the ZFS6000 type middle-level caving support for a full-height LTCC trial under hard coal conditions. The panel had a coal thickness of 10 m, with nearly horizontal coal seams, and achieved great results. In 1995, the ZFSG6000 low-level caving support was developed, along with the trial of top coal blasting and weakening technology, which showed initial success. From 1996 to 1998, a joint project team consisting of Datong Mining Bureau, China University of Mining and Technology (Beijing), Chinese Institute of Coal Science Taiyuan Branch, and Taiyuan University of Technology, under the Ministry of Coal Industry's "Ninth Five-Year Plan" technological research program, developed a new type of ZFS6000 low-level caving support. The team studied the movement laws of top coal and roof strata, implemented stepwise blasting for roof caving, and established a fractal energy release model for coal body blasting. Through pre-blasting of deep holes in the coal seam, the top coal blasting weakening effect was significantly improved. During the industrial trial in the Panel 8911, the monthly production reached 121,400 tons, with a mining recovery of 80.3%. The annual output of LTCC panel under hard coal conditions exceeded one million tons [49, 50]. The project received the second prize of the National Science and Technology Progress Award in 2000.

(2) LTCC in thick coal seams at Jincheng Mining Bureau

The main coal seam mined by Jincheng Mining Bureau is the No. 3 seam, with a coal thickness ranging from 4.25 to 7.67 m, with an average thickness of around 6 m. The coal seam is nearly horizontal, with some areas having a gentle dip. In 1990, Jincheng Mining Bureau began conducting LTCC trials at Fenghuangshan Mine. In 1994, a LTCC trial was conducted in the Panel 13310 at Gushuyuan Mine, where the average monthly production was 131,600 tons, with a top coal recovery of 79.6% and a mining recovery of 90.5%. The trial was successful. In 1998, LTCC was promoted at Chengzhuang Mine, where the main support model used was ZFS5200-16/32B. Due to the hardness of the coal seam ($f = 2$–4), the research focused on the drawing mechanisms of top coal and technologies to improve top coal recovery during the mining process.

3. LTCC in "three soft" high-gas thick coal seams

(1) LTCC at Weijiadi Coal Mine, Jingyuan Mining Bureau

In 1995, the Jingyuan Mining Bureau of Gansu initiated LTCC trials in high-gas coal seams at Weijiadi Coal Mine. The coal seam had an average thickness of 12 m, a dip angle of 20°, a coal hardness coefficient of 0.3, and a gas content of 9.3 m^3/t. The panel was 80 m long, utilizing the ZFSB3600-17/28 type low-level caving support, with a mining height of 2.5 m. To address the technical challenges of mining extra-thick and soft coal seams with high gas content and self-ignition potential, several measures were implemented. A gas drainage roadway was driven along the coal seam's roof to pre-drain gas, and a gas discharge roadway was established above the panel to remove gas from the upper corner during mining. Nitrogen and mud were injected into the goaf to prevent spontaneous combustion of residual coal in the goaf. The mining method employed a two-cutting and one-cutting technique, where coal was cut from the bottom up, with supports moved and caving performed in a sequence from bottom to top. This process enabled a monthly production of 73,000 tons and a mining recovery ratio of 87% [51]. The successful application of LTCC technology resolved the long-standing issues at Weijiadi Mine, including uneven roof conditions, thick parting, failed slicing mining, and low productivity in blasted mining. The cost per ton of coal was reduced by 15.27% compared to blasting mining, and overall production increased more than twofold, achieving direct labor efficiency of 32.63 tons per worker and turning the mine from a loss into profitability [52, 53].

(2) LTCC at Huaibei Mining Bureau

At Huaibei Mining Bureau, certain mines such as Zhuxianzhuang and Luling have extra-soft coal seams with a coal hardness coefficient of $f \leq 0.3$ and high gas content, with the risk of gas outbursts. The gas content is 16 m^3/t, and the average coal seam thickness is 8–12 m. Due to the complexity of the coal seam conditions, blasting mining has traditionally been used. In 1997, Zhuxianzhuang Mine trialed blasting and LTCC, and in 1998, they experimented with a grid-type support for LTCC. To improve mining efficiency and mechanization levels, Huaibei Mining Bureau collaborated with China University of Mining and Technology (Beijing), supported by the National Natural Science Foundation of China (Key Project 59734090), and in 1998, they conducted a successful LTCC trial at Zhuxianzhuang and Yangzhuang mines using the ZF2400-16/24 lightweight low-level caving support. This trial deepened their understanding of the drawing laws of loose top coal and yielded valuable experience in controlling leakage at the face, managing high-gas emissions, and roadway support.

In 2005, Huaibei Mining Bureau, in collaboration with China University of Mining and Technology (Beijing), Beijing Coal Mining Research Institute, and Chinese Institute of Coal Science Chongqing Branch, conducted pre-mining top layer LTCC trials at Luling Mine. The trials involved a systematic study of mining techniques, mine pressure, support design, equipment matching, and gas control. Gas drainage and control measures were implemented through floor-layer drilling, pre-mining top layer techniques, and comprehensive gas prevention using high-drainage roadways. To prevent coal wall damage and reduce dust, water was injected into the coal seam to increase coal wall cohesion. These measures were successful, and using the ZF3600-16/25 low-level caving support, the panel achieved an annual output of 1 million tons, with a recovery of 90% [50]. The project proposed the important academic concept that increasing the support resistance and stiffness effectively alleviates coal wall pressure and prevents coal wall spalling. Additionally, a calculation model for shear failure of soft coal walls and tensile failure of hard coal walls was developed [54]. These findings laid the foundation for the later proposal of a dual criterion for designing support resistance, which considers both balancing the main roof pressure and maintaining coal wall stability [55] Currently, the support resistance at Huaibei Mining Bureau typically ranges from 8000 to 10,000 kN, and the use of high-resistance supports has become an inevitable choice for efficient mining of extra-soft coal seams.

4. LTCC in inclined thick coal seams
(1) LTCC at Wulan Mine, Shitanjing Mining Bureau, Ningxia

In November 1992, Wulan Mine conducted LTCC in the Panel 5321 of the No. 3 coal seam. The average coal seam thickness was 6.79 m, with a dip angle ranging from 17° to 37°, and a coal hardness coefficient of 0.6–1.2. The panel was 44.5 m long, utilizing 29 ZFSB3200-16/28 type supports, 4 ZFG3400-20/30 transitional supports, and 2 ZTF3900-20/30 type end supports. The main mining process focused on preventing the support from tilting and the AFC from slipping. The mining method employed four "bottom-up" techniques: cutting coal from bottom to top, moving the supports from bottom to top, pushing the AFC from bottom to top, and caving coal from bottom to top. The panel achieved a mining recovery of 84.98%. This trial marked the initial success of LTCC in inclined coal seams.

(2) LTCC at Second Mine, Hebi Mining Bureau

In October 1999, Hebi Mining Bureau's Second Mine, in collaboration with China University of Mining and Technology (Beijing), conducted a LTCC trial in the Panel 3604 of the inclined thick coal seam. The average coal seam dip angle was 30°, with a coal thickness of 6.03 m and a panel length of 92 m. The mine used lightweight supports with a capacity of 2,400 kN and a weight of 7.5 tons. The trial highlighted the critical role of side guards for the supports and measures such as pressure-assisted support movement in preventing the supports and equipment from tilting or slipping. Additionally, the training of technicians and workers, as well as the standardized management of the panel, contributed to the success of the trial. The panel achieved a monthly production of 50,000 tons, with a mining recovery ratio of 93%.

Given the complex and diverse coal seam conditions across China, the required technical models and equipment vary greatly. The LTCC technology for complex conditions is the result of joint technological efforts between Chinese coal enterprises, research institutions, universities, and design organizations. Besides the four typical complex conditions mentioned above, LTCC technology has also been successfully applied in various other conditions, including in the extra-thick coal seams of Fushun, the multi-seam LTCC at Datun Yaoqiao Mine and Huainan Xieqiao Mine (with coal seam thicknesses below 6 m), as well as the residual coal pillar and corner coal mining at Tangshan Mine of Kailuan Bureau and Xingtai Mine of Xingtai Bureau. The technology has demonstrated significant adaptability in extracting coal pillars and corner coal.

2.5.2.6 Lightweight Support for LTCC

Lightweight supports for LTCC have played a significant role in the history of LTCC in China, closely related to the development of coal mining technology and the economic conditions of coal mining enterprises. In the early stages of LTCC during the 1980s and 1990s, the overall mechanization level of coal mining in China was low. Many mining bureaus still relied on blasting and conventional mining methods. As a result, lightweight supports (weighing $\leq$ 8 tons) were adopted to reduce the support area while maintaining support strength, thus saving on investments in panels and avoiding the problems associated with heavier, more difficult-to-handle supports (weighing 12–18 tons at the time). These lightweight supports also offered better stability and stronger support compared to simple supports like single units and sliding supports, making them a preferred choice for many coal mining enterprises.

In 1985, the Meihekou Mine in Jilin used the YFY1600-16/26 type high-level lightweight support (weighing 6.5 tons) for STCC in steeply inclined coal seams. In 1991, the Wangjiashan Mine in Jingyuan used lightweight supports for horizontal STCC, with a panel length of 40 m and a monthly coal production of 20,000 tons. In 1993, the Huating Mine in Gansu, followed by several other mines in the 1990s, including those in Xingtai, Fengfeng, Xu Zhou, Kailuan, and Jining, adopted lightweight supports for LTCC, achieving excellent results. The maximum monthly output in these panels reached 120,000 tons, though most panels produced between 50,000 to 80,000 tons. Compared to blasting caving method and conventional caving method, the output was significantly increased, while the investment in equipment for the panels remained relatively low. At the time, lightweight supports typically featured a four-post design with either single-joint or single-swing bar structures, with a rated working resistance between 2000–2800 kN and a support strength of 0.47–0.69 MPa. The support height ranged from 1.6 to 2.4 m, and the maximum applicable panel dip angle was 30° (Hebi No. 2 Mine).

However, with the rapid advancement of coal mining technology and equipment in China, and improvements in the economic efficiency of coal enterprises, there was a growing demand for high-output, high-efficiency mining. Additionally, the relatively low overall support capacity of lightweight supports, their susceptibility to damage, poor control over coal wall and end-face stability, and the relatively low production from panels led to the gradual phase-out of lightweight supports after 2005. Mining equipment tended to become larger and heavier. Despite this, lightweight supports for LTCC played a crucial transitional role in the development of LTCC in China and made significant contributions during that period.

2.5.3 Technological Innovation and Export Stage

After 20 years of exploration and development, the LTCC technology has formed a unique technological model and equipment system in China, showcasing its advantages in high productivity, efficiency, low energy consumption, and low cost. Over the past 20 years, in response to China's diverse coal seam conditions and the growing national demand for coal energy, LTCC has experienced further development and significant innovations. These advancements can be mainly observed in the following areas.

2.5.3.1 Large Mining Height LTCC

1. LTCC in the thick coal seams of Datong Tashan Mine

The technology for LTCC in 20-m-thick coal seams, represented by Datong Tashan Coal Mine, has made breakthrough progress. In 2008, as part of the national "11th Five-Year Plan" major science and technology support project, the ZF15000/28/52 four-column large mining height support system and associated equipment were developed. A binary criterion based on roof pressure and coal wall stability was established, along with a 3D top coal caving theory. Additionally, a "cantilever beam-hinged rock beam" structural model was proposed for overlying rock breakage, and preventive and treatment technologies for low-gas content and high-gas emission coal seams were developed. The project achieved an annual output of 10.85 million tons with a top coal recovery of 86.7% [56], and the results were awarded the first prize of the National Science and Technology Progress Award in 2014. In 2018, as part of the "13th Five-Year Plan" technology support project, the ZF21000/27.5/42D four-column high-resistance support was developed, with further research on intelligent technology for large mining height LTCC.

2. 7-m large mining height LTCC in Shandong Energy Group Jinjiantan Mine

In 2019, Shandong Energy Group developed 7-m large mining height LTCC at the Jinjiantan Coal Mine in northern Shaanxi. The ZY21000/38/70D two-column large mining height support system and associated equipment were developed, with a support strength of 1.65 MPa. The technology aimed to improve the top coal recovery by increasing the mining height and changing the top coal fracturing. The panel is 300 m long, with a maximum daily output of 71,700 tons, achieving a recovery of 90%. Intelligent technologies such as active sensing, automatic analysis, intelligent processing, and system coordination were applied to build an intelligent large mining height LTCC panel.

3. LTCC for shallow-buried hard coal at Pingshuo Mining Area

In 2002, China National Coal Group Pingshuo Company applied LTCC for shallow-buried hard coal beneath the non-mined areas of open-pit mines. The 4th coal seam is 7 m thick on average, and the 9th coal seam is 13.14 m thick on average, with coal buried at depths of 70 to 200 m and a hardness coefficient of $f = 2$–3. During initial assessments of the LTCC, reference was made to Datong Xinzhouyao Mine's LTCC methods, and a pre-blasting roadway for top coal was designed. However, due to the development of fractures in the coal seam, no pre-blasting was required despite the shallow depth. The 9th coal seam achieved an annual output of 6 million tons, with a recovery of nearly 90%. In 2010, the ZFY12000/23/40D two-column support system was trialed for large mining height LTCC in the 4th coal seam, with successful results [57].

4. Large mining height LTCC in other mining areas

In 2007, Huaibei Mining Industry's Guobei Mine implemented large mining height LTCC in extremely soft coal seams with parting. The coal seam had a thickness of 9–10 m, with an average parting of 1.54 m, and a coal hardness coefficient of $f = 0.1$–0.3. The ZF6800-19/38 low-level caving support system was used to address the challenges of efficiently mining soft coal seams with parting. In 2007, Lu'an Wangzhuang Mine constructed the first automated large mining height LTCC panel, employing an electro-hydraulic control system for the supports. This system primarily solved issues related to high gas outbursts and improved coal recovery rates, achieving automation for all processes except top coal caving. The results of this project were awarded the second prize of the National Science and Technology Progress Award in 2010. In 2013, Shendong Coal Group Corporation Limited's Huang Yuchuan Mine adopted the ZF21000/25/42D four-column positive four-link low-level caving support to mine the 6th upper coal seam, with an average thickness of 9.56 m, an dip angle of 0°–17°, and a mining height of 3.5 m. The panel achieved an annual output of 6 million tons. Additionally, large mining height LTCC was successfully applied in deep, thin main roof at the Zhaogu Mine in the Jiaozuo area of Henan Energy Chemical Group, and in deep shafts at the Xinjulong Mine in Shandong Energy, both resulting in high productivity and efficiency.

Large mining height LTCC offers several advantages, including high coal recovery rates, large ventilation cross-sections, and high efficiency. By increasing support resistance and rigidity and strengthening management, coal wall stability can be effectively controlled. In recent years, this technology has been widely applied across many mining areas and has even achieved excellent results in relatively soft coal seams.

2.5.3.2 Top Coal Caving Technology in Steeply Inclined Thick Coal Seam

1. LTCC in steeply inclined thick coal seams

(1) Wangjiashan Mine of Gansu Jingyuan Coal Group

The coal seam in the Wangjiashan Mine has an inclination angle ranging from 38° to 49°, with an average of 43.53°, and a thickness ranging from 13.5 to 23 m, with an average of 15.5 m. The coal seam hardness coefficient is $f = 1.0$. LTCC was trialed starting in 2003. The panel utilizes an upper roadway for advancing along the floor and a lower roadway advancing along the roof, with the panel configured horizontally at the bottom and an arc-shaped transition between the upper and lower faces. This arrangement aids in preventing equipment tipping and sliding. The panel is 95 m long, using ZFQ3600/16/28-type low-level support systems with a coal-cutting height of 2.6 m. The average monthly production is 97,100 tons, with a coal recovery ratio of 82.27% [58]. Due to the large inclination angle of the coal seam, the support stability was a challenge, so devices to prevent tipping and sliding were designed for the supports, along with measures to control the sliding of the scraper conveyors at the front and rear of the support.

(2) LTCC in Shanxi Dayuan Coal Mine of Jizhong Fengfeng Group

The mining area of Shanxi Dayuan Coal Mine is in the southern eastern wing of the Ningwu Coalfield's Nanjing Syncline. The panel has an average inclination angle of 53°, with a maximum of 62°, and an average thickness of 6.8 m. The coal seam has a hardness coefficient of 0.3, making it a typical "three-soft" coal seam. The panel is 80 m long, with a possible mining length of 680 m and a pseudo-tilted layout. The Fengfeng Group cooperated with China University of Mining and Technology (Beijing) to develop dynamic load impact and lateral extrusion pressure calculation methods for the support system. The ZFY4800/17/28 type wide-side protection, high-pressure anti-extrusion support was developed to prevent support overturning and equipment sliding. In combination with mining methods and equipment, solutions to prevent tipping and sliding of the equipment were developed, as well as a mining method of "downward dynamic section, upward caving within each section". This resulted in a monthly production of 80,000 tons and a coal recovery of 85% [59].

2. STCC in steeply inclined thick coal seams

For steeply inclined thick coal seams with a thickness greater than 20 m, STCC technology can be applied. The Wudong Mine of the China Energy Xinjiang Company achieved high production and efficiency using this technique in steeply inclined thick coal seams. The coal seams thicknesses are ranging from 20 to 50 m and inclination angles are of 67° to 87°. The Dahonggou, Xiaohonggou, and Jiangou coal mines adopted STCC in 1997–1998, with a sublevel height of 14 m, a coal cutting height of 2.4 m, and blasting methods for top coal treatment, using ZF2800/20/32 lightweight supports. After 2007, following the regulations of the Coal Mine Safety Regulations that stipulate a maximum mining ratio of 1:8 for steeply inclined thick seams, the Wudong Mine adjusted the sublevel height to 18–21 m, with a mining height of 3.2 m and ZF6500/20/40 type caving supports. This led to a significant increase in panel output and efficiency. In 2010, innovations were made to mining equipment, parameters, and processes. To improve top coal recovery and caving efficiency, a weakening top coal roadway was excavated along the strike of the coal seam, enabling pre-blasting and post-blasting water injection to weaken the top coal. The sublevel height was adjusted to 25 m, with a cutting height of 3 m, and ZFY10000/22/40D type caving supports were applied, allowing for an annual output of up to 4 million tons. Further in-depth research was carried out on roof movement, mining processes, top coal weakening technologies, dynamic pressure, and gas prevention [60].

The extraction of steeply inclined thick coal seams is difficult and inefficient, especially when the coal seam thickness is less than 20 m, where STCC results in short panels, high drift advance rates, difficulty in top coal caving, and low efficiency. When using LTCC, the equipment's anti-tipping and anti-sliding capabilities are harder to control, coal walls are more prone to spalling, and coal caving control is more challenging, leading to low efficiency. Over the past decade, significant progress has been made in steeply inclined thick coal seam LTCC technologies through collaborative research among enterprises, universities, and research institutes. A technical route has been developed that reduces the inclination angle of the lower panel through proper drift layout, prevents tipping and sliding of the supports with high-resistance wide-side protection, and controls the movement of the top coal through scientific mining processes. As a result, these technologies have been successfully applied. Although STCC was initially tested on steeply inclined thick seams in China, it only demonstrated the potential and adaptability of this technology at that time. Today, with the Wudong Mine in Xinjiang serving as a representative, this technique has undergone major innovations in terms of equipment, mining parameters, mining processes, output and efficiency, and disaster prevention, marking a fundamental shift from the early-stage technology.

2.5.3.3 Intelligent LTCC Technology

Since 2000, China's LTCC has gradually entered the era of automation, represented by hydraulic support systems with electro-hydraulic control, and has achieved automated caving primarily driven by sequential memory control. After initially introducing and adapting advanced automation technologies from countries such as the United States, Germany, and Australia, and combining them with a large amount of independent research and development, China has integrated automation technology in longwall mining with artificial intelligence and big data. This combination has led to the gradual formation of intelligent mining technologies for longwall mining faces. Corresponding technical standards have been established, greatly advancing China's coal mine intelligent mining technology, bringing it to an internationally advanced level. Certain technologies, such as remote control, automatic data processing, and vision-based intelligent control, are at the international leading level. The intelligent LTCC system consists of two parts. The first is the automation of the shearer and associated processes, which can directly utilize the automation technology from longwall mining faces. Compared with fully mechanized mining, the shearer can adopt a fixed cutting height, eliminating the need for coal-rock identification during the cutting process. The second part involves the automation of the top coal caving process, which has progressed more slowly. While efforts such as using sound frequency spectra, near-infrared spectroscopy, natural radiation, and vibration signals for coal and gangue identification during top coal caving have been conducted, this aspect remains at the experimental stage. In 2021, China University of Mining and Technology (Beijing) developed a breakthrough intelligent control LTCC technology based on image recognition, marking the first real application of intelligent control in top coal caving. This technology has already been applied in Huaibei and Kailuan mines and is undergoing industrial trials at the Caojiatan mine [38, 61, 62].

2.5.3.4 Technology Export

In October 2006, the first set of LTCC equipment provided by Yan Coal Australia was put into use at the Ausdrill mine, establishing Australia's first LTCC panel. In October 2013, the company provided a set of two-column caving supports and related equipment for the North Gunnedah Mine in Queensland, Australia. Apart from Australia, some coal mines in countries such as India, Turkey, Russia, and Vietnam have also adopted LTCC technology and conducted basic research [63–66]. In 2022, Zhengzhou Coal Mining Machinery Group supplied a full set of LTCC equipment to Turkey's Bozhai Coal Industry, including ZFY4600/18/30 hydraulic supports, MG300/730-WD shearer, SGZ730/500 front AFC, SGZ764/500 rear AFC, transloader, crushers, belt conveyors, self-moving machine tails, electrical equipment, pump stations, and centralized control facilities.

2.5.4 Technology Advancement and Development Direction in LTCC

2.5.4.1 Technology Advancements in China's LTCC

Over the past 40 years, China has made significant innovations in the technology, equipment, and theory of LTCC, which has become a landmark technology in the global coal industry and a brilliant calling card. It has evolved from early technology imports to later technology exports, driving the research and application of LTCC in other countries with thick coal seams. Due to its characteristics, LTCC offers advantages over other thick coal seam mining technologies, including low investment, low energy consumption, low cost, and high efficiency. Essentially, it is a low-carbon mining method, aligning with the current "dual carbon" strategy.

1. Technological R&D initiatives

Due to the low cost and high efficiency of LTCC, coal mining enterprises are highly motivated to adopt this technology. However, at the time, there was a lack of in-depth research and relevant regulations, and early practical experience was insufficient. To guide the healthy development of LTCC, the former Ministry of Coal Industry established the "Experts on Top Coal Caving" in 1994. The group consisted of 18 experts, including chief engineers from provincial coal bureaus and mining bureaus with successful applications of LTCC, experts from the Coal Science Research Institute, and professors from the Beijing Graduate School of China University of Mining and Technology, such as pioneers in China's top-coal caving practice and theoretical research, Researcher Fan Yunce and Professor Wu Jian. In 1995, the expert group developed the "Interim Technical Regulations for LTCC" to lay the foundation for the later provisions related to top coal caving in the "Coal Mine Safety Regulations".

In May 1996, the Ministry of Coal Industry released the "Reform and Development Outline of the Coal Industry during the Ninth Five-Year Plan", which explicitly stated: "New breakthroughs will be made in LTCC, with the goal of promoting this new process in all thick coal seams suitable for top coal caving." Minister Wang Senhao of the Ministry of Coal Industry noted: "LTCC is a revolution in the mining methods for thick coal seams". During the same year, the former Ministry of Coal Industry identified five key technological research directions for the coal industry in the Ninth Five-Year Plan. These included LTCC, anchor support technology, prevention and control of gas, coal dust, and water hazards, high-yield and efficient mine integration, and clean coal technology. Four research projects were established around LTCC, including research on safety assurance technology for LTCC faces, methods to improve recovery rates, laws of top coal movement and mine pressure, and development of hydraulic supports and associated equipment for top-coal caving. These projects were implemented in mining bureaus such as Yangquan, Datong, Zhengzhou, and Yanzhou. The Ministry of Coal Industry actively promoted the development and application of LTCC through organized research, enabling rapid development and significant innovations in this field in China.

In 2008 and 2018, the Ministry of Science and Technology listed large mining height LTCC as a key scientific research project for the Eleventh and Thirteenth Five-Year Plans. In 2008, the Ministry of Science and Technology included the "Research and Development of Full Set of Technology and Equipment for Large Mining Height LTCC in Extremely Thick Coal Seams" (2008BAB36B00) in the major National Science and Technology support programs since the founding of the country, investing a total of 445 million yuan in research funds.

Entering the twenty-first century, numerous research projects on LTCC emerged nationwide, yielding abundant results. Theoretical research mainly came from various projects supported by the National Natural Science Foundation, such as the key project on "Movement and Control of Rock Layers in Deeply Buried Weakly Cemented Thin Strata" (51934008) and "Rock Strata Control and Three-Dimensional Drawing Law of Top Coal in Shallow-Buried Thick Coal Seams" (U1361209). Technological research was largely spearheaded by coal mining enterprises, in collaboration with research institutions, universities, and coal machinery manufacturing companies, to jointly tackle new technology development and create mining technologies suited to different conditions. The results of LTCC technology research have accounted for a significant portion of China's Coal Industry Science and Technology Awards and other provincial coal mining science and technology prizes.

2. Technological advancements

China has established multiple LTCC technology models suitable for its thick coal seam conditions. These include high-output and high-efficiency LTCC technologies represented by Lu'an and Yanzhou, where the coal seams are near-horizontal, 6–8 m thick, with relatively simple gas conditions; large mining height LTCC for thick coal seams in the Datong, where the coal seams are near-horizontal, 15–20 m thick, with harder coal and more developed fractures; the "three-soft" coal seam LTCC represented by Huaibei Mining Area, where the coal seams are gently inclined, about 10 m thick, with a hardness coefficient less than 0.5 and high gas conditions; steeply inclined seam LTCC technology represented by Fengfeng Dayuan Coal Industry, with a coal seam dip angle of 60°, 6–10 m thick; steeply inclined thick coal seam STCC technology represented by Wudong Mine in Xinjiang, where the coal seams are steeply inclined and over 30 m thick; and deep-well LTCC represented by Shandong Energy Xinjulong Mine and Henan Zhaogu Mining Area, where the coal seams are gently inclined to nearly horizontal, 6–10 m thick, with mining depths of 600–1000 m. In recent years, LTCC technology has also been innovatively developed and widely applied in Shaanxi, Inner Mongolia, and other regions.

There have been breakthrough advancements in the development of top coal caving supports. The support frame design has evolved from the early four-column type to the current use of both four-column and two-column supports, with a strong trend toward replacing four-column supports with two-column types, especially in some intelligent panels where two-column supports are preferred. In four-column support designs, new supports have been developed with unequal strength in the front and rear columns based on the roof pressure characteristics. The support resistance has also increased from the initial 5000 kN level to the currently commonly used 10,000 kN level. For large mining height LTCC, the maximum resistance of the support has reached 21,000 kN, setting a world record for the highest resistance of comprehensive mining supports. China's research and manufacturing of mining supports have reached world-class levels, with exports to the United States, India, Turkey, Russia, and other countries.

Intelligent LTCC technology has developed rapidly, evolving from simple adaptations of intelligent mining technology to the development of LTCC specific intelligent mining technologies. There is widespread interest in and development of various intelligent LTCC technologies, with efforts directed toward industrialization. Significant advancements have been made in controlling the coal face and roof, including high pre-tightening, 700 MPa high-strength anchor support for all coal roadways, pillar-free roadway protection technology, hard roof water-pressure fracturing technology, far-near field collaborative roof control technology, and prediction and prevention technology for rock burst. Additionally, technologies for filling mining in LTCC panels have made important strides. Standardized technologies and measures for gas, fire, water, and dust prevention in LTCC have also been established.

3. Scientific and technological achievements

Since 2000, China has won 7 national science and technology awards related to LTCC, including 1 first prize and 6 s prizes. They are: key technology and equipment for large mining height LTCC in extra-thick coal seams (2014, first prize), key technology for LTCC in hard and thick coal seams (2000), key technology and equipment for automated LTCC and their application at home and abroad (2009), key equipment and automation technology for safe mining in extra-thick coal seams (2010), theory of top coal drawing in LTCC and control technology and application of surrounding rock in thick coal seam mining (2011), key technology of complete equipment for LTCC panel in coal seams with large dip angles (2012), and key theory and technology for LTCC in steeply inclined thick coal seams (2016). These national science and technology awards reflect some important scientific and technological progress made by China in LTCC technology in different periods, including mining theory, technology, and equipment.

According to incomplete statistics, China has published over 40 academic books on top coal caving mining [38]. Among these books, some focus on basic theories, such as "Drawing Laws and Intelligent Drawing", "Basic Theories and Applications in Top-coal Caving Mining", "Research and Application of Roof Activity Laws in Top Coal Caving Mining", and "Basic Theory of Top Coal Caving Mining"; some are inclined to engineering practices, such as "Technology of Large mining height Automatic Top Coal Caving Mining", "New Theory of Full Height Mining in Thick Coal Seams", and "Comprehensive Mechanized Top Coal Caving Mining Technology"; and some are summaries of top coal caving mining technologies in specific mining areas, such as "Theory and Technology of Top Coal Caving Mining in Extra-thick Coal Seams in Datong Mining Area" and "Practice and Understanding of Comprehensive Mechanized Top Coal Caving Mining in Yanzhou Mining Area".

Through a search and analysis on CNKI, there are 5316 papers (Chinese Social Sciences Citation index) published between 1982 and 2022 with the themes of "fully mechanized top coal caving" or "top coal caving". The SCI database reveals a total of 475 academic papers on the topic of "top coal caving" from 1982 to 2022, while the EI database shows 946 journal and conference papers on the same topic during the same period.

The aforementioned national awards, published books, and academic papers reflect the work and academic achievements made in the field of top coal caving mining in China and worldwide over the past 40 years. Although these statistics may not be comprehensive, they provide an overview of the academic research, technological progress, and innovation in top coal caving mining. The top 10 authors with the highest number of SCI and EI papers are all China's authors, indicating that China's academic research in top coal caving mining is internationally leading and dominant. In recent years, scholars from countries such as Australia, Russia, Turkey, and Vietnam have also published academic articles on top coal caving mining.

4. Academic exchange

Extensive and in-depth academic exchange is one of the important means to promote scientific and technological progress. In December 1990, the Graduate School of China University of Mining and Technology (Beijing) organized "the first national symposium on the theory and practice of top coal caving mining" in Beijing, with over 200 participants from the leadership, engineering technicians, and experts of the Ministry of Energy and the China National Coal Mining Corporation. Later, the second to fifth symposiums on the theory and practice of top coal caving mining were held in Yanjiao, Hebei; Dujiangyan, Sichuan; Jinggangshan, Jiangxi; and Hangzhou, Zhejiang in July 1995, September 1998, August 2000, and October 2005, respectively. In July 2022, China University of Mining and Technology (Beijing) organized the "2022 International Conference on Green Intelligent Mining for Thick Coal Seam", commemorating the 40th anniversary of LTCC in China. Over 50 academicians, experts, and graduate students from both domestic and foreign countries presented at the conference, including 17 speakers from 6 foreign countries.

Extensive and in-depth exchanges were made on the technical achievements in thick coal seam mining, actively promoting the achievements of China's LTCC technology.

To commemorate the 30th anniversary of LTCC in China, China Coal Technology and Engineering Group Tiandi Technology Co., Ltd. edited and published the book "Innovative Development of Theory and Practice in Fully Mechanized Top Coal Caving Technology" in October 2012. The book contains 101 papers, compiling research achievements and practical experiences in various fields over the past 30 years, including LTCC theory, caving technology, mine pressure manifestation and top coal motion laws, roadway layout and support, hydraulic support design and manufacturing, and supporting equipment for the panel, top coal caving properties, top coal recovery, overburden rock failure and ground protection, research and application of safety instruments for top coal caving mining, "one ventilation and three prevention" technology for LTCC panels, related technologies for top coal caving mining in high-gas coal seams, and water prevention and control technologies for top coal caving mining.

In the third edition of "Longwall Mining" written by Syd S. Peng, Chap. 15, titled "Longwall Top Coal Caving Mining", has been added, providing a detailed introduction to China's LTCC technology, equipment, and the BBR caving theory.

2.5.4.2 Future Development of LTCC Technology

1. Promoting comprehensive intelligent mining of LTCC panels

Intelligent mining has been the main technical direction promoted by China's coal mining industry in recent years, with rapid development and corresponding technical standards established. However, it is still in the initial stage of intelligent mining overall. The main manifestations of this initial stage are: (1) Certain mining process links have not yet been fully intelligentized, or even just in the initial stage of automation, such as the panel cutting process at the end, drawing process, precise coal-rock identification, equipment collision prevention, and personnel proximity warning technology, which still have significant room for improvement; (2) The reliability of the intelligent mining system needs to be significantly improved, and the applicable conditions need to be expanded as much as possible, especially with excessive manual intervention in complex geological conditions; (3) The compatibility of various subsystems is weak, with many data silos, and the big data models and processing technologies suitable for coal mine mining are not yet mature; (4) There is currently excessive reliance on vision-based control technology, with insufficient capabilities in the research, development, and manufacturing of multi-information high-precision sensing systems; (5) There is still a certain degree of blindness in introducing related industry technologies, with incomplete analysis of the complexity and specificity of coal mines, and the industry's own research and development capabilities need to be improved; (6) The research and development of intelligent drawing technology for LTCC panels is relatively backward, and current applications of intelligent drawing are not yet widespread. Achieving comprehensive intelligentization of LTCC panels requires the joint efforts of the entire industry, drawing on advanced domestic and international concepts and technologies, persevering in strengthening research and development capabilities, addressing the above challenges, and achieving comprehensive intelligentization and unmanned underground production of top coal caving mining.

2. Establishing a scientific mining system based on mining strata movement

LTCC involves a large mining height, resulting in more intense overburden movement compared to slicing mining, with a wide distribution of mining stress and a broad impact on the overburden and coal seam floor. Although there have been many studies on the overburden movement of LTCC over the past 40 years, corresponding law results have not yet been summarized and established. The combination of research on the control of surrounding rock in the stope and the movement of overburden in extra-thick coal seam is not close enough, and the research methods lack consistency. For example, the control of the roof in the stope is mainly based on mechanical analysis, while surface subsidence is mainly based on statistical methods. The combination of roadway surrounding rock control and overburden movement in the stope needs to be strengthened. Fundamentally, whether it is the movement of surrounding rock in the stope, the movement of surrounding rock in the roadway, or the movement of overburden up to the surface, they all belong to the problem of surrounding rock movement during mining excavation, which are different research scales and objects in the unified field of mining-induced overburden, interconnected and mutually influential. A unified field research model should be established. The unified field research of mining stress and overburden movement is the basis for developing technologies such as efficient mining, water-preserving mining, green mining, and disaster reduction mining based on mining strata movement, as well as the basis for the development of mining and tunneling equipment, mining engineering layout, and the construction of intelligent mining systems.

3. Accelerate the research on the low-carbon path of LTCC

The outstanding advantages of LTCC are low cost, low energy consumption, high efficiency, and low investment, which is also a strong support for the "dual carbon" strategy and energy supply security. However, in the process of developing LTCC technology, it is necessary to improve the recovery ratio by using free-pillar mining, and developing the technology of top coal caving at the end of the panel. To improve the top coal recovery, a certain amount of gangue needs to be drawn, which leads to a high gangue mixed ratio in the raw coal of the LTCC panel. Therefore, it is necessary to research and develop high-efficiency underground sorting technology and underground gangue high-efficiency treatment technology for LTCC to achieve energy conservation and emission reduction. Due to the technological characteristics of LTCC, it is difficult to conduct backfill mining on the panel, and the technological progress is slow. However, in the long term, it is of great significance to research and develop top coal caving backfill mining technology, especially backfill mining combined with underground sorting. Scientific mining layout, panel design, equipment matching, intelligent path selection, and new equipment research and development can further reduce mining energy consumption and costs.

References

1. Gazette of the state administration of work safety and the state administration of coal mine safety, 2016. Coal Mine Safety Regulations. Coal Industry Press, Beijing
2. Jin ZM (2001) Theory and technology of top coal caving mining. Coal Industry Press, Beijing
3. Wang JC (2009) Theory and technology of thick coal seam mining. Metallurgical Industry Press, Beijing
4. Wang JC, Lv HY, Wang ZH et al (2019) Technical principle and experimental study on fully mechanized top-coal caving mining after extracting the middle slice in extremely thick coal seam. J China Coal Soc 3:906–914
5. Du JP, Meng XR (2009) Mining science. China University of Mining and Technology Press, Xuzhou
6. Wang JC, Wang ZH (2017) Failure mechanism of fully mechanized top-coal caving mining under the composite effect of loading and unloading. Tong Mei Ke Ji 3:1–8
7. Priest SD (1993) Discontinuity analysis for rock engineering. Champman & Hall, New York
8. Hudson JA, Harrison JP (1997) Engineering rock mechanics. Elsevier Science Ltd., Kidlington
9. Wang JC, Zhang JW, Wang ZH (2018) Basic theories and application in top-coal caving mining. Science Press, Beijing
10. Liu XS, Ning JG, Tan YL et al (2016) Damage constitutive model based on energy dissipation for intact rock subjected to cyclic loading. Int J Rock Mech Min Sci 85:27–32
11. Li XQ, Yin GZ, Cai B (2010) Experimental study on deformation and seepage properties of outburst coal samples under cyclic loading. Chin J Rock Mech Eng 29(S2):3498–3504
12. Wang JC, Bai XJ, Wu ZS et al (2000) The research on the fractured blocks of the top-coal in the longwall top-coal caving technique of the hard coal seam. J Coal Soc 3:238–242
13. Wu J (1991) Theory and practice of sub-level caving method in China. J China Coal Soc 16(3):1–11
14. Yu B, Zhu DJ, Chen ZH (2017) Top-coal drawing law of LTCC mining based on stochastic medium theory. J China Coal Soc 42(6):1366–1371
15. Wang JC, Wei WJ, Zhang JW (2020) Theoretical description of drawing body shape in an inclined seam with longwall top coal caving mining. Int J Coal Sci Technol 7(1):182–195
16. Wang JC, Wei WJ, Zhang JW (2019) Effect of the size distribution of granular top coal on the drawing mechanism in LTCC. Granular Matter 21(3):70
17. Wang JC, Fu Q (2002) The loose medium flow field theory and its application on the longwall top coal caving. J China Coal Soc 4:337–341
18. Wang JC, Li ZG, Chen YJ et al (2004) The experimental study of loose medium flow field on the longwall top coal caving. J China Coal Soc 29(3):260–263
19. Wang JC, Zhang JW (2015) BBR study of top-coal drawing law in longwall top coal caving mining. J China Coal Soc 40(3):487–493
20. Wei WJ, Song ZY, Zhang JW (2018) Theoretical equation of initial top-coal boundary in longwall top coal caving mining. Int J Min Miner Eng 9(2):157–176
21. Zhang JW, Wang JC, Wei WJ et al (2019) Experimental investigation on the effect of size distribution on the flow characteristics of loose top coal. J China Coal Soc 44(4):985–994
22. Wang JC, Wei WJ, Zhang JW et al (2017) Stability analysis of support around the longwall top-coal caving mining in steeply thick coal seam. J China Coal Soc 42(11):2783–2791
23. Wei WJ, Wang JC, Zhang JW et al (2023) Drawing mechanisms of granular top coal considering the structure of hydraulic support in longwall top coal caving. Bull Eng Geol Environ Aims Scope 82:138

24. Yang SL, Wei WJ, Zhang JW (2020) Top coal movement law of dynamic group caving method in LTCC with an inclined seam. Min, Metall Explor 37(5):1545–1555
25. Zhang JW, Wang JC, Wei WJ et al (2018) Experimental and numerical investigation on coal drawing from thick steep seam with longwall top coal caving mining. Arab J Geosci 11(5):96
26. Zhang JW, Wang JC, Wei WJ (2018) Effect of face dip angle on the drawing mechanism in longwall top-coal caving mining. J China Univ Min Technol 47(4):805–814
27. Wang JC, Wei WJ, Zhang JW et al (2020) Numerical investigation on the caving mechanism with different standard deviations of top coal block size in LTCC. Int J Min Sci Technol 30(5):583–591
28. Wang JC, Wei WJ, Zhang JW et al (2021) Laboratory and field validation of a LTCC recovery prediction model using relative size of the top coal blocks. Bull Eng Geol Env 80(2):1389–1401
29. Wei WJ, Yang SL, Li M et al (2022) Motion mechanisms for top coal and gangue blocks in longwall top coal caving (LTCC) with an extra-thick seam. Rock Mech Rock Eng 55(8):5107–5121
30. Wei WJ, Pan WD, Zhang JW et al (2023) Dynamic sublevel caving technology for thick seams with large dip angle in longwall top coal caving (LTCC). Granular Matter 25:56
31. Melo F, Vivanco F, Fuentes C et al (2007) On draw body shapes: from Bergmark-Roos to kinematic models. Int J Rock Mech Min Sci 44(1):77–86
32. Zhang JW, Cheng DL, Wang JC et al (2023) Theoretical calculation model of top coal drawing body in horizontal sublevel top coal caving mining. J China Coal Soc 48(2):576–592
33. Yang SL, Wei WJ, Yang L et al (2023) Theoretical investigation and key caving technology development at the end area of longwall top coal caving (LTCC) panels. Comput Part Mech
34. Wei WJ (2021) Drawing mechanism of top coal considering the size distribution in longwall top coal caving mining. China University of Mining and Technology, Beijing, 2021
35. Wang JC, Huang GJ, Yang BG et al (2011) Top coal caving law tracker and method for measuring top coal caving law. Patent number: ZL200910080005.9. 2011-09-21
36. Wang JC, Yang SL, Huang GJ et al (2013) Research on top coal tracker of fully mechanized top coal caving mining and measurement of top coal recovery rate. Coal Sci Technol 41(1):36–39
37. Wang JC, Geng HL, Zhang JW (2014) Numerical simulation study on rational layout density and method of coal migration tracing mark. Coal Eng 46(2):1–3
38. Wang JC, Wei WJ, Zhang GY et al (2022) Top coal drawing mechanism and intelligent drawing. Science Press, Beijing
39. Wang JC, Zhang JW, Chen Y (2016) Research on technology of improving top-coal recovery in longwall top-coal caving mining based on BBR system. J Min Sci Technol 1(1):38–48
40. Yang SL, Zhang JW, Chen Y et al (2016) Effect of upward angle on the drawing mechanism in longwall top coal caving mining. Int J Rock Mech Min Sci 85:92–101
41. Wang JC (2023) 40 years development and prospect of longwall top coal caving in China. J China Coal Soc 48(1):83–99
42. China National Coal Association. Zheng Coal machine set up the international market again fruitful[EB/OL]. https://www.coalchina.org.cn/index.php?m=content&c=index&a=show&catid=16&id=141466
43. Wu J (1998) Review of fully mechanized caving mining technology for 15 years. China Coal, pp 1–9
44. Ren BG (2002) Comprehensive mechanized top coal caving mining in China. Coal Industry Press, Beijing
45. Wang JC (2018) Engineering practice and theoretical progress of top-coal caving mining technology in China. J China Coal Soc 43(1):43–51
46. Fan YC (2012) A revolution of mining methods in China's thick seam, innovative development of technology theory and practice of fully mechanized top coal caving. Coal Industry Press, Beijing
47. Fan YC, Shi WL, Zhang CG et al (1997) Quantification analysis of mining loss in fully mechanized caving face in Wangzhuang coal mine in Luan mining bureau. J China Coal Soc 22(6):606–611

48. Wu J, Yu HY, Zhang HG (1994) Three models of technology for mining gently inclined thick coal seam with sub-level caving mining method. J China Coal Soc 19(6):612–619
49. Xie HP, Wang JC, Chen ZH et al (1999) Study on top-coal blasting technique of full-mechanized caving in the hard thick coal seam. J China Coal Soc 24(4):350–354
50. Wang JC, Bai XJ, Wu ZS et al (2000) The research on the fractured blocks of the top-coal in the longwall top-coal caving technique of the hard coal seam. J China Coal Soc 25(6):238–242
51. Wang JC, Wu J, Li B et al (1997) Observation, Measurement and analysis on strata pressure of fully mechanized longwall caving face in special thick seam. Coal Sci Technol 25(9):2–5
52. Shang HT (1997) Prospect of adopting fully mechanized sublevel caving under complicated seam condition from the practice of Weijiadi mine in China. J China Coal Soc 22(3):242–247
53. Wang JC, Wu J (1998) Fully mechanized top coal caving mining technology under complex coal seam conditions. In: Theory and practice of the third national top coal caving mining proceedings, 1998, pp 86–90
54. Wang JC (2007) Mechanism of the rib spalling and the controlling in the very soft coal seam. J China Coal Soc 32(8):785–788
55. Wang JC, Wang L, Guo Y (2014) Determining the support capacity based on roof and coal wall control. J China Coal Soc 39(8):1619–1624
56. Kang HP, Xu G, Wang BM et al (2019) Forty years development and prospects of underground coal mining and strata control technologies in China. J Min Strata Control Eng 1(2):7–39
57. Zhang ZW (2011) Study on suitability of hydraulic powered two lag shield caving support in Pingshuo mining area. Coal Sci Technol 39(11):31–35
58. Xie JW (2005) High efficiency fully mechanized longwall top coal caving mining technology of steeply inclined thick coal seam. Coal Industry Press, Beijing
59. Wang JC, Zhao BW, Zhao PF et al (2017) Research on the longwall top-coal caving mining technique in extremely inclined and soft thick coal seam. J China Coal Soc 42(2):286–292
60. Chen JQ, Wang SB, Liu XD (2022) Study on influencing factors of gas emission in horizontal sub-level mining face of steeply inclined and extra-thick coal seam. Coal Sci Technol 50(3):127–135
61. Wang JC, Pan WD, Zhang GY et al (2022) Principles and applications of image-based recognition of withdrawn coal and intelligent control of draw opening in longwall top coal caving face. J China Coal Soc 47(1):87–101
62. Peng SS (2020) Longwall Mining, 3rd edn. Taylor & Francis Group, London
63. Kumar R, Singh AK, Mishra AK et al (2015) Undergrond mining of thick coal seams. Int J Min Sci Technol 25(6):885–896
64. Unver B, Yasitle NE (2006) Modelling of strata movement with a special reference to caving mechanism in thick seam coal mining. Int Coal Geol 66(4):227–252
65. Klishin VI, Klishin SV (2010) Coal extraction from thick flat and steep beds[J]. J Min Sci 46(2):149–159
66. Клишин ВИ (2018) ИННОВАЦИОННЫЕ ТЕХНОЛОГИИ И СПОСОБЫ ОБЕСПЕЧЕНИЯ ПОВЫШЕНИЯ ПРОИЗВОДИТЕЛЬНОСТИ И БЕЗОПАСНОСТИ ПОДЗЕМНОЙ УГЛЕДОБЫЧИ[J]. Горный информационно-аналитический бюллетень (специальный выпуск) 11:52–63. https://doi.org/10.25018/0236-1493-2018-11-48-52-63

Chapter 3
Large Mining Height

The full-seam mining method with a large mining height (hereinafter referred to as large mining height) is a longwall mining method that mechanically mines the full thickness of coal seams reaching and exceeding 3.5 m at once [1]. Since the application of large mining height technology in China, it has developed rapidly. Due to its advantages such as high production, high recovery rate, simple process, high efficiency, and ease of automation, it has become the preferred mining method for accelerating coal mine automation, informatization, and intellectualization, realizing the coal industry's goals of "reducing mines, reducing panels, reducing production, and reducing personnel". This method aligns with China's development trend of promoting coal industry upgrading and energy technology revolution in the direction of "safety, efficiency, environmental protection, and low carbon" [2]. It is favored in the mining of thick coal seams in suitable mining areas, and has been widely promoted and applied in mining areas such as Shendong, Huainan, Jincheng, Datong, Lu'an, and Yulin.

3.1 Large Mining Height Technology

3.1.1 Characteristics of Large Mining Height Technology

Large mining height is generally suitable for thick coal seams with simple geological structures, hard coal quality, a thickness of 3.5–10.0 m, stable occurrence, and an dip angle of less than 12° (maximum not exceeding 20°), as well as stable or relatively stable roof and floor. Compared with conventional fully mechanized mining panels, large mining height panels have greater production capacity, which is conducive to rational centralized production. They also boast a high resource recovery rate, low roadway excavation ratio and maintenance cost, relatively simple mining processes and roadway layouts, making them more efficient and economical. Compared with

J. Wang and W. Wei, *Thick Coal Seam Underground Mining*,
https://doi.org/10.1007/978-981-96-3874-1_3

LTCC, large mining height has a higher coal recovery ratio and simpler coal mining procedures. When compared to slicing mining, the production output of large mining height panels is significantly increased, with a noticeable reduction in the excavation ratio of mining roadways. Additionally, it reduces the need for false roof installation and the relocation of coal mining equipment, greatly enhancing mining efficiency. However, due to the increase in mining height, the power requirements for hydraulic support, shearer, and AFC have increased, leading to a larger investment in equipment. As the power increases, the equipment size also gets larger, resulting in an increase in the excavation size of mining roadways. Under traditional mine auxiliary transportation conditions, equipment relocation and installation are relatively difficult. Furthermore, the increase in mining height and equipment size can also lead to increased difficulties in preventing and controlling coal wall spalling, equipment tipping and slipping prevention, roof fall handling, and production management during the mining process.

3.1.1.1 Mine Pressure Behavior in Large Mining Height

After mining in the longwall panel with large mining height, the height of the caving zone increases with the mining height. If the collapsed immediate roof strata cannot fill the goaf and a large free space appears below the hard rock strata, the broken main roof strata will be difficult to form a "masonry beam" type balance. During its slewing motion, it often produces a dynamic load effect on the lower strata and the panel support, forming a higher supporting pressure, and causing strong periodic weighting. Therefore, the working resistance of the supports in the large mining height panel is generally relatively high, and the pressure from the main roof is more intense. The phenomenon of local roof fall and coal wall spalling is more serious, and the scope of coal wall spalling increases with the mining height.

3.1.1.2 Characteristics of Coal Mining Technology in Large Mining Height

(1) The recovery ratio of coal resources is relatively high. During the initial mining of the LTCC panel, to reduce the initial pressure intensity of the roof, the coal caving process is generally not carried out, and during the normal cyclic caving process, it is impossible to release all the top coal under the premise of considering the gangue mixed ratio, resulting in some coal resources being left in the goaf. Under the condition of suitable coal seam thickness, the coal resource loss in the large mining height panel is less, and the development of small coal pillar roadway protection technology, goaf-side entry retaining, and goaf-side entry driving technology in the large mining height panel is relatively mature, which can greatly reduce the loss of coal resources caused by the reserved section coal pillar and significantly improve the recovery ratio of coal resources in the mining area.

(2) The coal mining technology is simple and the production efficiency is high. When mining the top layer of the slicing mining panel, it is necessary to lay a false roof, and the equipment relocation is inevitable for the mining of the lower layer. While in LTCC, it requires a coal caving process. The increase in production processes leads to a relative reduction in production efficiency. The large mining height process is relatively simple, and with the advancement and application of remote electro-hydraulic control systems, shearer memory cutting technology, automatic hydraulic support following and moving technology, and intelligent geological support technology, the number of personnel and labor intensity on the panel have been significantly reduced, greatly improving the production efficiency of the panel.
(3) The initial mining height needs to be controlled. To facilitate the installation of equipment such as large mining height hydraulic supports, shearers, and AFCs in the set-up room, the height of the set-up room should not be too high, and the initial mining height is consistent with the height of the set-up room. Starting from the set-up room, the panel maintains the initial mining height and gradually increases the mining height to the normal mining height after the immediate roof collapses for the first time. To allow the roof of the goaf to collapse as soon as possible, depending on the roof conditions, measures such as retracting anchor rods and anchor cables can be taken within the set-up room, and blasting measures can also be adopted if necessary.
(4) The control difficulty of hydraulic supports is high. With the increasing mining height of the large mining height panel, hydraulic supports are more prone to slipping or tilting accidents during operation, seriously affecting the safety of coal mining. In the large mining height panel, the coal mining area is relatively large, and the standing posture of the hydraulic support cannot maintain a long-term stable state under complex conditions, greatly increasing the difficulty of controlling the hydraulic support, as shown in Fig. 3.1a.
(5) The difficulty of preventing and controlling rib spalling has increased. With the increase in mining height, the motion range and damage degree of the overlying strata on the panel also increase, the influence range of the advanced

Fig. 3.1 Coal wall rib spalling of large mining height panel: **a** pressed support; **b** local rib spalling; **c** large deformation of stope surrounding rock

supporting pressure on the coal wall becomes larger, the pressure from the main roof becomes more intense, and the mine pressure behavior in the stope will be more obvious. Additionally, the pressure and duration of the overlying strata acting on the coal wall and roof will also increase, reducing the stability of the coal wall. The phenomenon of large-area, deep, and multi-regional rib spalling of the coal wall will occur more frequently, as shown in Fig. 3.1b, c.

3.1.2 Layout of Large Mining Height Panel

3.1.2.1 Ordinary Large Mining Height Panel

Ordinary large mining height panel adopts a single strike longwall mining method, which refers to the longwall panel advancing along the strike, and the coal seam is mechanically mined in a single pass, as shown in Fig. 3.2. Similar to ordinary fully mechanized mining panels, large mining height mining requires the arrangement of a panel in the coal seam. The support stands on the floor and directly supports the roof. The shearer cuts the entire coal seam, and the coal falling from the coal wall is transported out of the panel by an AFC. Due to the large size of mining equipment in the large mining height panel, the size requirement for the mining roadway during the relocation of the panel also increases accordingly.

Considering the transportation of coal mining equipment and the support of the mining roadway, the mining roadway of the large mining height panel is generally excavated along the coal seam floor, and the height is generally not more than 5 m. This creates a certain height difference between the panel and the mining roadway, as shown in Fig. 3.3. According to different coal seam thicknesses, common end transition modes for large mining height panels include three types: smooth transition, stepped transition, and combined transition. Smooth transition refers to the method of leaving triangular top coal. When the shearer cuts the coal to a certain distance from the end, the shearer drum is gradually lowered to reduce the mining height, so

Fig. 3.2 Large mining height panel of Kouzidong Mine, Xinji, China Coal

that the mining height of the panel gradually decreases from the normal height to the mining roadway height, achieving a smooth transition from the panel to the roadway. See Fig. 3.3a. The stepped transition is a vertical transition. By adopting a special end support with a large side guard plate, a stepped transition can be achieved at the end of the panel. That is, when the shearer horizontally cuts coal from the panel to the special transition support at the machine head and tail, the large mining height can be vertically reduced to the mining roadway height and then cut through. See Fig. 3.3b. This transition mode can not only improve the recovery ratio of coal resources but also eliminate the problem of support eccentricity and improve the stress state of the support. Considering the support capacity of the large side guard plate, the combined transition method of combining one-level stepped transition and smooth transition is adopted for large mining height panels of 8 m and above, as shown in Fig. 3.3c. This significantly reduces the loss of triangular coal at the top of the transition sections at both ends of the panel, improving the recovery ratio of coal resources.

3.1.2.2 Extra-Large Mining Height Panel

The mining method of the extra-large mining height panel is the same as that of the ordinary large mining height panel. Due to the excessive size of mining equipment and the limited underground transportation capacity, it is impossible to transport the entire machine in the mining roadway. Therefore, a loading and assembly chamber must be arranged near the set-up room to transport the dismantled equipment to the chamber for assembly and then deploy it in the set-up room. Additionally, due to the large mining height, the fallen coal, gangue, and leaked gangue have a significant impact force, and coal blocks may be ejected during the coal wall spalling process, posing a significant threat to the safety of personnel and equipment at the panel. Therefore, protective measures such as gangue retaining nets are arranged between the supports and the coal wall (Fig. 3.4). During large mining height mining, shearer operators and support shifting workers operate the equipment behind the shield net to prevent flying gangue and leaked gangue between supports from injuring operators, ensuring safe production at the panel. Furthermore, the increase in mining height has led to more prominent coal wall spalling issues, placing higher demands on the support capacity of the rib protection device.

Extra-large mining height panels in China basically adopt high-strength multi-level rib protection device to support the coal wall. During normal operation of the panel, the grouped rib protection devices are generally retracted for no more than 3 supports; in severe cases of spalling, the grouped rib protection devices are retracted for no more than 2 supports; after the support shifting is completed, the telescopic beams on the upper part of the support are promptly extended; shearer operators generally lag behind the rear drum of the shearer by 2 supports to extend the rib protection devices, ensuring that the rib protection devices are close to the coal wall.

Starting from the set-up room, the panel maintains the initial mining height before the first roof weighting. After the first roof weighting is over, the mining height is gradually increased to the designed mining height. When the main roof of some

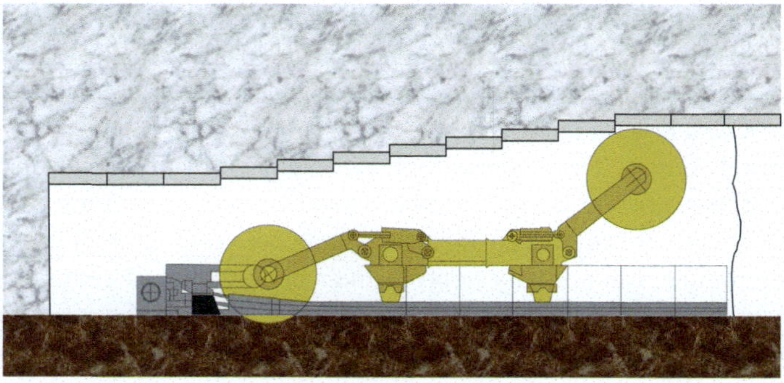

(a)

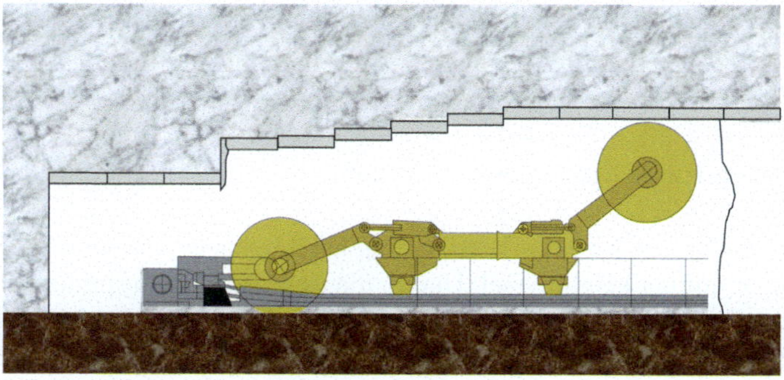

(b)

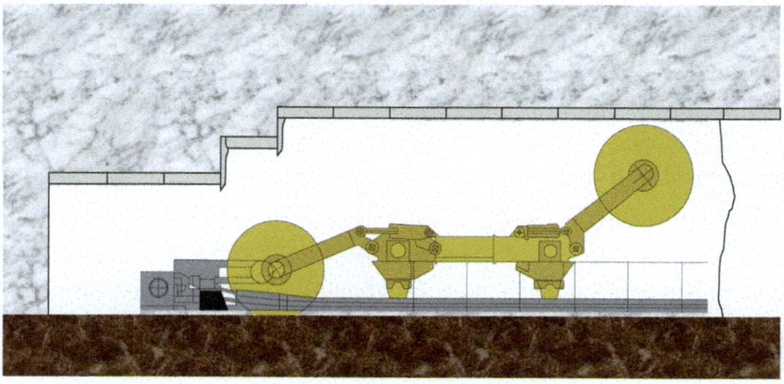

(c)

Fig. 3.3 Transition mode of large mining height panel: **a** smooth transition; **b** stepped transition; **c** combined transition

mines is relatively hard, artificial intervention will be taken to make the main roof break prematurely, preventing safety accidents caused by excessively large first roof weighting interval. Due to the difficulty and high cost of equipment disassembly and installation, to reduce the number of equipment relocations in extra-large mining height panels, the general panel has a long advancing length, up to 6000 m.

(a) (b)

Fig. 3.4 Caojiatan 10 m extra-large mining height panel: **a** installation of gangue retaining net; **b** three-level rib protection device

Extra-large mining height panels have a large mining height, a large one-time coal cutting height, and a large coal cutting volume, which cannot be met by the mining and transportation equipment of ordinary large mining height panels. The increase in mining height requires that the diameter of the shearer drum increases accordingly, which inevitably requires the shearer to have a larger arm length and power, and at the same time requires the shearer to have a stable heavy-load cutting speed to ensure stable and efficient production in extra-large mining height panels. For AFC, the imbalance of AFC load caused by the coal falling from the coal wall in extra-large mining height panels is more obvious, which causes some troubles to the stable operation of the AFC. Moreover, the increase in coal transportation volume poses higher challenges to the design of the AFC's size, machine power, scraper chain speed, and scraper stability.

3.1.3 Roadway Layout

Large mining height roadways generally adopt the retreating mining sequence, with the roadways including tailgate, headgate, and set-up room. These three components form and serve the panel, collectively referred to as mining roadway. The cross-section size of the mining roadway should meet the requirements for transportation, ventilation, pedestrian traffic, and safety. Due to the large size of equipment used in large mining height, the roadway cross-section size is also relatively large compared to the roadway of general fully mechanized mining, which increases the difficulty of the roadway excavation and support.

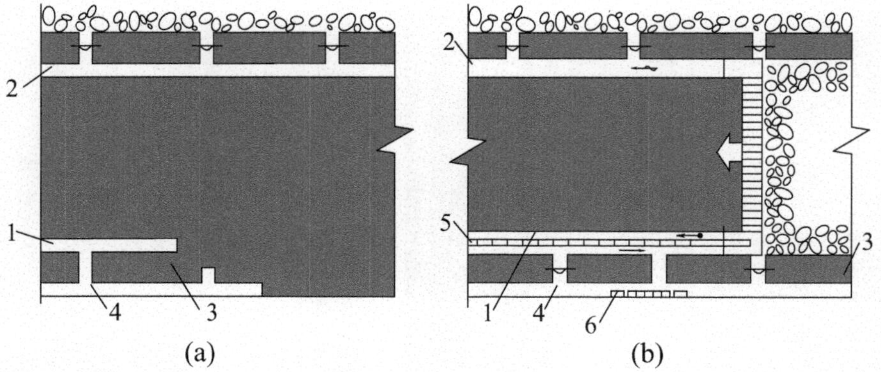

Fig. 3.5 Double-roadway layout: **a** excavation period; **b** mining period (1-headgate in this section; 2-railentry in this section; 3-section coal pillar; 4-tailgate in the next section; 5-belt conveyor; 6-Switch, pumping station and electrical equipment)

3.1.3.1 Double-Roadway Layout

The adoption of double-roadway layout facilitates the alternating mining and excavation of two adjacent panels. For panels with high gas content or long strike length, the use of double-roadway layout and excavation addresses the ventilation issues during long-distance excavation, which is conducive to ensuring the excavation ventilation and safety. As shown in Fig. 3.5, during the excavation of the roadway in the current section, both the headgate in this section and the tailgate in the next section are excavated simultaneously. A connecting roadway must be excavated between the two roadways to ensure ventilation safety during excavation. During the mining period of the panel, the belt conveyor and other electrical equipment can be arranged in the two roadways separately. The headgate is abandoned as mining progresses, while the roadway arranged electrical equipment is maintained to serve as the tailgate for the next panel.

3.1.3.2 Multi-roadway Layout

Multi-roadway layouts mainly include three-roadway and four-roadway. In high-yield and efficient mines with favorable mining conditions and the use of continuous miners for excavating entries, multi-roadway layouts and excavation methods are adopted for the section entries to fully utilize the advantages of continuous miners in roadway excavation and meet the ventilation and safety requirements under high-yield conditions. Due to the long advancing length of the large mining height panel, the multi-roadway layout, while satisfying the ventilation capacity of the panel, utilizes underground trackless tyred vehicle for auxiliary transportation, reducing the time for workers to enter the panel and increasing their average labor efficiency. The layout is shown in Fig. 3.6.

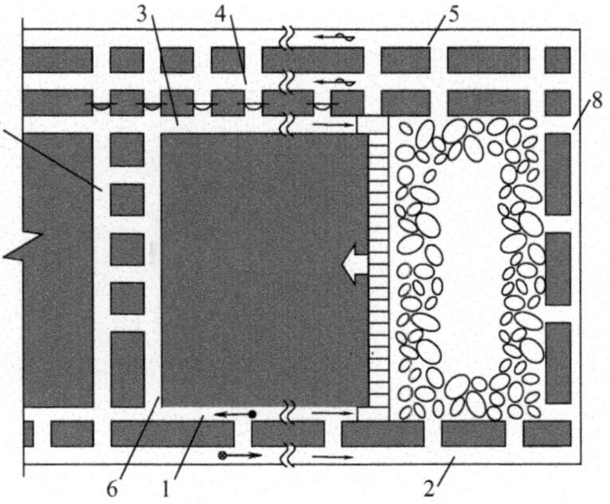

Fig. 3.6 Multi-roadway layout (1-headgate; 2-haulage roadway; 3-blowing roadway; 4, 5-tailgate; 6, 7-removal channels; 8-gas tail roadway)

3.1.3.3 Goaf-Side Roadway Retaining

The method of maintaining the original roadway along the edge of the goaf after the panel mining is known as goaf-side roadway retaining. Goaf-side roadway retaining technology is one of the important ways of non-pillar mining, which involves retaining the roadway of the panel through effective support techniques during the mining period [3]. As shown in Fig. 3.7, goaf-side roadway retaining can eliminate section coal pillars, contributing to an improvement in coal resource recovery. Due to the difficulty in maintaining the retained roadway during the retention period, additional support measures or other measures need to be taken alongside the roadway, as illustrated in Fig. 3.8. Due to technological limitations, this method is less commonly used in panels with large mining height.

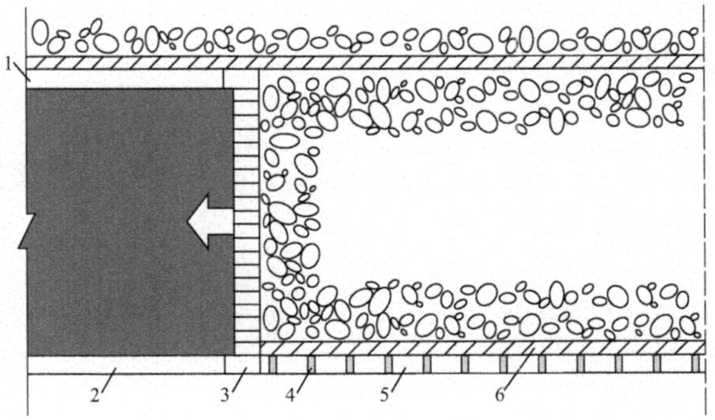

Fig. 3.7 Goaf-side roadway retaining (1-tailgate in this section; 2-headgate in this section; 3-end support; 4-portal support; 5-tailgate in next section; 6-roadway-side support)

Fig. 3.8 Temporary support with portal hydraulic support

3.1.4 Large Mining Height Technique

3.1.4.1 Equipment Layout with Large Mining Height

The main equipment in large mining height panel includes shearer, self-propelled support, and AFC. Additionally in the roadway, there are also transloader, crushers, belt conveyors, hydraulic pump stations, and auxiliary transportation equipment for transporting personnel and materials in the panel. The specific layout is illustrated in Fig. 3.9.

The 14030 panel of the Zhaogu No. 2 Mine, Henan Energy Group Co., Ltd., is a typical panel with large mining height. The main mining coal seam, known as Coal Seam No. 2, has an average thickness of 5.9 m. The coal seam structure is simple, with a coal seam dip angle ranging from 4° to 6°. The 14030 panel is equipped with a total of 102 hydraulic supports, including: 92 intermediate supports of the model ZY18000/30/65D, 5 transitional supports of the model ZYG18000/29/60D, 3 end supports of the model ZYT18000/26/53D, 2 end supports of the model ZYT13000/25/50D. The main equipment and technical parameters of the 14030 panel are listed in Table 3.1.

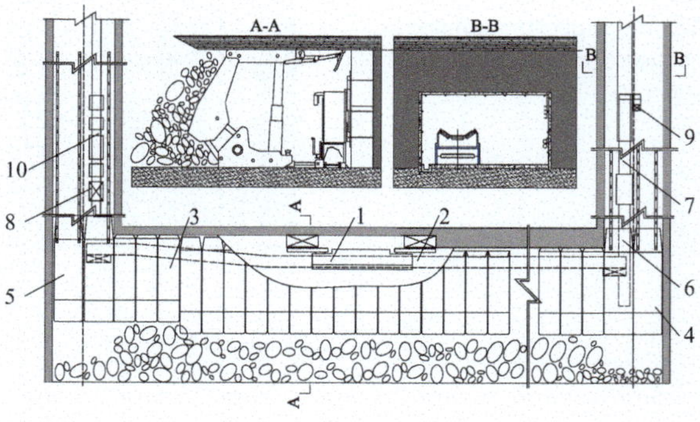

Fig. 3.9 Equipment layout of large mining height panel (1-shearer; 2-AFC; 3-hydraulic support; 4-lower end support; 5-upper end support; 6-t transloader; 7-belt conveyor; 8-centralized console; 9-pumping station; 10-equipment train)

Table 3.1 Main equipment and technical parameters of the 14030 panel

Equipment	Number	Technical parameter
Shearer	1	MG900/2320-GWD, power 2330 kW, Supply voltage 3300 V, drum diameter 3000 mm, truncated depth 865 mm, traction 1330/625 kN
AFC	1	SGZ1200/2000, motor power 2 × 1000 kW, motor voltage 3300 V, medium double-stranded form, scraper chain speed 1.59 m/s, scraper spacing 912 mm, chain center distance 280 mm, middle groove specifications (length × wide × high) 2050 × 1200 × 375 mm
Wheel crusher	2	PLM4000, crushing capacity 4000 t/h, maximum input blockage 1600 × 1200 mm, exclude granularity ≤ 300 mm, supply voltage 3300 V, spray out dust
Transloader	1	SZZ1200/525, length 55 m, conveying capacity 3000 t/h, supply voltage 3300 V, motor power 525 kW, monolithic box welding, climbing angle 10°
Intermediate support	92	ZY18000/30/65D, electro-hydraulic control system, support width 1900–2150 mm, setting load 12,370 kN, support center distance 2050 mm
Transition support	5	ZYG18000/29/60D, electro-hydraulic control system, support width 1900–2150 mm, setting load 12,370 kN, support center distance 2050 mm
End support	3	ZYT18000/26/53D, electro-hydraulic control system, support width 1900–2150 mm, setting load 12,370 kN, support center distance 2050 mm
End support	2	ZYT13000/25/50D, electro-hydraulic control system, support width 1680–1880 mm, setting load 8724 kN, support center distance 1750 mm
Pumping station	4	BRW500/31.5, rated pressure 31.5 MPa, rated flow 500 L/min, tank capacity 3000 L, motor power 315 kW
Belt conveyor	1	DSJ120/150/2 × 400, conveying capacity 1500 t/h, tape speed 4 m/s, motor power 400 kW, tape width 1200 mm
Belt conveyor	1	DSJ120/100/2 × 100, conveying capacity 1000 t/h, tape speed 3.15 m/s, motor power 110 kW, tape width 1200 mm
Transducer	1	BPJV-3 × 1250/3.3, rated power 3 × 1250 kW, rated input voltage 2 × 1700 V, rated input current 2 × 825 A, output frequency range 0–50 Hz
Spray pump	2	BPW516/16, rated pressure 16 MPa, rated flow ratio 516 L/min, motor power 160 kW, the working medium is clean neutral water
Spray pump	2	BPW315/6.3, rated pressure 6.3 MPa, rated flow ratio 315 L/min, motor power 160 kW, the working medium is clean neutral water
Backwash high-pressure filtration station	1	ZGLZ-2000II, rated pressure 31.5 MPa, rated flow ratio 2000 L/min, filter rating 25 μm

(continued)

Table 3.1 (continued)

Equipment	Number	Technical parameter
Return filter station	1	HGLZ-2000II, rated pressure 7 MPa, rated flow ratio 2000 L/min, filter rating 60 μm
Automatic dispensing station	1	ZMJ-KRPYZ-10, ratio range 1–6%, ratio capacity 10 t/h
Underground online self-cleaning comprehensive water supply purification station	1	JXGSZ-70JB-6A, water-yielding capacity 6 t/h, in-line self-cleaning
3300 V combination switch	2	QJZ1-1600/3300-8
1140 V combination switch	2	QJZ1-800/1140-4
Magnetic separator	1	RCBD-10DT2

3.1.4.2 Parameter of Panel

The parameters of the panel with large mining height mainly refer to the length of the panel, the advancing length of the panel, the height of the panel, and the cutting depth of the shearer, among others.

1. Length of the panel

The length of the panel is one of the crucial parameters in large mining height. A reasonable panel length should aim to achieve high average daily production, low cost per ton of coal, and compatibility with mining equipment, while minimizing the impact of significant geological formations. Since the mining equipment used in large mining height is large, relocating them incurs high costs. Therefore, increasing the length of the panel within a certain range can not only enhance production capacity but also improve mining efficiency and profitability. It can also reduce the ratio of roadway excavation and relatively minimize the loss of coal pillars between roadways. However, excessively long panels can reduce the advancing speed, increase the support time for the roof and coal walls using hydraulic supports, and elevate the risk of disasters such as coal wall spalling and roof falls at the ends. This is not conducive to preventing spontaneous combustion of residual coal in the goaf, which adversely affects production safety. The length of the panel is also constrained by factors such as equipment performance, coal seam geological conditions, production and technical management levels, and gas emission. Therefore, it is necessary to consider the weights of various influencing factors comprehensively, taking into account the actual production conditions and needs of the coal mine, to determine a reasonable panel length. Currently, the length of the panel with large mining height in China generally ranges between 200 and 400 m.

2. Advancing length of the panel

The advancing length of the panel is primarily influenced by geological conditions, the laying length of the belt conveyor, the overhaul period of equipment, the costs of installing and relocating equipment on the panel, as well as the maintenance and transportation costs of the section roadway. Geological conditions impose strong limitations on the advancing length of the panel. After the district is divided, the advancing length of a coal seam panel with fewer geological structures is mainly determined by the boundary of the coal seam distribution. However, for coal seams with numerous geological structures, especially those with large faults, the advancing length of the panel is often determined by the boundary of the geological structures. In recent years, significant progress has been made in domestic retractable belt conveyors, with multi-drive conveyors reaching a length of up to 6000 m. Limited by the level of technical equipment, the current upper limit for the advancing length of the panel is 6000 m. The overhaul period of the equipment on the panel is also a factor that affects the advancing length. Coal mine production should also consider economic factors while ensuring safety and efficiency. In general, increasing the advancing length of the panel can reduce the number of relocations and the length of roadway excavation, effectively reducing some economic expenditures. However, excessively long advancing lengths can cause difficulties in roadway maintenance, resulting in higher maintenance costs and transportation costs, while also posing challenges to power supply, ventilation, and other aspects.

3. Height of the panel

The height of the panel is also one of the important parameters in large mining height, which is mainly influenced by factors such as the level of equipment on the panel, coal quality, coal seam occurrence conditions, and geological structures. With the significant improvement in China's equipment manufacturing technology, the height of the panel has gradually increased under suitable coal seam conditions. By 2023, it has mainly undergone heights of 3.5, 4, 5, 5.5, 6.3, 7, 7.2, 8, 8.2, 8.8, 10 m, and so on. With the increase in mining height, it inevitably leads to excessive loading on the coal wall. Therefore, the working resistance of the hydraulic support on a large mining height panel is greater compared to a LTCC panel. The rib protection device is longer, and the resistance of the hydraulic cylinder is larger, to ensure the stability of the coal wall on the panel with large mining height.

4. Cutting depth of shearer

The cutting depth refers to the depth at which the shearer drum cuts into the coal wall, and it is a major factor determining the production capacity and installed power of the shearer. The cutting depth of the shearer is mainly influenced by the advancing speed of the panel, the support form, and the capacity of the AFC. As the coal wall stability is poor in large mining height, appropriately increasing the cutting depth of the shearer can improve the advancing speed of the panel, reduce the load time on the coal wall, decrease the weakening degree of coal body strength, and help lower the frequency and intensity of rib spalling. "Three equipment matching" is

Table 3.2 Parameters of some panel with large mining height

Coal mine	Panel	Mining height/m	Length of the panel/m	Advancing length of the panel/m	Cutting depth of shearer/mm
Gengcun	13031	4.0	195	1182	600
Bulianta	3202	5.0	240	5000	865
Sihe	2303	5.5	178	933.7	865
Shangwan	51202	6.3	301.5	4463	865
Bulianta	22303	7.0	301	4966	865
Wangzhuang	8101	7.2	270	546	865
Bulianta	12511	8.0	319.1	3139.3	865
Jinjitan	108	8.2	300	5538	865
Shangwan	12401	8.8	299.2	5254.8	865
Caojiatan	122104	10.0	300	5977	865

an important principle for selecting mining equipment. The difference between the maximum and minimum roof control distances of the hydraulic support should be the same as the cutting depth of the shearer, and the speed of the support moving should be consistent with the hauling speed of the shearer. The transportation capacity of the AFC needs to match the production capacity of the panel. Increasing the cutting depth of the shearer will enhance the production capacity of the panel while also increasing the transportation pressure on the AFC. Therefore, the shearer should determine a reasonable cutting depth based on the transportation capacity of the supporting equipment.

Table 3.2 shows the specific parameters of some representative panel with large mining height in different mining areas in China.

3.1.4.3 Entering Coal Wall Mode

Generally, double-drum shearer is used for the end oblique cutting of coal walls in large mining height, followed by the cooperation of the shearer operation and the advancement of the AFC to cut the coal wall. The specific process is shown in Fig. 3.10.

Figure 3.10a: When the shearer cuts to the end of the panel, the conveyor behind it, at a certain distance, has moved closer to the coal wall, leaving a section of bottom coal between the front and rear drums.

Figure 3.10b: Change the positions of the drums, lowering the front drum and raising the rear drum. Cut into the coal wall in the opposite direction along the curved section of the conveyor until reaching the straight section of the conveyor, and then straighten the conveyor.

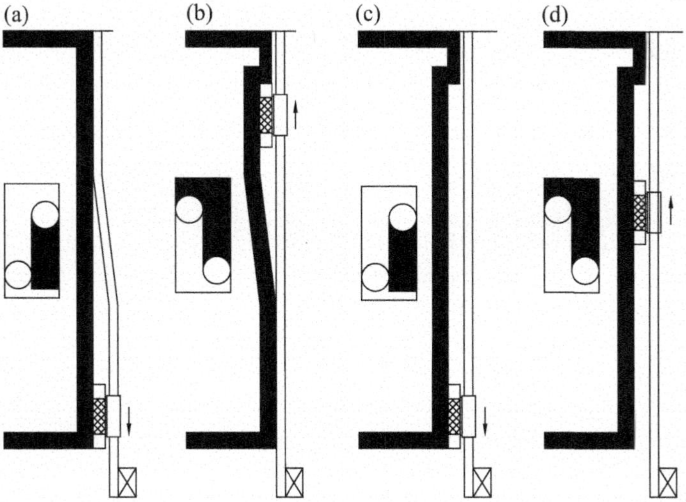

Fig. 3.10 End oblique cutting of coal walls by the double-drum shearer in large mining height **a** step 1; **b** step 2; **c** step 3; **d** step 4

Figure 3.10c: Swap the upper and lower positions of the two drums again, and return to cut the triangular coal to the conveyor head, leaving a section of bottom coal under the machine body.

Figure 3.10d: Swap the upper and lower positions of the drums again, and the shearer moves upward to cut off the bottom coal under the machine body. After cutting the coal wall straight, the shearer moves upward to cut coal normally.

3.1.4.4 Cutting Method

The process of coal cutting by the shearer and its coordination with other procedures is referred to as the coal cutting method of the shearer. There are two types of coal cutting methods: bidirectional coal cutting and unidirectional coal cutting. With the emergence of high-power, high-strength shearers, bidirectional coal cutting (two round trips) is mostly used in the panel with large mining height in China. Specifically, when the shearer starts to cut coal in a large mining height panel with double drums, it cuts coal upwards along the bottom from the lower cut, hangs beams randomly, pushes the conveyor, and simultaneously loads floating coal and props. After the shearer cuts to the upper cut, it repeats the same process after entering the knife and descending, and cuts another round of coal when returning. This method accelerates the advancing speed of the panel, reduces the occurrence of rib spalling and roof falling accidents, and reduces the non-functional wear between equipment.

3.2 Hydraulic Support System in Large Mining Height Panel

3.2.1 Development Process of the Large Mining Height Support

China's early coal mining was limited by the mining technology and mining equipment of that time, with the maximum mining height of coal seams being only 2.5 m. Single-drum shearers were mostly used for coal cutting at the panel, and single hydraulic props were used to support the roof. The single-drum shearer had to be divided into two cuts for large mining heights, with the top and bottom cuts. During the top coal cutting, only metal articulated top beams could be hung, and the props could only be set under the hung top beams after the bottom coal was cut and the AFC was pushed forward, thus increasing the suspended area and time of the roof. In addition, since the cutting drum was only arranged on one side of the shearer, the end of the panel without the drum side could only be blasted manually, and coupled with the short ranging arm of the cutting drum, both ends of the panel needed to be blasted manually to open the incision, resulting in the coexistence of two kinds of processes at the panel, reduced efficiency, and complicated management. China urgently needed a safe and efficient coal mining method at that time, and the large mining height emerged as the times required.

China's research and development of fully mechanized mining technology and equipment for thick coal seams with large mining heights began in the 1980s. From 1986 to 1990, the Beijing Mining Institute of the China Coal Research Institute, Pingdingshan Coal Mining Machinery Factory, Yima Mining Bureau, etc., jointly carried out the research and development test of China's first ZY3500/25/47 two-pillar shielded large mining height hydraulic support, which designed a limiting device between the roof beam and the shield beam of the hydraulic support for the first time to improve the adaptability of the hydraulic support to the roof. The complete set of equipment was successfully applied in Gengcun Coal Mine. During the "Seventh Five-Year Plan" period, the Beijing Mining Institute of the China Coal Research Institute, Xuzhou Mining Bureau, Xi'an Coal Mining Machinery Plant, Zhangjiakou Coal Mining Machinery Plant, etc., relied on the National Key Scientific and Technological Research Projects to carry out research and development and application of fully mechanized large mining height technology and equipment for thick coal seams with soft roof, soft coal and soft floor (4.5 m), deepening the understanding of the laws of fully mechanized mining in thick coal seams with large mining heights [4].

By the early 1990s, the technology of fully mechanized mining with large mining height was successfully promoted and applied in mining areas such as Tongchuan, Kailuan, Xishan, Yanzhou, Xuzhou, Xingtai, and Shuangyashan in China [5]. Starting from 1995, the Shendong Mining Area began to comprehensively introduce 4.0–5.0 m superior hydraulic supports and related equipment for large mining heights, achieving high yield and efficient mining of the panel with large mining height. As the coal seam thickness in the Shendong Mining Area is generally thick (6.0–8.0 m), and the maximum mining height of the imported superior fully mechanized mining equipment is only 5.0 m, the panel with large mining height can only be mined with top coal and bottom coal left, resulting in a significant waste of coal resources.

In 2003, the China Coal Research Institute and the Jincheng Mining Bureau took the lead in developing domestically produced superior hydraulic supports for large mining heights with a maximum support height of 5.5 m. They successfully developed 30 ZY8640/25.5/55 large mining height hydraulic supports for the first time and successfully applied them in the Panel 3302 of Jincheng Sihe Coal Mine in 2004, achieving a maximum daily yield of 30,000 tons and a maximum monthly yield of 760,000 tons. After a year of underground industrial testing, the overall condition of the hydraulic supports was good.

In 2005, driven by the national major technological equipment development project, the research and development of "6 million tons per year fully mechanized mining equipment" was carried out. Specifically, the ZY8600/24/50D domestic superior hydraulic support was developed for the 5-1 coal seam of Wanli No. 1 Mine of Shenhua Group. The support prototype successfully passed the 50,000-cycle loading type test, meeting the European CEN1804 standard. Through underground industrial testing, the equipment startup ratio of the complete set of equipment reached over 90%, with a maximum monthly yield of 804,000 tons and an annual production capacity exceeding 7 million tons.

From 2008 to 2011, with the support of the National "Eleventh Five-Year" Science and Technology Support Program, the research and development of "10 million tons per year mine fully mechanized mining equipment and key technologies for large mining heights" was carried out. A complete set of domestically produced fully mechanized mining equipment for large mining heights with a maximum mining height exceeding 6 m was developed, and the annual production capacity of the panel exceeded 10 million tons.

From 2009 to 2013, aiming at the needs of high-yield and efficient mining of hard and thick coal seams with a thickness of 6–8 m in coal mines in northern Shaanxi, China, such as Hongliulin Coal Mine, Jinjitan Coal Mine, Bulianta Coal Mine, and Shangwan Coal Mine in Shendong, taking the 5-2 coal seam of Hongliulin Coal Mine as the engineering background, the research and development of "7 m extra-large mining height fully mechanized mining technology and equipment" was carried out. The ZY18800/32.5/72D extra-large mining height hydraulic support and related equipment were developed, achieving an annual yield of 12 million tons for the panel. The hydraulic support developed this time first adopted a three-stage rib protection device and significantly reduced the triangular coal loss at both ends of the panel through the design of a "large gradient transition" supporting method. Figure 3.11 shows the ZY21000/38/70D two-column hydraulic support used in the 7 m large mining height panel of Jinjitan Coal Mine.

Fig. 3.11 ZY21000/38/70D extra-large mining height hydraulic support

From 2014 to 2016, to further enhance the resource recovery ratio of hard and extra-thick coal seams in western mining areas, targeting the occurrence conditions of the 2-2 coal seam in Jinjitan Coal Mine, the world's first set of extra-large mining height hydraulic support (ZY21000/38/82D) with a support height greater than 8.0 m was developed, breaking through a series of key technologies and matching challenges, as shown in Fig. 3.12. The developed complete set of equipment began underground industrial testing in October 2016, achieving a daily production of 66,000 tons and a monthly production exceeding 1.5 million tons in the panel. This innovative practice broke through the limit of mining height in thick coal seams and provided valuable experience for extra-large mining height.

In 2018, the Shendong Shangwan Coal Mine applied an extra-large mining height hydraulic support with a maximum support height of 8.8 m, as shown in Fig. 3.13, once again demonstrating the advantages of extra-large mining heights above 8 m.

Fig. 3.12 ZY21000/38/82D extra-large mining height hydraulic support

Fig. 3.13 ZY26000/40/88D extra-large mining height hydraulic support

Since 2020, targeting the engineering conditions of the hard and extra-thick 2-2 coal seam in Caojiatan Coal Mine (coal seam thickness of 8.08–12.36 m), an extra-large mining height hydraulic support (ZY29000/45/100D) with a maximum support height of 10 m and supporting equipment have been designed and developed, once again breaking through the extra-large mining height limit. Table 3.3 is a list of hydraulic support parameters for some representative large mining height panels.

Table 3.3 Hydraulic support parameters of representative panel with large mining height

Coal mine	Panel	Mining height/m	Support type	Setting load/kN	Support center distance/m
Gengcun	13031	4.0	ZY3500/25/47	2600	1500
Bulianta	3202	5.0	JOY8670-2.4/5.0	6056	1750
Sihe	2303	5.5	ZY8640/2550/5500	5888	1750
Shangwan	51202	6.3	ZY10800/28/63	7912	1750
Bulianta	22303	7.0	ZY18000/32/70D	12,370	2050
Wangzhuang	8101	7.2	ZY15000/33/72D	12,370	2050
Bulianta	12511	8.0	ZY21000/36.5/80D	11,025	2050
Jinjitan	108	8.2	ZY21000/38/82D	16,546	2050
Shangwan	12401	8.8	ZY26000/40/88D	19,782	2400
Caojiatan	122104	10.0	ZY29000/45/100D	22,368	2400

3.2.2 *Design and Management of Support for Large Mining Heights*

3.2.2.1 Support Design for Hydraulic Support

1. The principle of determining the support strength

The support strength of hydraulic support should be compatible with the geological conditions of the coal seam occurrence and the mining pressure manifestation strength of the roof. The mining pressure manifests intensely with the increase of mining height. The support strength of the large mining height hydraulic support should meet several principles, including that the initial support force and working resistance of the support should adapt to the pressure generated by the movement of the immediate roof and the main roof; the structure and support characteristics of the support should adapt to and protect the integrity of the roof; the base should adapt to the UCS of the floor rock; the support height should be compatible with the mining height or coal seam thickness, with good safety performance and equipped with equipment such as telescopic beams and bottom adjustment cylinders; the support cross-section should be compatible with ventilation requirements, and the three-machine matching should meet relevant requirements.

2. Verification of the support strength and working resistance of the support

The selection of hydraulic supports is based on the design requirements of the panel, and only the working resistance and support strength of the hydraulic supports are checked. The working resistance and support strength of the supports must be able to meet the support requirements.

(1) Calculation of working resistance of hydraulic support

$$F = N \times H \times S \times Z \times g \tag{3.1}$$

where F is the working resistance of the support, kN; N is the multiple of mining height; H is the maximum mining height of the panel, m; S is the maximum support area of the support, m^2; Z is the bulk density of roof rock, kg/m^3; K is dynamic load coefficient; g is the gravitational acceleration, at 9.8 m/s^2.

The basic support selected for the panel has a working resistance greater than the calculated working resistance, which can meet the design requirements for panel support.

(2) Calculation of support strength of hydraulic supports

$$p = N \times Z \times H \times g \tag{3.2}$$

where p is the support strength, kN/m^2.

The support strength of the support selected for the panel is greater than the calculated support strength, which can meet the design requirements for panel support.

3. Calculation of advanced support strength

$$p = M \times Z \times K \times g \tag{3.3}$$

where M is the height of the caving arch, m; K is the dynamic load coefficient.

The support strength of the selected support is greater than the calculated support strength, which can meet the design requirements for the advanced support of the main haulage roadway.

4. Calculation of overhanging distance at both ends

$$L_{\text{max}} = \sqrt{\frac{2M_Z[\sigma]}{\gamma}} \tag{3.4}$$

where L_{max} is the limit fracture step of immediate roof, m; M_z is the thickness of immediate roof, m; $[\sigma]$ is the tensile strength of strata, MPa; γ is the bulk density of strata, t/m^3. If the overhanging shape of the triangular area at the end is regarded as an isosceles right triangle, the maximum overhanging length of the triangular area behind the panel can be calculated using the following Equation:

$$L = \frac{L_{\text{max}}}{\sqrt{2}} - L_k \tag{3.5}$$

where L_k is the maximum roof control distance of the end support, m; the overhanging length of the triangular area at both ends of the panel should be controlled within L.

5. Design requirements for emulsion pump station

The liquid supply system of the fully mechanized mining panel is composed of a large-flow emulsified pump, emulsified liquid tank, large-flow valve, high-pressure backwash filter station, self-cleaning filter, and related high and low-pressure pipelines. A double-line ring-shaped supply and return system is adopted, requiring the emulsified liquid concentration to be maintained within an appropriate range. It is strictly prohibited to adjust the set value of the unloading valve randomly. It is necessary to strengthen the maintenance and repair of the pump station and pipelines, keep the hydraulic system intact, and eliminate any leaks, dripping, bubbling, and fluiding phenomena. There are strict requirements on the water supply pressure and water quality. The water outlet pressure of the spray pump must meet the water requirements of the equipment on the panel. The water quality must be clear, transparent, and free of impurities, with a neutral or weakly alkaline water quality. Every shift should inspect the filter, pipelines, and water quality to ensure the integrity of the equipment and the water quality meets standards. Inspect the automatic backwash filter to ensure reliable operation. The normal operation of the liquid supply system in the fully mechanized mining panel is crucial, and it must be used and inspected strictly according to regulations to ensure the safe and efficient operation of the equipment.

3.2.2.2 Management Method of Roof

During the initial mining, final mining, normal mining, roof pressure, rib spalling, passing through the caving area, and crossing geological structures in the panel with large mining height, special safety technical measures must be formulated based on the actual roof conditions and support conditions of the panel to strengthen roof management and ensure safe production. Taking the roof management method of Caojiatan Coal Mine as an example, a brief introduction is provided below.

1. Roof support methods during normal operation

To ensure the safe and efficient operation of the panel with large mining height, it is necessary to comply with the specified quality standardization requirements. Attention should be paid to the vertical transition of the roof at both ends to ensure smooth adjustment and avoid situations such as cutting beams, excessive beam end distances, or collapsing supports. The initial support force of the supports must meet the requirements, and during the process of moving the supports, the extension and retraction of the balance cylinders should be adjusted reasonably according to the flatness of the roof and floor to ensure tight contact between the supports and the roof. The rib protection device should be promptly extended behind the rear drum of the shearer, and the side guard plates should be used to adjust the support shape during the process of moving the supports to maintain the vertical state of the supports and the roof. The clearance between the supports must comply with the regulations, and the height of the supports should match the mining height. It is strictly prohibited to use the supports at excessive heights.

During coal cutting operations, the mining height must be strictly controlled, and the length of the canopy eaves of the panel is limited. The top beam of the support should be parallel to the roof, and the support should be vertical to the roof and floor to provide effective support for the roof. During production, it is strictly prohibited for personnel to observe the coal wall between two supports or walk in front of the large feet of the supports. After the machine is stopped, entering the AFC area requires ensuring that the equipment is locked, checking the safety of the roof and ribs, and assigning a dedicated person for on-site supervision.

Maintenance of hydraulic supports should be strengthened to ensure that the hydraulic system has no leakage or fluid motion, and the supports operate flexibly. The safety valves of the support columns and balance jacks should be checked to ensure that they reach the rated pressure when the safety valves are opened. When moving the AFC on the panel, a smooth transition should be ensured, and sharp turns should be avoided. Preventive maintenance should avoid the periodic roof pressure, and the shearer should be stopped at a reasonable position for maintenance.

2. Management of periodic roof pressure

During the roof pressure period in the panel with large mining height, on-site organization should be strengthened to achieve rapid progress and shed off the pressure. During normal production, all personnel must walk inside the supports and are strictly prohibited from walking in front of the supports. When the roof is separated or broken, the supports should be moved promptly behind the front drum of the shearer under pressure to ensure tight contact between the top beam of the support and the roof. When the roof is broken, the advance supports should be moved promptly. When choosing the shutdown time, it is necessary to reasonably select the area with intact roof conditions at both ends for shutdown maintenance.

3. Management of roof fall during caving

To ensure the safe production of a large mining height panel, it is necessary to promptly advance the supports when the panel is under excessive pressure, severe rib spalling occurs, or the roof is fractured. When there is a significant distance between the hydraulic support and the coal wall, the supports should also be advanced in a timely manner to reduce the unsupported roof distance. When the shearer passes through the roof fall area, if there is no large-scale roof fall, the roof should be controlled promptly by reducing the cutting height and advancing quickly. In case of a large-scale roof fall, unidirectional cutting at the head or tail of the machine should be adopted to gradually advance, accelerating the advancement of the panel, reducing the scope of the roof fall, and minimizing the unsupported roof area. When there is a partial roof fall on the panel, the supports on both sides of the roof fall area should be advanced promptly to prevent the expansion of the roof fall area. If the roof fall area is located at either end of the panel, additional support for the roof and ribs of the roadways is required.

3.2.2.3 Management of Roofs at Panel Ends and Roadways

1. Management of support at panel ends

To ensure safe production at the panel, the side and top shields of the end supports should be extended promptly after advancing the supports. Maintenance personnel should enhance equipment maintenance to ensure that there are no leaks in the end supports and that the hydraulic components such as safety valves and jacks are operating reliably. When retreating the supports at the ends of the panel roadways, roof weakening measures should be taken if there is a large overhang in the goaf lagging the cut-off line. During mining, appropriate reinforcement support methods should be adopted based on the condition of the roof and rib at the panel ends. If severe rib spalling occurs at the ends, pre-tensioned cable tray protection using iron wire should be employed, and pre-tensioned cable tray anti-fall devices should be used for protection, advancing in a timely manner.

2. Management of safety exits

The height of the safety exits at the upper and lower ends of the panel should meet safety requirements to ensure that personnel can evacuate the panel smoothly in case of an emergency. Any abnormal conditions in the roof and ribs must be addressed promptly to reduce safety hazards. If the safety exit tends to become smaller, the pseudo-inclination of the panel should be adjusted in a timely manner to control the size of the safety exit. During shift changes, dedicated personnel should inspect the roof and ribs, handling any abnormal changes, and confirm safety before allowing workers to start operations.

If personnel need to pass through or work on the motors at the head or tail of the machine, it must be ensured that no other personnel are working in the area. The end support shield should be extended promptly, the support should be closed, and the three-machine interlock should be implemented, with dedicated personnel assigned for supervision to prevent accidents. To ensure the smoothness of the safety exits at the panel, the supports at the upper and lower ends should be strengthened using end supports to increase the stability and safety of the support.

3. Advanced roof support in roadways

To ensure safe production at the panel, appropriate advanced supports should be set to control the roof distance, and the support configuration should be adjusted to ensure that the supports are perpendicular to the roof and floor. The top shields of the supports should fit tightly against the roof to ensure the safety of equipment and personnel. Near the panel of the crossheading, temporary closures or fences with conspicuous warning signs should be set up to prohibit personnel from entering. The roof should be monitored for any changes, and if any safety hazards are found, measures should be taken promptly to reinforce the support. During the operation of the advanced supports, the push–pull cylinders must be kept in a retracted state, and the surrounding personnel's safety should be observed. Only after confirming safety can the work proceed.

3.3 Coal Wall Stability Control

With the gradual improvement of comprehensive mechanized equipment manufacturing and panel management levels, large mining height technology has developed rapidly. Due to the significant increase in the height of single mining, the pressure exerted by the overlying strata on the coal wall and roof also increases. Additionally, changes in boundary conditions caused by mining-induced unloading actions decrease the stability of the coal wall, leading to frequent incidents of spalling and roof falls, which severely affect normal mining operations. Coal wall failures typically exhibit gradual, periodic, temporary support, and large deformation characteristics. The stability of the coal wall is significantly influenced by the characteristics of the coal seam and mining conditions. The main hazards include:

1. Impact on normal production: Coal wall failure affects the normal production of the mine. Large blocks of coal that fall and break are difficult to transport, severely impacting the normal advancement of the panel and deteriorating the stability of the surrounding rock. Additionally, large coal blocks falling onto the AFC reduce its performance and lifespan.
2. Threat to worker safety: Coal wall failure poses a serious threat to worker safety. Falling large coal blocks can injure workers, and hard coal wall failures can cause ejected coal blocks to injure people.
3. Worsening of support and surrounding rock relationship: Further spalling of the coal wall deteriorates the relationship between the support and the surrounding rock, increasing the unsupported space at the end panel. This prevents the support from functioning properly, triggering roof falls at the end panel and exacerbating coal wall spalling.

3.3.1 Mechanism of Coal Wall Spalling

A mechanical model for coal wall stability is established to identify the external influencing factors of coal wall failure at the panel. These factors include the load of the overlying strata on the coal wall, the equivalent concentrated force on the coal wall, the equivalent bending moment of the coal wall, the load on the rib protection devices, and the length of the rib protection devices. Internal influencing factors include the cohesion and the internal friction angle of the coal body. The stability coefficient of the coal body in front of the panel is defined according to the Mohr–Coulomb yield criterion. By combining the mechanical model of coal wall stability and the stability coefficient of the coal body, an analysis of coal wall failure factors is conducted. The Ritz method, based on the displacement variation principle in the energy principle, is used to solve the stress and displacement fields within the coal body of the panel [6].

3.3.1.1 Forms of Coal Wall Spalling

Through years of field observations and theoretical research, it has been found that the main types of coal wall failure at the panel are shear and tensile failures [7], as shown in Fig. 3.14. However, actual coal wall failure forms are not limited to these two basic types and can be more diverse. Figure 3.15 shows the statistics of coal wall failure types at the Panel 2612 with a 6.5-m mining height at the Dongpang Mine of Hebei Jizhong Energy Group. In this figure, types (a) and (d) account for over 80% of the failures, while type (c) accounts for 16%. The failure forms in (a) and (d) belong to shear failures and can be represented by Fig. 3.14a.

The failure type shown in Fig. 3.15c is due to the presence of a relatively stronger rock layer in the middle of the coal seam, which controls the failure mechanism. Its failure principle still belongs to shear failure. Shear failure occurs above the rock,

Fig. 3.14 Basic forms of coal wall failure: **a** shear failure; **b** tensile failure

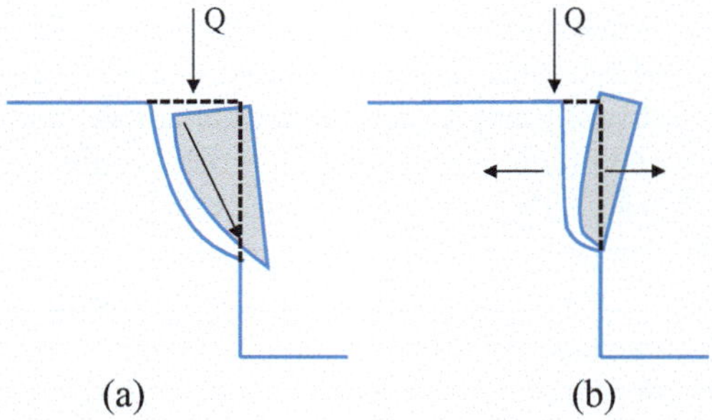

while tensile failure occurs below the rock due to compression. The failure type shown in Fig. 3.15b is a tensile failure, with extrusion occurring in the middle of the coal wall. Generally, tensile failure often occurs at the panel of hard coal seams, and this type of failure is rare. The above analysis indicates that shear failure is the predominant type of coal wall failure at the panel.

3.3.1.2 Mechanical Model of Coal Wall Shear Failure

In Fig. 3.16, the mechanical model for solving shear failure in the coal wall of a large mining height panel mainly includes two methods: (1) Exact solution: This involves calculating the shear failure surface of the coal wall as a curved surface. However, this method is complex in terms of calculation. (2) Approximate solution: This simplifies the curved failure surface to a plane *abc* and the coal wall pressure to a uniformly distributed load q (in reality, the coal wall pressure q is not uniformly distributed) [8]. According to the Mohr–Coulomb strength criterion, the stability coefficient K is defined as the ratio of the shear resistance T along the shear slip surface to the sliding force S on that surface: $K = T/R$. If this value is less than 1, shear failure occurs in the coal wall; otherwise, the coal wall remains stable.

From Fig. 3.16, using the limit equilibrium method, the following can be derived:

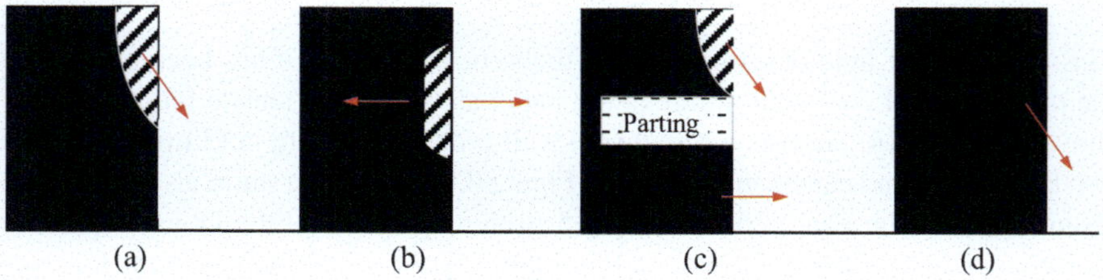

Fig. 3.15 Types of coal wall failure in the 2612 panel of Dongpang: **a** upper section damage; **b** middle section damage; **c** upper and lower section damage; **d** overall damage

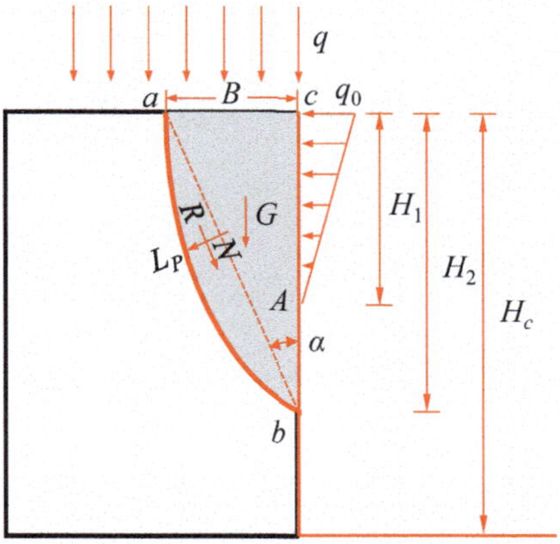

Fig. 3.16 Mechanical analysis of coal wall shear failure

$$
\begin{aligned}
R &= (Q + G)\cos\alpha - Q_0 \sin\alpha \\
T &= cL_p + N\tan\varphi = \frac{cH_2}{\cos\alpha} + [(Q + G)\sin\alpha + Q_0\cos\alpha]\tan\varphi
\end{aligned}
\tag{3.6}
$$

Among them,

$$
\begin{aligned}
&L_p = \frac{H_2}{\cos\alpha}; Q = qB = qH_2\tan\alpha; Q_0 = \frac{q_0 H_1}{2}; G = \frac{1}{2}H_2^2\gamma_c\tan\alpha; \\
&N = (Q + G)\sin\alpha + Q_0\cos\alpha
\end{aligned}
\tag{3.7}
$$

where T is the shear resistance provided by the failure surface; R is the sliding force along the failure surface ab; N is the normal pressure on the failure surface; L is the length of the failure surface; q is roof load concentration; q_0 is the load density of the coal wall pressure applied by the rib protecting device; H_c is the height of the coal wall; H_1 is the height of the rib protecting device; H_2 is the height of the failure body; G is the weight of the failure body; γ is the unit weight of coal; α is the angle between the failure surface and the coal wall; Q is the coal wall pressure acting on the failure body; Q_0 is the horizontal force exerted by the rib protecting device on the coal wall; c is the cohesion of the coal body; φ is the internal friction angle of the coal body. Therefore, the stability coefficient of the coal wall at the panel is:

$$
K = \frac{T}{R} = \frac{[(Q + G)\sin\alpha + Q_0\cos\alpha]\tan\varphi + cH_2\sec\alpha}{(Q + G)\cos\alpha - Q_0\sin\alpha}
\tag{3.8}
$$

3.3.1.3 Sensitivity Analysis of Coal Wall Failure

To verify the impact of various factors on the failure of coal walls in large mining height panels, a sensitivity analysis is conducted based on the mechanical model of

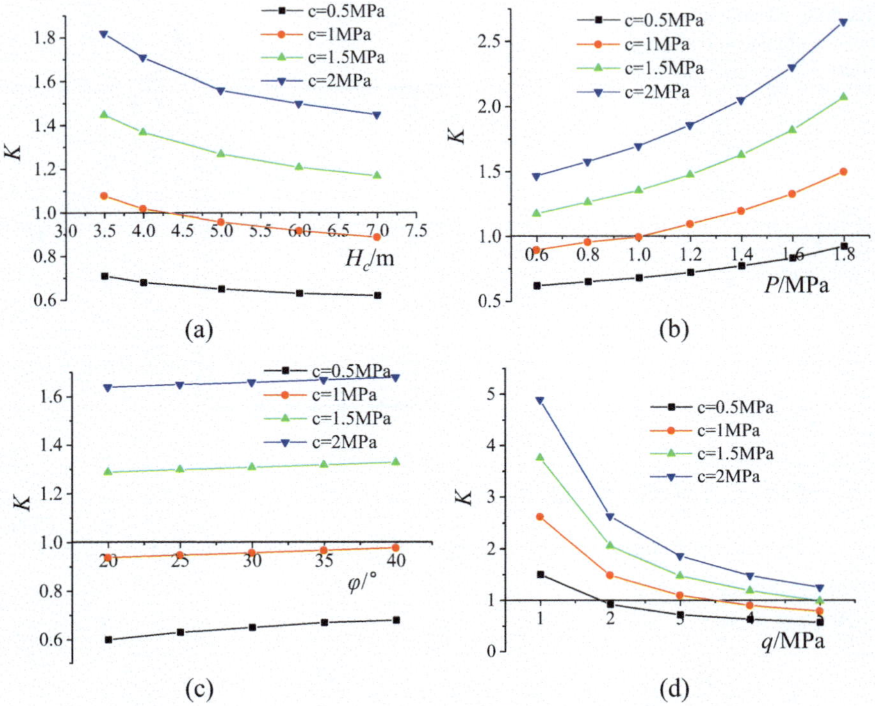

Fig. 3.17 Stability coefficient K variation curves: **a** with Hc; **b** with P; **c** with φ; **d** with qs

coal wall shear failure. The basic data for the coal wall in the panel are as follows: $H_c = 5$ m, $c = 0.5$ MPa, $\varphi = 25°$, $q_0 = 0.1$ MPa, $H_1 = 2$ m, $\alpha = 45° - \varphi/2$, $\gamma_c = 14$ kN/m^3. These values are substituted into Eq. (3.8). By setting $H_2 = H_c$, and varying P, H_c, φ, and c, the stability coefficient K values are obtained, as shown in Fig. 3.17.

Figure 3.17 illustrates the relationship between the stability coefficient K and the changes in coal body cohesion c, mining height H_c, support intensity P, coal body internal friction angle φ, and roof load concentration q. From the figure, the stability coefficient K increases significantly with an increase in coal body cohesion c and support intensity P, indicating high sensitivity. This means that enhancing the coal body cohesion c or increasing the support intensity P significantly improves the stability of the coal wall. The stability coefficient K decreases with an increase in coal wall height H_c, that is, higher mining heights increase the likelihood of coal wall failure. The stability coefficient K increases with an increase in the internal friction angle φ of the coal body, however, the effect is not very sensitive. This indicates that increasing the internal friction angle of the coal body has a limited impact on improving coal wall stability. An increase in the roof load concentration q rapidly decreases the coal wall stability coefficient K. Therefore, reducing the pressure on the coal wall is an important measure to improve coal wall stability.

The coal wall pressure is the primary cause of coal wall failure. The larger the mining height the greater the pressure on the coal wall. Conversely, the greater the support intensity of the hydraulic supports, the smaller the roof pressure acting on the coal wall, making it less prone to spalling. The rib protection devices of the supports play a significant role in preventing the sliding and collapse of the failing coal wall, however, they have a limited effect in controlling the further failure of the coal body in the plastic zone.

3.3.1.4 Displacement and Stress Fields of the Coal Body at the Panel

To solve for the displacement and stress fields of the coal body in front of the panel, the mechanical parameters of the coal body are taken as follows: elastic modulus of the coal body $E = 30$ MPa; poisson's ratio $v = 0.35$; Cohesion $c = 1$ MPa; internal friction angle $\varphi = 36°$; rib protection device load $q_0 = 0.1$ MPa; roof load acting on the upper coal $q = 0.8$ MPa; shear stress $\tau = 0.4$ MPa; equivalent concentrated force on the coal wall $P = 2000$ kN; equivalent moment on the coal wall $M = 10{,}000$ kN m. The model size parameter is the length of the coal body in front of the coal wall $L = 10$ m. The coal seam thickness $H_c = 6$ m, rib protection device height $H_1 = 3$ m. Under the above basic parameter conditions of the coal wall stability mechanical model, the horizontal and vertical displacement distribution cloud maps of the coal body in front of the panel, obtained using the Ritz method, are shown in Fig. 3.18.

In Fig. 3.18, the vertical axis represents the coal wall height, and the horizontal axis represents the distance from the model's origin, with $x = 10$ m being the coal wall. The closer to the coal wall of the panel, the greater the deformation of the coal body. At the upper part of the coal wall, both the horizontal and vertical displacements reach their maximum values, indicating that the upper part of the coal wall is the most prone to spalling, consistent with field observations. The maximum horizontal displacement is 47.25 cm, and the maximum vertical displacement is 61.57 cm. The relationship for the stress state at any point on the Mohr's circle and the conversion between the maximum and minimum principal stresses is:

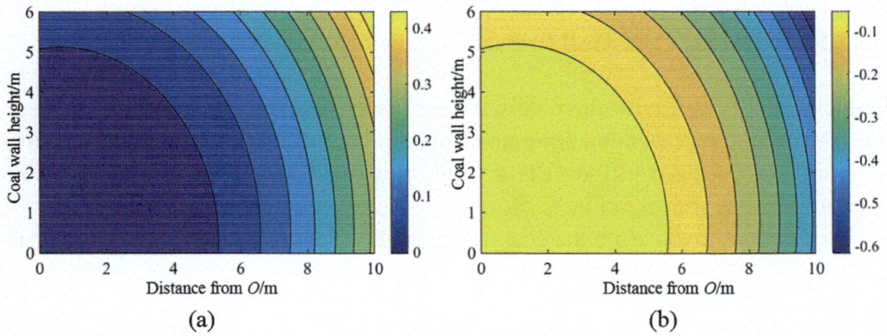

Fig. 3.18 Displacement of the coal body in the panel: **a** horizontal displacement; **b** vertical displacement

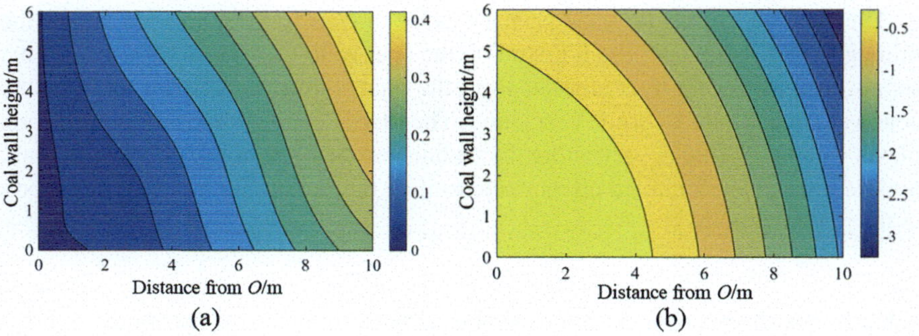

Fig. 3.19 Stress distribution in front of the panel coal body: **a** maximum principal stress; **b** minimum principal stress

$$\frac{\sigma_1}{\sigma_3} = \frac{\sigma_x - \sigma_y}{2} \pm \sqrt{\left(\frac{\sigma_x + \sigma_y}{2}\right)^2 + \tau_{xy}^2} \tag{3.9}$$

Substituting the horizontal stress σ_x, vertical stress σ_y, and shear stress τ_{xy} obtained using the Ritz method into Eq. (3.9), the principal stress distribution within the coal body in front of the panel can be determined, as shown in Fig. 3.19. The coordinates in the figure correspond to the coal wall stability mechanical model in Fig. 3.18, with $x = 10$ m indicating the position of the coal wall. The maximum principal stress within the coal body of the panel is predominantly tensile, with the maximum value occurring at the upper part of the coal wall. The minimum principal stress is compressive, with its absolute maximum value also appearing at the upper part of the coal wall. Therefore, under the action of the maximum and minimum principal stresses, the upper part of the coal wall is prone to tensile-shear and compressive-shear failures.

3.3.2 Coal Wall Stability Control

3.3.2.1 Reducing Coal Wall Pressure

Reducing the pressure on the coal wall and increasing the coal wall strength are the main ways to prevent spalling and improve coal wall stability. Enhancing the working resistance and stiffness of the supports is an effective method to reduce coal wall pressure, as shown in Fig. 3.20.

The relationship between the coal wall pressure Q and the support resistance P is given by Eq. (3.10). The pressure on the coal wall decreases as the support working resistance P increases.

$$Q = \frac{L^2(q + h\gamma_g) + 2(T_1 h - T_1 \tan\varphi L - PL_s)}{2L_R} \tag{3.10}$$

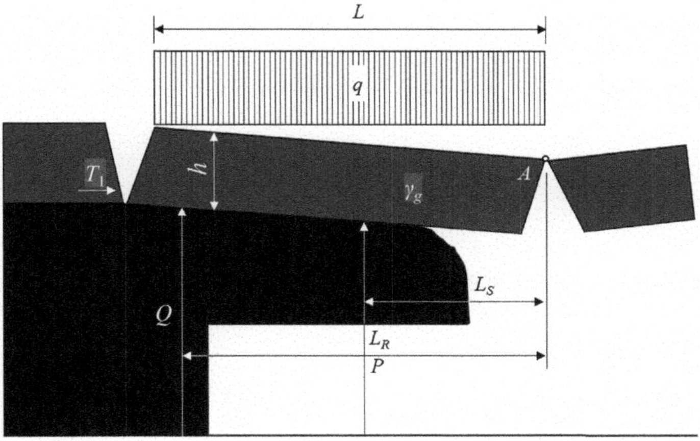

Fig. 3.20 Illustration of the balanced structure "block beam" formed by the fracture of the roof at the coal wall

where R is the pressure on the coal wall; P is the working resistance of the support; L_R is distance from Point A at the coal wall to the roof breakage; L_S is the distance from Point A at the support to the roof breakage; L is the length of the main roof broken rock block; γ_g is the bulk density of the main roof broken rock block; h is the height of the main roof broken rock block; q is the load on the upper part of the main roof; T_1 is the horizontal force acting on the main roof broken rock block; φ is the friction angle between the broken rock blocks.

3.3.2.2 Coal Wall Water Injection

1. Basic principle of coal wall water injection

Currently, to effectively address the problem of coal wall spalling during the mining process in large mining height panels, the coal wall is generally reinforced using advanced deep water injection technology. The principle of this technology is that the pressure water diffused into the joint fractures of the coal seam causes a large number of water molecules to be adsorbed on the capillary pores inside the coal body, thereby enhancing the integrity of the coal body through the mutual attraction between water molecules. To ensure the effectiveness of advanced deep water injection, quick-setting cement can be used to inject grout into shallow holes in the roadway side wall to enhance its compressive strength. The water injection pressure can be obtained from Eq. (3.11), and the water injection volume per single hole can be obtained from Eq. (3.12) [9, 10].

$$\begin{aligned} P &= P_0 + K_y V_H \\ P_0 &= 156 - 78/(0.5 + 0.001 H_m) \\ K_y &= 6.75f - 3 \end{aligned} \tag{3.11}$$

where P_0 is the minimum water injection pressure, MPa; P is the initial pressure of coal wall water injection, MPa; K_y is the coal seam property coefficient; f is the coal hardness coefficient; V_H is the unit water injection rate, m^3/h; H_c is the seam thickness.

$$Q = K_Z \cdot L_Z \cdot B_j \cdot H_c \cdot \gamma \cdot q_d \tag{3.12}$$

where Q_Z is the water injection volume per single hole, m^3; L_Z is the length of the water injection hole; K_Z is the water injection coefficient, taken as 1.2; q_d is the water injection volume per ton of coal; γ is the density of the coal; B_j is the spacing between water injection holes.

2. Experimental study on coal samples with different moisture content

Injecting water into the coal wall can enhance its cohesion, especially for soft coal seams. When the moisture content ranges from 12 to 18%, a significant improvement in cohesion can be observed. Compression experiments on coal samples with different moisture content were conducted using a strain-controlled triaxial apparatus. All coal samples were taken from the No. 8 coal seam of the Luling Coal Mine in Huaibei, Anhui Province. Cylindrical coal specimens with a diameter of 61.8 mm and a height of 150 mm were artificially prepared, as shown in Fig. 3.21. Figure 3.22 shows the before and after comparison photos of the specimens.

Figure 3.23 shows the limit Mohr circles and strength curves under different moisture content conditions. The variations of cohesive strength c and the internal friction angle φ with moisture content was derived, as depicted in Figs. 3.24 and 3.25, respectively. With an increase in moisture content, the cohesive strength c shows a tendency to increase. At a moisture content of 16.84%, the cohesive strength c reaches

Fig. 3.21 Artificially prepared coal specimen

Fig. 3.22 Comparison before and after specimen failure

its maximum value of 384 kPa, marking a significant increase of 120%. The internal friction angle φ, on the other hand, generally decreases with increasing moisture content. At a moisture content of 16.84%, the internal friction angle φ reaches its minimum value of 34.50°, representing a decrease of 12%, albeit relatively small. These results indicate that for soft panels, a substantial increase in cohesive strength c after reasonable water infusion plays a considerable role in preventing coal wall collapses.

Figure 3.26 depicts the relationship between triaxial compressive strength and confining pressure, as well as moisture content. With an increase in confining pressure, the triaxial compressive strength significantly increases. Conversely, as the moisture content increases, the triaxial compressive strength decreases.

3. Application at Huozhou Coal and Electricity Xinzhi Mine [11]

The 2-208 panel of Huozhou Coal and Electricity Xinzhi Mine is located in the 310 horizontal second mining area, with a long axis of 175 m and an inclined length of 579.5 m, mainly extracting coal seam 2#. The thickness of the coal seam ranges from 3.8 to 4.3 m, with an average thickness of 4.1 m, and an inclination angle 2°–6°, with an average inclination of 4°, classified as low-sulfur fat coal. The coal seam has a complex structure, typically containing two layers of gangue. The immediate roof consists of mudstone and sandy mudstone, with a uniform thickness of 3.0 m; the main roof is composed of fine sandstone (K8), with a uniform thickness of 7.2 m; the immediate floor is mudstone, with a uniform thickness of 4.5 m; and the main floor is composed of medium sandstone (K5), with a uniform thickness of 6.5 m. The 2-208 panel adopts a fully mechanized mining method with a single-pass full-height longwall retreating technique, the advancing interval is 0.8 m, while the coal wall caving occurred during the mining process, requiring effective measures to control it.

Based on the parameters required for water injection construction in the 2-208 large mining height panel of the Xinzhi Mine and the coal quality conditions, the final water injection pressure is determined to be 2–3 MPa, with a single-hole water

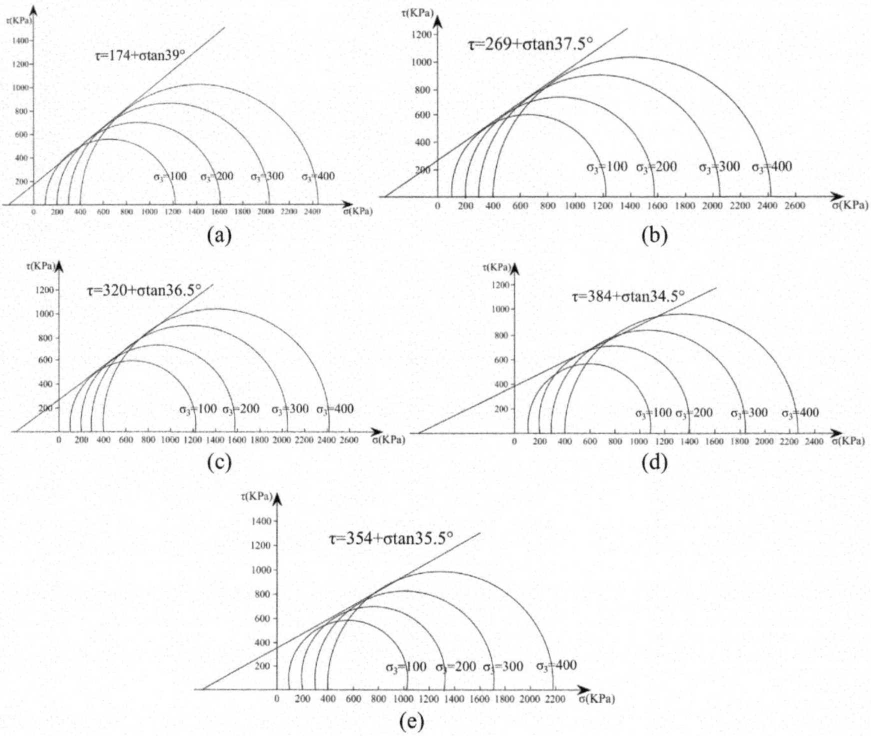

Fig. 3.23 Mohr circles and strength curves under different moisture content conditions: **a** 3.48%; **b** 8.73%; **c** 12.27%; **d** 16.84%; **e** 18.32%

Fig. 3.24 Cohesive strength change

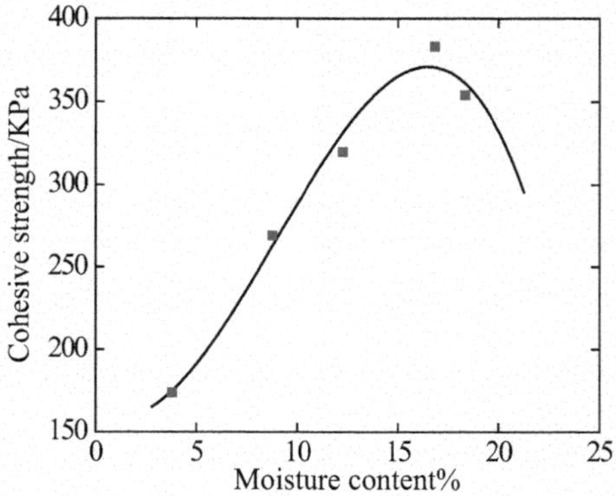

injection volume of 89.4 m^3 and a water injection ratio of 0.9 m^3 per hour, requiring a total of 85 h for water injection construction. When the distance between the panel and the water injection borehole is in the range of 8–20 m, water injection should be stopped. Considering that the panel advances by 2.5 m per day, the minimum distance

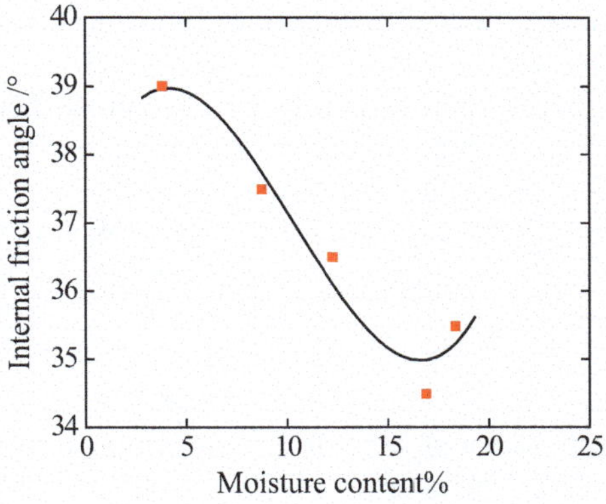

Fig. 3.25 Internal friction angle change

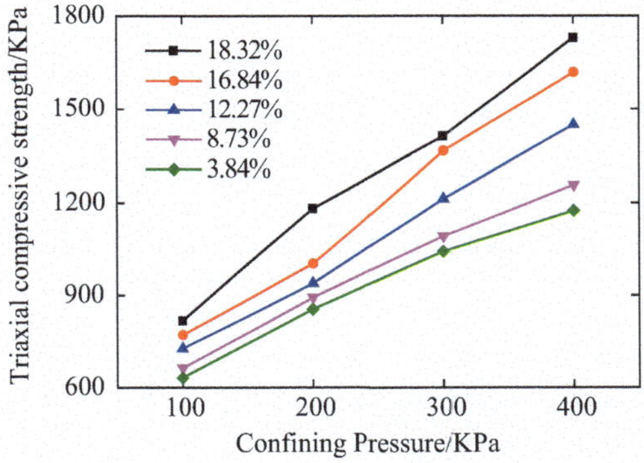

Fig. 3.26 Variation of triaxial compressive strength with confining pressure

between the panel and the water injection borehole should be at least 17–29 m. To facilitate on-site construction, the specific location for advanced water injection is determined to be 25 m. After the implementation of this technology, the frequency of coal wall caving reduced by about 80%, effectively controlling the occurrence of coal wall caving and ensuring the safe and efficient mining of the 2-208 panel.

4. Application at Xishan Coal and Electricity, Duerping Mine [12]

The main seam of 72909 panel at Duerping Mine is 2# coal seam. The coal seam thickness ranges from 1.90 to 3.50 m, with an average thickness of 2.93 m and an average dip angle of 6°. The coal seam has a complex structure and contains an unstable gangue layer with a thickness ranging from 0 to 0.45 m and an average thickness of 0.05 m. The immediate roof consists of fine sandstone with an average thickness of 1.94 m, while the main roof is composed of K6 fine sandstone with an average thickness of 3.93 m. The immediate floor is composed of sandy mudstone with an average thickness of 0.6 m, and the main floor consists of K5 coarse sandstone with an average thickness of 4.67 m. The 72909 panel adopts comprehensive mechanized coal mining technology and has exposed 3 collapsed pillars and 2 faults during exploration and excavation. Coal rib spalling and roof caving occurred during the mining process, significantly impacting the safety of the panel, necessitating effective measures to ensure the stability of the coal rib.

After considering the site conditions, the depth of the injection holes for grouting was determined to be 6 m, with hole positions located 2.1–2.6 m above the panel floor. The spacing between injection holes was set at 5.3 m. The grouting pressure ranged 5–8.2 MPa, with a single-hole grouting flow ratio of 3.45 m^3 and a minimum grouting time of 41.4 min per hole. Following the implementation of shallow-hole grouting technology, on-site measurements revealed an increase in the moisture content of the coal body from the original 1.16 to 4.53%, reaching the ideal moisture content for the coal body. Furthermore, the frequency of coal rib spalling decreased by approximately 80% during the mining process. Compared to the pre-grouting depth of 1 m, the average depth of coal rib spalling was significantly reduced to 0.4 m after grouting, effectively controlling the occurrence of coal rib spalling and greatly improving the stability of the coal wall.

3.3.2.3 Coal Wall Grouting

Coal rib grouting is an effective technique for improving the stability of coal wall. By increasing the working resistance and stiffness of the supports, it can effectively alleviate coal wall pressure and reduce the occurrence of coal wall damage. Currently, grouting reinforcement materials are mainly divided into three categories: cement-based grouting materials, chemical grouting materials, and cement-chemical composite grouting materials. Cement-based grouting materials have the advantages of high strength, good durability, no pollution, and low cost, making them suitable for loose and clearly separated broken coal and rock bodies with less stringent requirements on aging properties. Chemical grouting materials are mostly high polymer materials, characterized by low viscosity, good injectability, strong permeability, and fast solidification, and are widely used in projects with tight schedules. Common coal wall grouting schematics can be seen in Fig. 3.27.

Grouting materials are important factors affecting the effectiveness of grouting, and grouting parameters mainly include grouting pressure, grouting volume, and grouting time. Grouting pressure is the driving force for the flow, diffusion, and penetration of grout in the surrounding rock, depending on factors such as the permeability of the rock, the required reinforcement range, and the properties of the grout.

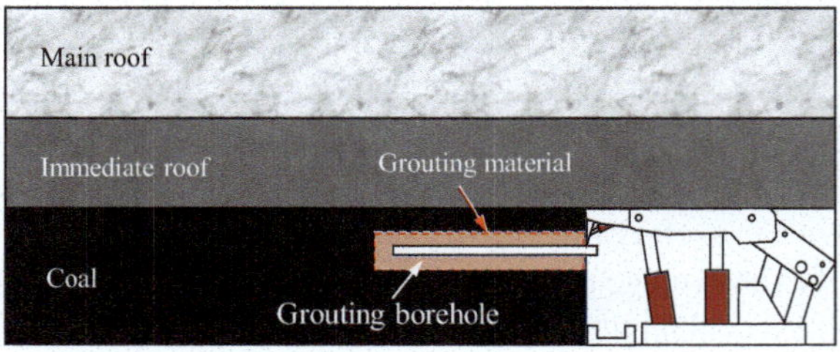

Fig. 3.27 Schematic diagram of coal wall grouting

Grouting volume is closely related to the lithology and fragmentation degree of the surrounding rock, the range of grouting pressure and time, etc. Through the rational setting of testing experiments, the grouting performance can be effectively improved.

1. Engineering application at Zhaogu No.1 Mine

The coal seam mined at the 11071 panel of Zhaogu No.1 Mine is the 2-1 coal seam, which is stable and has an average thickness of 6.09 m. The coal seam has a dip angle of 0.2°–3.36°, with an average of 1.8°, and has a simple structure. The 2-1 coal seam is located at the bottom of the Shanxi Formation, and the entire area is exploitable. It belongs to a relatively stable type of thick coal seam with developed jointing and bedding, and locally contains partings. The coal seam mainly consists of lumps of coal. However, due to the large mining height and the development of longitudinal fractures, coal wall collapse occurs, which adversely affects production. To address this issue, a combined reinforcement method of advance reinforcement of upper and lower roadways along with in-face combined anchor cable grouting reinforcement is adopted. The layout and grouting plan of the grouting boreholes are shown in Fig. 3.28.

Along the direction of the coal wall, grouting holes are drilled towards the roof, with openings made at a distance of 0.5–0.8 m from the roof, and the spacing between holes is 1.5 m. Grouting is first carried out in odd-numbered holes, such as the first, third, fifth… with an upward inclination of 10°–15°, a diameter of 28 mm, a depth of 3 m, and a final grouting pressure of 3 MPa. Subsequently, even-numbered holes such as the second, fourth, sixth… are drilled for grouting, with an upward inclination of

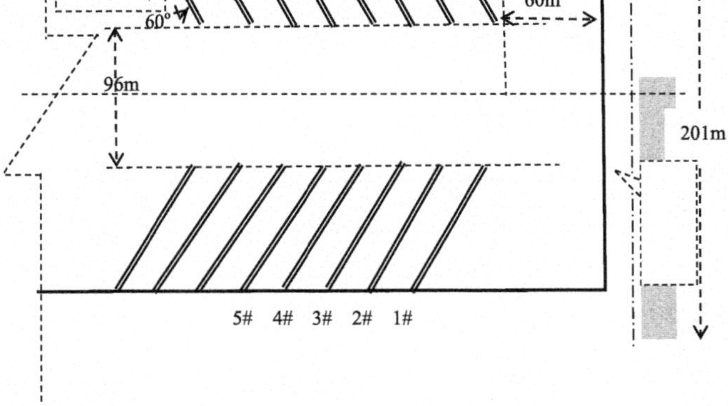

Fig. 3.28 Layout plan of grouting drillholes in headgate and tailgate of the panel

15°–20°, a diameter of 42 mm, a depth of 6 m, and a final grouting pressure of 6–8 MPa. Arrangement of grouting holes is shown in Fig. 3.29.

The high-strength composite material grouting anchor cable is composed of three high-strength composite material anchor cable strands combined with components such as aluminum-plastic grouting pipes, guide caps, and grouting joints. It features high tensile strength, easy cutting without damaging the cutting teeth of the coal shearer, flame retardancy, antistatic properties, safe use, and spark-free cutting. It is suitable for use in high-gas and high-dust mines. Additionally, it is lightweight, chemically stable, resistant to corrosion, and unaffected by acidic or alkaline environments. Its length can be adjusted according to engineering design, and the anchor cables can be cut and assembled on-site. They can be bent into a disc with a radius of approximately 1 m, facilitating transportation and on-site construction. The technical parameters are listed in Table 3.4.

2. Coal wall grouting through fault in Shengtai Coal Mine [13]

Grouting technology plays a great role in crossing faults and controlling leakage in front of the support. When the 15203 panel of Shengtai Coal Mine passed through the F7 fault, due to the high working resistance of the support and the large deformation of the surrounding rock in the dynamic pressure area, the triangular coal pillar at the end of the tail end collapsed seriously, and it was difficult to move the support at the end and head. After technical demonstration, it was decided to adopt flexible grouting reinforcement technology for the coal wall on the panel and deep hole grouting reinforcement technology for the coal wall on the side of the roadway. The grouting drilling layout is shown in Fig. 3.30.

After the grouting of the coal wall, UCS of the coal rock has been improved, the instability of the coal wall in the later mining process of the panel has been

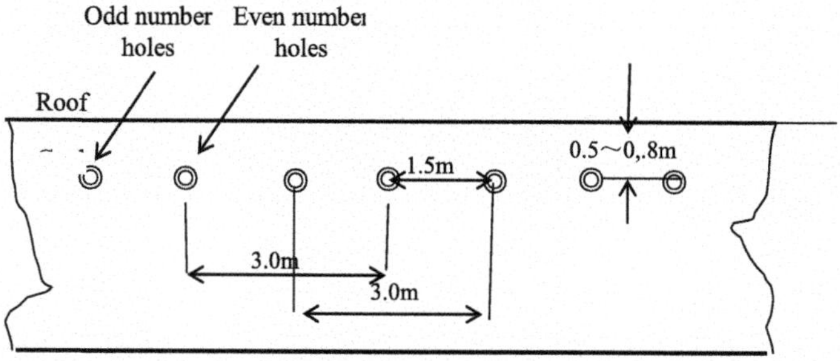

Fig. 3.29 Layout plan of grouting holes for reinforcement on the panel

Table 3.4 Technical parameters

Anchor rod outer diameter/mm	Bending radius/mm	Tensile force/ KN	Tensile strength/MPa	Elongation at break/%	Diameter of the anchor bundle/mm
20	1000	105	1200	2.5	6

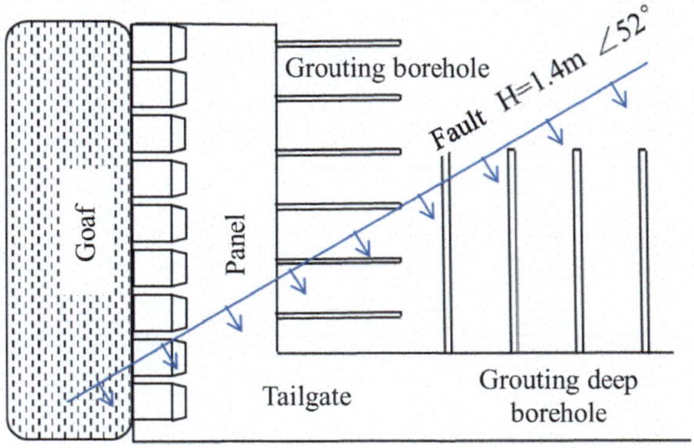

Fig. 3.30 Plan view of grouting boreholes layout on the panel

significantly improved, and the collapse of the tail triangular coal pillar has been significantly controlled. After the surrounding rock of the coal wall in the fault area is reinforced with grouting, the pressure on the support is significantly reduced, and the mining speed is significantly increased.

3. Coal wall grouting over collapsed pillars in Xiadian Coal Mine [14]

The 3119 panel of Xiadian Coal Mine is located under the 3 mining area. The coal seam thickness is 6.5–7.1 m. According to the comprehensive geophysical prospecting results of the panel, there are multiple geological structures in the mining space. Among them, the JDX16 collapse column is 600 m away from the first set-up room, with a length of 60 m and a width of 50 m. It is basically located in the middle area of the panel. During the mining process, it will serious coal wall rib spalling problems occur. Therefore, when the panel passes through the collapse column, the coal mass in the collapse column is reinforced by grouting. The grouting scheme is shown in Fig. 3.31.

According to the stress evolution analysis of the surrounding rock, the cracks in the coal and rock mass are relatively developed within the range of 40–20 m from the panel to the collapse column, which is conducive to the diffusion of slurry. Grouting is carried out currently, and the maximum pump pressure for borehole grouting is set to 10–13 MPa, ensuring that the slurry diffuses and penetrates the coal and rock mass around the entire collapse column. A single row of drilling holes is arranged in the headgate, and horizontal grouting holes and downward-inclined drilling holes are set by numbering to respectively reinforce the surrounding rock above the support and the coal wall within the machine mining height; "Sanhua drilling holes" are arranged in the tailgate, all are constructed at an elevation angle. After the grouting of the coal wall, the number of coal wall fragments decreased, the roof of the panel was relatively complete, and there was no roof collapse, which effectively ensured that the panel could pass through the collapse column normally.

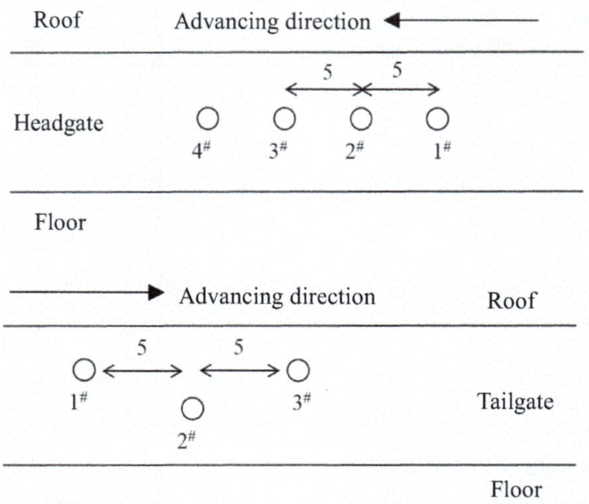

Fig. 3.31 Grouting drilling layout diagram

3.3.2.4 Coal Wall Reinforcement

Reinforcing the coal wall with anchor rods is a common technology. However, because the shearer cuts the coal wall, metal anchor rods cannot be used in the coal wall. Instead, fiberglass anchor rods, bamboo anchor rods or wooden anchor rods are commonly used. However, since the deformation of these anchors often cannot be coordinated with the coal wall, the reinforcement effect is not ideal. To solve this problem, a flexible reinforcement technology for coal walls has been developed in recent years [15], as shown in Fig. 3.32.

The flexible reinforcement technology of the coal wall refers to the "brown rope and grouting" full-length anchoring technology. The brown rope used in reinforcement technology itself has a certain support stiffness, however, in view of its flexible characteristics, materials with properties like brown rope are defined as flexible

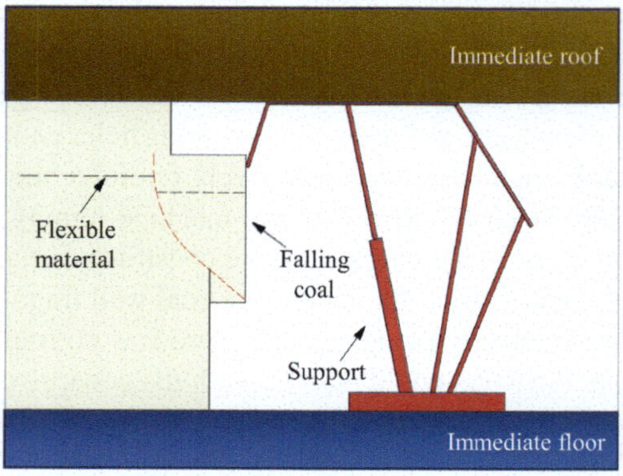

Fig. 3.32 Coal wall flexible reinforcement technology

support materials. The principle of the "brown rope and grouting" flexible reinforcement technology for the coal wall is to use polymer materials such as chemical slurry to integrate the brown rope with the coal body of the coal wall, thereby achieving the full-length anchorage of the coal wall of the panel. In addition, the deformation of the coal body before the coal wall of the panel is large, the larger elongation of the brown rope can adapt to this, and can inhibit the further fragmentation of the damaged large coal body. Whether this new type of coal wall flexible reinforcement technology of "brown rope and grouting" can better control the damage of coal wall fragments requires the tensile strength and elongation of the flexible brown rope, as well as the control of flexible materials and slurry materials. The mechanism of action and the coordinated deformation principle of "brown rope and grouting" and the coal wall of the panel were systematically studied to provide a basis for the prevention and control of coal wall flakes on the panel.

1. Engineering application in Zhaogu No. 2 Mine

To control the occurrence of coal wall damage and roof collapse in the large mining height panel of Zhaogu No. 2 Mine and ensure the safety of the stope working space, the "brown rope and grouting" reinforcement technology was used to strengthen the coal wall of the 6.3 m large mining height stope. Porain slurry material (materials A and B are mixed at a ratio of 1:1) is used for grouting reinforcement. The grouting pipe is not taken out during the mining process. The grouting pipe used is a plastic pipe with small deformation and low shear strength. It does not affect normal production, as shown in Fig. 3.33.

By optimizing the brown rope, borehole diameter and chemical slurry dosage, the coal wall reinforcement process and parameters of the complete section, the supporting pressure section and the geological structure section were finally established, as shown in Table 3.5.

The large mining height coal wall of Zhaogu No. 2 Mine adopts "brown rope and grouting" for flexible reinforcement and control, combined with a series of stope roof management policies such as giving full play to the efficiency of the use of support panels, timely removal of the support and improving the support strength of the support. Through these measures, the damage to the coal wall of the panel has been well controlled, the time for dealing with flakes and roof collapse has been significantly shortened, the startup ratio has been greatly improved, and the daily output of the panel has exceeded 10,000 tons.

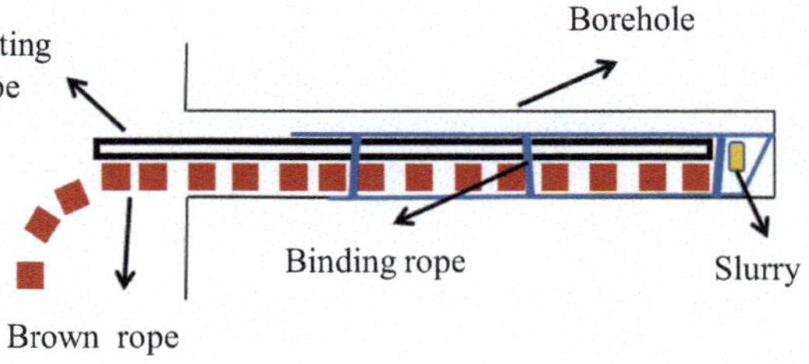

Fig. 3.33 "Brown rope and grouting" reinforcement process

Table 3.5 Coal wall reinforcement parameters

Location	Drilling diameter/cm	Drilling depth/m	Drilling angle/°	Drilling row spacing/m	Distance from roof/m	Brown rope diameter/cm
Complete section	42	8	13	4	2	25
Mining pressure influence section	28	6	50 ± 2	2	4.5	15
Construction section	76	30	3–4	4	3.5	30

2. Ruilong mine panel application

According to the geology and mining conditions of the undulating coal seam in Ruilong Mine, the "brown rope and grouting" technology is used to strengthen the coal wall to control the coal wall rib spalling and the roof falling off of the panel. The reinforcement plan is shown in the Table 3.6.

Use a drill bit with a 42 mm bolt machine to drill holes into the coal wall, keeping the drilling angle at 90° with the direction of the coal wall, and the distance between the drilling hole and the floor is about 2/3 of the mining height. Since the width of the plastic zone in front of the coal wall is about 5 m and the advancement distance of the two production teams on the panel is 3 m, the drilling depth is determined to be 5 m. The diameter of the brown rope selected is 15 mm. Use tie wire to tie the brown rope to the grouting pipe (to facilitate cutting by the shearer, the grouting pipe is made of plastic), and then use a bolt machine to drive it into the borehole, and finally use a grouting pump to evenly mix the chemical slurry (the two materials are mixed one to one) and inject it into the plastic pipe tied with brown rope. The amount of slurry required for each drilling hole is about 25–35 barrels. The deformation and damage of the coal wall before and after "brown rope and grouting" reinforcement are shown in Fig. 3.34.

When the panel of Ruilong Mine was advanced upward, the coal wall of the panel was flexibly reinforced with "brown rope and grouting". Furthermore, the on-site stope roof control measures were further strengthened, which effectively prevented and cured the coal wall fragments of the panel. Side and end panel caving, thus

Table 3.6 "Brown rope and grouting" reinforcement scheme and parameters for the panel

Location	Drilling row spacing/m	Drilling diameter/cm	Drilling depth/m	Drilling angle/°	Distance from roof/m	Brown rope diameter/cm
Coal wall seriously damaged	42	5	90	3	2	15

(a) (b)

Fig. 3.34 Before and after flexible reinforcement of the coal wall at the panel: **a** before flexible reinforcement; **b** after flexible reinforcement

further ensuring high-safety and high-efficiency mining of the inclined top-caving panel.

3. Application of 8101 large mining height panel in Wangzhuang Coal Mine

Due to the soft coal quality and large mining thickness at the 8101 large mining height panel of Wangzhuang Coal Mine, the problem of coal wall rib spalling on the panel is serious. Especially when the panel advances to the vicinity of the F286 fault, the rib spalling and roof collapse accident cannot be effectively solved, and the production of the panel was seriously affected, resulting in the suspension of production at the panel for a time. Corresponding measures were taken to prevent and control coal wall rib spalling, and the stability of the coal wall at the panel was significantly improved. The geological structure of the 8101 soft coal seam panel with large mining height is complex, especially in the section affected by the geological structure, the surrounding rock fragments of the panel need to be grouted and reinforced with Malisan materials (a polyimine gum grease material) to reinforce the coal wall in front of the panel. The slurry solidified body can improve the mechanical properties of coal and form a load-bearing structure and network skeleton structure. After adopting coal wall chipping prevention and control measures, Wangzhuang Coal Mine improved the stability of the coal wall at the panel. The number of coal wall rib spalling, depth, and height were all significantly reduced. Accidents caused by large pieces of coal wall, the rib spalling depth being greater than 1.2 m, and the rib spalling height being greater than 1.7 m, at the panel are no longer occur. The damage to the coal wall at the panel has been effectively controlled. The start-up ratio of the shearer has been increased, ensuring the safe and efficient advancement of the panel. The production capacity of the panel has been restored and improved.

3.4 Large-Section Roadway Support Technology

Due to the substantial increase in the mining intensity and output of large-scale mining, the required equipment continues to become larger, and the roadway cross-section size gradually increases. The maintenance of large-section roadways has become a difficult problem restricting the safe and efficient mining of coal mines.

3.4.1 Principle of Roadway Anchor Support

Anchor support promotes the transformation of surrounding rock from a load body into a bearing body. Although the mechanism of action of anchor bolts is somewhat different under different geological conditions, they all reinforce the surrounding rock inside the surrounding rock around the roadway to form a surrounding rock bearing body, which is beneficial to the stability of the surrounding rock. Traditional anchor support theories are based on certain hypotheses and explain the mechanism of anchor support from different angles and conditions. In recent years, theoretical research on anchor support has further developed. The theory of strengthening the surrounding rock of roadway using anchor support has been proposed, and anchoring technology has been studied as a system as a whole, further revealing the essence of anchor support [16].

3.4.1.1 Suspension Theory

Suspension theory holds that the function of anchor support is to suspend the weaker rock stratum of the roadway roof on the upper stable rock stratum to enhance the stability of the weaker rock stratum. For the layered rock mass often encountered in mining roadways, the suspension effect of the anchor rod is shown in Fig. 3.35a. If the surrounding rock in the shallow part of the roadway is soft and broken, and there is a loose and cracked area in the roof, the suspension function of the anchor rod is to anchor this part of the rock mass that is prone to falling to the deep rock layer that is not loose, as shown in Fig. 3.35b.

3.4.1.2 Composite Beam Theory

If there are several layers in the roof rock layer, the combined beam theory believes that on the one hand, the role of the anchor rod is to provide anchoring force to increase the friction between the rock layers, prevent the rock layer from continuing to slide along the layer, and avoid the phenomenon of separation; on the other hand, the anchor rod body can increase the shear stiffness between rock layers and prevent horizontal dislocation between rock layers, thereby locking several thin rock layers

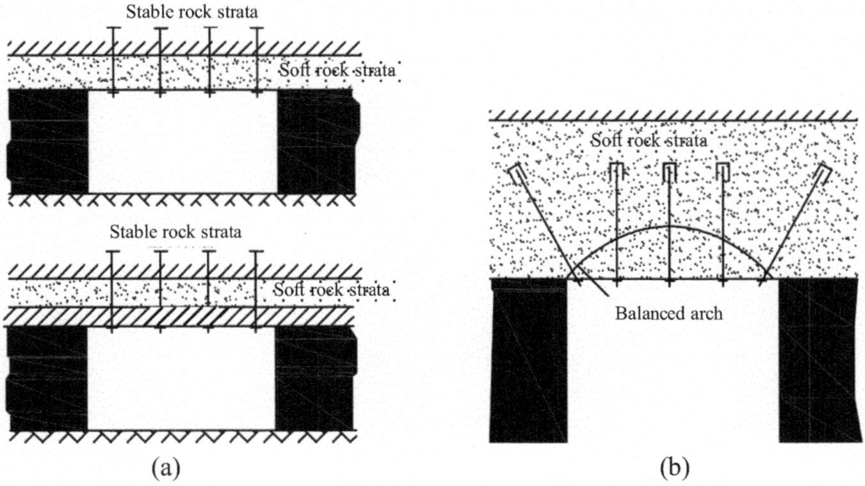

Fig. 3.35 Anchor support and suspension function: **a** hard roof anchor, **b** weak roof anchor

Fig. 3.36 Layered roof anchor composite beam

within the anchorage range of the roadway roof into one thicker rock layer, as shown in Fig. 3.36.

3.4.1.3 Combined Arch (Compression Arch) Theory

The combined arch theory holds that prestressed anchor rods are installed in the rupture zone of the surrounding rock of an arched roadway to form a conical distributed compressive stress zone from both ends of the rod body. If the distance between anchor rods is small enough, the compressive stress cone formed by each anchor rod will be interleaved, forming a uniform compression zone in the rock mass, that is, a compression arch. The rocks in the compression arch are compressed in both radial and tangential directions and are in a 3D stress state. The strength of the surrounding rock is improved and the supporting capacity is increased accordingly, as shown in Fig. 3.37.

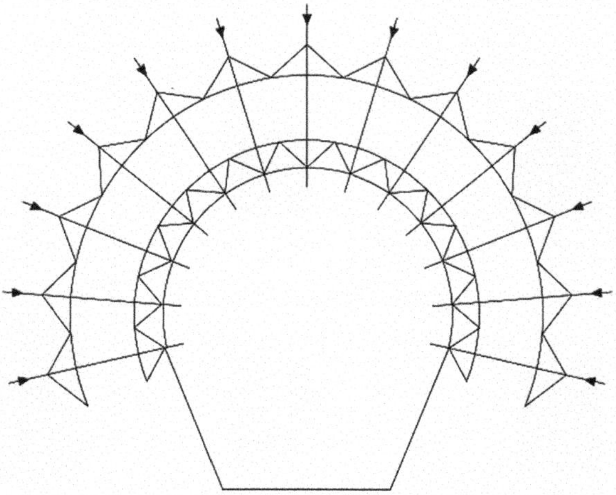

Fig. 3.37 Anchor rod combined arch (compression arch) theory

3.4.1.4 Maximum Horizontal Stress Theory

This theory holds that the horizontal stress of the mine rock formation is usually greater than the vertical stress, and the stability of the roadway roof and floor is mainly affected by horizontal stress. When the stress is redistributed after the excavation of the mining roadway with prominent layered characteristics of the surrounding rock, the vertical stress will transfer toward the two sides, and the horizontal stress is transferred to the roof and floor. The influence of the vertical stress mainly appears on the two sides and leads to the destruction of the two sides, while the influence of the horizontal stress mainly appears on the roof and floor rock layers, as shown in Fig. 3.38. The function of the anchor is to restrain the expansion of the rock formation along the axial direction of the anchor and to restrain the shear dislocation of the rock formation in the direction perpendicular to the axial direction of the anchor.

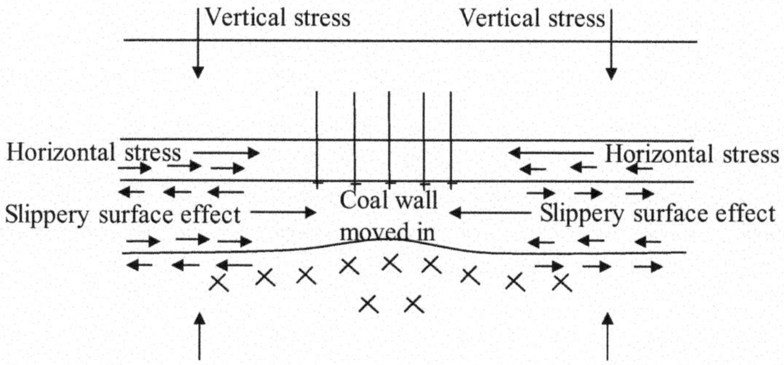

Fig. 3.38 Principle of maximum horizontal stress

3.4.1.5 Surrounding Rock Strength Strengthening Theory

The theory of strengthening the surrounding rock strength is proposed based on the characteristics of the surrounding rock in soft-rock coal roadways. Through similar material simulation tests and theoretical analysis before and after roadway surrounding rock anchoring, the key points of the surrounding rock strength strengthening theory are as follows:

(1) The essence of roadway anchor support is that the anchor rod and the rock mass in the anchoring area interact to form a unified load-bearing structure.
(2) Roadway anchor support can increase the mechanical parameters (E, c, φ) of the anchor and improve the mechanical properties of the anchored rock mass.
(3) The surrounding rock of the roadway has the broken zone, plastic zone, and elastic zone. The rock mass in the anchor bolt anchoring zone is in the broken zone or in the above 2–3 zones. The rock strength in the corresponding anchoring zone is at post-peak strength or residual strength. Anchor support makes the strength of surrounding rock of the roadway, especially the surrounding rock in the post-peak area being strengthened, increasing the peak strength and residual strength.
(4) Coal roadway anchor support can change the stress state of the surrounding rock and increase the confining pressure, thereby improving the bearing capacity of the surrounding rock.
(5) After the strength of the anchorage of the surrounding rock of the roadway is increased, the scope of the broken zone and plastic zone around the roadway and the surface displacement of the roadway can be reduced, and the development of the broken zone and plastic zone of the surrounding rock can be controlled, which is beneficial to maintaining the stability of the surrounding rock of the roadway.

3.4.2 Roadway Support Technology

3.4.2.1 Anchor Support

Anchors are rod-shaped structures anchored in coal and rock bodies to maintain the stability of roadway surrounding rocks. Anchor rods are used as the main component of the support system for roadway surrounding rocks to form an anchor rod support system. Its essence is to make poles from metal parts, wood parts, polymer parts or other materials, and drill them into pre-drilled holes in the surrounding rock. It uses the special structure of its head, pole body and tail plate, or relies on the bonding effect, combining the surrounding rock and stable rock mass to produce suspension effects, composite beam effects, and reinforcement effects to achieve the purpose of support [17].

1. Mechanical anchor

Mechanical anchors are anchored in the borehole through a mechanical locking device at the root of the anchor, including expansion shell anchors and slotted wedge anchors.

(1) Expansion shell anchor. The expansion shell anchor consists of a solid rod with screws at both ends, an expansion shell at the root of the anchor, a tray and a nut. The expansion shell in Fig. 3.39 is just one of several types currently in use. The expansion shell is an anchoring device composed of wedges and 2–4 shell leaves. The wedge is screwed to the screw at the root of the anchor rod. When the screw is rotated, the wedge is pulled toward the drilling direction. The blades of the housing open and press against the drilling wall, establishing a contact force between the housing and the hole wall. The anchoring force at the shell-rock interface is proportional to the contact force. Anchors can easily be firmly anchored in hard rock as long as a high enough torque is applied to the anchor. However, in soft rock, when the torque is too high, there is a risk that the hole wall rock will be crushed by the expansion shell. If the rock is crushed, the anchoring force of the anchor will be significantly reduced. Therefore, the torque must be appropriate when installing expansion shell anchor in soft rock. After the expansion shell is installed, tighten the nut and apply pre-tightening force to the anchor rod. The relationship between the pre-tightening force P and the applied torque T is shown in Eq. (3.13). The failure of expansion shell anchors is rarely due to breakage of the rod body, usually due to failure of the screw and pallet, or due to loose anchorage slippage at the expansion shell.

$$P = BT \tag{3.13}$$

where B is a constant related to the anchor diameter and installation parameters. According to experience, the anchor rod with a diameter of 16 mm has $B = 164$; the anchor rod with a diameter of 19 mm has $B = 131$. The anchor pre-tightening force is usually 50–60 kN, and the pre-tightening force should not exceed 60% of the anchor yield load.

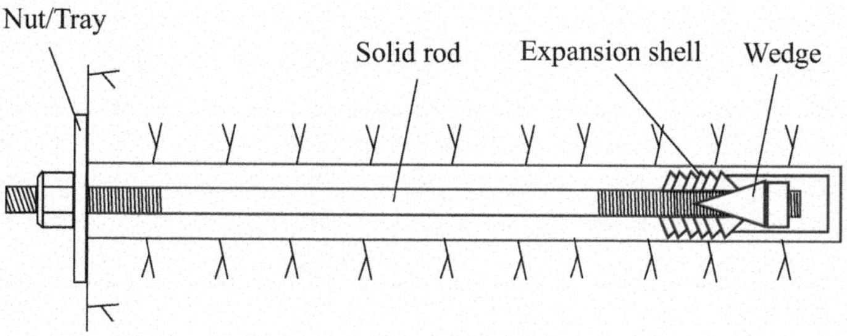

Fig. 3.39 Expansion shell anchor

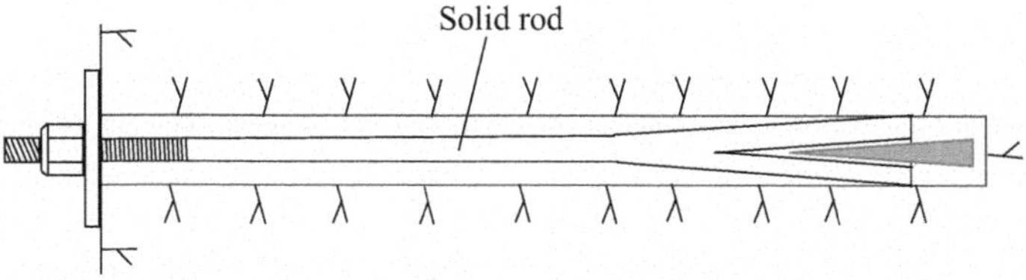

Fig. 3.40 Slotted wedge anchor

(2) Slotted wedge anchor. The slotted wedge anchor has a longitudinal slot at the root, and a wedge is inserted into the slot (Fig. 3.40). During installation, the anchor rod is pushed into the drill hole. The wedge hits the bottom of the hole and then hits the anchor rod head. The wedge is squeezed into the groove to open the bifurcation at the root of the anchor rod, and the bifurcation of the rod body is squeezed into the hole wall. Then install the tray and tighten the nuts to complete the installation. The slotted wedge anchor is fixed in the borehole through the friction between the root of the rod and the rock. The pre-tightening force of this type of anchor is not large, and the maximum pre-tightening force of a steel anchor with a diameter of 20 mm is 30–50 kN. To make the body to contact the bottom of the drill hole, the drillhole length must be less than the anchor length. Due to the low pre-tightening force and unreliability of the anchor, slotted wedge anchors are not widely used at present.

2. Grouting anchor

   (1) Full-length grouting steel anchor. Full-length grouting steel anchor refers to an anchor that is filled with cement slurry or resin so that the entire length of the anchor is wrapped in the grouting body. The ribs on the surface of the anchor rod mesh with the solidified grouting body to fix the anchor rod in the rock body, as shown in Fig. 3.41. Pre-tightening force cannot be applied immediately after the cement grouting anchor is installed. If pre-tightening is required, the pre-tightening force can be applied after the cement slurry has hardened (such as one week after grouting). Applying pre-tightening force too early will cause the bonding between the anchor rod and the grouting body to be damaged.

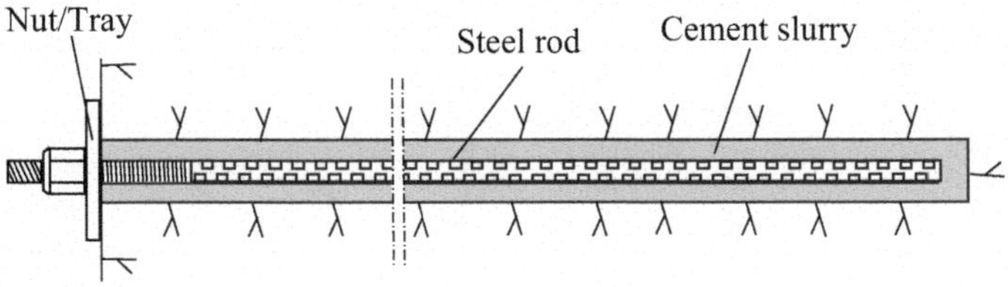

Fig. 3.41 Full-length grouting steel anchor

(2) Full-length grouting rebar anchor. Rebar anchors are similar to steel anchors, except that their surface has coarse-pitch threads, as shown in Fig. 3.42. Rebar anchors are hot-rolled to form coarse threads (pitch 10 mm) along their length. There is no need to process additional threads at the head of the rebar anchor rod. The nut is directly screwed onto the coarse thread of the rod body. The strength of rebar anchors is the same along their entire length. Anchors can be installed using either resin or cement grout.
(3) End grouting anchor. The end grouting anchor uses resin to bond a section of the anchor root into the drill hole, as shown in Fig. 3.43. The anchor rod body can be made of steel bars or rebar, and pre-tightening force can be applied immediately after the end grouting anchor is installed. In terms of inhibiting rock mass displacement, the role of end grouting anchors is similar to that of expansion shell anchors. Since the end grouting anchor is bonded with resin grouting, its anchoring is more reliable than the expansion shell anchor. The technical parameters of end grouting anchors are the same as those of steel anchors and rebar anchors.
(4) Full-length cement grouting groove wedge anchor. The full-length cement grouting groove wedge anchor is made of steel bars with wedges at the far end of the anchor. As shown in Fig. 3.44, the anchor length usually does not exceed 3 m. During installation, first pump the cement slurry into the drill hole, and then insert the anchor rod. When the anchor rod touches the bottom of the hole, an impact thrust is exerted on the anchor rod head, and the wedge is squeezed into the rod end groove and splits. The end of the rod is squeezed against the wall of the drilled hole. Then tighten the nut and pallet and apply pre-tightening force. The maximum preload force for a steel anchor with a diameter of 20 mm is 30–50 kN. After the grouting body has solidified, the full-length cement grouting slot wedge anchor performs the same function as the traditional full-length grouting steel anchor. Wedges

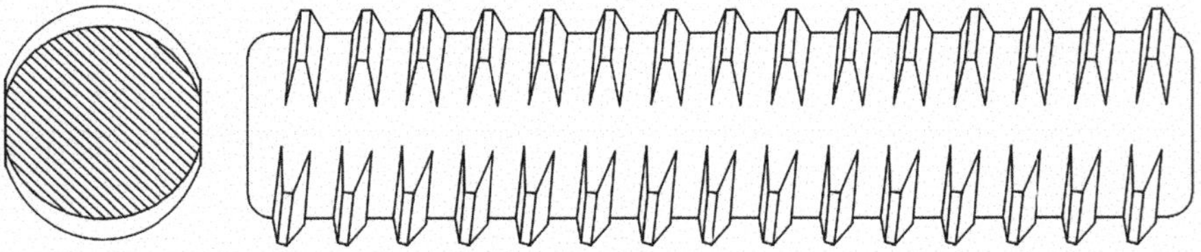

Fig. 3.42 Rebar anchor

Fig. 3.43 End grouting anchor

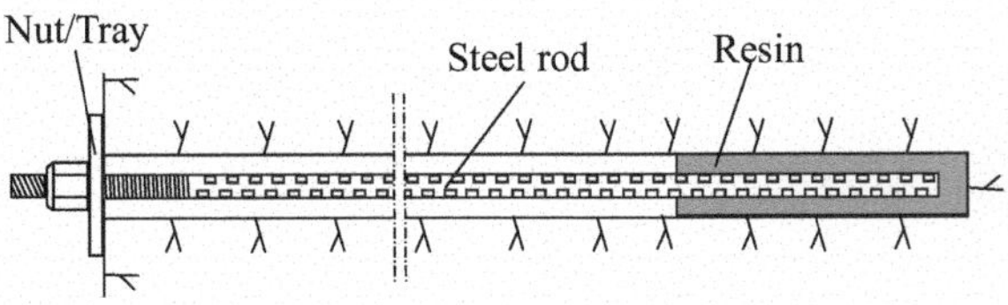

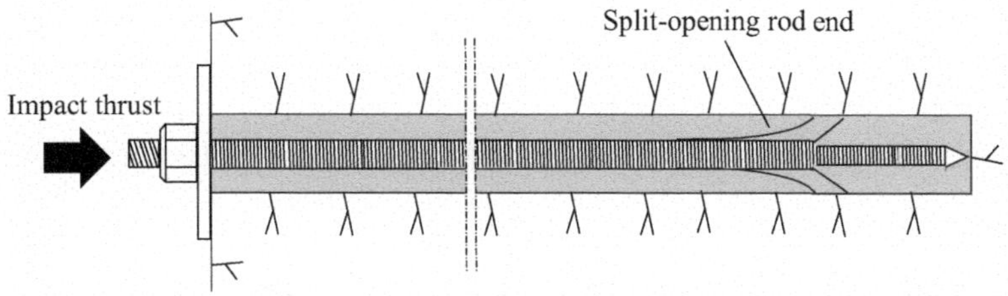

Fig. 3.44 Full-length cement grouting groove wedge anchor

are added to the end of the rod to apply pre-tightening force to the anchor rod.

3. Self-drilling anchor

The general installation process of cement grouting anchors is to drill holes first, then pump cement slurry into the holes, and finally push the anchors. When the coal and rock mass is extremely fragmented, hole wall collapse may occur during or after drilling, making it difficult or completely impossible to inject slurry into the hole, or to insert anchor rods into the hole. Self-drilling anchors can be used in this case. The self-drilling anchor consists of a hollow rod body and a drill bit at the far end of the anchor rod. The cylindrical surface of the anchor rod is rolled into a coarse thread, which is not only used to connect the bolt of the nut, but also to enhance the bond between the rod body and the cement slurry. During installation, the rod body is drilled into the formation with the drill bit. When drilling is completed, the rod body remains in the hole, and then cement slurry is injected into the hole through the center hole of the rod body. When the cement slurry flows out from the annulus between the anchor rod and the drill hole, the grouting stop. When drilling, the anchor rod is the drill pipe, and the anchor rod must be strong enough to withstand the drilling torque. Several anchors can be connected through joints to form extended anchors.

3.4.2.2 Anchor Cable Support

Anchor cables are divided into single strands, double strands and multiple strands. The anchor cables are usually bonded in the borehole with cement slurry, as shown in Fig. 3.45. Some shorter lengths of anchor cable can be installed using resin adhesive. An anchor rope can be composed of multiple strands of steel wire. Each strand of steel wire is composed of multiple steel wires wound around a central steel wire. The cross-section of steel wire is generally round and is cold drawn through a mold. A strand of steel strand usually has 7, 19 or 37 steel wires, and several strands of steel strands are twisted together to form an anchor cable. Figure 3.46 shows the cross sections of the 1 × 7 type anchor cable and the 3 × 7 type anchor cable. The first number marked on the anchor cable is the number of steel strands in the anchor cable, and the second number is the number of steel wires in each strand. For example, a 1 × 7 type anchor cable means a single-strand steel strand anchor cable, with each strand having 7 steel wires. The more wires in each strand or cable, the greater the flexibility. The 1 × 7 type anchor cable is harder and less flexible than the 3 × 7 type anchor cable.

Anchor cables are made from different grades of stainless steel or carbon steel. Stainless steel anchor cables have great corrosion resistance, however, the cost is higher. Common anchor cables are usually made from less expensive carbon steel. Anchor cables made of galvanized carbon steel have a certain degree of corrosion resistance and have a relatively good anti-corrosion effect. If necessary, corrosion-resistant materials can be coated on the surface of the anchor cable to increase corrosion resistance. Anchor cables made of steel strands without special treatment are called ordinary anchor cables. To improve the anchoring force of the anchor cable in the mortar, the diameter of the steel strand can be enlarged at certain intervals or fixings can be installed on the steel strand.

The advantages of the anchor cable are that the length is adjustable, the load-bearing capacity is high, and it can be wound on a reel for easy transportation. Since the anchor cable wire is made by cold drawing, the disadvantage of the anchor cable is its poor deformation ability. The maximum elongation of the anchor cable is 3–5%, which is much lower than the elongation of ordinary hot-rolled steel anchors. To improve the deformation ability of the anchor cable, the middle section of the anchor cable is often wrapped with a plastic sleeve to separate the anchor cable from the grouting body, seeing Fig. 3.47. Using this debonding technology, the deformation capacity of the anchor cable is significantly improved.

The pallet is connected to the anchor cable through the cylinder wedge locking piece, seeing Fig. 3.47. When installing the pallet, first loosen the cylinder and

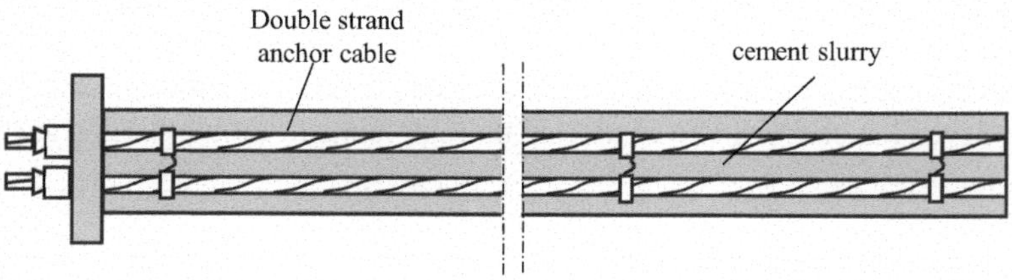

Fig. 3.45 Anchor cable

Fig. 3.46 Cross-section of **a** 1 × 7 type anchor cable and **b** 3 × 7 type anchor cable

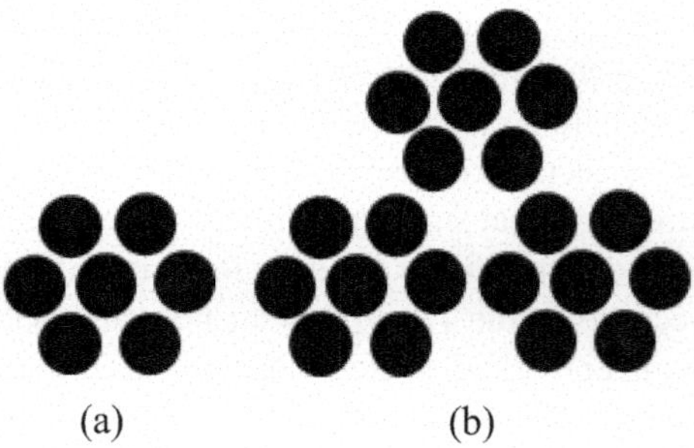

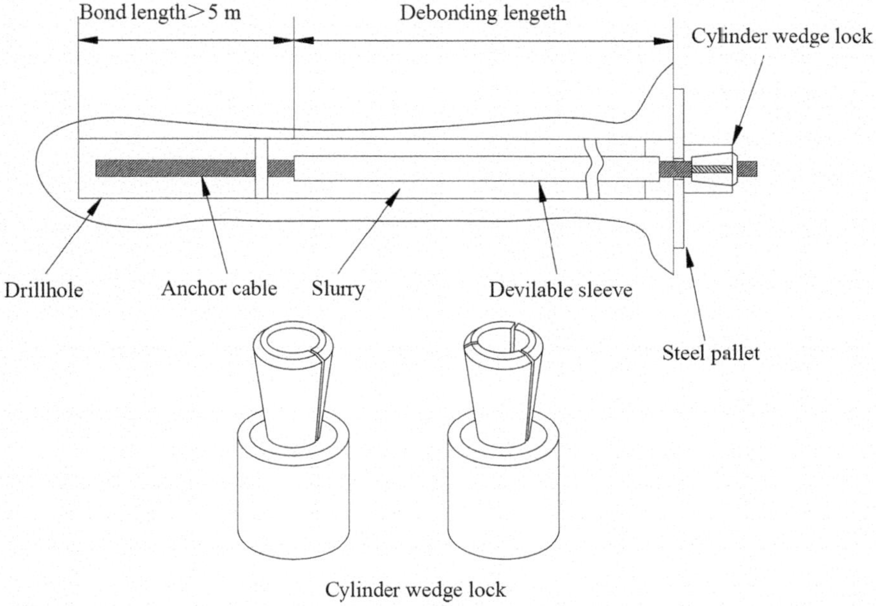

Fig. 3.47 Design drawing of debonding anchor cable

wedge of the cylinder wedge lock on the head of the anchor cable, tighten the anchor cable, press the wedge into the cylinder, relax the anchor cable, and the anchor cable contracts to lock the wedge inside the cylinder. When grout is used to install anchor cables in vertical holes in the vault, there must be a suspension device at the far end of the cable to prevent the cable from slipping under the influence of gravity. A commonly used end suspension device is the "fishbone barb". The "fishbone barb" is fixed to the end of the anchor cable with a steel wire. After the anchor cable is inserted into the bottom of the hole, the barb hangs the anchor cable on the hole wall, and then the cement slurry is pumped into the hole. If a resin roll is used for bonding, a resin mixing device is required at the end of the anchor cable to stir the adhesive in the resin roll. There are two commonly used resin mixing device methods. One is to bulge the steel wire at the far end of the anchor cable, and the other is to wrap a thick steel wire around the end of the anchor cable in addition to the expanded steel wire.

3.4.2.3 Joint Support

1. Anchor beam network combined support

When a single anchor cannot effectively control the surrounding rock, combined anchor support is often used. Anchor beam net combined support is the combined use of anchor rods, shielding nets and joists. Anchors are generally resin anchors, and metal diamond mesh, metal warp and weft mesh, plastic mesh, and steel mesh can all be used as shielding mesh. There are mainly four types of beams and steel strips used in China.

(1) W-shaped steel strip

The W-shaped steel strip is made of thin steel plates with a thickness of 2.5–3.0 mm and a width of 250–300 mm tied into a W-shaped cross section, as shown in Fig. 3.48. The W-shaped steel strip has high bending strength and a large roof protection area. When the cracks in the roof are relatively developed and the pressure is not high, it can be used in the roof support system. However, when the pressure on the roadway roof is high, the W-shaped steel strip is prone to shearing destroyed and unfit for use.

(2) Reinforced ladder beam

Reinforced ladder beams are generally welded from two round steels with a diameter of 12–18 mm, and the shape is similar to a ladder, as shown in Fig. 3.49. Reinforced ladder beams are easy to process and low in price; however, their bending strength and roof protection area are not as great as those of W-shaped steel strips. When the roof pressure is high, the welding joints of the ladder beams are prone to damage and their strength is low.

(3) M-type steel strip

The main features of the M-type steel strip are that the concave bending section modulus of the steel strip is 2.75 times the convex bending section modulus. It is easy to stick to the roof rock surface during installation. When the roof sinks, the

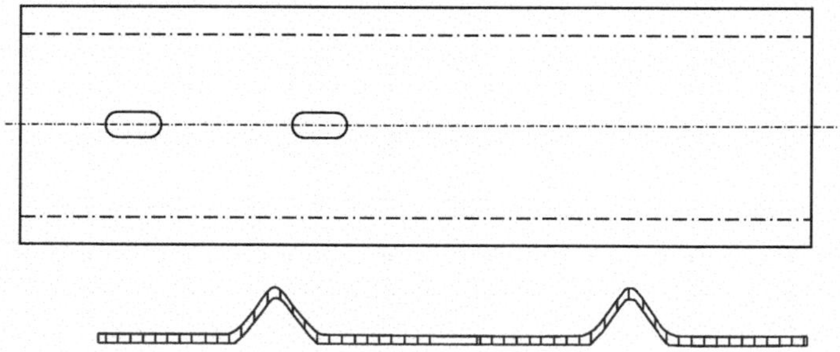

Fig. 3.48 W-shaped steel strip shape

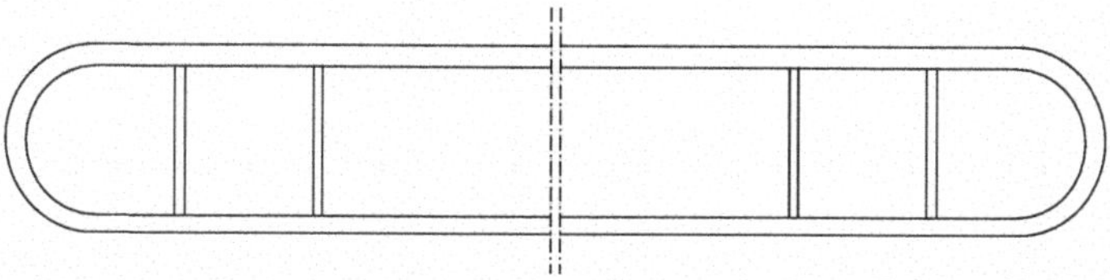

Fig. 3.49 Shape of reinforced ladder beam

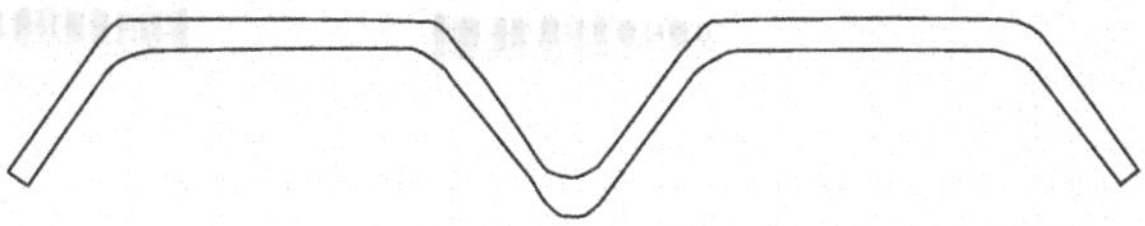

Fig. 3.50 M-shaped steel strip

steel strip and the surrounding rock will form in one piece, it can also solve the problem of easy tearing of W-shaped steel strips. Its cross-sectional shape is shown in Fig. 3.50.

(4) Type II light duty steel strip

To address the issues of unreliable welding, poor integrity, and poor contact with the surrounding rock in reinforced ladder beams, a Type II light duty steel strip has been developed. This steel strip has good overall performance, is economical in usage, and is suitable for roadway side support.

The combined support of anchor beams and nets can form an integrated support structure. The integral structure formed by the anchors, beams, and nets enhances the integrity of the surrounding rock of the roadway. The high tensile effect of steel strips and joists transforms the anchoring force of the anchors into a top support force for the steel strips and joists. This makes the anchoring force on the surface of the surrounding rock more evenly distributed and increases the supporting area of the anchors, preventing the fall of small rocks between the anchors.

2. Truss anchor support

The common truss anchor is composed of four parts: the anchor, the tie rod, the tensioner, and the bearing plate. The main types of truss anchors include the single-type truss anchor (Fig. 3.51); duplex truss anchor, which consists of 2 to 3 single truss anchors arranged in the span direction of the roadway; cross truss anchor, where single truss anchors cross at the intersection of roadways to form a cross shape; continuous truss anchor, which connects single or duplex truss anchors along the axial direction of the roadway to form an integrated whole.

Whether it is a single or duplex truss anchor, horizontal and vertical compressive stress zones will form within the roof rock layer, as shown in Fig. 3.52. The anchoring force of the anchor and the pre-tightening force of the tie rod cause the "neutral axis" of the roof to move downward, enhancing the roof rock layer's resistance to bending and reducing the tensile stress inside and on the surface of the roof. For the fractured roof, the horizontal pressure provided by the truss anchor increases the friction factor along a group of fractures in the transverse direction of the roadway, improving the integrity of the "fracture beam" and facilitating the formation of an arch structure in the roof beam. Through the tie rod, multiple anchors of the same tie rod form an integral support structure. On the other hand, when the roof undergoes bending deformation or bending subsidence, the combined action of the tie rod and inclined anchor will generate greater compressive stress and friction within the roof and its fracture body, reducing or even offsetting the tensile stress that may occur in the middle of the roadway roof, preventing further bending and subsidence of the roof.

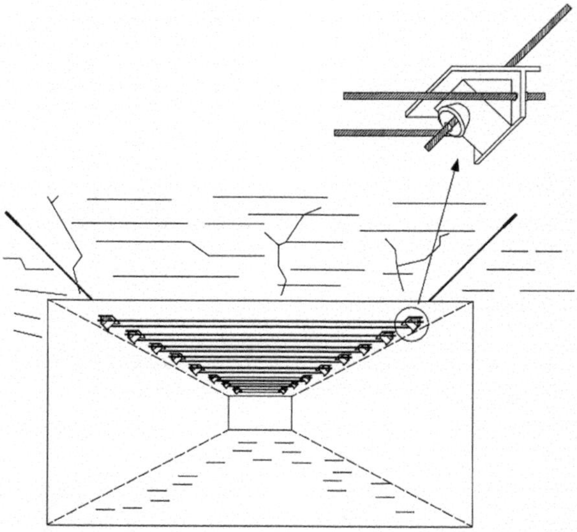
Fig. 3.51 Single-type truss anchor

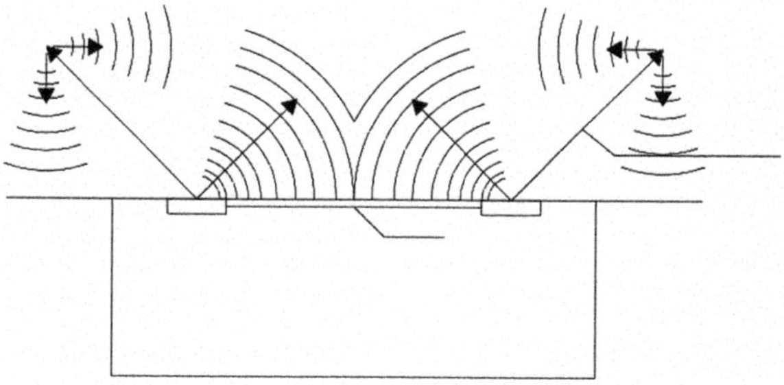
Fig. 3.52 Truss anchor support force

The key to truss anchor support is the pre-tightening force generated in the tie rod, making the design of the tie rod crucial. Tie rods are generally made of round steel, and their beam-like structure can provide pre-tightening force, which can be divided into two types: top-contacting and non-top-contacting. Truss anchors are highly adaptable and can be used to support intact roofs or those with developed joint fractures, broken roadways, and large cross-sections or areas with large hanging roofs, such as chambers or roadway intersections. Currently, single truss anchors have been used in roadways with a maximum span of 7 m, and the maximum width of roadways supported by duplex truss anchors has reached 9 m at the abroad. In the room and pillar mining method, the maximum width of the mining rooms supported by duplex truss anchors has reached 12 m. In China, the maximum width of roadways supported by duplex truss anchors has also reached 9 m.

3.4.3 Large-Section Roadway Support Cases

3.4.3.1 A Case of Wangzhuang Coal Mine [18]

Wangzhuang coal mine primarily exploits the No. 3 coal seam, which has an average thickness of 6.52 m and a dip angle ranging from 2° to 12°. The coal seam has a relatively low gas content, classifying it as a low-gas mine. The 9105 panel has a width of 339 m, and the transport and return roadways are 3432 m long. The roadway cross-section dimensions are 5.6 m in width by 4.2 m in height, which qualifies it as a large-width condition roadway.

Roof Support: The "reinforcing mesh, threaded steel anchor rod, and prestressed anchor cable" combined support system is employed. The reinforcing mesh is first laid on the roadway roof for support, ensuring it is in firm contact with the roof, with wooden backing boards used under the mesh to tightly back against the forward exploration beams. Then, Φ22 mm × 2700 mm threaded steel anchor rods are installed with an inter-row spacing of 0.75 m × 0.75 m. The anchor rods in the middle of the roadway are arranged vertically to the roof and sides, while those on the sides of the roadway are angled at 45° to the roof and sides. The installation proceeds from the middle to the sides in a stepping fashion. When installing the first central top anchor rod, the corresponding forward exploration beam is retracted to not affect the positioning of the anchor rod spacing. Lastly, Φ18.9 mm × 8000 mm low-relaxation prestressed steel strand anchor cables are installed with a spacing of 1.7 m × 1.2 m, arranged in a "2-3-2" staggered pattern. The installed anchor cables should be capped, and the cables are bound with No. 10 wire under the anchor cable locks, wrapped three times in a cross pattern to bind and fix the cables and bearing plates to the top metal mesh.

Roadway side Support: The "I-beam shed, threaded steel anchor rod, and prestressed anchor cable" combined support system is used. When drilling anchor rod holes on the sides, the arrangement of the pillar side and the mining side anchor rods is consistent with the top anchor rods. Due to significant deformation of the pillar side during the original roadway layout, two additional Φ18.9 mm × 5000 mm steel strand anchor cables are installed per row on the pillar side, with a spacing of 1.8 m × 1.2 m. I-beam sheds are also used for support, made from No. 11 mine-use I-beams into trapezoidal sheds, with a shed spacing of 0.8 m, beam length of 3500 mm, and leg length of 2839 mm each, with a footing of 450 mm. Four Φ18 mm round steel tie rods are used to link between the legs of the shed, with a spacing of 1 m between the upper and lower tie rods, and a tie rod length of 1.1 m each, four per shed. The tie rod seats are welded onto the legs of the shed, with one seat welded at 1 m and another at 2 m on the leg pad. The large roadway cross-section support scheme is shown in Fig. 3.53. The results show that the convergence of the roadway roof and floor and the sides is minimal, effectively controlling the deformation of the surrounding rock in the large cross-section roadway, meeting the safety requirements for the panel's mining operations.

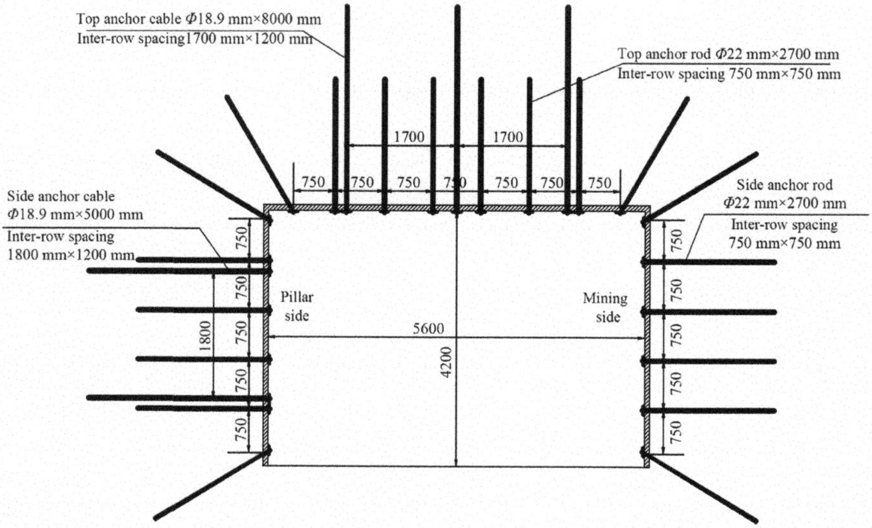

Fig. 3.53 Large-section roadway support in 9105 panel of Wangzhuang coal mine

3.4.3.2 A Case of Caojiatan Coal Mine [19]

The Caojiatan Coal Mine's panel 122109, situated in the 12th panel of the 2-2 coal seam, boasts a floor elevation ranging from + 957 to + 972 m and a surface elevation between + 1261 and + 1300 m. The 2-2 coal seam, characterized by its simplicity and stability, has a density of 1.32 tons per cubic meter and a Protodyakonov scale of hardness less than 3. The headgate of the 122109 panel, with a cross-sectional dimension of 6.5 m in width by 4.35 m in height, paneld substantial issues due to the inadequacy of the fiberglass anchor rods and accompanying trays and nuts in the original support scheme. This discrepancy in strength, especially when compared to the threaded steel anchor rods arranged alongside the coal pillars, led to widespread dislodging of the fiberglass anchor rods and fragmentation of the trays within a 0–15 m range ahead of the terminal support in the retreat mining area. Consequently, this resulted in extensive damage to the surrounding rock, expansion of fractures, and even slabbing.

To ensure safety, a high-performance anchor rod support system was established, centered around high strength, high pre-tightening, and high system rigidity, tailored for rapid excavation in coal alleys with wide spacing between rows.

1. Pillar side support: threaded steel anchor rods and reinforcing mesh

The anchor rods used are left-handed, non-longitudinal threaded steel rods with a specification of Φ20 mm × 2000 mm, made of BHRB335 steel. The spacing between anchor rods is 1000 mm × 1000 mm, with 4 rods per row. The topmost anchor rod is positioned 300 mm from the roof, inclined at a 15° angle, while the remaining anchor rods are installed vertically against the rock panel. The anchor rod holes have a depth of 1950 mm. Each anchor rod is secured using an MSK2380 resin cartridge anchor, with a minimum pre-tightening force of 120 N m and a minimum anchoring force of 10 t. The anchor rod trays are butterfly-shaped iron trays with dimensions of 150 mm × 150 mm × 10 mm. The reinforcing mesh is arc-welded and has dimensions of 3600 mm × 1100 mm, with 4 mm steel bars and a grid size of 100 mm × 100 mm. The top of the mesh is bent by 100 mm, and it overlaps the roof mesh by 100 mm and adjacent mesh by 100 mm.

2. Mining side support: fiberglass anchor rods and plastic-coated steel mesh

The anchor rods used are GQN60 high-strength, torsion-resistant fiberglass rods with a specification of Φ22 mm × 2400 mm, along with accompanying tray nuts. The spacing between anchor rods is 1000 mm × 1000 mm, with 4 rods per row. The topmost anchor rod is positioned 300 mm from the roof, and the anchor rod holes have a depth of 2300 mm. Each anchor rod is secured using an MSK2380 resin cartridge anchor, with a minimum pre-tightening force of 50 N m and a minimum anchoring force of 10 t. The mesh is made of plastic-coated steel, with dimensions of 3600 mm × 5000 mm, laid in the longitudinal direction. The mesh overlaps the roof mesh by 100 mm and adjacent mesh by 100 mm. The plastic-coated mesh in the rib section does not contain steel wires, and the mesh openings are 50 mm × 50 mm.

3. Roof support: anchor rods, anchor cables, and reinforcing mesh

   (1) Anchor rods support: The roof anchor rods are BHRB335 grade left-hand non-longitudinal threaded steel anchor rods with specifications of Φ22 mm × 2600 mm. The spacing between rows of anchor rods is 1200 mm × 1000 mm, with six rods per row. The outermost anchor rod is 250 mm from the side, installed at a 15° outward angle, while the four middle anchor rods are installed vertically to the roof, with a hole depth of 2550 mm. Each anchor rod is secured with one MSK2380 type resin cartridge, requiring a pre-tightening force of no less than 200 N m and an anchoring force of no less than 10 t. The anchor tray used is an arched high-strength tray with specifications of 150 mm × 150 mm × 10 mm. The reinforcing mesh is an electric arc-welded steel mesh with specifications of 6500 mm × 1100 mm, a steel diameter of 4 mm, a grid size of 100 mm × 100 mm, and an overlap width of 100 mm.
   (2) Anchor cables support: Two cables are arranged per row on the roof, made of steel strand wire with specifications of Φ17.8 mm × 6250 mm, and a hole depth of 6000 mm. The spacing between cables is 2400 mm × 3000 mm, installed vertically against the rock. Each cable is secured with two MSK2380 type resin cartridges, with a pre-tightening force of 14 t and an anchoring force of no less than 24 t. The cable tray is a high-strength arched adjustable centering tray with specifications of 250 mm × 250 mm × 20 mm. The cross-sectional view of the roadway support system is as shown in Fig. 3.54.

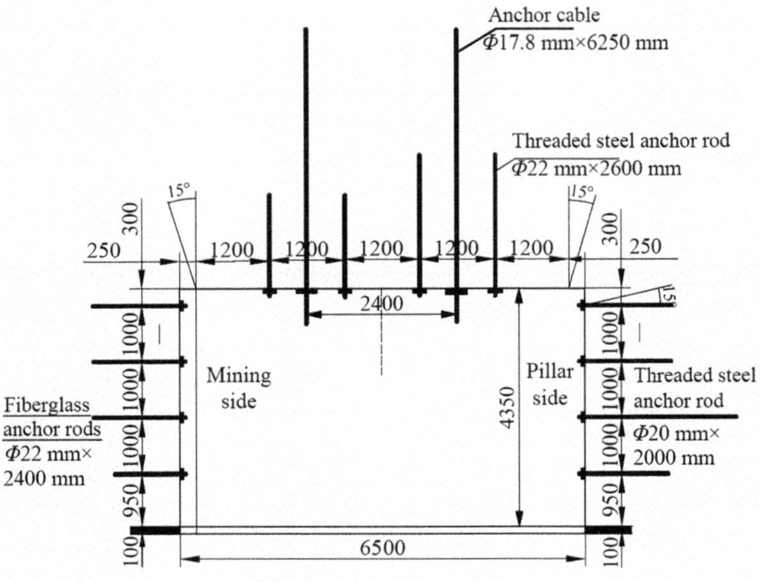

Fig. 3.54 Support for the large-section of the headgate at Caojiatan coal mine

3.4.3.3 A Case of Ganhe Coal Mine [20]

The structure of the 2# coal seam in Ganhe Coal Mine is relatively simple. The average thickness of the coal seam is 3.75 m, with approximately 0.3 m of parting. The average dip angle of the coal seam is 3°, and its hardness coefficient is $f = 1.5$. The cross-section of the 2-1081 drivage roadway measures 5 m × 3.8 m. Due to the larger roadway section, the coal body of the 2# seam is relatively soft, and the strength of the coal above and below the parting varies. Additionally, the horizontal stress in the roadway exceeds the vertical stress, leading to a tendency for wall deflection during excavation. Moreover, the roof and floor rock layers of the roadway exhibit well-developed bedding planes, making the rock mass prone to fragmentation during excavation and mining. The combination of high ground stress and the influence of fault zones in the roadway direction may result in roof collapse or floor heave during excavation and mining. Conventional support designs often lead to excessive deformation of the roadway surrounding rock, affecting normal mine production. These challenges pose significant difficulties for support design. The roadway support scheme, as shown in Fig. 3.55, adopts a combined support system using anchor nets and beams.

Each row of roof rods consists of seven rods, using high-strength threaded steel rods with specifications of Φ22 mm × 2500 mm, spaced 800 mm × 800 mm apart. Each rod is secured with one CK2340 resin anchoring agent and one Z2388 resin anchoring agent. The shoulder angle rods are 100 mm from the roadway side, with an angle of 75° to the roof, fitted with Φ130 mm × 8 mm × 25 mm wedge washers,

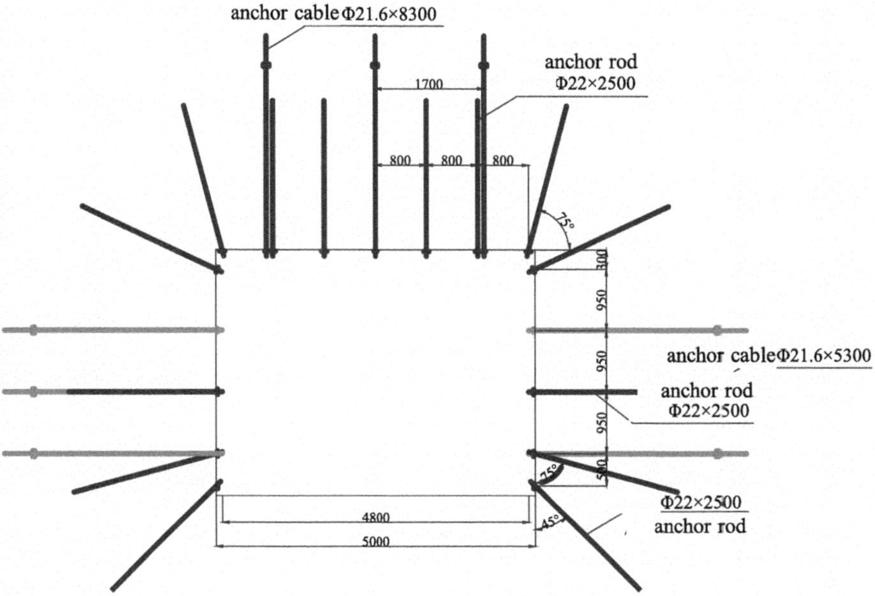

Fig. 3.55 Support design for the roadway at Ganhe coal mine

while the rest are perpendicular to the roof, with a pre-tightening applied to the rods of no less than 50 kN.

For the side support, each row on each side has four rods, using high-strength threaded steel rods with specifications of Φ22 mm × 2500 mm, spaced 950 × 800 mm apart. The topmost rod is 300 mm from the roof, at an angle of 70° to the roadway side, while the rest are perpendicular to the roadway side. The lowest rod is 650 mm from the floor, at an angle of 75° to the roadway, with a bottom corner rod placed 150 mm from the floor, using high-strength threaded steel rods with specifications of Φ22 mm × 2500 mm, with a pre-tightening applied to the rods of no less than 50 kN.

The anchor cables are arranged in a "3 × 3" pattern, spaced 1700 mm × 3200 mm apart, with specifications of Φ21.6 mm × 8300 mm, secured with two CK2340 and three Z2388 resin anchoring agents, and steel plates with specifications of 400 mm × 400 mm × 10 mm. Three medium-length anchors are added to the middle part of the roadway side, spaced 950 mm apart, with a row spacing of 1600 mm, with specifications of Φ21.6 mm × 5300 mm, secured with two CK2340 and three Z2388 resin anchoring agents, with a pre-tightening applied to the anchors of no less than 200 kN.

The strengthening support scheme effectively controlled the convergence of the two sides and the roof and floor of the roadway, ensuring the stability of the surrounding rock. The main purpose of the strengthening support is to form a solid thick wall in the side of the surrounding rock, reducing the plastic zone of the surrounding rock in the roadway. Due to the enhanced stability of the side, the subsidence of the roof and the heaving of the floor were also reduced, ensuring the stability of the surrounding rock in the roadway.

References

1. Du JP, Meng XR (2009) Mining science. China University of Mining and Technology Press, Xuzhou
2. Yin XW (2019) Research status of strata control and large mining height fully-mechanized mining technology in China. Coal Sci Technol 44(8):37–45
3. Hou CJ's Team (2013) Ground control of roadways. China University of Mining and Technology Press, Xuzhou
4. Wang GF, Pang YH, Xu YX et al (2023) Development of intelligent green and efficient mining technology and equipment for thick coal seam. J Min Saf Eng 40(5):882–893
5. Kang HP, Xu G, Wang BM et al (2019) Forty years development and prospects of underground coal mining and strata control technologies in China. J Min Strata Control Eng 1(2):7–39
6. Song GF, Yang SL, Wang ZH (2018) Longwall face stability analysis using Ritz method and its 3D physical modelling study. J China Coal Soc 43(8):2162–2172
7. Wang JC, Wang ZH, Kong DZ (2015) Failure and prevention mechanism of coal wall in hard coal seam. J China Coal Soc 40(10):2243–2250
8. Wang JC, Yang SL, Kong DZ (2016) Failure mechanism and control technology of longwall coalface in large-cutting-height mining method. Int J Min Sci Technol 26(1):111–118
9. Cheng C (2014) Study on rule and technological parameters of water injection to strengthen stability of soft coal wall. China University of Mining and Technology
10. Zhou DW (2009) Study on technology of soft coal seam rib spalling preventing by water infusion with shallow holes. Anhui University of Science and Technology
11. Li XF (2019) Application of deep hole pre-injection technology in improving coal wall stability of large mining height face. Shandong Coal Sci Technol 7:1–3
12. Chen PP (2019) Research and application of coal wall water injection prevention technology for 72909 working face in Du'erping Mine. Coal Chem Ind 42(5):4–7
13. Li YB (2023) Application of grouting reinforcement technology in crossing fault area at fully-mechanized working face. Jiangxi Coal Sci Technol 2023(4):36–38+42
14. Li SH (2022) Prevention and control technology of coal wall slope passing through collapse column in Xiadian coal mine. Shandong Coal Sci Technol 40(5):48–50+54
15. Yang SL, Kong DZ (2015) Flexible reinforcement mechanism and its application in the control of spalling at large mining height coal working face. J China Coal Soc 40(6):1361–1367
16. Qian MG, Xu JL, Wang JC et al (2021) Mine pressure and ground control. China University of Mining and Technology Press, Xuzhou
17. Li CL (2021) (Sweden); Rockbolting principles and applications (Zhao TB, Li YR trans). Science Press, Beijing
18. Du XG (2024) Research on control technology of surrounding rock in the large section roadway under structural stress disturbance in Wangzhuang coal mine. Shandong Coal Sci Technol 42(1):1–5
19. Han CD, Xu XL, Lei YJ et al (2020) Optimization of surrounding rock control technology of large section roadway in Caojiatan coal mine. Shaanxi Coal 39(6):32–35+42
20. Li GB (2013) Research on the surrounding rock stability and control techniques of large section roadway in Ganhe coal mine. China University of Mining and Technology (Beijing)

Chapter 4
Slicing Mining

Slicing mining technology was one of the primary methods used in early China for extracting thick and extra-thick coal seams [1]. Typically, slicing mining involves dividing nearly horizontal, gently inclined, inclined, and steeply inclined thick coal seams into several layers with inclined planes parallel to the coal seam. Each layer is approximately 2.0–3.0 m thick, and mining proceeds layer by layer. To ensure the safety of mining in lower layers, artificial false roofs are often installed above upper layers to separate the falling gangue from the coal seam below [2]. Slicing mining has several advantages. The smaller mining height results in less pronounced ground pressure, making it easier to control the surrounding rock. Additionally, the equipment requirements are lower. However, there are also drawbacks, such as lower coal recovery efficiency due to the smaller mining height, higher economic costs, and poorer economic benefits [3, 4]. In the 1980s, the rapid development of LTCC technology led to its widespread adoption in thick coal seams [5]. Simultaneously, advancements in hydraulic supports and other mining machinery increased the height of large mining height technology, resulting in a parallel development of LTCC and large mining height technologies. Consequently, the application of slicing mining technology in thick coal seams has significantly decreased in recent years. However, for extra-thick coal seams (20 m or more) and steeply inclined coal seams in western regions, slicing mining techniques (either slicing mining or STCC) are still employed.

J. Wang and W. Wei, *Thick Coal Seam Underground Mining*,
https://doi.org/10.1007/978-981-96-3874-1_4

4.1 Slicing Mining Roadway Layout

4.1.1 Slicing Mining Parameters and Roadway Layout

4.1.1.1 Slicing Mining Types

Slicing mining can be categorized into two types. The first, known as simultaneous layered mining, occurs within the same segment: once the upper layer has been completely mined or advanced a certain distance, the excavation of the return roadways for the lower layer commences, followed by its extraction, as depicted in Fig. 4.1a. The second approach, sequential layered mining, involves completing the mining of the upper layer before proceeding to excavate the return roadways and extract the lower layer within the same mining area or field, as illustrated in Fig. 4.1b. The former method shortens the erosion time of the artificial false roof, while the latter extends it.

In the case of simultaneous layered mining, the interval between mining the upper and lower layers is brief. If an artificial false roof is employed, this can diminish the erosion time of the false roof. However, if a regenerated roof is used, the fragmented roof from the upper layer's mining cannot form a regenerated roof in a short period, or the bonding quality of the regenerated roof may be poor, compromising the safety of the lower layer's mining. Conversely, with sequential layered mining, the longer interval between minings leads to a stronger erosive effect on the artificial false roof by the fragmented roof of the upper layer, which may jeopardize the safety during the lower layer's mining. Similarly, a longer interval is conducive to the formation of a regenerated roof, enhancing its stability.

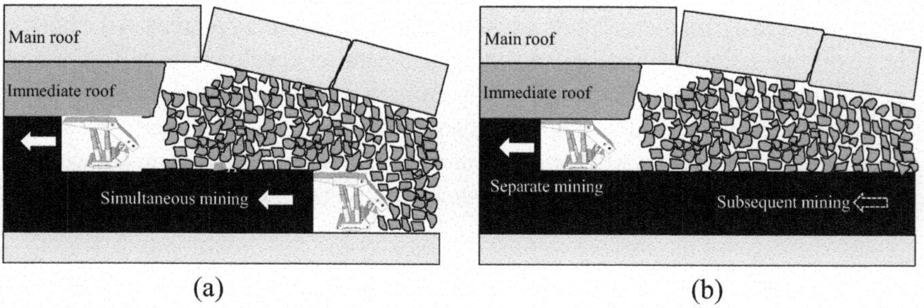

Fig. 4.1 Slicing mining types: **a** simultaneous layered mining; **b** sequential layered mining

4.1.1.2 Coal Mining Parameters

1. Inclined layer thickness

When employing the comprehensive mechanized mining process, the typical thickness for layered mining is approximately 3 m. However, with advancements in shearer, the layered thickness can be correspondingly increased. Given the frequent variations in coal seam thickness, and considering the significant subsidence of artificial false roofs or regenerated roof layers, it is essential to ensure sufficient mining height for the lower layers, allowing the shearer to operate effectively in that layer. To control the slicing mining height, some mines drill boreholes every 30–50 m in the upper and lower gate roads and panel toward the bottom coal seam during the mining of the first layer. These boreholes help determine the number of layers and the appropriate mining height.

2. Panel length

The determination of panel length in slicing mining follows the same principles as the single longwall mining method. However, due to the added complexities of installing mesh and conducting underground operations, panel lengths are generally kept relatively short. Additionally, thick coal seams often exhibit a natural tendency toward spontaneous combustion. To prevent coal self-ignition, panels must accelerate their advance speed to ensure sufficient monthly progress. Consequently, panel lengths should not be excessively long. For specific parameters related to slicing mining panels in different mining regions, referred to Table 4.1.

4.1.2 Roadway Layout

The judicious position of mine roadway is a critical aspect of roadway layout planning. By positioning roadway strategically, one can ensure that the forces exerted on the surrounding rock are minimal and uniform. This prevents severe deformation of the rock body due to significant differences in the forces acting on both sides of the roadway, which could otherwise complicate roadway support and maintenance, and potentially trigger other geological disasters, adversely affecting fluent production operations. Furthermore, optimal roadway layout allows for the coal pillars that protect the roadway to be of a reasonable width, thereby minimizing coal loss, maximizing coal recovery ratio, reducing production costs, and enhancing economic benefits. In the mining of thick coal seams using the slicing mining, there are three methods for the roadway layout: inclined, horizontal, and vertical.

Table 4.1 Partial slicing mining panel parameters

Number	Coal mine	Layers	Panel	Thickness/m	Panel length/m	Advance length/m
1	Xieyi	Upper	5111B11b	1.9	172	831
		Lower	5111B11b	3.5	182	845
2	Panyi	Upper	1331(3)	2.0	125	600
		Lower	1331(3)	2.2	125	600
3	Dongda	Upper	3310	2.9	176	1248
		Lower	3310	3.2	176	1248
4	Luwa	Upper	2303	3.0	160	860
		Lower	2303	4.3	125	720
5	Yuecheng	Upper	1308	3.1	116	870
		Lower	1308	3.0	175	2129
6	Dongchuan	Upper	5103	3.4–3.8	200	1696
		Lower	5103	3.0	200	1696
7	Daliuta Huojitu	Upper	12208	4.0	287	4156
		Lower	12208	6.0	287	4156
8	Gubei	Upper	13121	4.5	205	1049
		Lower	13121	3.1	205	1049
9	Zhangji	Upper	1613A	4.5	200	1503
		Lower	1613A	2.0–4.0	200	1503
10	Panbei	Upper	11113	5.3	129	401
		Lower	11111	3.6	164	401
11	Makou	Upper	8401	6.0	120	500
		Lower	8402	2.0	120	500

4.1.2.1 Inclined Layout

As illustrated in Fig. 4.2, the mining roadways for each layer are arranged at an angle of 25°–35°, typically suitable for coal seams with an inclination of less than 15°–20°. A coal pillar is often left between the upper segment's headgate and the lower segment's tailgate. The size of this pillar depends on factors such as the thickness and angle of the coal seam and the softness of the coal quality, generally not less than 15 m, or larger.

Relative to the length changes of the middle and lower layer panels, the inclined layout is further divided into inward staggered and outward staggered types. The inward staggered layout refers to the lower layer's roadways being positioned inside the upper layer's panel, that is, below the goaf area of the upper layer's panel, at a certain distance from the upper layer's coal pillar, forming a right trapezoidal coal pillar, as seen in Fig. 4.2a. This arrangement, due to the lower layer's roadways being on the inside of the upper layer's panel, results in larger coal pillars from top

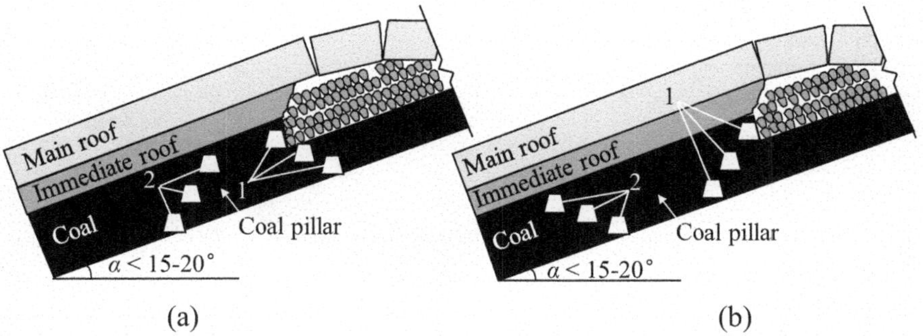

Fig. 4.2 Inclined layout of layered roadway: **a** inward staggered; **b** outward staggered (1-headgate in the upper section; 2-tailgate in the lower section)

to bottom, shortening the lower layer's panel, inevitably reducing the coal recovery ratio and increasing coal loss. However, this method places the mining roadways in the stress relief zone, facilitating roadway excavation and production maintenance, and reducing roadway maintenance costs. The inward staggered layout of the middle and lower layer mining roadways is inward staggered by half to one roadway width, shortening the panel, enlarging the coal pillar, and allowing the roadway to be excavated along the false roof below the goaf area, making it easier to maintain and also prone to upward air leakage.

In the outward staggered layout, the lower layer roadway is situated on the exterior side of the upper layer panel, nestled within the upper layer sectional coal pillar, thus forming an inverted trapezoidal coal pillar, as shown in Fig. 4.2b. This configuration elongates the panel, with the coal pillar dimensions diminishing as one descends. The roadway's position falls within the influence range of the lateral mining pressure exerted by the upper layer coal pillar. This method can complicate the maintenance of the roadway, especially the absence of an artificial false roof at the upper and lower exits of the middle and lower layer panels makes both coal mining and support more challenging. Consequently, this layout is less frequently employed.

4.1.2.2 Horizontal Layout

As depicted in Fig. 4.3, each layer mining roadway is arranged at the same horizontal elevation, with the sectional coal pillars forming parallelograms. This layout is conducive to material transportation, pedestrian movement, and ventilation. The headgates, located beneath the upper layer goaf, are subject to less pressure and are easier to maintain. However, the tailgates, positioned under the sectional coal pillars, bear the brunt of lateral mining pressure, making their maintenance more arduous. It is crucial to ensure that the vertical distance between the upper and lower segment mining roadways is no less than 5 m. Therefore, this layout is generally reserved for coal seams with an inclination exceeding 20°–25°, lest the sectional coal pillars become excessively large.

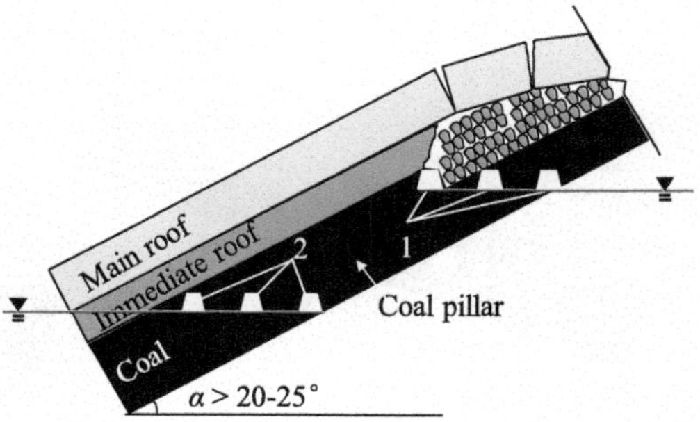

Fig. 4.3 Horizontal layout of layered roadway (1-headgate in the upper section; 2-tailgate in the lower section)

4.1.2.3 Vertical Layout

As depicted in Fig. 4.4, the mining roadways for each layer are arranged in a vertically overlapping fashion, with the sectional coal pillars forming an approximate rectangle. Under conditions where the coal seams dip angle is less than 8°–10° (particularly in nearly horizontal thick coal seams), this layout can reduce the dimensions of the sectional coal pillars. Concurrently, the lower layer mining roadways, excavated along the artificial roof laid out by the upper layer mining roadways, facilitate easier directional control. By analyzing the stress distribution in the surrounding rock of the top layer mining area, the bottom layer panel roadway are positioned directly beneath the corresponding roadway of the top layer, leading to the development of vertical roadway layout technology for slicing mining [6]. The movement and redistribution of stress in the rock layers surrounding the mining area, caused by the top layer mining process, significantly impact the layout and support methods of the bottom layer roadway. Under the influence of mining activities, along the direction of advancement of the retreating panel, the stress in the coal body and coal pillars on both sides of the mining space undergoes considerable changes with varying distances and times from the panel, typically manifesting in three distinct stress zones. These include the original stress zone, unaffected by mining activities, far from the panel; the stress increase zone near and around the panel, affected by mining activities; and the stress reduction zone within the panel where the mining influence stabilizes. The stress increase zone comprises three parts: rising, intensifying, and weakening stress. After the mining process, as the rock layers above the goaf settle and the gangue in the caving zone progressively collapses and compacts, the pressure in the goaf and on both sides of the coal pillars stabilizes.

Vertical layout, ideally situated at the periphery of the goaf within the stress attenuation zone, offer a prime location for roadway layout. However, during the mining of the top layer panel, the absence of meshing in both roadways leads to the naturally formed regenerative roof having poor cementation strength. This can result in debris falling and roof collapse accidents if roadway support is not timely or if the control distance to the roof is too great. Additionally, one side of the roadway falls within a stress concentration zone, which necessitates extensive later repair work. The presence of secondary collapses in the roof of the top layer roadway further

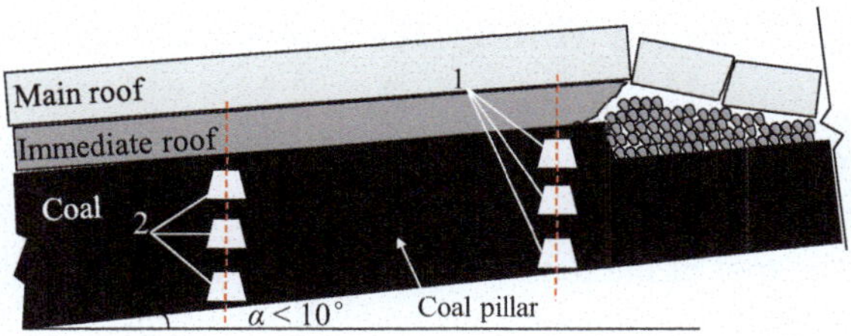

Fig. 4.4 Vertical layout of layered roadway (1-headgate in the upper section; 2-tailgate in the lower section)

complicates roadway maintenance. Xingtai Mine employs shelf support to address the significant side pressure and severe deformation characteristic of certain sections of the vertical roadways. By utilizing anchor rods support on both sides, combining proactive and passive support methods, the bearing capacity of the coal walls on both sides is further reinforced. This, coupled with optimized construction techniques, has facilitated the vertical layout of roadways in the slicing mining of thick coal seams, as shown in Fig. 4.5. Here, the maximum stress concentration coefficient is denoted by K, γ represents the unit weight of the overlying rock strata, and h denotes the mining depth.

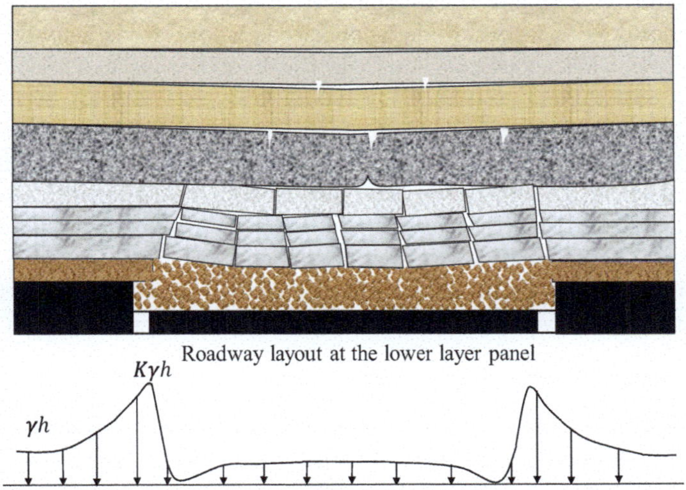

Fig. 4.5 Stress distribution of vertical roadway layout

4.2 Slicing Mining Technology and Equipment

4.2.1 Slicing Mining Technology

The roof of the top-layered coal seam is the original roof and the floor is the coal seam. The coal mining method is basically the same as the longwall mining method of medium-thick coal seams. To isolate the coal seam below from the collapsed gangue after the top layer mining and provide a safe environment for the mining of the lower coal seam, it is necessary to lay an artificial roof or form a regenerated roof during the upper layer mining.

4.2.1.1 Artificial Roof

Artificial roofs mainly include bamboo or thorny roofs, metal or plastic mesh roofs.

1. Bamboo or thorny artificial roof

Some mining areas in China, especially some mining areas in the south, use the bamboo basketry or thorny as artificial roof materials, seeing Fig. 4.6. Bamboo basketry is made of bamboo or thin bamboo rod by wire woven into the basketry, about 0.7–1.0m wide, about 2.2–2.4m long. The thorny is a piece of basketry woven with chaste strips. Bamboo basketry or thorny false roof can only be laid on the floor, generally tilted along the panel from the bottom to top, basketry between each other, the lap with wire connection, the joint should be fixed on the bottom beam to prevent basketry from sliding. When the gangue in the goaf has better cementation properties, it can also be replaced with double basketry instead of bottom beam. Each layer shall be laid except the lowest one. The integrity of bamboo basketry or thorny false roof is poor, the strength is low, the suspended area allowed under the false roof is small, and it is easily perishable and ineffective, thus it is gradually used less.

2. Metal mesh artificial roof

Metal mesh artificial roof is generally made of 12–14 galvanized iron wire, of which the high-strength mesh edge is woven with 8–10 iron wire. The mesh shapes of the metal mesh artificial roof are square, diamond, honeycomb, etc., as shown in Fig. 4.7. Metal mesh artificial roof has the advantages of high strength, high flexibility, small size, light weight, easy transportation and laying, corrosion resistance and long service life. Therefore, it is widely used in slicing panels.

The metal mesh on the panel is automatically laid by hydraulic supports, and there are two laying methods: laying along the top and laying along the bottom. Laying along the top generally involves adding a bracket under the front exploration beam or top beam of the hydraulic support, as shown in Fig. 4.8a. The metal mesh is installed on the bracket, and the mesh is passed over the front end of the support beam and pressed against the top plate. When the support moves forward, the mesh roll will unfold automatically. After one roll of mesh is laid, a new roll of mesh is

Fig. 4.6 Bamboo artificial roof

Fig. 4.7 Metal mesh artificial roof

installed, and the edge of the new mesh is connected to the edge of the old mesh. The work of connecting the metal mesh is done manually under the support beam, and the long side of the laid top mesh is perpendicular to the panel direction. The main disadvantage of this method is that the efficiency of connecting the metal mesh is low. Since the mesh sags near the coal wall, when the mining height is low, there is not enough space on the support to place the metal mesh roll, or the metal mesh roll hinders the smooth passage of the shearer.

Laying along the bottom means that the mesh roll brackets for the inter-frame mesh and the mid-frame mesh are installed under the shield beam at the rear end of the support (some supports are at the front end of the base), and the front and rear rows of mesh rolls are placed alternately, with the long side of the mesh overlapping by 150–200 mm and the short side overlapping by about 500 mm. When the support moves forward, the mesh rolls unfold automatically on the bottom plate, as shown

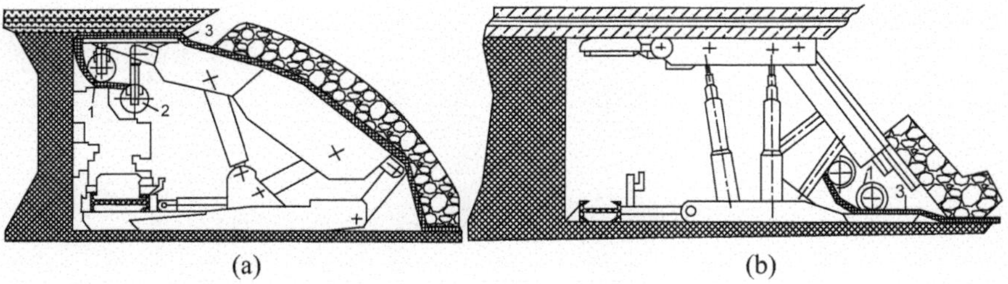

Fig. 4.8 Hydraulic support mechanized **a** roof net and **b** bottom net laying (1-mid-frame mesh roll; 2-inter-frame mesh roll; 3-metal mesh)

in Fig. 4.8b. The work of connecting the metal mesh rolls is carried out under the protective beams and does not interfere with the coal mining operations.

3. Plastic mesh artificial roof

The plastic mesh artificial roof is woven from polypropylene resin plastic belts, and its laying method is basically the same as that of the metal mesh artificial roof. The mass of the plastic mesh is only about 1/5 of the metal mesh of the same area. It has the advantages of light weight, odorless and non-toxic, flame retardant, anti-static, high flexibility and corrosion resistance. It has the value of promotion and application after further reducing costs. At present, high-strength polyester fiber flexible nets for mining are also used as artificial roof materials. The net laying has strong integrity, high strength, and simple construction, which reduces the labor intensity of manual operations and achieves high production and high efficiency.

4.2.1.2 Regenerated Roof

During the top layer mining, after the immediate roof with high argillaceous content collapses, under the pressure of the overlying rock strata and the injection of water or grouting into the goaf, the falling gangue can be re-cemented to form a regenerated roof with certain stability and strength after a period. The lower layer can be mined directly under the regenerated roof without laying an artificial false roof. The formation time and integrity of the regenerated roof are influenced by factors such as the composition of the strata, water content, and the magnitude of the roof pressure. Generally, this process takes at least 4–6 months, and in some cases, it may require up to 1 year. The delay time of upper and lower layered coal mining panels should be greater than the above time.

When the regenerated roof is affected by mining, the roof is easy to break and the lower layer maintaining is difficult, grouting can be used to improve the strength and stability of the regenerated roof and ensure the safety of the lower layer mining [7]. Roof grouting reinforcement technology can utilize either underground grouting techniques or surface grouting techniques. Grouting operations underground are constrained by site conditions, resulting in the use of relatively simple grouting equipment and inconvenient construction processes, making it challenging to employ high-pressure grouting. Surface grouting techniques can increase grouting pressure, however, selecting appropriate grouting areas presents difficulties [8].

When mining under an artificial false roof or regenerated roof, the roof is collapsed rock, so the periodic pressure is not obvious. The key to roof management is to protect the broken roof. The fully mechanized mining panel under the artificial roof should use shielding or supporting shielding supports, and the distance between roof beam end of the support and the coal wall should be minimized. After the shearer cuts the coal, the support should be moved immediately and supported in time. The distance between the drum and the roof mesh should not be less than 100 mm to avoid the formation of mesh bags at the coal wall or cutting of the top mesh. If the metal mesh is found to be damaged, it should be repaired in time.

4.2.2 Slicing Mining Equipment and Support Selection

4.2.2.1 Principles for Selecting Equipment for Slicing Mining

When selecting equipment for the slicing mining panel, there are no special requirements for shearers and AFCs. The focus of selection is the hydraulic support. When mining the top, middle and lower layers, the supports have different requirements [9–11].

1. Basic principles for selecting hydraulic supports:

    (1) The weight of the hydraulic support should be reasonable, which should be controlled within the transportation capacity of the mine auxiliary transportation equipment. If the weight of the hydraulic support is too heavy and exceeds the capacity of the mine's auxiliary transportation equipment, the support needs to be dismantled and transported, which consumes a lot of manpower and material resources. Therefore, the weight of the support should be reduced as much as possible, and high-strength materials should be used when necessary.
    (2) The transport height of the support should be as small as possible. The minimum height of the support is usually its transport height. A support with low height, low center of gravity and stable loading has no special requirements for the height of the transport roadway. The dimensions of the main roadways, air gates, etc. of the production mine have been fixed and can meet the requirements of support transportation without the need for modification.
    (3) The support height has a large range of variation. The hydraulic support has a large height scalability and can adapt to a wide range of mining heights. The hydraulic support can not only meet the needs of supporting panels with different mining heights, but also meet the needs of mining under conditions of faults and changing coal thickness.

(4) The support structure has strong adaptability. The design of the support should meet the requirements of the support structure under different geological conditions in the mine, so that the support can adapt to the needs of top layer, middle layer, lower layer and single coal seam mining as much as possible.

2. Requirements for hydraulic supports in top layer mining:

(1) Since the roof of the top layer panel is rock, the top layer mining has a greater requirement for the support strength of the hydraulic support than the lower layer. The support strength of the support can be determined based on the measured mine pressure or the roof type. If the main roof periodic pressure is obvious, the safety valves of the columns and balance jacks should use medium or large flow, otherwise the speed of unloading and releasing pressure of the support will be low, causing damage to the support components.
(2) The floor of the top layer panel is the coal seam, which is a soft floor. Therefore, the smaller the relative pressure of the front end of the support base to the floor, the better, which can prevent the support from piercing the bottom.
(3) The gas emission quantity of top layer mining is larger than that of lower layers, so the larger ventilation volume requires the panel support to have a larger ventilation section.

3. Requirements for hydraulic supports in middle and lower layer mining:

(1) The roof of the lower layer panel is a regenerated roof, and the floor is a coal seam or soft rock layer. The front end of the support base is required to have a lower relative pressure to the floor to prevent the support from piercing the bottom.
(2) The middle and lower layers mining requires that the maximum support height of the support should be much higher than the mining height. This will effectively support the roof when it leaky or collapses, and prevent the roof from being empty or collapsing.
(3) Hydraulic supports are matched with AFCs and shearers, and are carried out in a timely roof protection manner. In actual production, the support can be moved forward to make the support top beam close to the coal wall to prevent the roof from leaking. The hydraulic support can also be moved with pressure to prevent roof leakage and collapse.
(4) The support should have complete side protection devices.

4.2.2.2 Slicing Mining Equipment Case

1. Panel geological conditions

The 11012 panel of Zhaogu No. 2 Mine is located in the west of the first panel. It is the bottom layered panel of the 11011 panel that has been mined. The design length of the tailgate is 2038.2 m, the design length of the headgate is 2056.7 m, and the design length of the set-up room is 167.5 m. The lithology of the roof and floor of the panel is as follows:

False roof: It is composed of gray mudstone and carbonaceous mudstone in the solid coal section, with a thickness of 0–0.5 m, uneven distribution, and falls with mining. The solid coal section is between the stop mining line of the 11011 and 11012 panel.

Immediate roof: The immediate roof near the solid coal section between the stop mining line of the 11011 panel and the designed stop mining line of the 11012 panel is mainly composed of mudstone and sandy mudstone, with a thickness of 1–6.5 m and good stability.

Artificial false roof: The artificial false roof in the bottom layer section is 0.5–12 m thick and is composed of gray-black mudstone and sandy mudstone that collapsed after the top layer mining. The roof is poorly rusted and relatively broken.

Main roof: It is composed of gray sandstone with an average thickness of 6.68 m. It is mainly composed of quartz, star-shaped minerals and white mica fragments, with obvious horizontal stratification and good stability.

Immediate floor: It is composed of gray-black sandy mudstone and gray mudstone, with an average thickness of 14.06 m. It is rich in plant root fossils and easily expands when exposed to water.

Main floor: L9 limestone, dark grey, average thickness 1.33 m, locally developed calcite veins, dense and hard.

2. 11012 panel equipment selection

The 11012 panel uses the ZZ9000/16/32D hydraulic support (Fig. 4.9), with a working resistance of 9000 KN. It is estimated that 112 supports will be installed in 11012 set-up room. The technical parameters of the hydraulic supports are shown in Table 4.2, and the equipment configuration of the 11012 panel is shown in Table 4.3.

Fig. 4.9 ZZ9000/16/32D hydraulic support

Table 4.2 Technical parameters of ZZ9000/16/32 hydraulic support

Item	Unit	Parameter
Type	Four-column shielded	
Height (min/max)	mm	1600/3200
Width	mm	1430
Center distance	mm	1500
Initial support force (P = 31.5 MPa)	KN	7752
Working resistance (P = 34 MPa)	KN	9000
Pressure on the floor	MPa	3.17
Support strength	MPa	> 1.22
Pump station pressure	MPa	31.5
Operation method	Electro-hydraulic control	
Total weight	t	24
Shipping dimensions (length x width x height)	mm	5051 × 1.43 × 2685

Table 4.3 11012 panel equipment configuration

Number	Name	Models and specifications	Unit	Quantity	Note
1	Shearer	MG300/720-AWD	Unit	1	1140 V
2	Hydraulic support	ZZ9000/16/32	Unit	112	–
3	AFC	SGZ764/630	Unit	1	1140 V
4	Transfer conveyor	SZZ730/110	Unit	1	1140 V
5	Crusher	PCM-110	Unit	1	1140 V
6	Belt conveyor self-moving	ZY2300	Unit	1	–
7	Belt conveyor	DTL100/63/2 × 110	Set	2	1140 V
8	Belt conveyor	DTL100/63/2 × 75(single drive)	Set	2	1140 V
9	Emulsion pumping station	BRW400/31.5	Unit	1	1140 V
10	Clean water pump	BPW315/12.5 (two pumps and one tank)	Unit	1	1140 V
11	Combination switch	QJZ-1600/1140(660)-8	Unit	2	1140 V

4.3 Roof Control in Slicing Mining

Compared with general large mining height, the roof movement laws and fracture characteristics in the top layer mining of slicing mining are quite similar, and there is no significant difference in roof control. However, the lower layer mining possesses distinct characteristics. As mentioned above, the roof of the lower layer mining is mainly composed of artificial false roof or regenerated roof. Compared with the roof of the upper layer mining, the cementing layer of the regenerated roof has low strength and poor stability, and is prone to plastic deformation, causing the roof to delaminate and sink, which can easily lead to local roof falls and large-scale roof falls induced by local roof falls, which makes it more difficult to control the roof [7, 12]. The mine pressure behavior law of lower layer mining has the following characteristics.

(1) The static load of the support increases and the dynamic load decreases. Since the immediate roof during the lower layer mining is the regenerated roof, its overhang and integrity are poor, and the gangue has good follow-up performance, which is manifested as timely pressure. Therefore, the initial support force and working resistance of the support are both smaller than those of the top layer mining under the same geological conditions.
(2) The influence range of the front abutment pressure is reduced. Due to the pre-mining roof stratification, the strength and stiffness of the immediate roof are reduced, thereby reducing the degree of stress concentration in front of the coal wall and the range of stress concentration.
(3) The load distribution of the support changes. According to the measured results, the loads on the left and right columns of the support are relatively even, with almost no eccentric load, and the load on the front column of the support is greater than that on the rear column.

Therefore, the key technology for roof control in thick coal seam slicing mining is mainly roof control during lower slicing mining. At present, the commonly used roof management technology for slicing mining is to determine a reasonable panel layout. The roof control principle of taking protection as the main method and support as the auxiliary method should be adopted for the roof of the panel, to determine the reasonable initial support force and the control distance of the roof [13], strengthen the safety management of the panel, and improve the stability of the roof.

4.4 Roadway Support Technology for Slicing Mining

4.4.1 Roadway Support Technology

Compared with the roadway support for large mining height, the roadway support for slicing mining has a smaller cross-section, a lower degree of mine pressure, and is generally easier to support. Compared with the general large mining height roadway support, the roadway cross-section of the top layer mining roadway support technology is smaller, the roadway deformation is smaller, the support principle is similar, and there is no obvious difference in technical characteristics. Regarding the roadway support technology for lower-layer mining, the mining movement of the top-layer mining affects the stability of the roadway surrounding rock of the lower layer, making the surrounding rock of the lower layer roadway more broken and lower in strength, and the support difficulty is greater than that of the top-layer mining. The surrounding rock failure of the lower layer panel roadway in slicing mining is mainly caused by stress concentration. The coal pillars left in the upper layer to protect the roadway form stress concentration, so that the layer roadway is affected by the concentrated pressure of the coal pillars and the mining pressure during the mining process.

Therefore, the roadway support technology for slicing mining focuses on the roadway support for lower slicing mining. The commonly used support idea is to study the characteristics, advantages, and disadvantages of different roadway layout methods during slicing mining and determine the size of the roadway offset. It can also solve the problem of surrounding rock crushing in the lower layer roadway by adopting the roof pressure crushing zone grouting technology, and then select a more suitable lower layer roadway location [14]. At present, roadway support technology mainly focuses on the following aspects: (1) determining the reasonable size of coal pillars; (2) improving roadway support and improving support quality; (3) reasonably determining the order and time arrangement of slicing mining; and (4) reasonably unloading roadway pressure [15].

4.4.2 *Cases of Roadway Support*

4.4.2.1 11012 Panel of Zhaogu No. 2 Mine

(1) Cross-sectional shape: regular trapezoid, top net width 3330 mm, bottom net width 4930 mm, centerline net height 2900 mm.

(2) Support form: bolt-mesh-cable, 12# industrial steel trapezoidal shed, and fork shed combined support.

(3) The side is supported by anchor cables, with specifications of Φ21.6 mm × 4250 mm, row spacing of 1100 mm × 1000 mm, anchor length of not less than 1500 mm, arranged in a staggered pattern, and anchor cable trays of 12 mm × 200 mm × 200 mm steel plates. The side anchor cables are laid in three rows from top to bottom. The first row of anchor cables is 600 mm from the top beam, with a row spacing of 1000 mm; the second row of anchor cables is 1700 mm from the top beam, with a row spacing of 1000 mm; the third row of anchor cables is 2800 mm from the top beam, with a row spacing of 1000 mm.

(4) The roof uses a 2100 × 3000 mm metal diamond mesh, which is laid parallel to the centerline of the roadway on the long side. The mesh size is 70 mm × 70 mm, and the mesh overlap width is not less than 3 grids (210 mm). 14# electro-galvanized wire is used for double-strand double-track binding every 2 grids (140 mm) in a three-flower arrangement. The overlap between the diamond mesh on the top beam and the steel mesh on the side is tied with double-strand 14# electro-galvanized wire, with the galvanized wire head facing the coal wall side. The metal mesh on the side is welded with Φ6 mm steel bars, with a mesh width of 1100 mm × 1700 mm. The short side is laid parallel to the centerline of the roadway, with the longitudinal reinforcement facing inward and the transverse reinforcement facing outward. The mesh overlaps 100 mm,

and 14# electro-galvanized wire is used for diagonal binding in each grid, with the galvanized wire head facing the coal wall side.

(5) I-beam sheds are uniformly made of 12# industrial steel. The roadways are supported by 3600 mm (beam) × 3200 mm (column) regular trapezoidal industrial steel single sheds, with a shed distance of 500 mm, shed leg feet of 800 mm, and column socket depth of not less than 200 mm.

(6) Wooden pads must be added at the interface between the I-beam roof beam claw and the shed leg, and at the interface between the I-beam roof beam head and the shed leg. The specifications of the wooden pads are 10 mm × 90 mm × 120 mm. The roof is backed with thorn sticks and diamond-shaped metal mesh. The spacing between the thorn sticks is 500 mm and the diameter is not less than 30 mm. All thorn sticks should be evenly and firmly laid, with both ends penetrated.

(7) Six connecting plates are arranged on each shelf, one at 500 mm from both ends of the top beam, and one at 200 mm and 2600 mm from the top of the column legs. The connecting plates are made of 12 mm thick steel plates, and the U-shaped clips used to fix the connecting plates are made of 20 mm diameter round steel. 12# channel steel is welded on the upper end of the bracket column leg as the limit groove of the industrial steel top beam, with a length of 125 mm. 12 mm thick steel plate is welded on the lower end of the bracket column leg, with a specification of 150 mm × 120 mm, and the long side is consistent with the construction direction of the roadway. The support section is shown in Fig. 4.10.

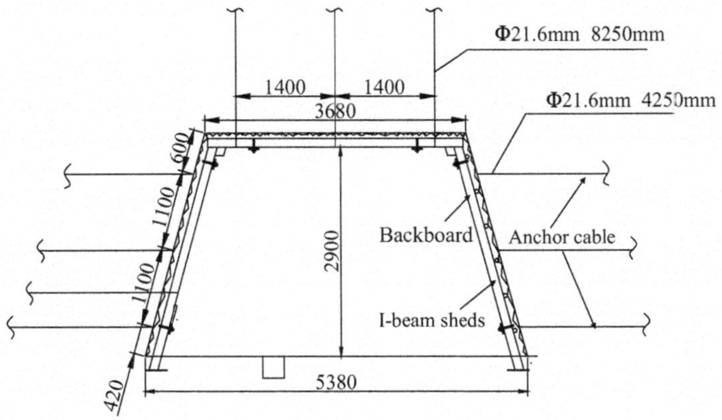

Fig. 4.10 Roadway support scheme for the panel of Zhaogu No. 2 Mine

4.4.2.2 Laogongyingzi Coal Mine [16]

1. Project overview

The average thickness of the 5# coal seam in Laogongyingzi coal mine is 14.2 m, the average inclination is 4°, the coal type is lignite, the structure of the coal seam is simple, the gangue lithology is mainly mudstone and siltstone, and carbonaceous mudstone can be seen in some parts. The roof is mainly composed of mudstone and fine-grained sandstone, and locally coarse-grained sandstone. The roof contains water, the lithology is relatively loose, mud cemented, and it swells when it encounters water and is easy to break. The floor is mainly fine-grained sandstone and siltstone, which are loosely cemented, swell when it encounters water, and are easy to break. The 5# coal seam is divided into three layers and mined in layers using a vertical roadway layout. Each layer adopts comprehensive mechanized coal mining, and a 2.0 m coal seam is left between the layers as a false roof.

2. Support method

The I05(8)2 panel is the middle layer panel of Sect. 4.8, with a panel length of 196 m and an advancing length of 1241 m. Above it is the goaf of the I05(8)1 panel, and to the west are the goafs of the upper layer I05(7)1 panel and the middle layer I05(7)2 panel. The section coal pillar is 9 m. The roadway cross-section is trapezoidal, with dimensions of 3000 mm in upper width, 5010 mm in lower width and 3160 mm in height. The original roadway adopted the support method of "frame anchor cable and steel strip", with 6 rows of I-beams laid on the roof with a spacing of 500 mm, and then the backboard was laid. Six rows of I-beams are laid on both sides with a spacing of 500 mm. The backboard is laid and then anchor cables are used for support. The anchor cables are Φ22 mm × 4200 mm high-strength prestressed low-relaxation steel strands. Three strands are laid in each row with a spacing of 700 mm, as shown in Fig. 4.11.

3. Roadway deformation:

   (1) Severe bulging and spalling of the coal pillars in the roadway occurred. The bulging occurred over a range of more than 15 m along the roadway direction, with a displacement of about 1500 mm. The bulging caused the I-beams to undergo significant bending deformation. After the bulging reached a

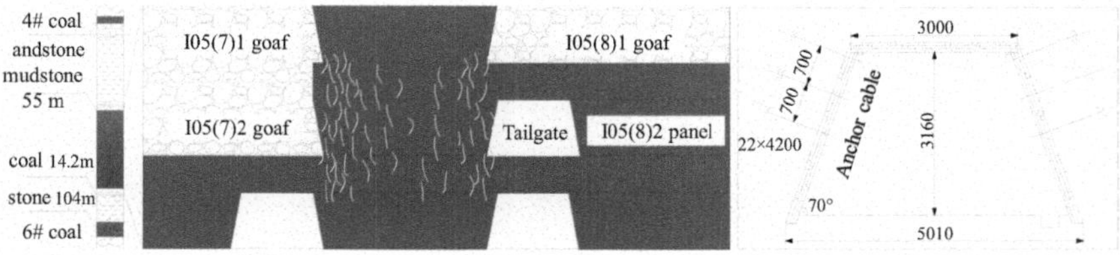

Fig. 4.11 Roadway layout and support scheme of panel in Laogongyingzi mine

certain extent, it developed into large-area spalling, resulting in large cavities between the I-beams and the coal pillars. The spalling depth in severe areas reached more than 1000 mm, causing the failure of support bodies such as anchor rods, I-beams, and backboards. Moreover, large deformation of the coal pillars continued to occur after the expansion and repair.

(2) The deformation degree of the solid coal wall is lower than that of the coal pillar wall. During the roadway excavation stage, the deformation is small. When affected by the disturbance of the panel mining, the roadway is severely deformed. The displacement of the bulge wall ranges from 500 to 750 mm. Some I-beams are bent and deformed, and the axial damage range along the roadway is about 3 m.

(3) The amount of roadway roof subsidence is small, mainly occurring within 20 m in front of the panel, with a subsidence of about 300 mm, and the steel strip slightly bending and deforming. There is a bottom bulge in the roadway within 20 m ahead of the panel, with the maximum measured bottom bulge of about 200 mm, which has little impact on the safe production of the panel.

4. Roadway reinforcement scheme

(1) Coal pillar side support: Based on the original support, a 100 mm thick C20 concrete layer is sprayed on the surface of the roadway wall, and then grouting is carried out behind the wall. The grouting holes are arranged perpendicular to the coal wall at a spacing of 1200 mm. The diameter of the grouting holes is 50 mm, and the grouting depth is 4500 mm. The grouting material is Malisan materials, and the grouting pump is a ZBQ-5/12 pneumatic double-liquid grouting pump. The pump station pressure is set to be greater than 6 MPa. Once leakage or slurry leakage is found, grouting is stopped. In addition, it takes a certain amount of time for the slurry to solidify. Considering the influence of overburden movement on the grouting effect, it is chosen to carry out grouting reinforcement in the roadway 150 m in advance. Anchor rods are installed on the side of the roadway. The parameters of the anchor rods are Φ20 mm $\times$ 1800 mm, and the spacing between rows is 800 mm $\times$ 1000 mm. The anchor rods are arranged vertically to the surface of the side of the roadway, as shown in Fig. 4.12.

(2) Roof support: On the basis of the original support, a C20 concrete layer with a thickness of about 80 mm is sprayed on the roof.

(3) Solid coal side support: Based on the original support, a C20 concrete layer of about 100 mm is sprayed on the surface of the roadway wall. No grouting is performed on the coal body.

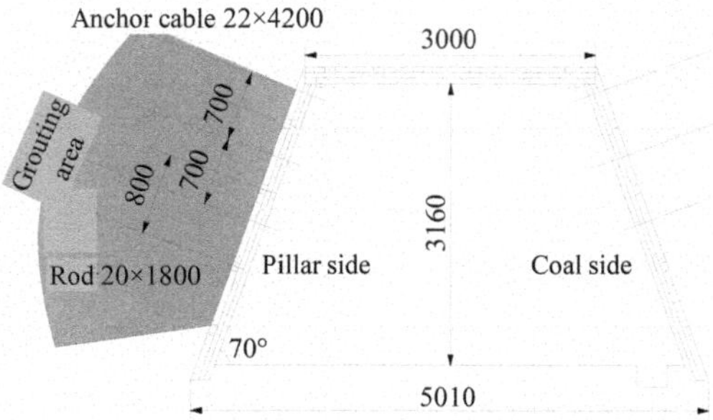

Fig. 4.12 Reinforcement and support scheme for roadways at the panel of Laogongyingzi Mine

4.4.2.3 Yuecheng Coal Mine [17]

1. Project overview

Yuecheng Coal Mine is affiliated to Qinxiu Coal Industry Co., Ltd. of Shanxi Jincheng Coal Industry Group, with a designed production capacity of 1.5 million t/a and a mine area of 13.81 km^2. Currently, the main coal seam mined is No. 3, with a coal seam thickness of 5.04–7.16 m and an average of 6.11 m. Since the mine is an integrated mine of Jinmei Group, if the large mining height technology or LTCC technology is adopted, the shaft, roadways, mining equipment and other facilities of the original mine cannot be used anymore, and a large amount of funds must be invested to transform the mine. Since Jinmei Group has rich experience in slicing mining and mature mining technology, after comprehensive consideration, the mine still adopts the slicing mining method, which helps to reduce initial investment and shorten the construction period.

2. Support schemes

Roof support: The top anchor rod adopts left-handed threaded steel anchor rod with a diameter of 22 mm and a length of 2200 mm. The spacing between anchor rods is 800 mm × 900 mm. They are all arranged vertically on the roof. The tray specification is 150 mm × 150 mm × 10 mm. One anchor with a specification of K2335 and one anchor with a specification of Z2360 are used. A thickened steel strip with a length of 3.8 m is used. 6# steel mesh is selected for roof protection with a mesh specification of 80 mm × 80 mm. The diameter of the roof anchor cable is 21.6 mm, and the length of the steel strand anchor cable is 7300 mm. It is arranged in a rectangular shape with a spacing of 2000 mm × 1800 mm.

Two-side support: The side anchor rods are made of fiberglass anchor rod with a diameter of 18 mm and a length of 2000 mm. The spacing between the anchor rods is 900 mm × 900 mm. Three anchor rods are arranged in each row of each side, and they are arranged vertically on both sides. A low-viscosity resin anchoring agent with a specification of K2335 is used. The side is protected by a diamond mesh with a mesh size of 50 mm × 50 mm and a mesh size of 3300 mm × 1100 mm. The support section is shown in Fig. 4.13.

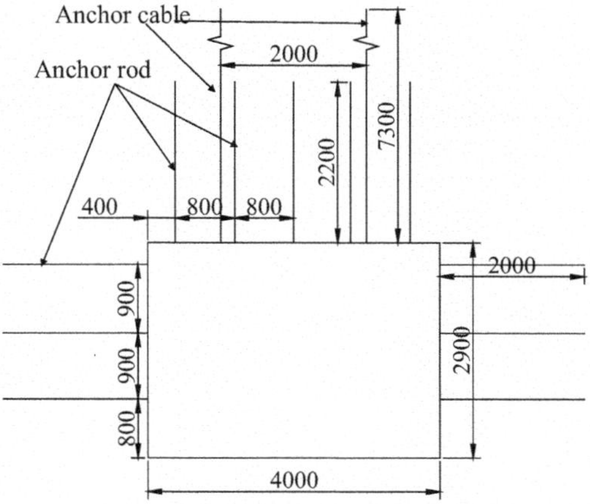

Fig. 4.13 Roadway support scheme for panel of Yucheng coal mine

References

1. Wang JC, Zhong SH (2008) The present status and the key issues to be resolved of thick seam mining technique in China[J]. China Science paper Online 3(11):829–834
2. Wang JC (2009) Theory and technology of thick coal seam mining[M]. Metallurgical industry press, Beijing
3. Qian MG, Xu JL, Wang JC et al (2021) Ground pressure and strata control[M]. China University of Mining and Technology Press, Xuzhou
4. Meng X R, Wang H P, Liu C H, et al. Selection principles and development status of thick seam mining methods in China[J] Coal Science and Technology, 2009, 37(1): 39–44.
5. Wang JC (2023) 40 years development and prospect of longwall top caving mining in China[J]. Journal of Coal Society 48(1):83–99
6. Yan WG (2007) Vertical gateway layout technology for thick seam slicing mining in Xingtai Mine[J]. Coal Science and Technology 35(3):13–15
7. Chi XL, Yang K, Fu Q et al (2023) Mechanical behavior and stability control of regenerated roof in long wall stratified mining of thick steeply dipping coal seam[J]. Coal Science and Technology 51(6):1–10
8. Zhai XR, Wu JW, Hu R et al (2022) Study on surface pre-grouting reinforcement layer of stratified regenerated roof under bifurcated coal seam[J]. Coal Science and Technology 50(11):30–39
9. Wang JG (2015) Steeply inclined thick coal seam slicing mining working face equipment selection research[J]. Coal Mining Machinery 36(9):228–231
10. Yang SL, Zhang P, Li FS et al (2010) Load determination of powered supports for fully mechanized top coal caving mining working face in horizontal slice of steep inclined thick seam[J]. Coal Science and Technology 38(11):37–40
11. Yang SZ, Zhai XX, Kang QY et al (2003) Study on reproducing roof strata and the equipment forming a complete set of thick coal seam slicing[J]. Journal of Jiaozuo Institute of Technology (Natural Science Edition) 22(1):5–8
12. Chi XL, Yang K, Fu Q et al (2022) Study on stress distribution law of regenerated roof in fully-mechanized Slicing Mining of steeply dipping coal seam[J]. Journal of Mining and Safety Engineering 39(5):891–900
13. Zhao H S. Constitution and control of regenerated roof[J]. Coal Science and Technology, 1993, 5: 2–5+63.
14. Zhang N, He YN (1994) The influence of shear stress on the layout of lower slice roadways[J]. Journal of Mining & Safety Engineering 1:37–40

15. Zhou L X. Layout and supporting of roadways in slice mining of thick coal seam[J]. Coal Mining, 2002, (4): 19–20+26.
16. Li GZ, Sun LJ (2020) Instability mechanism and support technology of coal pillar roadway in slice mining[J]. Coal Engineering 52(12):46–50
17. Li ZQ (2018) Research on Layout and Supporting of Slice Mining in Thick Coal Seam[J]. Energy Technology and Management 43(6):51–53

Chapter 5
Intelligent Mining

The automated mining of fully mechanized mining panel started in 1984 in Mogali County coal mine in West Virginia, the US. The landmark technology is the first use of electric and hydraulic control support, realizing the automatic propelling of hydraulic support, reducing the operating time of the hydraulic support, and guaranteeing the support of the initial support force and the propelling of the norms in place. 1990 later, the fully mechanized mining panel of the shearer, hydraulic support, AFC, and "three machines" automation technology is a mature and reliable operation. In the past 10 years, China has also made significant advances in intelligent mining technology in coal mines through the adoption of advanced automation technologies from the United States, Germany, and Australia, as well as through extensive independent research and development. This has entailed the integration of automation technology for the fully mechanized mining panel with artificial intelligence and big data technologies. Additionally, the country has formulated corresponding technical standards, which have greatly accelerated the advancement of intelligent mining technology. Among these developments, remote control and intelligent mining based on machine vision have made the most rapid progress [1].

5.1 Development History of Intelligent Coal Mining in China

5.1.1 Basic Concepts of Intelligent Mining

With the comprehensive development of the fully mechanized mining technology and the application of a new generation of information technology, intelligent mining technology has come into being, promoting China's major technological change from the fully mechanized coal mine to coal mine intelligence. Intelligent mining

J. Wang and W. Wei, *Thick Coal Seam Underground Mining*,
https://doi.org/10.1007/978-981-96-3874-1_5

refers to the application of the internet of things, cloud computing, big data, artificial intelligence, and other advanced technologies, so that the panel shearer, hydraulic support, AFC and hydraulic power equipment, etc. to form an intelligent system with independent perception, independent decision-making and automatic control of the operation function, to realize the panel of the falling coal (cutting or top coal caving), support, coal transport operation working conditions of the self-adaptive and process synergistic control of the mining method [2, 3].

For the thick coal seam LTCC panel, intelligent control and manually assisted top coal drawing mode can be adopted. Since the cutting height of the shearer is not limited by the thickness of the coal seam in LTCC panel, it is not necessary to use the intelligent height adjustment technology, however, it is still necessary to intelligently control the amount of the lower drum of the shearer according to the change of the undulation of the bottom plate of the coal seam. Due to the complexity of the site environment, the intelligent coal drawing technology is in the preliminary application stage of research and development.

For the large mining height panel, the large mining height intelligent coupling man–machine cooperative and efficient comprehensive mechanized mining mode can be adopted. Due to the large mining height panel being mostly heavy equipment, and mining height increase leads to equipment stability deterioration, the heavy equipment group is prone to interference. The control precision of the existing large mining height mining panel intelligent mining equipment, intelligent cooperative control precision is still difficult to meet the requirements of unmanned mining, thus the intelligent mining of the fully-seam mining panel should be based on intelligent manipulation, supplemented by man–machine cooperative control.

5.1.2 Development of Intelligent Mining

In the 1960s, China began to do coal mine automation control related research and experiments. Coal mines used a special cable for simple equipment control, the empty contacts as sensors, small bulbs or other objects to show the equipment on and off, the alarm clock for timekeeping. This stage cannot be strictly called automation, this stage of the automation development of China's coal mining process is slow.

In the 1980s, with the development of microcontroller technology, initial attempts were made in the application of computerized centralized monitoring of local production links in mines, such as the monitoring technology of the belt conveyor in the No. 2 shaft of Yangquan No. 3 Coal Mine, and the centralized control of the clean coal loading station of the coal washing plant of the Jiahe Coal Mine, and so on. In addition, Xi'an Coal Mine of Liaoyuan Mining Bureau actively promoted the application of electronic computers with the great assistance of various departments, and made some progress in gas monitoring. With different types of gas telemetry alarms and disconnecting devices emerging, the individual mines in mine ventilation automation have been installed with remote monitoring and remote-control devices, which enable the staff to remotely control the starting and closing of the downhole air doors and the opening and stopping of the local fans from the control room on the ground. With the increasing mechanization of coal mine production, research in coal mine automation technology has made significant strides, achieving promising results. For instance, advancements have been made in centralized control devices for AFCs, communication signals, and the CK-2 machine in mining panels. Developments also include shaft telephones, signaling devices for slanting shaft manned vehicles, and automatic telephone dispatching communication systems for mines. Notable innovations that reached the international level in the 1980s, such as the KJ-1 machine type for mine environment monitoring and production monitoring systems, have been particularly impactful. These scientific research achievements have been applied in production, yielding substantial economic and social benefits. However, despite these advancements, the overall level of coal mine automation in China remains significantly behind that of advanced coal-producing countries worldwide.

In the 1990s, China began to promote the programmable logic controller (PLC) automation system in coal mines and achieved better results. Subsequently, large-scale control systems represented by distributed control system (DCS) and supervisory control and data acquisition (SCADA) were gradually applied.

Since entering the twenty-first century, China's coal mining automation has been further developed, and China's safe and efficient fully mechanized mining technology and equipment have been rapidly developed. With the development of coal mining technology, the requirements for coal mine automation have been gradually improved, and the number of underground equipment is large. The cooperation and interaction between various systems are very important, which also brings big problems for the development of coal mining automation. In recent years, with the development of the industrial Internet, and coal mines in the realization of multi-system aggregation, the establishment of a comprehensive automation platform has been the rapid development of the realization of the interaction between the information.

With the transformation of China's economic development mode, the coal industry is transforming from the rough production mode to intensification and refinement, and intelligent mining has become the development direction and inevitable trend of safe and efficient coal mining. "Action Plan for Energy Technology Revolution and Innovation (2016–2030)", "State Administration of Safety Supervision on the implementation of "mechanized replacement, automation reduction" technology strong safety special action notice" both list the intelligent coal mining technology as a key research direction.

In February 2020, The National Development and Reform Commission, the National Energy Administration and other eight ministries and commissions jointly issued the "Guidance on Accelerating the intelligent Development of Coal Mines", pointed out the need to accelerate the reform of the supply structure and high-quality development of the coal industry, which was a milestone for the development of China's coal industry, and also formally drew the curtain on the construction of coal mine intellectualize.

In November 2020, China Coal Industry Association, China Coal Science and Industry Group and Coal Mine Intelligent Innovation Alliance jointly released the

"China Coal Mine Intelligent Development Report (2020)", which systematically summed up the basic situation of China's coal mine development and informatization construction, elaborated on the progress of research on the basic theories and key technologies of coal mine intelligence, described in detail the construction practice of the intelligent demonstration coal mines, and laid out the standard system of coal mine intelligent construction.

At the end of 2020, the National Development and Reform Commission, National Energy Administration launched the first batch of 71 national intelligent coal mine construction demonstration projects, making every effort to promote the demonstration and cultivation of intelligent construction, and accelerating the enhancement of the industry's intelligent level.

Since 2020, the National Energy Administration, together with seven departments, has accelerated the intelligent construction of coal mines, focusing on the use of 5G communications, artificial intelligence, and other new generation of information technology to transform and upgrade the traditional coal industry, which can help coal mines to achieve the reduction of personnel, increase safety, improve mining efficiency, and the role of coal as the bottom of the pocket in energy security is further consolidated. Up to now, Chian has built a total of 1043 intelligent coal mining panel and 1277 intelligent driving panel. In recent years, coal-producing provinces and regions, in collaboration with relevant state departments, have initiated a comprehensive program of intelligent construction in coal mines. This initiative has entailed a significant acceleration in the mechanization of manpower replacement, automation to reduce manpower, and the introduction of intelligent technologies that require fewer personnel. The objective of this program is to optimize and enhance the quality of coal mine production and safety, while simultaneously strengthening the resilience and elasticity of the coal supply chain. This approach has laid a solid foundation for guaranteeing the safe and stable supply of coal. The National Energy Group, China Coal Group, and seven other key coal enterprises have constructed an intelligent production capacity of 1.393 billion t/a, representing 74.7% of total production capacity and an average staff labor efficiency of the comprehensive mechanized mining panel has increased by 27.7%, while the average level of digging for a comprehensive single roadway has increased by 32.8%. Over the past two years, intelligent coal mines have played a pivotal role in enhancing coal production and ensuring a reliable supply of coal. Furthermore, the development of coal mine intelligence has strongly promoted the transformation and upgrading of coal upstream and downstream industries. Preliminary statistics, at present, China's coal mine intelligent construction total investment of nearly 200 billion yuan, the investment completion ratio of more than 50%, the increase in market demand for intelligent, effectively driven by the rapid development of mine Internet of Things, coal machinery and equipment manufacturing, intelligent control systems, safety monitoring and early warning and other new industries and new business models.

5.2 Intelligent Mining Equipment and Control Technology

Shearers, hydraulic support, and AFC are the three most important mining equipment in the fully mechanized mining panel, commonly known as "three machines". The left and right traveling wheels of the shearer ride on the pin row of the AFC, which is in a state of mutual engagement, and the hydraulic support is arranged on the top of the AFC, which protects the mining workspace as well as isolates the airspace. These three kinds of equipment cooperate and work together to complete the coal mining process, that is, "three machines" matching. The automation and intelligent linkage control of these three pieces of equipment is the foundation of intelligent coal mining.

5.2.1 Shearer Memory Cutting and Coal-Rock Recognition Technology

5.2.1.1 Shearer Memory Cutting

The shearer memory cutting technology refers to setting up a set of memory systems on the shearer, which can record every step of the shearer's operation in the working process, and optimize the cutting parameters of the operation according to the actual situation. In this way, in the follow-up work, the parameters can be adjusted adaptively for the coal mining operation of the same type of coal seam to improve the operation efficiency and cutting accuracy. In the working process of the shearer, there are three directions of movement, which are the reciprocating traction movement along the AFC direction, the pushing movement to the coal wall direction, and the up and down movement of the drum. In a cutting cycle, because the thickness of the coal seam is changing, the shearer driver needs to constantly adjust the height of the drum to make the cutting edge of the drum close to the top of the coal seam as much as possible. After a cycle of cutting, the hydraulic support pushes the AFC towards the coal wall, and then the hydraulic support follows up, completes the pushing action, and is ready to start the next cutting. Furthermore, because the change of coal seam thickness is slow, in each cutting cycle, for the same position of the shearer, the suitable height of the drum is always not much different, so the path of the drum cutting the next knife can be determined by the information of the position and inclination of the shearer. Only an experienced shearer driver is required to carry out a manual demonstration operation, and the subsequent operation can be carried out automatically according to the cutting path of the first knife.

Its working principle is shown in Fig. 5.1. When the driver operates the shearer for the first demonstration cutting, the various types of sensors on the panel collect the position and attitude data of the shearer in real-time and store them in the computer. The information includes traction direction and velocity, drum position, panel length, and transverse and longitudinal inclination. In the automatic cutting

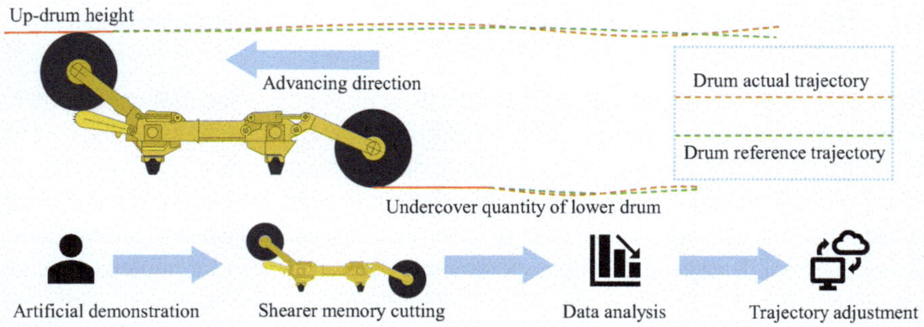

Fig. 5.1 Memory cutting principle [2]

mode, the computer determines the drum height of the next cutting cycle through historical data, to realize the path memory. Taking the direction of the panel as the positive direction of the *x*-axis, the advancing direction of the shearer as the positive direction of the *y*-axis, and the upward adjustment direction of the drum as the positive direction of the *z*-axis, a spatial rectangular coordinate system is established. During the first cutting, *N* sampling points are evenly set on the *x*-axis, and the corresponding shearer pose information and the drum height of each point at that time are recorded as the demonstration cutting path, recorded as $\{A_1, A_2, \ldots, A_n\}$. During the second cutting, the real-time pose information of the shearer and the cutting path of the previous cutting can be combined. Repeat the first cutting path $\{B_1, B_2, \ldots, B_n\}$ as a cycle for coal mining operations. During the cycle, if the shearer driver finds that the drum height adjustment is not appropriate, it can be manually adjusted, and the adjusted path will be remembered as a new cutting path. It should be noted that due to the problem of control accuracy, the number of cycles of memory cutting should not be too many, generally 4–5 cycles are appropriate. After the end of the cycle, the shearer needs to be set to manual operation mode again, the benchmark parameters are set again, and then the demonstration cutting is carried out.

The infrared transmitter and inertial navigation system are installed on the shearer. The inertial navigation system is composed of an inertial navigation component, airborne antenna, wireless base station, shearer, and centralized control center communication. The inertial navigation component is installed on the shearer and protected by an explosion-proof hood. The shearer inputs 3300 V power to transmit the inertial navigation data to the centralized control center. The shearer is equipped with CPE (wireless CPE is a mobile signal receiving mobile signal and wireless WIFI signal forwarding mobile signal access equipment, which can replace wireless client equipment such as wireless network card) and an antenna. It cooperates with the radio base station fixed on the rack for the communication system and the support electro-hydraulic control system. After the shearer learns 1–2 cutting cycles through the set learning mode, it realizes automatic memory cutting, and the support automatically follows the shearer by setting the following mode. And through the inertial navigation system automatic alignment, as shown in Fig. 5.2.

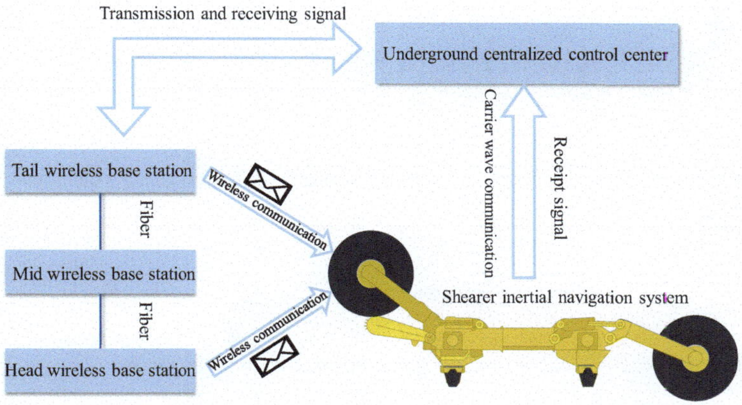

Fig. 5.2 The inertial navigation system of the shearer

Generally, the wireless base station is installed at the head, the tail, and the middle of the panel to cover the whole panel. If the panel is too long, it can be increased by several in the middle according to the specific situation. The wireless base station is connected to the panel video switch, which is transmitted to the roadway network by Gigabit cascaded Ethernet. It is mainly used for high-speed data communication between the centralized control equipment and the inertial navigation system. The automatic straightening of the shearer mainly includes four steps: alignment, measurement, calculation and straightening, and repeated execution.

5.2.1.2 High-Frequency Electromagnetic Wave Detection Technology of Coal-Rock Interface

Memory cutting is generally applicable to the panel with stable coal seam occurrence. The United States and Australia all use this technology. When the thickness of a coal seam changes greatly, it is necessary to identify the coal and rock in the panel in real-time. At present, the coal-rock interface cut by a shearer is mainly detected by image, drum output power, hyperspectral, and other technologies. In recent years, the China University of Mining and Technology-Beijing has developed a technology for detecting coal-rock interfaces by using high-frequency electromagnetic wave methods. As shown in Fig. 5.3, the shearer is equipped with a special geological radar, which transmits high-frequency broadband electromagnetic waves through the signal-transmitting antenna, and the receiving antenna receives the reflected wave from the target medium interface. When the electromagnetic wave propagates in the medium, its path, electromagnetic field strength, and waveform will change with the electromagnetic properties and geometric shape of the medium. Therefore, according to the travel time (also known as two-way travel time), amplitude, and waveform data of the received electromagnetic wave, the distribution of the target medium and the position of the coal-rock interface can be inferred.

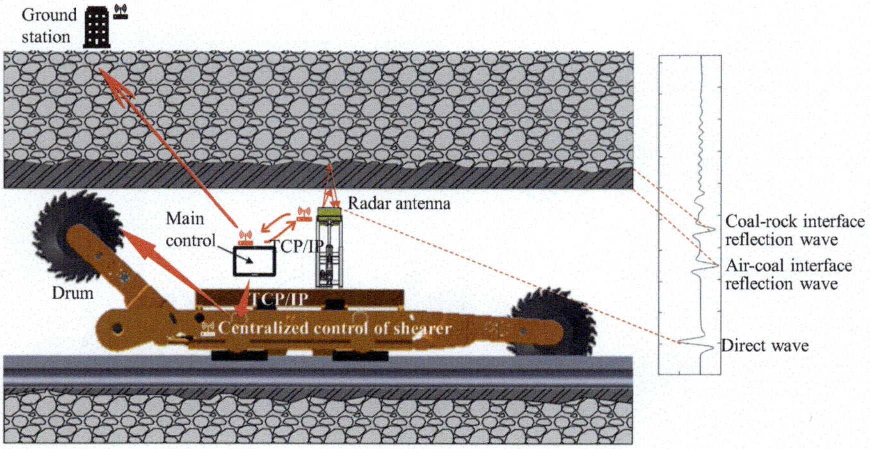

Fig. 5.3 High-frequency electromagnetic wave detection diagram

As shown in Fig. 5.3, using the airborne non-contact detection radar antenna of shearer, the information on top coal thickness or the position information of the next coal-rock interface can be obtained in real-time. The technical core lies in the design of the non-contact detection radar antenna, the characteristics of antenna energy reflection loss, and the spatial distribution characteristics of electromagnetic radiation. The antenna radiation surface and shielding device with good impedance matching are designed to improve the antenna radiation directivity, electromagnetic energy focusing, and emission efficiency, and shield against the interference of external electromagnetic wave signals. Based on obtaining the radar data of the coal-rock interface with a high signal-to-noise ratio, the information on the air-coal interface and coal-rock interface is obtained by combining the intelligent recognition and tracking algorithm of the interface, and the discrimination mechanism of the interference source is established to ensure the robustness of recognition and tracking algorithm in complex mine environment. The antenna intelligent support uses the air-coal interface information as the control signal to realize its attitude adaptive control, maintains the antenna suspension distance, and sends the coal-rock interface information and position information to the shearer control system to realize the dynamic update of the 3D geological model. Furthermore, it can realize the independent operation of the shearer on the coal seam cutting operation [4–7].

1. Air-coupled radar antenna

As the basic unit of radar radiation or receiving electromagnetic waves, the performance of the antenna directly affects the positioning accuracy, detection resolution, and target recognition level of the whole radar system. It is an important direction in the research and development of radar systems to make the antenna have broadband, high gain, and high emissivity. To realize the rapid and real-time detection of the coal-rock interface in a mine environment, the radar antenna is required to adopt the non-contact detection method. After the antenna is placed in the air, its coupling mode will change, which will produce strong standing wave interference and be easily interfered with by the surrounding environment. Therefore, the design of an air-coupled radar antenna with excellent focusing characteristics is the key to the detection of coal-rock interface with mining [8]. The center frequency of the antenna can be designed according to the depth requirement of the measured target body, usually 900–1500 MHz.

2. Adaptive control of intelligent hydraulic support

The fluctuation of the coal seam roof in the panel changes, and the distance between the air-coupled radar antenna fixed on the upper part of the shearer and the roof also changes. If the distance is too large, it will directly affect the quality of the radar echo signal. If the distance is too small, it will directly affect the safety of the air-coupled antenna. In the process of mining, the safe and effective detection distance between the detection device and the top coal surface is set. The air coupling distance between the antenna and the roof surface is picked up by radar data, and the variation of the air coupling distance is detected. According to the geometric relationship of the intelligent hydraulic support device, the expansion and contraction of the support structure are adjusted accordingly, to realize real-time intelligent adjustment, avoid interference damage between the instrument device and the environment, and ensure the effective coupling distance of radar detection [9].

3. Intelligent identification and tracking of coal-rock interface

In radar data, it is the key to accurately identify the coal-rock interface by defining the data zero time and the accurate echo time of the reflection interface (i.e., the corresponding depth position). Through forward modeling and physical simulation, the reflection characteristics of the 'air-coal-rock' horizon under the condition of the whole space roadway in the mine are analyzed. Then, the interface detection accuracy and error under different suspended height antennas and different coal seam thickness models are analyzed respectively. According to the 2D map of the relative error of the layer thickness in the range of the main lobe extreme point and the least square method, the relationship between the sequence numbers of the layer samples is obtained, and then the coal and rock layer location selection scheme is obtained, which ensures the accuracy of the layer identification and the stability of the system [10, 11].

In the process of tracking the radar map of the coal-rock interface, there will be some gangue mixed in the coal-rock medium, or some interference sources such as fracture zone and gas-bearing zone, these phenomena can easily cause wave diffraction, that is, some waveforms similar to hyperbolic curves are formed on the interface, which affects the tracking effect. When a tracking error occurs at a certain position, the result of the horizon track that leads to subsequent errors is called the 'string layer'. The horizon tracking algorithm is established based on the idea of a 'multi-level window', and the interference source judgment and error correction mechanism are established to reduce the occurrence of the 'string layer' phenomenon and improve the robustness of the horizon tracking algorithm [12].

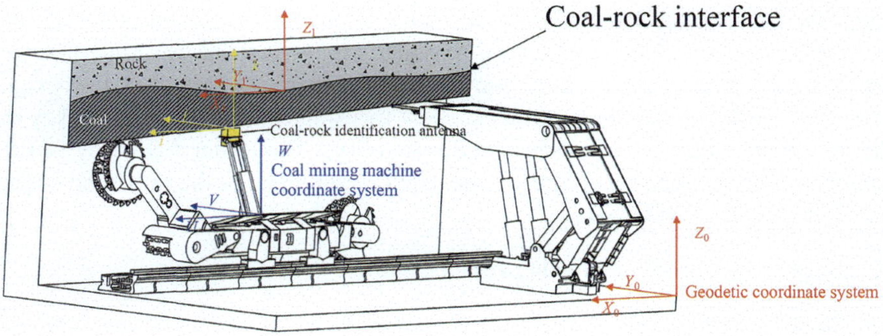

Fig. 5.4 Spatial coordinate transformation of coal and rock strata position

4. Coal-rock interface positioning and information interaction

In the application of coal-rock recognition, the air-coupled radar antenna is installed in the shearer body through intelligent hydraulic support, and the coal-rock interface is detected in real-time during the mining process of the shearer. Combined with the inertial navigation unit of the shearer, the position and posture of the intelligent hydraulic support, and the measured coal thickness data of the radar antenna, the spatial relationship model of the geodetic coordinate system (X, Y, Z), the shearer coordinate system (U, V, W) and the antenna coordinate system (I, J, K) is established (Fig. 5.4), and the absolute coordinate position of the coal and rock strata is obtained.

The radar data of the coal-rock interface is transmitted to the main control unit in real-time through wireless WiFi or Ethernet through detection. The main control unit processes the accurate position data obtained by identification tracking and space conversion and sends it to the centralized control system of the shearer in real-time through the local area network. The centralized control system of the shearer analyzes and guides the ranging arm action to realize the position adjustment of the cutting drum. Furthermore, the underground 5G network is used to transmit the position information of the coal-rock interface to the ground centralized control center in real-time, which is used for the intelligent platform to dynamically update the 3D geological model of the coal mine to meet the intelligent, dynamic and real-time requirements in the production and construction of the mine.

This technology is used to identify the coal and rock strata in the Jinjie Coal Mine of Shendong Coal Group Company by detecting the thickness of coal left after roof cutting. There is 10–20 cm coal seam under the roof to avoid soft immediate roof leakage. Through the coal-rock recognition system, the thickness of the remaining coal seam can be detected as the basis for guiding the next coal cutting, to ensure the maximum recovery of coal resources under the condition of setting the reserved coal thickness. In the process of mining, some areas are selected to excavate and measure the reserved coal seam of the roof, which is verified by comparison with the detection results. As shown in Fig. 5.5, the results show that the average error of the detection data of the coal-rock recognition system is $\pm$ 1.23 cm, and the average error percentage is 8.632%.

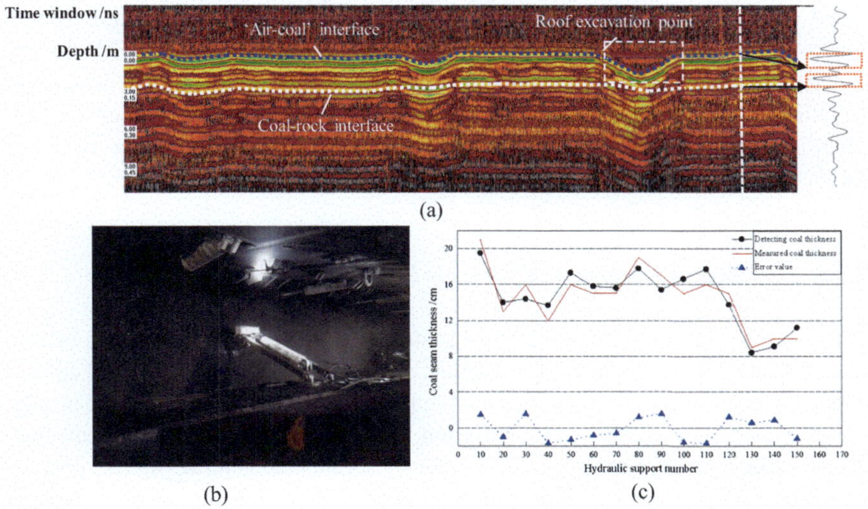

Fig. 5.5 **a** radar map of coal-rock recognition; **b** filed application of coal-rock recognition system; **c** error analysis of coal-rock interface identification

5.2.2 Hydraulic Support Electrohydraulic Control System

5.2.2.1 Intelligent Development Process of Hydraulic Support

In the 1950s, the UK developed the world's first stack hydraulic support, which was successfully applied in the Coleman coal mine in Australia. In 1985, the second-generation electro-hydraulic control system was developed. The system was designed to install a console in the transportation roadway, and the control of the connected support was completed through remote control, thus realizing the centralized control of the entire panel support. Furthermore, Germany, the United States, and other countries began to study the electro-hydraulic control system. PANZERMATIC-E system developed by Westphalia and Siemens is the first electro-hydraulic control system of hydraulic support in Germany. From the 1990s to the beginning of the twenty-first century, electro-hydraulic control technology has made rapid progress and achieved large-scale application. Taking the application of electro-hydraulic control systems in the United States as a reference, at the end of 1994, there were a total of 81 fully mechanized panels in the United States, of which 73 were equipped with electro-hydraulic control systems. The application of electro-hydraulic control has reached more than 90%, and the proportion is close to 92.7% after two years. Compared with western countries, China's research on hydraulic support and its electro-hydraulic control technology started late. In the 1970s, hydraulic supports from Germany and the United Kingdom were introduced. In 1991, Zhengzhou Coal Machinery Plant developed the electro-hydraulic control system DYZK-l and began to test and apply it. In 1996, the YLT electro-hydraulic control system was developed by the Coal Academy of Sciences, and the test application was carried out. In March 2008, the electro-hydraulic control system developed by Zhengzhou Coal Machinery Plant obtained the relevant qualification certificate through the test and certification of the National Test and Testing Center. At present, China's hydraulic support electro-hydraulic control system technology has matured, which can meet the needs of intelligent mining of various coal seams.

5.2.2.2 Intelligent Control of Hydraulic Support

At present, the electro-hydraulic control system of coal mine hydraulic support is a system that integrates mechanical, electronic computer technology, industrial control technology, network communication technology, and other technologies. Through the mutual parameter setting of the program, the position of the shearer is detected, and the support is driven to complete various actions, to achieve the purpose of automatic coal mining. Realize the automation of coal mining to reduce the number of mining panel workers, and reduce the labor intensity of coal mining workers.

The core component of the electro-hydraulic control system of the coal mine support is the controller. All the function keys of the hydraulic support control are integrated into the panel of the controller. The core executive component of the electro-hydraulic control is the pilot valve, which converts the electrical signal into a hydraulic control component. The hydraulic support is the core equipment for the safety support of the mining panel. There are usually hundreds of hydraulic supports in the long-wall fully mechanized panel to complete the roof timbering, coal wall protection, AFC, and other processes at the same time. Since the maturity of electro-hydraulic control technology, the movements of hydraulic support, such as lifting, lowering, moving, protecting sides board, and stretching the beam, have been able to realize program control. At present, in the intelligent fully mechanized panel, the key technologies of intelligent control of hydraulic support mainly focus on three aspects: hydraulic support attitude perception, self-moving support following the shearer, and straightness adjustment of the panel. Figure 5.6 is a diagram of the automatic straightening of hydraulic support in a fully mechanized panel.

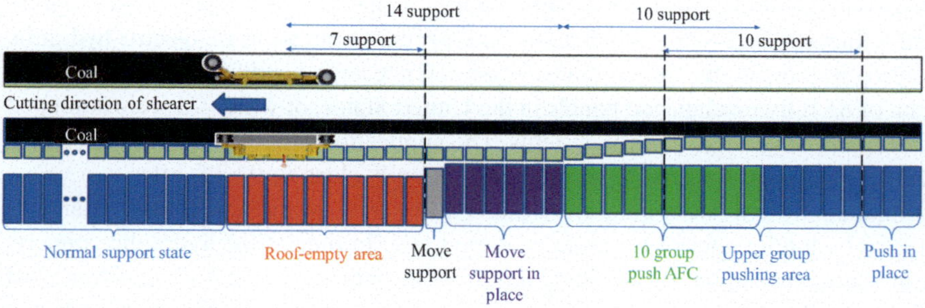

Fig. 5.6 Fully mechanized panel hydraulic support automatic straightening diagram

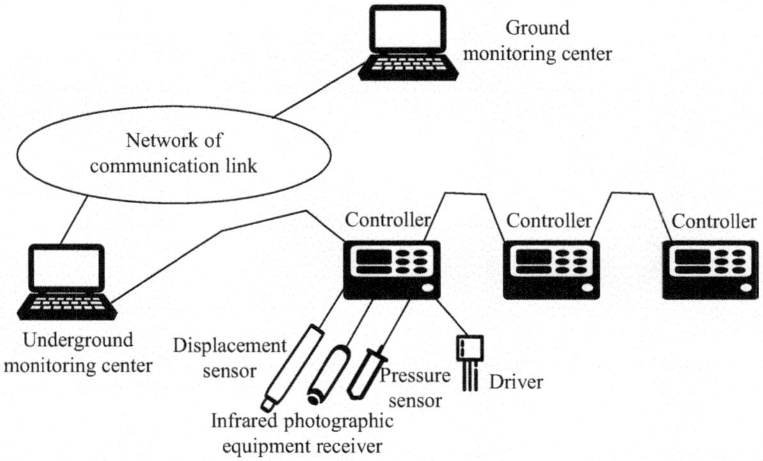

Fig. 5.7 Electro-hydraulic control system diagram of hydraulic support

The electro-hydraulic control system of coal mine hydraulic support is shown in Fig. 5.7. It mainly includes a ground dispatching monitoring center, underground monitoring center, controller, various sensors (infrared sensor, pressure sensor, displacement sensor, etc.), driver, etc.

In the electro-hydraulic control hydraulic support, a single support is equipped with a set of electro-hydraulic controllers, an infrared receiver, a displacement sensor, and two pressure sensors (Fig. 5.8). Complete configuration: the panel hydraulic support is equipped with an electro-hydraulic control system (the end hydraulic support has no electro-hydraulic controller), and a set of explosion-proof and intrinsically safe voltage-stabilized power supply, shielding coupler, repeater, and switch is set up for each six supports. The repeater supplies power to the corresponding group controller. The isolation coupler is used to isolate the electrical signal, transmit the signal, realize the grouping control, and identify the position of the shearer in real-time through the infrared receiver. The displacement sensor enables the user to accurately grasp the action stroke of the pushing jack. The pressure sensor, meanwhile, allows the monitoring of the initial support force of the front and rear columns of the support in real time. The controller, in turn, sets the parameters in accordance with the specified operation procedure and spacing, thereby enabling the user to carry out actions such as group operation and automatic push–pull movement with the hydraulic support. The electro-hydraulic main control valve maintains the manual operation button. When the electrical control system fails, the button of the solenoid valve can still be used to control the work of the hydraulic support. The main functions of the electro-hydraulic control system of the hydraulic support are as follows: (1) Group control of single support, adjacent support, and partition support. (2) local group locking, all electro-hydraulic control hydraulic support emergency stop. (3) Real-time online display of shearer position, push rod stroke, and support pressure. (4) with automatic pressure function, regular automatic backwashing, or

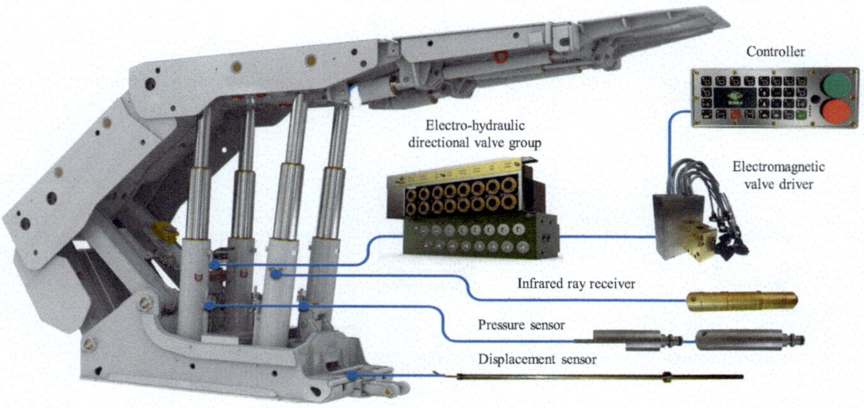

Fig. 5.8 Single electro-hydraulic control hydraulic support

manual backwashing. (5) automatic pull support, step-down shift supports, automatic spray. (6) Automatic follow-up movement of the hydraulic support moving support.

5.2.2.3 Self-Adaptive Support of Hydraulic Support Group

The group adaptive support of hydraulic support means that the hydraulic support can adjust the support resistance, support state, and support mode independently according to the load change and surrounding rock deformation.

1. Adaptive adjustment of hydraulic support resistance

The hydraulic support resistance of hydraulic support includes initial support force and working resistance. The adaptive control model of the surrounding rock-hydraulic support system of the panel is established. The adaptive adjustment of the hydraulic support resistance is based on the observation results of the lower chamber pressure of the column, based on the column pressure adjustment method of the column rapid pressurization and stress relief device, and the adjustable opening pressure and large flow safety valve. Through the position information of the hydraulic support, the column pressure control strategy of the roof and floor lithology and structure, the supporting attitude of the hydraulic support is adjusted in real-time, and the adaptive control of the support characteristics (support strength, stiffness, and stability) of the hydraulic support is realized.

2. Self-adaptive hydraulic support of supporting method

When the panel fluctuates and changes, through infrared perception, high-definition video image automatic capture, combined with the precise positioning system of the panel equipment, the position information of the shearer and the drum cutting trajectory line is automatically extracted, the deviation between the cutting curve and the geological model scanning line is analyzed in real-time, the memory cutting template is corrected, the drum cutting path is planned, and the hydraulic support is combined with the hydraulic support to prevent the bite frame and the maximum load of the left and right supports. The height of the drum is automatically adjusted to realize the adaptive support of the hydraulic support according to the change of the coal seam in the panel.

3. Support state adaptive principle and method

The angle sensor is installed on the base, the top beam, and the shield beam. The pressure sensor is set on the column to collect information on the support attitude and the support resistance parameter, and the geometric parameters of the support are obtained. The multi-dimensional attitude controller of the support is used to sense the attitude of the support in real-time and judge the balance state of the support. The electro-hydraulic control system regulates the support height of the column and the related jack stroke, changes the attitude of the top beam and the shield beam, and adjusts the support state of the hydraulic support.

5.2.3 Automatic Control of Transportation Equipment

5.2.3.1 Automatic AFC

AFC is the core equipment of underground material transportation in coal mines, which is directly related to the efficiency and economy of underground material transportation. In mine production activities, a AFC is a traction equipment for material transportation. In the specific opespeedn, the scraper is firmly fixed on the scraper chain, and the chain is driven by the hydraulic coupler and the transmission device. The chain movement drives the scraper to realize the transmission of materials. In the process of equipment opespeedn, the AFC is driven to circulate the sprocket to improve the material transportation of mining enterprises. At present, most of AFCs mainly adopt the constant rotate speed operation mode. Due to the influence of the fully mechanized mining rate, the conveyor is in a light load state in most cases, so the energy waste is large, which cannot meet the needs of economic and linkage operations. Moreover, the AFC works in harsh environments for a long time, and the failure ratio is high. The existing control system cannot self-test the working state of the AFC, the manual detection efficiency is low and the safety is poor.

1. Function of the automatic AFC

To realize the automatic operation of the coal mine AFC, it is necessary to clarify the function of the control system. According to the needs of intelligent mining in coal mines, the main functions of the intelligent control system of the AFC are automatic position adjustment, real-time monitoring of running status, and automatic fault alarm.

(1) Automatic position adjustment

During the operation of the AFC, it is necessary to ensure that the shearer is parallel to the hydraulic support and no collision occurs. To realize the intelligent control of the AFC, it is necessary to realize automatic position adjustment. When the relative position of the AFC and the hydraulic support deviates greatly, it should be adjusted to avoid a collision between the AFC and the hydraulic support. The position parameters of the AFC mainly include dip angle, horizontal degree, and azimuth angle. In the actual operation process, the AFC is usually driven by the hydraulic support. However, the hydraulic support is moved in groups. If the AFC has a large bending in the local area, it will not only lead to serious material spreading during transportation but also lead to local fracture due to the large shear force at the bending position. This requires that the position of the hydraulic support should also be detected when detecting the position of the AFC.

(2) Real-time monitoring of operating state parameters

The goal of intelligent control of the AFC is to control the running state of the AFC in real-time, to achieve the best performance. To realize the control of the running state, it is necessary to monitor the main running state parameters. The main operating state parameters of the AFC are motor power, motor rate, motor current, and motor temperature. These main state parameters are related to time, so it is necessary to monitor these parameters in real-time. During the operation of the AFC, when the mass of the loaded rock or coal is different, the rotate speed and power of the motor are also different. By taking these key quantities as control quantities, the running state of the motor is adjusted in real-time to achieve energy-saving operation. In the process of realizing optimal control, some corresponding algorithms need to be adopted.

(3) Auto fault alarm

Due to the poor working environment in the coal mine, the AFC has some faults. If these faults cannot be found in time, serious electromechanical safety accidents may occur. Some state parameters of the AFC are monitored. Once it is found that the operating parameters deviate from the predetermined value, an alarm should be issued immediately, to take the best treatment plan in time. For example, when the motor running current deviates from the predetermined value, it can be considered that the motor of the AFC is faulty or overloaded. When the motor running temperature is too high, it can be considered that the motor of the AFC is faulty, and the machine should be stopped for processing. There are many forms of alarm, which can be alarm lights or alarm sounds.

2. Realization of an automatic AFC

(1) Position detection

The position detection is mainly divided into horizontal degree and azimuth angle. A laser measuring instrument or infrared measuring instrument can be used in azimuth angle measurement. Considering that the AFC and the hydraulic support are linked controls, it is only necessary to measure the relative position of the AFC and the hydraulic support. The trajectory of the hydraulic support is determined by the shearer. To facilitate control, the position information of the shearer walking is usually recorded. Once the position of the shearer deviates, it can be adjusted immediately according to the calibration position. When correcting the running track of the shearer, the head is usually selected as the adjustment point.

In the actual operation process, the position adjustment process of the AFC is as follows: The position of the AFC is collected in real-time by the sensor installed on the head of the AFC. The collected location information is processed, mainly compared with the set location to see the degree of deviation. The results of the comparison are transmitted to the shearer, and the overall straightening of the AFC is realized by adjusting the position of the shearer.

(2) Working state control

To realize the optimal state of the AFC operation, it is necessary to control the running state of the AFC. The key to running state control is to accurately control the running state of the AFC motor. In the actual production process, due to the change in the coal or gangue loading amount of the AFC, the output power requirements will also change. By monitoring the loading capacity of the AFC, the relationship between the running speed and the loading capacity is calibrated, and then the running speed of the AFC is quantitatively adjusted.

The frequency conversion control system is usually used to realize the intelligent adjustment of the speed of the AFC. Motor variable frequency speed regulation is mainly achieved by changing the frequency of alternating current. When the running speed is fast, a larger power frequency is input, otherwise, a smaller frequency is used. To realize the intelligent adjustment of the speed, the feedback control system should be adopted. The feedback is to compare the output speed with the given speed to determine whether it needs to be adjusted. Through feedback, the speed can be automatically adjusted. It should be noted that the reference speed of the feedback needs to be given in advance according to the needs, that is, there must be a certain correspondence between the speed and frequency, which is determined according to the needs of the operation. After realizing the intelligent control of the speed, it can not only realize the unattended operation of the AFC but also realize energy saving.

(3) Automatic tension function

The construction of the control system for the automatic tensioning device of the AFC is based on the PLC programmable controller, which is used to control the sensor that detects parameters such as displacement and pressure. Additionally, the PLC monitors the change in bottom pressure of the telescopic cylinder in real-time and combines the chain tension change to automatically control the telescopic cylinder, thereby facilitating the transportation of materials. It is possible to make dynamic adjustments to the chain tightness of the AFC within a reasonable range, thus meeting the operational requirements in an optimal state of tightness and improving overall operational efficiency. As far as the control system of the automatic tensioning device of the AFC is concerned, it includes two parts: the hydraulic control unit and the electronic control unit. The electronic control unit includes a PLC programmable controller, DC power supply, explosion-proof digital input module, explosion-proof digital output module, explosion-proof analog input module, and sensor. The hydraulic control unit encompasses the safety valve, the hydraulic control expansion valve, the expansion cylinder, the electromagnetic directional valve, and the hydraulic control pilot valve. It receives the expansion signal transmitted by the control system, which enables the electromagnetic operation, opens the hydraulic control pilot valve, and injects oil into the expansion cylinder, thereby actuating the system and facilitating the automatic adjustment of chain tension, thus enhancing the level of production automation. The production environment of coal mine panel is relatively complex. In the panel of increasing production requirements, it is an inevitable choice to actively promote the optimization and improvement of production technology and equipment. To better meet the production needs of mining enterprises, actively promote the optimization and improvement of the control system of the automatic tensioning device of the AFC, which is helpful to improve the reliability of the production process, improve the operation performance of the equipment, and prolong the performance of the equipment. In this way, the production process is actively optimized, and the fault problems in the operation of the AFC are found and solved in time, which brings more considerable economic benefits to the enterprise.

(4) Fault diagnosis and alarm

To realize the safe operation of the AFC, the fault diagnosis and alarm function should also be realized. Fault diagnosis is conducive to the fault maintenance of the AFC, and the alarm is conducive to ensuring the safe operation of equipment. In general, fault diagnosis and alarm are achieved by monitoring the operating parameters of the AFC. The alarm function is usually relatively simple to implement, only by setting the threshold of the relevant parameters. Once a certain state parameter monitored exceeds the threshold, a special acousto-optic signal should be sent out immediately. The realization of this function depends on the sensors installed on the AFC and the host computer connected to these sensors.

5.2.3.2 Automatic Belt Conveyor

The underground belt conveyor is an important part of coal mine production, and it is also a key link that directly affects production efficiency. However, due to the harsh environment and high load conditions in coal mines, various faults often occur in coal mine belt conveyors. To improve the stability of the coal mine belt conveyor, the operation of the belt conveyor is automatically controlled, and the unmanned operation of the belt conveyor is realized. In the process of coal mine production, the underground belt conveyor consumes energy and supports coal mining and transportation. The transportation efficiency and speed are inextricably linked with coal mine production. Production efficiency is the key to the sustainable development of coal mines. The traditional underground belt conveyor is limited by the design and technical level, and the energy consumption is large, which is contrary to the goals of energy conservation, emission reduction, and green environmental protection. Coal mines must align themselves with the developmental trends of the contemporary era. This necessitates the pursuit of energy conservation, emission reduction, and expenditure reduction as core developmental objectives. Furthermore, there is a pressing need to optimize the underground belt conveyor and integrate advanced automation control technology into the aforementioned system. This integration should be guided by the overarching principle of intelligent and automated construction. Using modern technology to achieve upgrading and optimization, and provide more support for the sustainable development of coal mines.

The traditional underground belt conveyor is also automatically controlled. With the help of electronic technology, the scientific and orderly transmission of the belt is realized, and the overall system is driven to achieve continuous operation and complete production and transportation. In the process of system operation, parameters such as speed are extremely important, and automatic control is difficult to separate from the support of various parameters. However, due to technical limitations, there is a gap between some details and the ideal goal. In the new era, automation technology has been developed. Relying on the solid foundation provided by modern theory, the underground belt conveyor has been optimized and upgraded, and the overall intelligence and automation level of the system has been greatly improved. In particular, the signal distribution of the conveyor belt of the underground belt conveyor in the new era is more scientific, and the connection between the construction signal and the central system is closer and smoother. This provides more accurate information for the rate-regulating motor, which is convenient for the next action command to complete the automatic monitoring. This method can not only effectively reduce energy consumption, and meet the requirements of energy conservation and emission reduction targets, but also prolong the service life of traditional components and reduce maintenance costs.

The realization of an automatic belt conveyor needs to go through the following aspects:

(1) Scientific use of computers and a variety of sensor advantages. Relying on the advantages of computer and sensor technology development, an efficient working condition monitoring and control system is constructed to make up for the shortcomings of the traditional underground belt conveyor, realize continuous monitoring, and further improve monitoring accuracy and efficiency. Through the system, the parameters of the whole system can be accurately and comprehensively understood, and human labor can be effectively reduced.
(2) Achieving automatic control. After the application of automatic control in the underground belt conveyor, the process can be greatly simplified, the manpower workload can be reduced, and the technical advantages can be relied on to enable the staff to realize remote multi-site automatic monitoring management and data acquisition. Relying on the Internet, the Internet of Things, 5G technology, etc., a more complete and huge monitoring network can be constructed to maximize

work efficiency and provide a guarantee for the safe and stable operation of the system.

(3) Having a strong communication networking capability. The application of automatic control provides communication channels for the underground belt conveyor. Relying on advanced technology, multi-level linkage, and information interactive communication and sharing can be realized.
(4) Relatively perfect protection system. When the underground belt transportation automation control system is running, it will support the preparation of more perfect rules and regulations and management processes, and scientifically establish a management system. When the sensor is integrated into the system, the operation data of the belt conveyor is collected in advance, enabling risk prediction to be carried out and potential safety risks to be identified and mitigated. Additionally, the system is equipped with a fault quantity voice prompt, safety prompt, and alarm signal, which collectively serve to establish a foundation for the safe operation.

5.2.4 "Three Machines" Linkage Automation Technology

The "three machines" linkage control system of fully mechanized mining is the premise of realizing an unmanned panel. According to the current situation of production and use, the three-core equipment of the panel, namely shearer, hydraulic support, and AFC, are generally produced and provided by different manufacturers. To realize intelligent control, there are some problems, such as the agreement of each manufacturer is not open to the outside world, only monitoring, and centralized control cannot be realized. The basis of the linkage of the "three machines" is the acquisition of information [13], which mainly includes some aspects as follows.

5.2.4.1 Intermodal Control Information Provided by the Shearer

(1) Cutting method: oblique cutting at the end of the panel, no triangle coal left (cutting triangle coal), round-trip cutting 2 knives, pre-set and unchanged, to determine the traction trajectory of the shearer;
(2) The relative position of the shearer-hydraulic support to determine the starting position of the hydraulic support;
(3) Operation direction. The operation direction of the shearer is used to determine the moving hydraulic support and pushing AFC and its related steps.
(4) Traction speed: shearer traction speed directly determines the hydraulic support chase machine speed and way, but also the hydraulic support can reach the chase machine speed constraints.

5.2.4.2 Information Provided by the Hydraulic Support

(1) The relative position of the hydraulic support and the shearer is used to determine the starting position of the moving hydraulic support and pushing AFC.
(2) The way of moving hydraulic support is preset, which affects the speed of the hydraulic support;
(3) The speed of moving hydraulic support is set in advance and does not participate in the "three machines" linkage;
(4) The advance step distance is set in advance, which is used to determine the stroke of the hydraulic cylinder of the hydraulic support, and does not participate in the "three machines" linkage;
(5) The stroke of the hydraulic cylinder is preset, which is only related to the automatic movement of the hydraulic support, and does not participate in the "three machines" linkage.

5.2.4.3 Information Provided by the AFC

(1) The load of the AFC (load current) is related to the linkage of the coal falling amount (cutting rate) of the shearer;
(2) The chain speed of the AFC is related to the linkage of the shearer.
(3) The coal flow information of the head unloading port is related to the cutting speed of the shearer or the chain speed of the AFC.

The core function of the "three machines" linkage control system of fully mechanized mining is to realize the automatic retraction and extension of the hydraulic support, the automatic moving hydraulic support, and the pushing AFC by detecting the position and traction direction of the shearer. The "three machines" linkage control system mainly solves the coordinated operation among the shearer, hydraulic support, and AFC. The main functions of the control system include the following aspects:

(1) The control system detects and collects the relevant parameters of the "three machines" linkage, and sends them to the PC of the monitoring center for processing, storage, display, and other operations. The monitoring center and the "three machines" realize the interactive communication of information. Through the "three machines" linkage controller, the fully mechanized mining equipment of the panel is fully controlled, and the functions of sequential start-up and sequential shutdown are completed.
(2) The shearer automatically pulls and cuts coal. Furthermore, the position of the current shearer and the direction of the traction speed are transmitted to the control system. According to the received information, the control system controls the hydraulic support to automatically retract and stretch the rib protection device and the front detection beam. In the rear of the shearer, the hydraulic support is automatically lowered, moved, lifted, and pushed AFC. If the moving speed of the hydraulic support cannot keep up with the traction speed of the

shearer, a signal can be sent to request the shearer to slow down, to realize the automation of the hydraulic support and complete the electro-hydraulic control.

(3) The AFC transmits its current transport capacity and load information to the control system to adjust the traction speed of the shearer;

(4) The panel equipment has the function of emergency stop locking, the panel conveyor and the shearer have the function of interlocking, and the shearer and the spray pump have the function of linkage.

5.3 Automatic Positioning and Correction of Roadway Driving

The automatic control technology of roadway driving equipment has evolved through four stages: 'section monitoring, sight distance remote control, memory cutting', 'working condition monitoring, remote control, automatic cutting', 'autonomous navigation, intelligent cutting, single machine intelligence', and a complete set of equipment collaborative propulsion'. Currently, the length of newly excavated roadways in China's state-owned key coal mines exceeds 12,000 km per year, highlighting the significant scale of these operations. The primary method for roadway driving support is cantilever roadheader driving combined with single anchor drilling machine support, accounting for over 90% of operations. In recent years, to address the challenges of continuous roadway driving operations, various rapid roadway driving equipment has developed rapidly. These advancements center around the integrated digging and anchoring machine, equipped with bolt transloader, flexible belt conveyors, and stepping self-moving tails. This integration enables parallel operations of driving, anchoring, and transportation, greatly improving roadway driving efficiency [14].

5.3.1 Development of Roadway Driving Technology

5.3.1.1 Drilling and Blasting Roadway Driving [15, 16]

The drilling and blasting method involve excavating rock through a sequence of drilling, charging, and blasting. This technique dates to the early industrial revolution in the sixteenth century. The advent of explosives significantly advanced modern roadway driving technology. The invention of the pneumatic rock drill marked a pivotal leap in drilling and blasting technology. With the widespread use of nitroglycerin explosives and pneumatic rock drills, the drilling and blasting construction method gradually evolved. Over more than a century, this method has seen rapid development. The main advantages of the drilling and blasting method include quick operation preparation, strong adaptability to different rock types, flexible movement, easy handling of technical and geological faults, high efficiency, and low cost. In recent years, there have been significant improvements in construction machinery and technology, particularly in countries like the United States, Germany, Sweden, and South Africa. For instance, in Sweden and the United States, the method can achieve an efficiency of more than 30 m^3 per shift. However, some shortcomings remain, such as a low average roadway driving rate, difficulty in organizing multi-process cross-operations, inconsistent technical levels across processes, frequent over-blasting, and the harmful effects of blasting fumes.

Rock drilling and blasting are the primary processes in roadway driving, constituting 30–50% of the total cycle time. Traditionally, rock drilling in roadways utilizes a jackhammer with an air-leg. Electric rock drills can also be employed for low-strength rock. The jackhammer with an air-leg is flexible, easy to operate, and covers a small area for each drill, enabling simultaneous multiple rig operations and parallel drilling and installation. However, the widely used jackhammer with an air-leg has drawbacks, including high labor intensity and limited drilling speed and depth. Since the 1980s, hydraulic rock drills have gained popularity due to their numerous advantages: (1) High drilling speed and efficiency. In medium-hard rock, the drilling speed can reach 1.5–2.0 m per minute. The impact efficiency ranges from 45 to 55%, and the energy utilization ratio is between 30 and 40%. (2) Improved working environment. Hydraulic drills maintain constant thrust without rebound, produce low noise, and emit less oil mist, thereby enhancing the working environment. (3) Enhanced drilling capabilities. Hydraulic drills offer large torque, allowing for deeper and larger hole drilling, which is advantageous for deep-hole smooth blasting.

In 1979, Montabert of France produced the first practical hydraulic rock drill, weighing approximately 70 kg, typically mounted on a drill carriage. Over the past few decades, various types of drill carriages have been developed both internationally and domestically. Among these, hydraulic drill carriages have become the most widely used. The predominant walking modes for these carriages are trackless crawlers or rubber wheels.

The cost of rock breaking using explosives is significantly lower than that of mechanical rock breaking. When comparing these two methods, the energy consumption indicators of drilling and mechanical roadway driving are primarily considered. For example, when a roadheader is used in hard rock, the power consumption is 9–22 kW·h/m^3. In contrast, pneumatic rock drilling equipment consumes 0.9–1.6 kW h for every 1 m of borehole drilled. Typically, if 4 boreholes are drilled per square meter of the panel, the power consumption under normal conditions is 4 kW h/m^3.

5.3.1.2 Comprehensive Mechanized Roadway Driving Technology [17]

The comprehensive mechanized roadway driving in coal mines involves the integration of cantilever roadheaders and single-anchor drilling machines. The primary equipment used in this process is the cantilever roadheader. This method is widely applicable and is the most common technology used in coal mine roadway driving.

Fig. 5.9 EBZ200H roadheader

The comprehensive mechanized roadway driving system includes a cantilever roadheader, single anchor drilling machine, transfer machine, retractable belt conveyor (or AFC), power supply system, and ventilation and dust removal equipment, as shown in Fig. 5.9. Among these, the cantilever roadheader is the key equipment, and its performance is crucial for improving roadway driving efficiency and excavation footage.

The development and application of cantilever roadheaders in China began in the 1960s. Initially, traditional cantilever roadheaders had limited power. In the early 1980s, models such as the AM50 and S-100 roadheaders were introduced. Building on these, China developed mechanized roadway driving equipment tailored to its coal storage characteristics, which significantly advanced the country's roadheader technology. In recent years, China has developed alternative equipment, such as the EBJ-120TP, which has reached international standards in overall technical performance. The new generation of roadheaders in China features a reasonable design, compact structure, stable operation, high production capacity, strong rock-breaking ability, good adaptability, and high reliability. Additionally, these machines come equipped with industrial and mining detection and fault diagnosis functions.

The cantilever roadheader has been widely used in Chinese coal mines, playing a crucial role in maintaining stable and high coal production. However, the current practice of single roadway driving combined with the use of single anchor drilling machines for anchor support limits the efficiency of operations. This approach prevents the parallel execution of roadway driving and support activities, thereby restricting the further improvement of the roadway driving speed.

5.3.1.3 Continuous Miner Roadway Driving Technology [18]

A continuous miner is a type of combined mechanized roadway driving equipment with a large cutting width, integrating collection, transportation, and movement. It finds extensive use in the rapid driving of double or multiple roadways in rectangular section coal mines, as well as in room-and-pillar or room-and-board mining, corner coal mining, residual coal mining, and coal pillar recovery (Fig. 5.10). It has become pivotal in modern, high-yield, and high-efficiency mines. The introduction of continuous miners in China commenced in 1976, progressing through two stages of single machine and complete set introductions. Presently, continuous miners are employed in China's Shendong, Yanzhou, Jincheng, and Huangling mining areas. Shendong, with its nearly horizontal coal seams constituting over 90% of its coal roadways, boasts stable seams, favorable roof and floor conditions, and low gas content. The anchor rods support and rectangular section roadway driving are adopted, providing ideal conditions for continuous miner utilization. In the Shendong mining area, continuous miners have achieved significant milestones in rapid coal roadway driving. Monthly excavation footage exceeds 2000 m, reaching a peak of 3273 m in a single month (including coal transportation). The produce experience demonstrates that continuous miners, particularly in double roadway driving scenarios in longwall mining panels, meet the monthly advancement requirements of highly productive and efficient fully mechanized mining panels. Moreover, their roadway driving costs are lower compared to cantilever roadheaders.

In the continuous miner roadway driving process system, the roadway driving coal loading operation and roadway anchor support operation are completed by two different pieces of equipment: the continuous miner and the anchor drilling machine. Therefore, after the continuous miner completes the roadway driving (to a certain cutting depth) over a length of roadway, it must be moved to another area of operation. This allows the anchor drilling machine to enter the completed roadway to perform the anchor support operation. This necessitates continuous cross-operation between the two machines.

Fig. 5.10 EML340 continuous miner

5.3.1.4 The Integrated Digging and Anchoring Machine [19]

To enhance roadway driving efficiency and achieve simultaneous operation of driving and bolting in single roadway setups, the ABM20, the first integrated digging and anchoring machine, was developed by VAI Mining Equipment Co., Ltd. in 1991 and successfully tested at Tahmoor Colliery Mine in Australia. The machine features a main and auxiliary frame with relative sliding capability. The main frame is equipped with multiple sets of rod drilling machines and temporary support, while the auxiliary frame incorporates a cutting mechanism matching the roadway width, sliding relative to the main frame for coal cutting. This design enables parallel operation of roadway driving and anchoring, reduces roadway damage from equipment rolling, and facilitates one-pass roadway formation. Additionally, the equipment has a small empty roof distance (≤ 2 m), making it suitable for challenging rock conditions. Currently, the integrated digging and anchoring machine is extensively deployed in fully mechanized mining for preparation roadway driving, totaling over 400 units globally. They are predominantly used in the United States, South Africa, Australia, Russia, China, and other countries, with nearly 100 units in China alone. Over nearly three decades of development, manufacturers in Austria, the United States, Japan, Germany, and other countries have developed a comprehensive series of integrated digging and anchoring machines. These machines cater to mining heights ranging from 1.2 to 5.5 m, cutting widths from 4.0 to 7.2 m, cutting powers from 200 to 340 kW, and weights from 60 to 115 t, offering over 30 models adaptable to diverse working conditions. Significant theoretical and experimental research has been conducted in related fields such as intelligent anchoring and digging, integration of driving, anchoring, and detection, electro-hydraulic control of drilling rigs, and automated network laying.

In the research and development process of the integrated digging and anchoring machine, combined with the actual situation of China's coal mine underground roadway engineering, a significant amount of related innovative technology research has been conducted to enhance the adaptability of the integrated digging and anchoring machine and meet the requirements of various geological conditions in China. The key innovations include the following.

(1) Wide-track low-specific pressure chassis. The development includes a light, wide crawler chassis, and auxiliary support mechanism, reducing the ground pressure during the machine's movement and cutting and supporting operations by approximately 20%. This enhancement improves the machine's adaptability to different ground conditions, as illustrated in Fig. 5.11. Additionally, AC variable frequency walking drive technology is employed to mitigate the 'lying machine' phenomenon, utilizing large starting torque and high-precision quantitative speed regulation.
(2) Double-drive high-speed confluence heavy cutting reducer. A high-reliability cutting reducer has been developed, featuring a built-in gear pump and cooling circuit for active lubrication and forced cooling. Various sensors monitor and provide early warning of the lubrication status of the cutting reducer. The reducer

Fig. 5.11 EJM340/4-2H low-specific pressure types integrated digging and anchoring machine

is driven by double motors with high-speed confluence, addressing maintenance challenges associated with traditional single-motor systems while enhancing cutting power and capability.

(3) Pre-exploratory temporary support. Addressing the challenge of traditional temporary supports failing to accommodate broken roofs due to significant empty roof distances, a front exploration airborne temporary support has been developed. This innovation reduces the empty roof distance of the temporary support from 1.0 to 0.4 m, enhancing safety during anchor support operations and ensuring timely and effective active support from the anchor rods.

The coal roadway driving using the integrated digging and anchoring machine adopts a one-time roadway forming process. During each operation cycle, roadway driving and support are conducted simultaneously. After completing a section of anchor support, the system advances to initiate the next cycle, often relying on shuttle cars or bridge transloader for subsequent support, whether discontinuous or continuous. The Shendong mining area in China exemplifies the successful application of this integrated technology for roadway driving and anchoring. The average monthly excavation footage using the one-time roadway forming process reaches approximately 800 m. Compared to the continuous miner double roadway driving method, this approach offers a broader application scope, delivers effective support, achieves notable efficiency in roadway driving, and ensures high safety standards.

5.3.1.5 'The Integrated Digging and Anchoring Machine, Anchor Transfer Machine, and Large Span Transfer' Remote Control Intelligent Rapid Roadway Driving Technology [20]

Zhangjiamao coal mine is situated in the Jurassic coalfield in northern Shaanxi, characterized by favorable conditions, straightforward support requirements, and conducive environments for rapid roadway construction. Embracing the concept of equipment integration, digital monitoring, and automated control, the mine advocates an intensive support approach involving the integrated digging and anchoring machine, rod transfer machine, and double-span transition transportation, and it referred to as the "three machines" method.

To ensure continuous and reliable roadway development, the MB670-1 integrated digging and anchoring machine is selected. It facilitates swift continuous cutting, reducing single cycle times to under 10 min, which is known for its high reliability and simultaneous operation capabilities in roadway cutting and anchoring. The rod transfer machine serves multiple functions including coal transport, large block crushing, and anchor rod (cable) support, also referred to as the integrated anchor transport and breaking machine. It features three top anchors and two side anchor drilling arms, with horizontally movable top anchors on both sides capable of adjusting up to 1200 mm to support full-section top anchors. This flexibility ensures alignment with support design orientations and angles, facilitating concurrent roadway cutting and anchoring operations. Enhancing the effective overlap length of the long-span bridge transfer machine and belt conveyor, while minimizing rigid frame connections, is a key strategy to boost roadway development rates. Adopting a double-span transfer extends carrying distances to 100 m, further enhancing operational efficiency, as depicted in Fig. 5.12.

The integrated navigation technology has been developed to leverage the stability of laser guidance error and real-time online monitoring provided by inclination sensors (or inertial navigation systems). These technologies complement each other's weaknesses, enhancing overall performance and establishing a new navigation system. Downhole measurements demonstrate that the laser receiver achieves a resolution of 1 mm and accuracy within 3 mm over 100 m. Replacing the inclination sensor with an inertial navigation system enables distance ranging and integration with geodetic coordinates.

To address the key technology of multi-machine cooperative control in mining operations using high-precision extrasonic and laser sensors, a precise multi-machine positioning system and cooperative control algorithm have been developed. This system enables the automatic operation of the integrated digging and anchoring machine, anchor transfer unit, and breaking and rear bridge transfer machine. It incorporates a total of 10 laser ranging sensors, 14 extrasonic sensors, 2 encoders, and 6 travel switches distributed among the roadway equipment. The integration of extrasonic and laser ranging sensors enhances sensing accuracy. Utilizing equipment

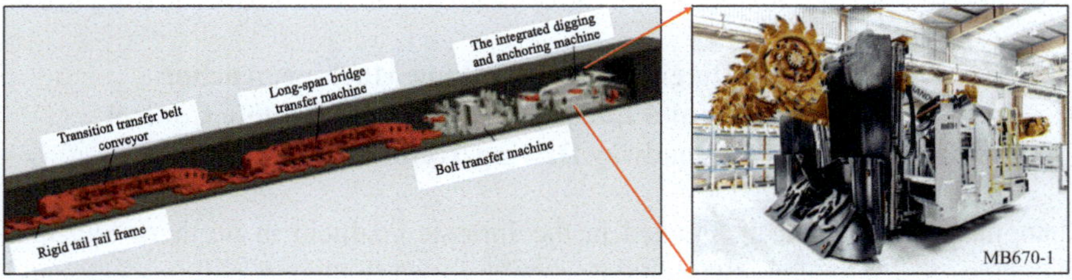

Fig. 5.12 Rapid roadway driving equipment supporting model [20]

location and status information, the system facilitates signal interaction and interlocking control across multiple devices, monitors equipment operational status, and enables one-key start and shutdown for all equipment.

A remote centralized control visualization platform was developed, equipped with the capability to construct a 3D geological model of the heading panel. This model is continuously updated based on actual geological data uncovered during roadway construction, integrating advanced detection information, roadway formation quality, and the 3D geological model to accurately depict the heading panel scene. The monitoring system features intelligent functions for environmental monitoring (such as dust, gas, and water) at the heading panel, with an added UWB personnel precise positioning system for proximity identification and alarm in hazardous areas. The system enables intelligent coordination across the processes of roadway driving, support, anchoring, transportation, and breaking. It includes positioning and cutting functions facilitated by the integrated navigation and positioning system, calculating the spatial position of cutting heads. Operations like starting, stopping, and cutting with equipment such as the integrated digging and anchoring machine and conveyors at the heading panel can be remotely controlled from underground centralized warehouses and long-distance control centers on the surface.

5.3.2 Automatic Roadway Driving Technology

5.3.2.1 Roadheader Composition and Working Principle

Usually, the cantilever machine consists of an execution structure, loading structure, walking structure, loader, and related electrical control and hydraulic drive systems. In mining roadway construction, the roadheader primarily operates through its control system to maneuver the front-end cutting machine within the mine roadway, shaping different sections according to operational requirements. The roadheaders used in mining can be categorized into vertical axis types (Fig. 5.9) and horizontal axis (Fig. 5.11), based on the installation orientation of the cutting head's driving shaft. Currently, compared to other roadheaders, the cantilever roadheader offers advantages such as a compact structure and advanced technology. It facilitates continuous, efficient, and safe operation across various coal seam conditions, effectively enhancing the working environment for mine operators. As a result, it finds widespread application in practical mining operations.

In the operation of the mining heading panel, the roadheader relies on the hydraulic structure of the entire mechanical system for power. The hydraulic system drives the cutting mechanism's rotation. As the shaft and cutter rotate within the cutting structure and the walking mechanism operates, the roadheader progresses along its designated path. This continuous operation enables the cutting, loading, and removal of mined rock, which is then transported out of the mine. Furthermore, to facilitate automatic roadway construction in mining operations, the roadheader's mechanical system is equipped with an automatic control system. This system autonomously interprets pre-set 3D motion trajectories to effectively manage the roadheader's advancing and boom-cutting positions during operation. It promptly captures real-time operational data and system commands, ensuring operational requirements are met through automated comparative analysis and error correction. This integration enables the automated driving of driving panels.

Compared with traditional mining roadway construction, in specific operations, integrating ground-issued mining instructions with automated control of each mechanical structure enhances the capability to meet unmanned roadway driving requirements, thereby significantly improving roadway safety. This approach offers notable advantages. In the mining heading panel operations, the diverse nature of coal seam deposits and complex environmental conditions present challenges. Coal seam thickness variations are significant. Roadway construction involves various equipment such as cantilever roadheaders, transloader, and rod drilling machines, often necessitating single roadway drives. Therefore, advancing automation technology in mining heading panels requires continuous development and testing of diverse technical features and applications to accelerate automation progress in mine roadway construction [19].

5.3.2.2 Key Technologies of Automatic Control of Driving Panel

The work of a roadway roadheader typically encompasses tasks such as roadway driving, swing cutting, and backward coal collection, which involve processes like alignment, positioning, cutting, clearing, and coal collection. Automating a roadway roadheader requires leveraging intelligent sensing and adaptive technologies to achieve the automated formation of roadway sections. Intelligent perception involves pose, formation, and state perception, while adaptive operations encompass adaptive correction, diagnostics, cutting adjustments, and more.

1. Automatic driving technology

In the mining production process, roadway driving operations are crucial tasks. The support technology employed during these operations forms a vital foundation for overall mine production. The roadway driving and support operations establish essential channels necessary for mine production. The progress of roadway driving operations in mines is influenced by the external environment and regional climatic conditions. Hence, to meet the demands of roadway driving in complex environments, enhancing the application of automation systems and equipment becomes imperative. For instance, deploying automatic mining equipment facilitates efficient mining, transportation, and loading operations, thereby enhancing overall mining efficiency. The roadheaders play a pivotal role as indispensable equipment in roadway driving and in facilitating the transport of mined minerals. Typically, automatic control systems in roadheaders for mine operations include features like automatic cutting, error correction, and positioning. Integrating hydraulic, electronic, and mechanical technologies within equipment systems effectively supports roadway driving and enables automatic control of roadheaders. Due to China's expansive geographical terrain and diverse regional environments, these automated control systems and technologies are crucial for efficient mine roadway driving. They not only streamline operations but also support personnel in executing simpler operational tasks, thereby promoting the advancement of intelligent mining and production capabilities in mines.

2. Safety control automation technology

The safety of mining operations not only directly affects the regular production and operation of mines but also significantly impacts national production safety. Ensuring the safety and feasibility of roadway driving in mining production is a fundamental requirement to promote safe operations and normal functioning of mining enterprises. Typically, the working environment and production equipment influence safety risks during mining roadway operations. Therefore, enhancing environmental monitoring and analysis of mining roadway driving, along with timely maintenance and upkeep of application equipment, are crucial steps to ensuring the safety of mining roadway operations.

In the monitoring of the working environment and equipment operation during mine roadway driving, automatic monitoring, and control equipment are utilized to oversee parameters such as air quality and humidity. Real-time analysis of this data informs the development of roadway driving plans and specific implementation strategies, ensuring both safety and operational feasibility. Due to the inherent dangers and uncertainties of the mining environment, continuous real-time monitoring and analysis of environmental conditions are crucial to maintaining safety during roadway driving operations. The primary focus areas for real-time monitoring during mine roadway driving include dust, gas levels, coal dust accumulation, and the impact of roadway driving on surrounding strata. It is essential to maintain these parameters within acceptable limits to achieve safe and efficient mine roadway operations. Moreover, monitoring the operational status, feedback systems, and aging of equipment throughout the roadway driving process ensures the safe and reliable operation of equipment, thereby guaranteeing the safety and efficiency of mine roadway driving operations under various conditions. Figure 5.13 illustrates the safety monitoring and early warning system for coal driving panels based on automation technology.

3. Automated transport technology

The soil, rock, and waste generated from roadway driving must be transported out through the roadway to ensure smooth operation of the driving panel and maintain the roadway's integrity for continuous operation. Additionally, in driving panel roadway driving, meeting the supply demands of materials and equipment requires efficient roadway transportation systems, including automated transport systems. Compared to traditional methods, these systems not only cater effectively to the material transportation needs of mine roadway driving but also significantly enhance efficiency. In recent years, advancements in mine roadway driving technology have seen the increased application of centralized belt control systems in practical engineering,

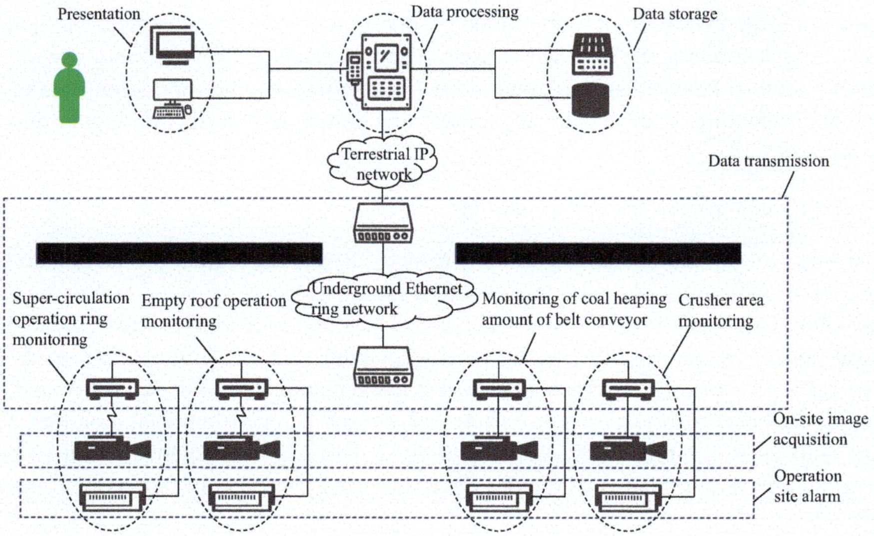

Fig. 5.13 Safety monitoring and early warning system of roadway driving panel based on automation technology

providing robust support for material transportation in driving panel roadway driving. As the workload of mine roadway driving grows, the introduction and implementation of new technologies and systems in belt centralized control transportation further enhance efficiency. Continuous improvements in the design and processing of belt conveyors at driving panels ensure they meet increasingly efficient and safe transportation demands. Strengthening support materials for mine roadway driving and enhancing the effectiveness of material transportation through bottom belt systems are essential to meet safety standards in mining operations.

4. Information transmission automation and its technical analysis

In the roadway driving of driving panels, the operational site is characterized by sudden changes and unpredictability. Failure to understand and control on-site operations during construction increases safety risks and reduces operational efficiency. To address this challenge, automation technology is essential in a driving panel operation. It enhances and optimizes real-time transmission and control of operational information, improving feedback and communication between ground control and heading panel operations. This automation meets the demand for efficient information transmission in a driving panel operation. Currently, Ethernet-based automatic transmission and control systems are increasingly applied in practice, especially in harsh working environments. These systems effectively establish information transmission and control, ensuring timely and rapid data transmission during operations. They enable the transmission of construction site data and equipment statuses to engineering management and command systems, facilitating efficient project development and scheme formulation [21].

5.3.2.3 Application of Automation Technology in Coal Mine Roadway Roadheader

1. Location technology

The roadheader typically operates underground where space is limited, relying on indoor navigation and positioning. Various positioning technologies are utilized based on detection signal differences, including mechanical visual positioning, extrasonic positioning, and radio positioning technologies. Table 5.1 illustrates the advantages and disadvantages of these technologies.

From Table 5.1, it is evident that many commonly used positioning technologies are influenced by the roadway environment and roadheader operating conditions. A single positioning technology cannot guarantee accurate detection or efficiently automate the roadheader's pose detection. Consequently, foreign scholars have begun exploring combined positioning detection technologies. One such technology is strap-down inertial navigation combined positioning, which addresses cumulative error issues and ensures accuracy even in confined roadway spaces. The principle of strap-down inertial navigation/extra-wideband (UWB) integrated positioning is illustrated in Fig. 5.14.

The positioning system comprises four UWB mobile base stations, four UWB fixed base stations, and a strap-down inertial navigation system installed on the roadheader. The simulated roadway coordinate system's origin is selected, and the four modules behind the roadheader are fixed to calibrate the 3D coordinates. The laser range finder is used for quick measurement of the fuselage's coordinates, constructing the attitude estimation Equation, and calculating the roadheader's pose based on the

Table 5.1 Analysis of the advantages and disadvantages of common positioning technology

Type	Advantages	Disadvantages
Mechanical visual	The position judgment is accurate, the spectral response range is wide, and it can run for a long time	The roadway environment can affect judgment
Extrasonic	The system structure is simple and easy to implement, and the detection timeliness is high	It's expensive and requires highly trained staff to operate
Wireless	The system can be shared by using the network. The transmission protocol is perfect and can be integrated	The signal may be interfered with, affecting the positioning accuracy
Inertial navigation	It has strong adaptability, high timeliness, and dynamic detection	The roadway space is narrow, and there is no calibration reference for inertial navigation. The device may have a cumulative error problem of positioning
Laser target	The system structure is simple and easy to implement, the detection results are accurate when the surface finish of the roadway section is good	The detection efficiency is low, and it may be affected by the noise of the roadway, so the section needs to have a certain flatness

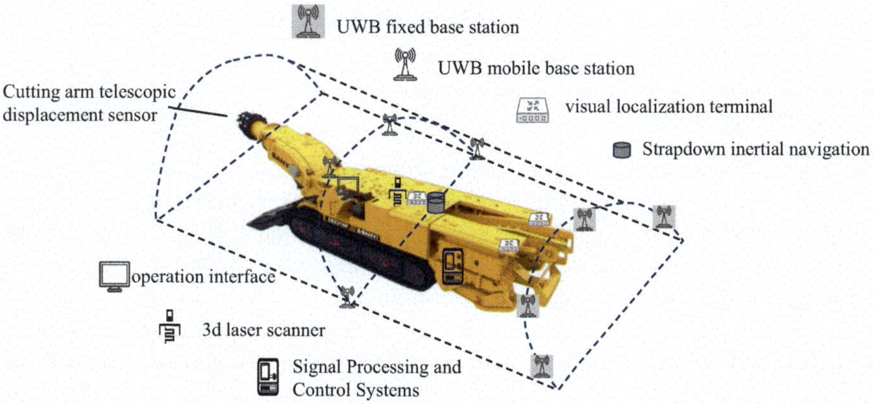

Fig. 5.14 Principle of SINS/UWB integrated positioning [22]

fuselage node coordinates. Pose deviation of the roadheader significantly impacts section quality; any deviations occurring during cutting must be promptly corrected. The roadheader posture correction primarily involves adjusting posture and elevation under cutting load and directional walking. Pitch posture correction is achieved using the front shovel and rear support hydraulic cylinders, while directional walking is adjusted via the crawler hydraulic motor. However, correcting the roadheader's walking pose can still encounter challenges such as control difficulty and insufficient accuracy in correction control.

2. Section automatic forming and adaptive cutting technology

The cross-section shape of a roadway typically includes trapezoidal, rectangular, semi-arched, and other configurations. To ensure compliance with section shape specifications and enhance operational efficiency, it is common practice to control the cutting head's path in an S-shape trajectory. Based on the roadheader's pose, the relative position of the section detection system in the coordinate system is calculated. Cloud mapping is used for error analysis, and designing cutting and walking trajectories to enable automatic section formation. Section cutting involves sweeping, brushing, and S-shaped cutting. Sweeping facilitates section cutting by managing deviations caused by coal and rock accumulation on the floor. Brushing corrects section errors, improving formation quality. S-shaped cutting ensures sublevel height and width, minimizing errors within a reasonable range. In the roadway section automatic forming technology, automatic cutting, and brushing technologies are advanced. Scholars have developed an automatic cutting path planning system, which can meet accuracy requirements for roadway cross-sections. Adaptive cutting technology studies aim for balanced cutting parameters and loads. Artificial neural networks control cutting heads automatically, or PID neural networks design adaptive cutting controls. Roadway sections often contain gangue, posing load variation issues during cutting. For small gangue areas, cutting rates adjust or yaw rates control cutting. Large gangue areas in roadway sections require avoiding gangue belts initially, prioritizing cutting other coal and rock. This challenge prompts new directions in adaptive cutting methods, yet detecting gangue belts remains problematic.

3. Intelligent cutting technology

If the supporting conditions are fixed, roadheader intelligent theory can achieve automatic cutting, loading, and transportation goals. Implementing automatic transportation and walking is less challenging, so practical efforts should focus on automating the cutting process. The roadheader's strong cutting force generates significant reaction forces, potentially compromising overall equipment stability and causing substantial vibration during operations. Hence, enhancing roadheader stability and chassis safety during cutting is crucial. Many manufacturers integrate section monitoring functions into roadheaders. During cutting operations, understanding section monitoring involves analyzing cutting machine and cutting head position data, body vibration, and force matching to adjust cutting forces and rates to meet production requirements. Dynamic pose measurement equipment can be installed behind roadheaders to adjust their operations in real-time. Prompt adjustments correct any inclination or deviation issues, ensuring smooth operations, improving roadway cutting cleanliness, and sustaining continuous roadway driving. Roadway driving equipment requires stable support structures, especially for mines with robust roof conditions that may allow for three-row excavations. Achieving long-section cuts with roadheaders is challenging and requires additional support after each cutting cycle.

In automatic intelligent cutting technology, compared to traditional methods, small-distance section cutting exhibits several distinct characteristics. Firstly, it offers higher cutting efficiency. Intelligent roadheaders optimize cutting paths, shorten total travel distances effectively, and regulate material feed using fuselage vibrations and retractions. By analyzing the relationship and synergy between cutting and traction forces, the utilization of cutting roadways can be enhanced, thereby improving overall cutting effectiveness. Compared to traditional methods, well-regulated cutting forces can prevent issues like empty cutter heads, reduce manual rounding efforts in roadway sections, and enhance cutting efficiency. Secondly, automated cutting yields smoother roadway sections. Through programmatic design, cutting paths align seamlessly with rear positioning operations. Regardless of the roadheader's position, the cutting segment follows a predefined route, ensuring the one-time formation of the roadway section with superior flatness. This approach minimizes over-excavation risks and eliminates challenges associated with manually controlling cutting volumes and limited visibility, helping to control costs. Automated intelligent cutting reduces labor intensity significantly. Traditional manual cutting operations require operators to assist in roadheader cutting tasks. In contrast, automated intelligent cutting achieves these tasks with a single click, quickly forming sections without human intervention. Operators monitor roadheader operations remotely, simplifying cost management by reducing labor requirements. In the end, higher security. Automatic intelligent cutting enables remote operation, separating operators from machinery and preventing damage from tool and gravel splashing during cutting. Furthermore, it minimizes operator exposure to dust pollution, enhancing overall operational safety.

4. Mechanized support technology

In the traditional roadway driving operation, the slow progress of the support process is the primary reason for the slow excavation footage, with the support phase often consuming 65–70% of the total excavation footage time. Achieving mechanized support is essential for realizing the goal of mechanized support. The application of intelligent technology in roadheaders must include mechanized support. Supporting equipment, such as drill pipes and anchor rods, requires regular replacement, making full mechanization challenging. Currently, most roadheader support work relies on manual intervention. Mechanized support can effectively synchronize with roadheader construction operations, enhancing support efficiency. Currently, only some mining areas in China use roadheaders or four-arm rod machines to achieve mechanized support goals. However, many mines fail to meet the standard where the roof span exceeds the length of one roadheader (10 m). In response, integrating rod machines into roadheader designs could improve support operation efficiency. After completing cutting operations, rod machines integrated within roadheaders can be deployed to perform support tasks. Depending on mine conditions, roadheaders can also be configured to support rod machines, utilizing a rod conveyor for top support while rear rod configurations handle side and anchor cable support tasks. During rod conveyor support operations, roadheaders can concurrently conduct cutting and support tasks, optimizing support timing to enhance roadway driving efficiency. The rod conveyor facilitates material transport from roadheaders and can configure rod machines for subsequent rod and anchor cable support tasks. Figure 5.15 depicts a double-arm hydraulic rod drilling rig for coal mines.

Fig. 5.15 Double-arm hydraulic rod drilling rig for coal mine

5.4 Intelligent Mining Information Technology

5.4.1 Mine Information System

Mine information systems represent the culmination of information technology integration within the mining sector, embodying years of innovation and evolution. These systems are tailored to meet the unique demands of safe mining operations in challenging underground environments. Typically, a mine information system comprises several layers: perception execution, transmission, storage analysis, control, and decision-making.

With the steady advancement of information technology and its transformative impact, network technologies leveraging broad connectivity and responsive capabilities have revolutionized traditional vertical hierarchical structures and centralized decision-making paradigms. This evolution has reshaped data flows and decision-making frameworks, shifting from master–slave hierarchical architectures to flat architectures supported by distributed decision-making and microservices. A key feature of this architecture is its ability to reengineer processes and management based on cutting-edge technologies such as AI, cloud edge computing, and the Internet of Things (IoT), tailored to meet the core requirements of mining and safety. By harnessing IoT and edge intelligence, it ensures agile responsiveness at the terminal system level while leveraging cloud intelligence and big data analytics to achieve overall optimization and efficient management of intelligent mining operations. The holistic architecture of the new generation information system is illustrated in Fig. 5.16.

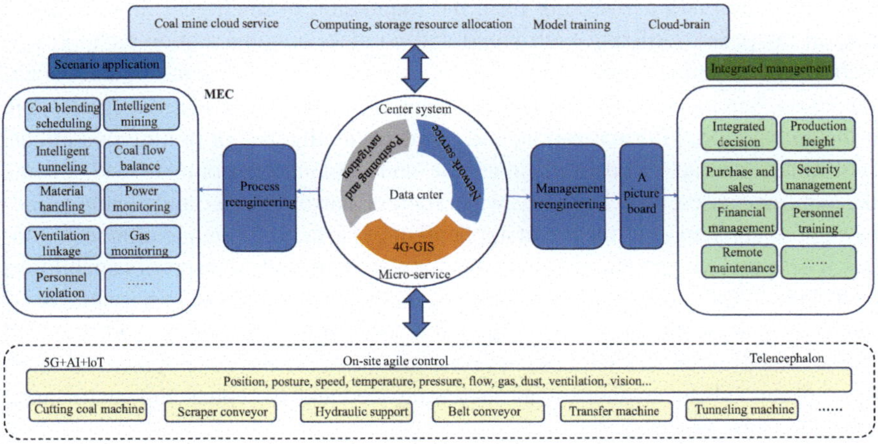

Fig. 5.16 The overall architecture of the information system [2]

Mine basic perception primarily involves comprehensive information gathering across various aspects such as geological conditions, hazard sources, mining environment, equipment, and personnel, achieved through the extensive use of multi-sensor technology. This approach enables the holistic perception of diverse information within different mining systems. Building upon this foundation, temporal synchronization establishes a unified time benchmark, while spatial information is constructed through temporal correlation, forming a unified space–time information system for the mine. For instance, in coal mines, comprehensive information perception regarding humans, machinery, and environmental factors forms the basis for disaster prediction and early warning systems. Real-time perception, fusion analysis, and predictive modeling of disaster-related data are crucial for effective disaster prevention and control. Information perception methods can be categorized into contact and non-contact methods. Contact methods involve sensors that directly interact with measured objects for perception and measurement purposes, including common sensors for mine gases, wind rates, flow rates, temperatures, dust concentrations, and mine pressures. Non-contact sensing techniques encompass infrared spectrum gas sensors, extrasonic wind speed sensors based on time differences, extrasonic flow sensors utilizing frequency differences, infrared temperature sensors, extrasonic level sensors, and laser radar sensors, among others.

5.4.2 Mine Communication and Network

5.4.2.1 Intelligent Mining Communication Requirements

Communication and networking form the foundation of intelligent mining operations at the panel. Utilizing both wired and wireless communication technologies enables functionalities such as information exchange, real-time control, and collaborative operations among personnel and equipment, thereby playing a crucial role in the automation, informatization, and eventual intelligence of mining operations. Coal mines, characterized by their unique underground environment with factors like explosive gases, dust, humidity, water ingress, narrow spaces, and winding roadways, pose challenges to the direct application of surface communication technologies. These include specific requirements for electrical explosion-proof capabilities, limited range for wireless communications, increased interference in wireless transmissions, lack of infrastructure support for positioning and timing, and the need for integrated wired and wireless network solutions. As the development of intelligent panels advances, there is an increasing demand for enhancements in mine communication systems.

(1) Highly reliable multi-source data transmission is essential, accommodating real-time needs for data, file, and video transmission. It supports integrated network services such as underground voice calls, data transfer, video surveillance, and personnel positioning.

(2) Robust broadband backbone network: To handle high-definition video transmission and extensive monitoring data across the mine, a backbone network with $\geq$ 10 Gbit/s transmission capacity and a redundant ring structure is crucial. Core equipment includes three-layer switches with self-diagnostic, traffic management, and network topology generation capabilities, ensuring network self-healing within 50 ms.
(3) High-capacity wireless broadband communication. Due to the abundance of coal mine production equipment, safety sensors, surveillance cameras, and other devices, amounting to tens of thousands of data nodes, effective access and transmission demand large-capacity and high-bandwidth capabilities. The accessible base station capacity must meet the full coverage requirements of the mine, with each station supporting no fewer than 32 users.
(4) Extra-low latency and precise clock synchronization: The underground communication network must achieve stringent real-time performance for automated driving, remote equipment control, and other critical systems.
(5) Network convergence caters to diverse application scenarios. It integrates various wired and wireless communication networks and network management systems, enabling heterogeneous network integration, interconnection, and efficient management of vast mine information resources from different origins. Intelligent coal mining applications span mining, roadway driving, transportation, lifting, drainage, communication, equipment maintenance, monitoring, scheduling, control, and washing. These scenarios require suitable network communication technologies that support high-bandwidth, high-reliability, low-latency applications, and multi-user access, facilitating information sharing and comprehensive utilization.

5.4.2.2 Intelligent Mining Backbone Network

At present, the backbone network of intelligent mining mines primarily utilizes high-bandwidth industrial Ethernet (Gigabit/10 Gigabit and higher industrial Ethernet). Industrial Ethernet technology has been tailored for adaptability beyond commercial Ethernet standards. It integrates additional control functionalities to address industrial production safety and stability requirements, offering tailored solutions for specific industrial applications. Industrial Ethernet satisfies demands for industrial production efficiency, stability, real-time operation, cost-effectiveness, intelligence, scalability, and other critical aspects. It extends down to control field equipment in actual enterprise production processes, making it a pivotal technical tool that comprehensively manages and controls industrial production processes. The main technical characteristics of an industrial Ethernet backbone network are as follows.

(1) Supports communication rates of 10,000 Mbit/s and above. Core equipment utilizes three-layer switches that incorporate routing and redundancy functionalities.

(2) Implements port-based Virtual Local Area Network (VLAN), IEEE 802.1Q VLAN, and generic VLAN registration protocol (GVRP) to streamline network planning.
(3) Features a comprehensive array of industrial network security functions.
(4) Supports standards like RSTP/STP/MSTP for redundancy protocols, including MRP (Media Redundancy Protocol) ring networks or multiple sub-ring protocol applications. Includes network traffic management and automatic network topology generation, with network self-healing time under 50 ms.
(5) Adopts a port modular design; media modules support hot swapping.
(6) Incorporates Quality of Service (QoS), such as IEEE 802.1p/Q and TOS/DiffServ to enhance network stability.
(7) Supports time synchronization protocols like IEEE 1588.
(8) Compatible with Ethernet/IP, PROFINET, MODBUS-TCP, and other industrial Ethernet protocols.

5.4.2.3 Intelligent Mining Wireless Communication Technology

In terms of mine wireless communication, 4G and WiFi, capable of handling calls and high-definition video, have been widely adopted, while the next generation 5G and WiFi 6 are also making their way into mining operations. Among the wireless transmission technologies suitable for low-power sensing terminals, LoRa, ZigBee, Bluetooth, and others are prevalent, each with its distinct technical advantages and limitations. Communication schemes such as UWB, NB-IoT, and Wave Mesh have found limited applications in some products, primarily due to constraints related to network node capacity, throughput, and real-time performance, which are still undergoing technical refinement.

1. WiFi wireless communication

At present, the mine WiFi wireless communication system is predominantly based on WiFi 4 (802.11n), while WiFi 5 (802.11ac) is less utilized in mines due to its reliance on the 5 GHz high-frequency band. The adoption of WiFi 6 (802.11ax), which supports the 2.4 GHz band, will promote the upgrade of the mine WiFi system.

The WiFi 6 wireless communication system consists of a dispatching switch, recording server, wireless management controller, touch screen dispatching station, ground wireless base station, mine intrinsically safe wireless base station, mine intrinsically safe mobile phone, mine flameproof and intrinsically safe power supply, and other supporting equipment. The system is integrated into the industrial Ethernet ring network deployed in the mine, eliminating the need for dedicated voice transmission lines and thereby reducing investment in line construction funds.

In addition, the mine WiFi6 system has the advantages of high transmission bandwidth, simple system, and low construction and maintenance cost, however, it also has the disadvantages of poor mobility, large transmission delay compared with 5G, and low call quality compared with 4G/5G system. It is suitable for the scene with a large number of concurrent traffic and low mobility and call quality requirements in the mine. In addition to providing voice communication, it can also be used for mobile broadband transmission, high-definition video transmission, emergency communication, underground wireless interconnection, etc. Furthermore, it can also be deeply integrated with personnel positioning, scheduling communication, and other systems to build a multi-system integrated network and promote the construction and development of intelligent mines.

2. 5G wireless communication technology

5G technology, that is, 5th generation mobile communication technology. The International Telecommunication Union (ITU) defines three application scenarios for 5G. That is enhanced mobile broadband (eMBB), extra-reliable low latency communications (URLLC), and massive machine-type communications (MMTC). It has the characteristics of large broadband (user experience speed up to 1 Gbit/s), low latency (as low as 1 ms), and wide connection (user connection capacity up to 1 million connections/km^2).

5G technology has the following characteristics:

(1) Continuous wide-area coverage. The most traditional communication scenario, aiming at ensuring user mobility and business continuity, provides users with a seamless high-rate business experience of more than 100 Mbit/s.
(2) Hot spot high capacity. It is mainly oriented to local hot spots, providing users with a data transmission speed of 1 Gbit/s to meet the needs of network traffic density.
(3) Low-power large connection. Aiming at the application scenario of deploying a large number of terminals in the Internet of Things, it meets the density index requirements of millions of connections per square kilometer, and the terminal power consumption is low.
(4) Low delay and high reliability. It is mainly oriented to the extreme performance requirements of vertical industries such as the Internet of Vehicles and industrial control, providing users with millisecond end-to-end delay or close to 100% transmission reliability.

The application of 5G technology in coal mine private networks provides an integrated wireless, high-reliability broadband transmission platform for coal mines. This platform enables mobile voice communication and scheduling, mobile high-bandwidth transmission, high-definition video transmission, real-time remote control, emergency disaster relief communication, underground unmanned driving, underground industrial interconnection, and other functions. It ensures reliable communication support for mine informatization, automation, and intelligent construction, meeting the communication needs of underground people-to-people, people-to-things, and things-to-things interconnections. The mine 5G communication system adopts Standalone (SA) architecture networking, with 5G core network localized deployment. It employs a 5G wireless private network separated physically from the public network. The technical core network and base stations are deployed locally within coal mines, ensuring that data does not leave the mining area. This setup enhances the security and reliability of data transmission.

5.5 Thick Coal Seam Intelligent Mining

5.5.1 Intelligent Caving Technology of LTCC

The coal mining process mainly involves two types of coal-rock identification issues. One is the identification of the coal-rock interface during the coal cutting process of the fully mechanized mining panel, and the other is the identification of coal and gangue during the top coal caving process of the LTCC panel. Both the coal-rock interface identification during the coal cutting process of the fully mechanized mining panel and the coal and gangue identification during the caving process of the LTCC panel are highly challenging, and there are no mature technologies yet. Although the coal-rock identification during the coal cutting process has been studied for more than 30 years, it still relies primarily on memory-based coal cutting. The identification of coal and gangue from the rapidly moving coal-gangue accumulation on the rear AFC during the caving process is even more difficult than the coal-rock identification during the coal cutting process of the panel, requiring higher accuracy and reliability. At present, the LTCC panel still adopts manual caving methods. Caving workers observe the coal-gangue flow on the rear AFC through the gap of the hydraulic support, observe whether the fallen gangue is drawn, and the amount of coal and gangue drawn. Then, they decide whether to close the DO and stop top coal caving. This method has high labor intensity, low production efficiency, high environmental dust, and is prone to misoperation. It cannot accurately distinguish between the gangue (broken immediate roof) and parting in the top coal.

Intelligent LTCC is based on intelligent fully mechanized mining technology, which realizes intelligent caving and ultimately achieves intelligent control of the LTCC panel. In the field of intelligent LTCC related technologies, preliminary exploratory experiments have been conducted in China in recent years. For example, the Xinglongzhuang Coal Mine of Yankuang Group experimented with a caving method combining programmed control and manual replenishment. The Wangzhuang Coal Mine of Lu'an Group tested sound frequency spectrum coal-gangue identification technology. China University of Mining and Technology conducted experimental research on an automatic control system for the caving process based on acoustic waves, near-infrared spectra, natural rays, etc. It uses high-performance multiple echo signal reflection lidar scanning technology to monitor the coal transportation volume on the rear AFC. Henan Polytechnic University proposed an active coal-gangue identification method based on microwave heating and infrared detection. Shandong University of Science and Technology analyzes the tail beam vibration signal to identify coal-gangue on the LTCC panel. Shandong Technology and Business University and other institutions have proposed the idea of using a variety of monitoring methods to achieve precise caving control. Beijing Tiandi Mako Electro-Hydraulic Control System corporation tested the memory top coal caving method. Tongxin Coal of Jin'neng Group Mine uses coal-gangue impact vibration sensors to identify the time when gangue reaches the DO.

The core of the above technology research focuses on how to identify and detect coal and gangue during the caving process, i.e., identifying whether the drawn material is a coal flow or a gangue flow during the caving process. However, as for the gangue ratio in the mixed flow of coal and gangue, there is no research. It is not possible to calculate the gangue ratio in the gangue flow and to distinguish whether the gangue caved is from the parting or the roof. To identify whether the drawn gangue comes from the parting or the roof, based on the independently developed top coal motion tracking instrument (invention patent number: ZL200910080005.9). In addition, a top coal motion time measurement system has been developed, and an intelligent caving control system based on this technology has been constructed, which can achieve precise control of multi-round sequential intelligent coal caving. It was initially applied in the 8105 panel of Zhuxianzhuang Mine in Huaibei. In recent years, deep learning has developed rapidly and has been widely applied in various industries and fields to solve scene-understanding problems in complex environments, providing technical support for image recognition intelligent caving technology. Through semantic segmentation, the coal-gangue boundary in the coal flow on the rear AFC of the LTCC panel can be extracted, then the gangue mixed ratio can be obtained and guide intelligent caving.

Foreign LTCC is only applied in a few countries such as Turkey, Bangladesh, Australia, and so on, most of them are the output of domestic LTCC, such as Yankuang Group Co., Ltd. applies LTCC to Australia's Ausda mine and explores the automated caving method based on time control. In 2018, Caterpillar Inc. put forward intelligent caving technology based on memorizing the stent positional state of the stent. In addition, the integrated mining panel automation technology represented by inertial navigation technology, extrasonic, thermal infrared, terahertz spectroscopy and other coal and rock identification technologies, virtual reality technology, and multi-sensor technology makes the intelligence of the integrated mining panel possible, which also provides a certain technological reference for the research of intelligent top coal placing mining technology.

The core of realizing intelligent caving is to correctly grasp the timing of opening and closing DO, however, no key breakthrough has been made so far. Following over a decade of collaborative scientific and technological research, the authors' team, through continued investigation and analysis of image, sound, vibration, and other forms of gangue recognition technology, ultimately narrowed their focus to image recognition technology. We devised a novel method for rapidly recognizing images based on the greyness difference characteristics of the gangue pile. A gangue image acquisition system was devised for the rear AFC of the top coal panel, which is subject to high dust conditions. This system enables the intelligent caving from the LTCC panel, which is a crucial step in achieving intelligent coal caving. The intelligent coal discharge of the panel has resolved the key technical issues associated with intelligent mining of the LTCC panel. This has been demonstrated through on-site applications at the Huaibei mine and the Kailuan mine [23, 24].

5.5.1.1 Effect Mechanism of Illumination on Coal-Gangue Image Features and the Selection of Coal-Gangue Recognition Features

At present, in LTCC panel, it is usually a caving worker judges whether the gangue has been drawn or estimates the amount of drawing top coal by observing the color (or gray scale) of the coal flow on the rear AFC near DO position and then determine whether the DO needs to be closed. The 40 years of top coal caving production experience also shows that in most coal seam occurrence conditions and production environments, it is feasible and effective to identify coal-gangue by the artificial naked eye and then control the switch of DO position. Therefore, it is entirely possible to use mature image recognition algorithms instead of manual identification of coal and gangue to achieve the purpose of reducing people and increasing efficiency. There are two main ideas about image recognition.

The first research idea is to use the classical image processing algorithm to calculate the gangue mixed ratio. Gray scale and texture are the two most used feature indicators for image recognition. The fractal dimension can be used to reflect the complexity of the coal-gangue surface texture. The principle of this method is simple, and it has been widely used in the field of image recognition of coal-gangue separation, and its reliability has been verified.

The second research idea is to use the deep learning algorithm to semantically segment the coal-gangue image, and then calculate the gangue mixed ratio. Whether it is a classical image processing algorithm or a semantic segmentation algorithm based on deep learning, the object of analysis is an image, and the image is the result of a variety of factors such as the target object, the light source, the environment, and the image acquisition system. The same object is placed in different light environments, or different sensors are used for acquisition, and the obtained images are also different. The results of further recognition through the algorithm may also be different, especially for the recognition of coal and gangue, such problems are more significant. Therefore, the concept of illumination is introduced into the research of image recognition automatic caving technology to provide the optimal illumination or optimal feature for coal-gangue image recognition, which is also the difference between coal-gangue image recognition and image recognition in other fields.

5.5.1.2 Accurate Segmentation of Coal-Gangue Image and Calculation of Gangue Mixed Ratio

Compared with other recognition technologies or means, one of the major advantages of image recognition automatic caving technology is that it can realize the recognition of gangue mixed ratio. Gangue mixed ratio refers to the ratio of the volume of the gangue (the surface area of the gangue or the projection area in the 2D image) to the total volume of the coal and gangue (the total surface area or the total projection area in the 2D image), which is caved from the DO and falls on the rear AFC. The value range is 0–100%.

At present, some commonly used automatic caving technologies, such as technologies based on sound or vibration signals, can only distinguish the two caving stages of coal caving and gangue caving, however, cannot distinguish the gangue mixed ratio. This is a typical principle of seeing the gangue closing DO. Recent studies have found that when the gangue mixed ratio is 10–15%, the top coal recovery can be maximized, which requires a quantitative and accurate judgment of the gangue mixed ratio when identifying coal-gangue, otherwise, it will cause a large loss of top coal.

The accurate judgment of the gangue mixed ratio is inseparable from the high-precision image segmentation results. The image segmentation in the field of top coal caving is different from other fields. For example, in the process of coal-gangue separation, the coal and gangue to be separated are paved on the belt conveyor and through the sensor, while the LTCC panel, the rear AFC drives the coal flow to move rapidly, and the coal and gangue accumulate and stack each other (Fig. 5.17), which is not conducive to boundary identification and gangue mixed ratio calculation.

To achieve accurate image segmentation, two ideas of classical algorithm and deep learning algorithm are studied. In terms of classical algorithms, a marker-controlled watershed method based on multi-scale reconstruction and marker control for coal-gangue image segmentation is proposed. It realizes the segmentation of coal-gangue mixed image (Fig. 5.18), however, the ability to identify the shadow area in the image is poor.

The deep learning method is introduced into the research of gangue mixed ratio recognition, and a lightweight gangue recognition and boundary measurement model for LTCC panel is proposed (Fig. 5.19, where r is the void rate). The low-level features of the image are extracted by a depth-separable lightweight convolutional structure to improve the speed of feature extraction. High-level image information is extracted by

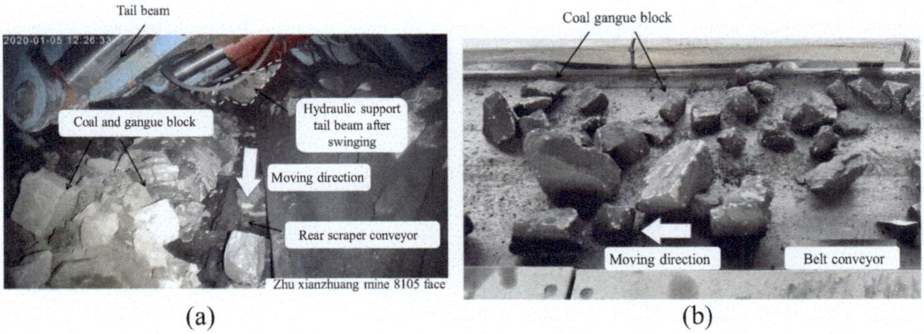

Fig. 5.17 Coal-gangue identification in different research fields: **a** LTCC panel; **b** Coal-gangue separation

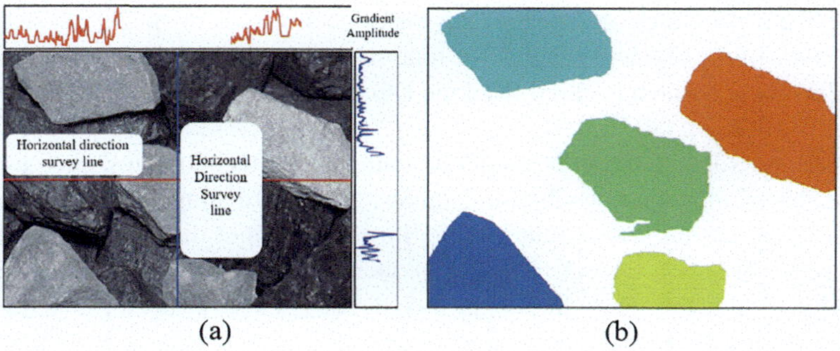

Fig. 5.18 Segmentation results of classical algorithm: **a** modified gradient amplitude indication; **b** segmented image

multi-scale modules. After low-level and high-level multi-scale information fusion, the complete boundary of the gangue target in the image is obtained. By labeling the coal-gangue image data set training, the accurate boundary of the gangue in the top coal image is quickly obtained (Fig. 5.20).

Gangue mixed ratio is a generalized concept, which can be expressed not only by projection area ratio but also by surface area ratio or volume ratio. By counting the number of pixels in the 2D image, the gangue mixed ratio represented by the projection area can be obtained. The coal and gangue in the coal flow on the AFC at the back of the LTCC panel are 3D blocks, and they are accumulated and stacked with each other. Therefore, the 3D stacking shape of the coal and gangue blocks is inverted by the 2D image of the coal and gangue, and the volume of the coal flow surface is obtained. The gangue mixed ratio is predicted, and the volume of gangue

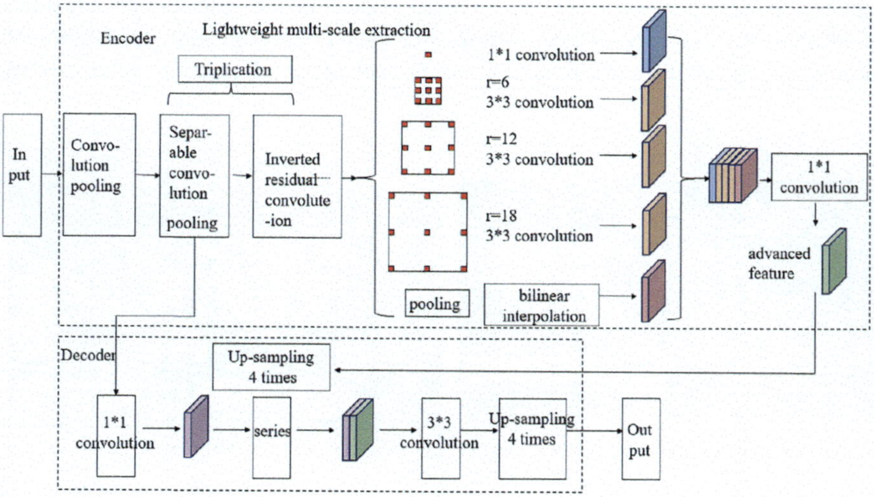

Fig. 5.19 Gangue identification and boundary measurement model in the LTCC panel

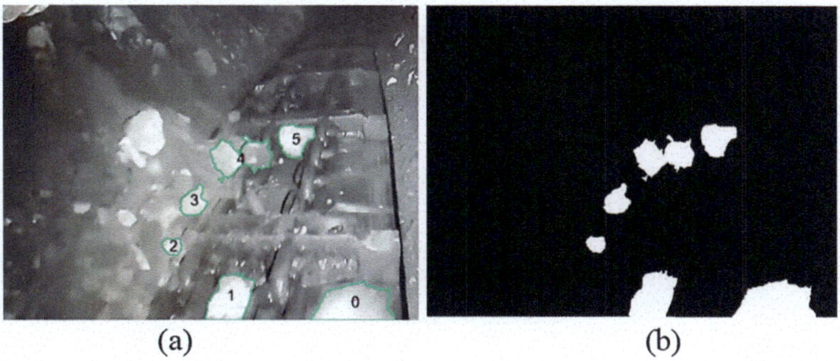

Fig. 5.20 Image acquisition and segmentation: **a** image segmentation results; **b** binarization processing

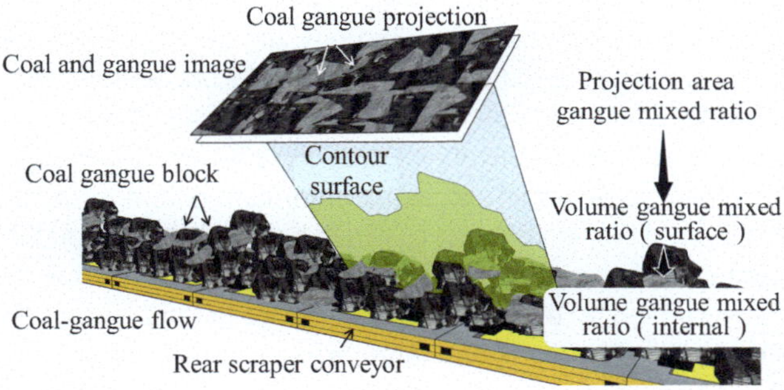

Fig. 5.21 Gangue mixed ratios involved in image recognition automatic caving technology

mixed ratio superimposed inside the coal flow is predicted (Fig. 5.21). This is a method to improve the measurement accuracy of the gangue mixed ratio, and it is also a place where the image recognition automatic caving technology is different from and ahead of other monitoring methods.

5.5.1.3 Illumination Intelligent Monitoring and Adaptive Adjustment Based on Stereo Vision

Illumination is an important factor to be considered in image recognition intelligent caving technology. Accurate measurement and control of illumination is the premise of obtaining high-quality images and achieving high-precision recognition. In the process of top coal caving, the drawn coal and gangue move rapidly on the rear AFC, and the illuminance measurement is difficult. Therefore, a new method for monitoring illumination using stereo vision depth measurement technology is proposed. The illumination value of the drawn coal-gangue surface is related to the power of the

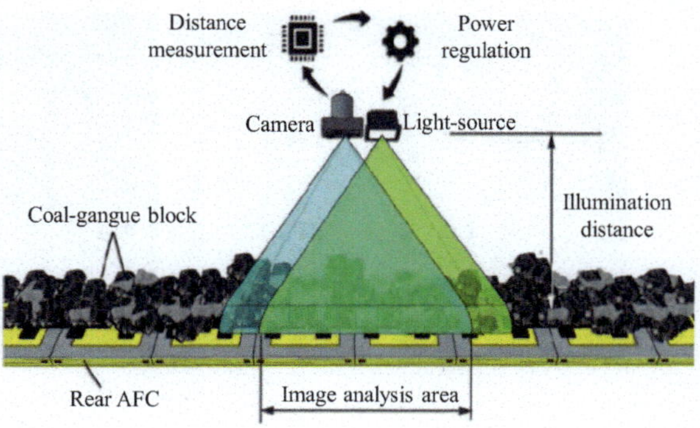

Fig. 5.22 Principle of automatic illumination monitoring and adaptive adjustment

light source and the distance between the light source and the coal-gangue surface (illumination distance). The light source power can be measured by a power meter. In the image recognition automatic caving technology, the stereo vision depth measurement technology can complete the image acquisition work without increasing the equipment and workload. While obtaining the distance data, it can also extract the gray, color, or texture features of the object to achieve semantic segmentation.

Therefore, this book proposes an automatic illumination monitoring and adaptive adjustment method based on stereo vision, which provides the optimal illumination for image recognition automatic caving technology. In the LTCC panel, the illumination distance is determined by the depth information of the image, and the real-time illumination is determined by the power monitoring, and then the optimal illumination is obtained by adjusting the power, and the image is collected and identified under the optimal illumination (Fig. 5.22), and the adaptive adjustment of illumination in the process of image recognition automatic caving is realized. If the camera and the light source are placed side by side, the image depth can be used directly instead of the illumination distance. If the layout is scattered, the illumination distance can also be determined by geometric relationship conversion.

5.5.1.4 Relationship Between Projection Area Gangue Mixed Ratio and Volume Gangue Mixed Ratio of Coal-Gangue Accumulation Body

2D and 3D shape characteristics are important indicators to characterize the shape of irregular coal-gangue blocks. 2D shape features are easily obtained from 2D images, however, they cannot fully reflect the shape of the block. The 3D shape characteristics contain many indicators, and the description of the block is more comprehensive, however, the 3D shape characteristics are difficult to measure, especially in the environment where the LTCC panel requires real-time monitoring of the coal-rock flow on the rear AFC. Therefore, fast, direct, and accurate estimation of 3D shape features from 2D images is an important issue. Based on laboratory experiments and numerical simulation, the relationship between the 3D shape characteristics and the 2D shape characteristics of the coal-gangue block is revealed.

The 2D shape characteristics are related to the perspective of image acquisition, that is, the 2D shape characteristics of coal-gangue blocks from different perspectives are different (Fig. 5.23). Therefore, when analyzing the 2D characteristics of coal-gangue blocks, it is necessary to first determine the observation perspective. In the LTCC panel, the coal-gangue block from the DO tends to accumulate on the rear AFC in the preferred orientation after falling and collision. Therefore, it is necessary to determine the observation angle of the coal-gangue block on the rear AFC, rather than observing at will, and the shape features extracted under the dominant orientation are meaningful. Based on this assumption, the coal-gangue block model is accurately reconstructed by multi-view image sequence, the numerical calculation of free fall is carried out by discrete element method, and the dominant orientation of blocks with different shape characteristic parameters is determined (Fig. 5.24).

On this basis, the 2D shape characteristics and 3D shape characteristics of the coal-gangue block in the preferred orientation are calculated respectively, and the correlation analysis is carried out to reveal the relationship between the 2D and 3D shape characteristics. The results show that the 2D and 3D shape characteristics have a high correlation (Fig. 5.25). Therefore, through the analysis of 2D images,

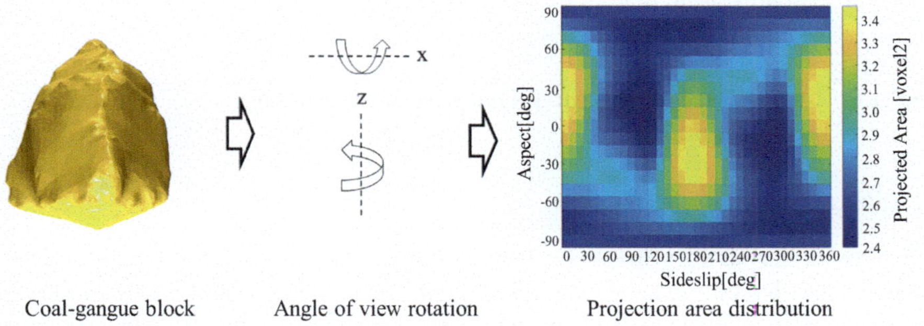

Fig. 5.23 The relationship between the projection area and the viewing angle

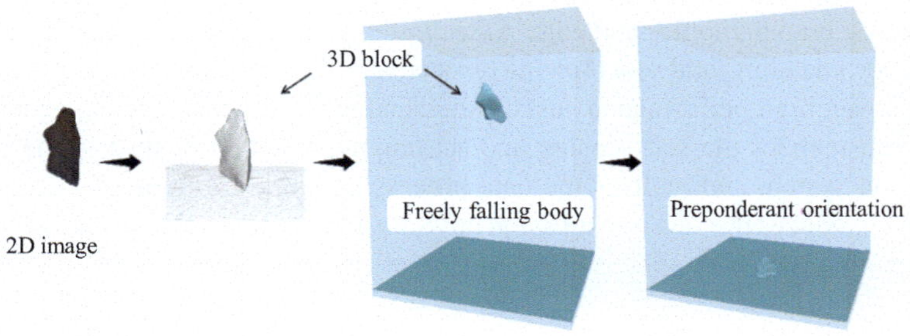

Fig. 5.24 Dominant orientation determination process

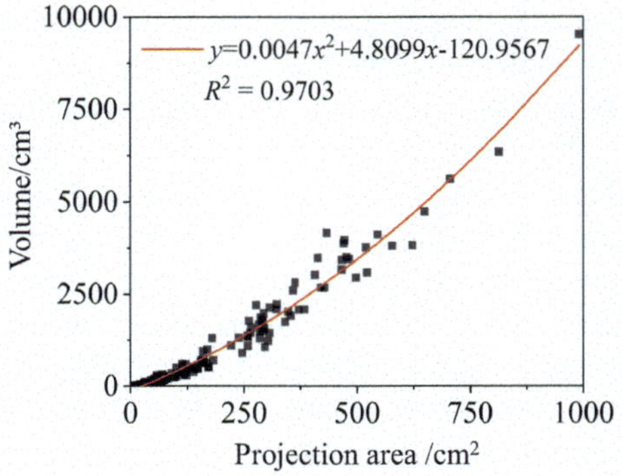

Fig. 5.25 The relationship between projection area and volume of coal-gangue block

the 3D shape characteristics of coal-gangue blocks can be estimated, and then the gangue mixed ratio data can be corrected. As the first step of the two-step strategy for high-precision prediction of S2I gangue mixed ratio, the transition from projection area gangue mixed ratio to surface volume gangue mixed ratio can be realized. On this basis, the accumulation characteristics of coal-gangue blocks are further analyzed, and the relationship between the surface volume gangue mixed ratio and the internal volume gangue mixed ratio of coal rock flow is obtained, to achieve the purpose of predicting the 3D gangue mixed ratio by using 2D images and realize the high-precision measurement of gangue mixed ratio.

5.5.1.5 Image Recognition Intelligent Caving Technology and Equipment

1. Image acquisition system

The dust in the caving process is extremely large. How to ensure the cleaning of the camera and obtaining clear and high-quality images has become the key to restricting the realization of image recognition intelligent coal caving. Figure 5.26 and Fig. 5.27 are the first-generation prototype of the image acquisition system with an independent data processing function, Insight-I, which can adapt to the high concentration dust and water mist environment in 2020.

Based on human bionics and edge AI technology, the system imitates the actions of blinking, kneading, and soot blowing respectively. Through high barrier pneumatic sealing cover, high performance dust cleaner, and high pressure blowing dust knife, the system realizes the self-sensing and scavenging function of dust in the image acquisition system and realizes the data processing work at the image acquisition end, which reduces the data transmission pressure and improves the system reliability and response rate.

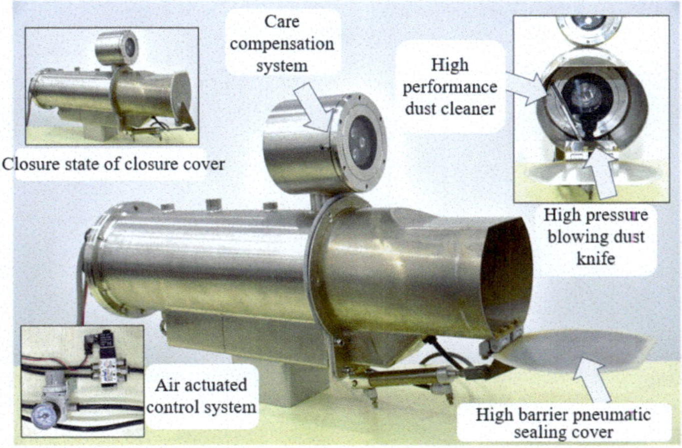

Fig. 5.26 Self-cleaning camera

Fig. 5.27 Installation of the camera on the support

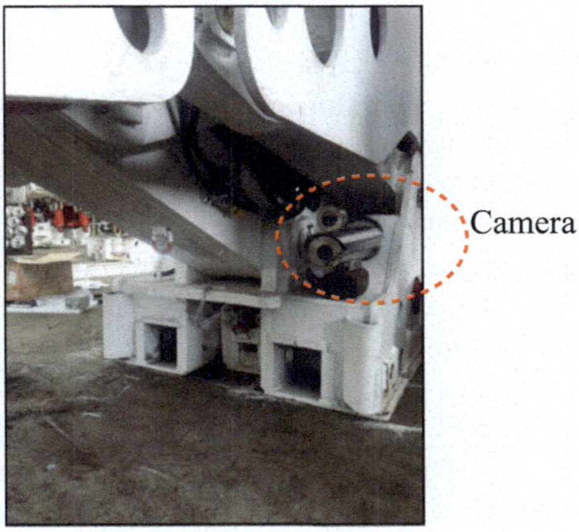

2. Intelligent caving online detection software

(1) Software architecture

The image recognition intelligent caving online detection software mainly includes four parts: an acquisition module, a scene interaction module, an image recognition module, and a communication module. It covers camera data acquisition, human–computer interaction interface, coal-gangue target recognition and measurement algorithm, graphical display of detection results, data communication, and other core contents. It can collect video/import video data in real-time, process image data acquisition, pop-up window prompt, remote transmission, and real-time storage of gangue mixed ratio data in a caving link. The real-time detection results of the gangue mixed ratio are displayed in a graphical form, and the serial communication

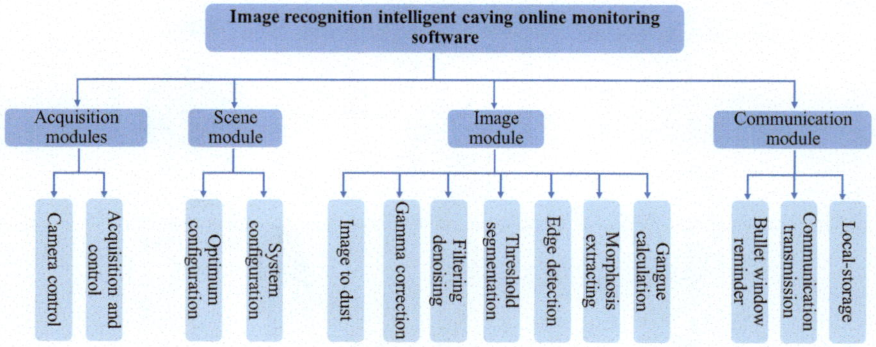

Fig. 5.28 Image recognition intelligent caving online monitoring software architecture

transmits the content to the controller to implement the intelligent control of the top coal caving. The software architecture is shown in Fig. 5.28.

(2) Software interface design

The main interface includes the following parts: menu bar, file opening and saving operation, camera opening and closing operation. Manually select a single picture or video under the folder for coal-gangue recognition, or choose to open the login interface and configure the camera to collect video images. The four image areas of the main interface from left to right from top to bottom are collecting video, video image detection area selection, detection result display, and gangue mixed ratio curve. The right side of the interface is the communication parameter setting of test results and real-time output of gangue mixed ratio.

To reduce the interference of complex background on the recognition process, the method of manually selecting the area is used to locate the area of the AFC, and the image recognition range is constrained in the working area of the AFC to improve the image processing ratio and recognition accuracy. To facilitate the staff to delineate the area of the AFC in the image, the mouse hardware device and Qt are used to realize the area selection task. The implementation of this task includes two steps:

(1) Initial rectangular frame delineation. The Qt mouse event is associated with the mouse hardware device to realize the left mouse button triggering, moving, and releasing functions, and area selection and positioning. Trigger the left mouse button, the trigger point is the starting point of the box; move the mouse to reach the desired end position; release the left mouse button, and the release point is the end of the selection box.
(2) Rectangular frame adjustment. When the initial selection box is completed, if the selection box needs to be further adjusted, the mouse cursor can be placed at the end of the selection box, and the left mouse button can be triggered to adjust the position of the corresponding selection box. After the selection frame is completed, the position of the starting point of the selection frame can be

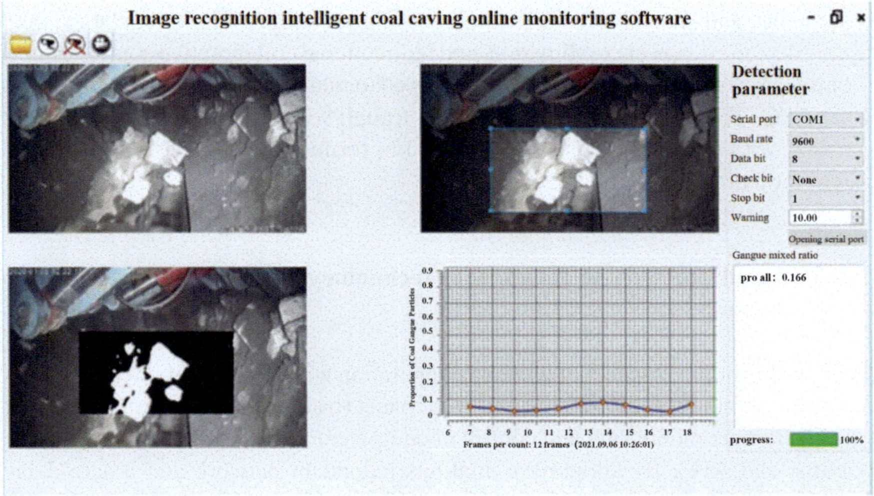

Fig. 5.29 Coal-gangue recognition and detection human–computer interaction interface

determined, to calculate its width and height. The image area selection of the AFC is shown in the upper right image box in Fig. 5.29.

The related achievements of image recognition automatic top coal caving technology have been tested in 8105 LTCC panel of Zhuxianzhuang Mine of Huaibei Mining Co., Ltd., 824 LTCC panel of Yuandian No.1 Mine and 0291 LTCC panel of Tangshan Mine of Kailuan Group, and good results have been achieved. The top coal recovery is expected to increase by more than 5%. In the past two years, it has been applied in the Shendong Baode Coal Mine and Wangzhuang Coal Mine of the Lu'an Group.

5.5.2 *Key Technology of Intelligent Large Mining Height*

5.5.2.1 Intelligent Extra-Large Capacity Transportation Technology [25]

The SGZ1400/3 × 1500 extra-large capacity central double-chain AFC has been designed, using F-grade chains with specifications of Φ56 mm × 187 mm, achieving a conveying capacity of 5000 t/h. For the first time, a new structure featuring a double middle plate with a cross side-discharge headframe and a large-stroke electro-hydraulic controlled automatic retractable tail with a double-layer box structure has been adopted. This new structure can automatically adjust the cylinder retraction amount based on changes in cylinder pressure, maintaining the scraper chain at an optimal tension at all times. Utilizing high-voltage variable frequency soft start technology, the system monitors the transportation status of the scraper chain, drive components, and chains. It features intelligent start, intelligent ratio regulation, coal volume detection, power coordination, and bidirectional collaborative control. A new 172 mm large-pitch pin row has been developed to match the 2050 mm specification of the AFC. This is paired with a 1600 mm trough width high-power reloader and a crusher with a crushing capacity of 7000 t/h, forming a complete transportation system for extra-large mining panels.

5.5.2.2 Intelligent System Integration Technology of Fully Mechanized Mining

Based on the mining concept of unmanned operation with supervised safety, a visualized remote control monitoring center in the panel roadway has been established for real-time remote monitoring and control of equipment such as the shearer, hydraulic supports, and AFC. By adopting a dual-bus redundant network and a zero-delay signal repeater (with a system response time of less than 300 ms) adapted to the electro-hydraulic control system of long and narrow panel supports, the technical challenge of signal transmission delay has been resolved.

The implementation of new equipment such as inclination sensors, pressure sensors, cameras, and monitors enable real-time monitoring of the coal wall spalling, hydraulic support posture, and equipment operating status at the panel. Proximity sensors can effectively determine the retraction status of the support's shield guard, preventing interference between the shearer and the shield guard, and reducing the overall support moving time. Laser rangefinders, pressure sensors, and displacement sensors are used to identify the panel height and the conditions of the roof and floor, enabling intelligent decision-making for corresponding actions. The installation of displacement and pressure sensors on the primary shield guard of the hydraulic support allows real-time detection and intelligent control of the coal wall support pressure, significantly improving the management of coal wall spalling at the panel.

Based on the sensory information from the aforementioned sensors, intelligent linkage control between the equipment at the panel has been achieved, addressing the challenge of coordinated control among the shearer, hydraulic supports, and AFC.

5.5.2.3 Surrounding Rock Support Coupling Control Technology

According to the characteristics of the intelligent panel with a large mining height, it is necessary to realize the coupling control of surrounding rock support in the panel. The coupling control can realize the functions of support attitude monitoring, roof surrounding rock coupling, sidewall surrounding rock coupling, and support toppling control. Specifically, as follows:

1. Support attitude monitoring

According to the inclination sensor components of the support base, top beam, shield beam, and front connecting rod, the working attitude of the support is detected in real time to prevent the inclination or bite of the support and stop or lock it in time.

2. Roof surrounding rock coupling

Based on the pressure sensors on the support columns, pressure and displacement sensors on the primary shield guard, and proximity switches on the tertiary shield guard, the parameters of the surrounding rock of the panel roof are monitored in real-time. This system effectively controls roof falls and collapses, allowing for the timely detection of dangerous situations in the surrounding rock of the roof.

3. Sidewall surrounding rock coupling

Through the stroke sensor, pressure sensor, and proximity sensor of the support, effective protection behavior is adopted after the deep intelligent perception of the change of the surrounding rock of the panel. By installing a stroke sensor on the support rib protection device, the shape of the group rib protection device is gradually opened, to prevent accidents such as the collapse of large coal after the rib protection device, damage to the column cylinder, or the stuck between the support and the cable groove, which makes the support unable to move. By installing a pressure sensor on the rib protection device, the rib protection device can perceive the support effect of the coal wall, and prevent the spalling accident in the fully mechanized intelligent panel with large mining height. By installing the proximity sensor on the three-level rib protection device, the equipment damage caused by the interference of the action of the rib protection device is prevented. In the case of failure of the proximity sensor, the system automatically locks the action of the rib protection device and alarms the support that has not recovered or cannot recover the rib protection device. The field application effect of surrounding rock coupling is shown in Fig. 5.30.

4. Support toppling control

When the panel has an inclination or is in an uphill or downhill mining state, large-height fully mechanized mining supports may experience jamming or tipping accidents during the support movement. To address this, angle sensors are used to detect

Fig. 5.30 Wall rock coupling field application effect

the working state of the hydraulic supports. Posture control is implemented during the movement of the supports. If the angle of any support component exceeds a specified value or if the angle between the support and adjacent support components exceeds a set limit, the system locks the support actions. This measure prevents jamming and tipping accidents, ensuring the stability and safety of the supports.

5.5.2.4 High-Definition Visualization Technology of Panel

Based on the special application environment of fully mechanized mining panels, a new intrinsically safe high-definition pan-tilt camera specifically for mining use has been developed through a comparative analysis of existing cameras both domestically and internationally. This camera features a 720p HD lens that supports automatic or manual switching between color and black-and-white modes, providing optimal video quality under low-light conditions and meeting the monitoring requirements of the panel. Additionally, external communication controls the internal pan-tilt motor, achieving a rotation precision of 1°. The stepper motor encapsulation process has been optimized to meet general mining design standards. Research on video image relay technology and image tracking technology through an underground main control computer has enabled real-time tracking of fully mechanized equipment such as shearers. This allows the live operational view of the panel to be automatically pushed to the operators, enabling immersive remote manual intervention at the panel.

5.5.2.5 Rapid Support Movement Control Technology for the Panel

Intelligent large mining height panels require rapid support movement control technology to effectively avoid the hazards of periodic pressure and rock bursts. The use of direct-drive rapid liquid supply control technology can enhance the response ratio of hydraulic supports. The FHD500/31.5 high-flow electro-hydraulic control reversing valve, with 2 DN25 inlet ports, 5 DN25 working ports, and a rapid return liquid function, paired with a 1000 L/min high-flow hydraulic control check valve for columns and a high-flow reverse pull push check valve, meets the needs for rapid movement of 7-m supports, achieving a support movement ratio of 8 s per support.

Through the simulation analysis of hydraulic support movement actions, the column action distance is set at 0.1 m, and the push–pull cylinder action distance is set at 0.96 m. The lowering, moving, and raising times for single support are 3 s, 3 s, and 2 s, respectively, with a total action cycle time of 8 s. The simulated movement curve of the hydraulic support lowering, moving, and raising actions is shown in Fig. 5.31, where L represents the stroke distance and t represents the time.

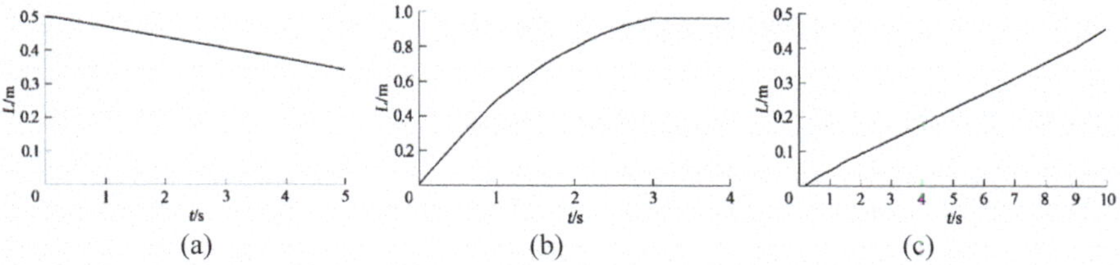

Fig. 5.31 Simulated movement curves of hydraulic support lowering, moving, and raising actions: **a** lowering column curve; **b** moving support curve; **c** raising column curve

5.5.2.6 3D Virtual Reality Technology for Equipment in the Panel

Using 3D animation technology, a 3D model of the equipment in the panel is created to integrate seamlessly with operational data. This approach achieves a 3D monitoring system driven by real-time operational data. The established 3D monitoring system for fully mechanized mining panels allows visualization of processes such as coal cutting by the shearer, coal collapse, coal transportation, and support movement. The virtual reality production scene of intelligent large mining height panels is illustrated in Fig. 5.32 [26].

Fig. 5.32 3D virtual reality panel of the intelligent large mining height

5.5.3 Intelligent Mining Case of Thick Coal Seam

5.5.3.1 Key Technology of Intelligent Caving

In Tongxin Coal Mine, the key technology of intelligent caving in extra-thick coal seam and the equipment of the LTCC panel have realized automation. The 8102 panel of Tongxin Coal Mine is the first automatic caving panel. Compared with the previous panel, it adds intelligent equipment such as the main roof beam, shield beam, rear swing beam, and height measurement sensor based on the original pressure sensor, push stroke sensor, support infrared receiving sensor, and so on. According to the thickness of top coal, top coal caving time, and caving method, the way of remote caving operation on the ground console is selected. It adopts three rounds of sequential caving with three support intervals per round. The time from receiving the signal to the end of the coal drawing for each support is 40 s. By selecting the caving sequence (ascending or descending of the unit, coal drawing range, etc.), the optimum size of the coal drawing port is determined to be 210–280 mm (return high stroke is 260 mm) according to the one-month test of inclination sensor. Furthermore, combined with the altimetry sensor, according to the support stroke, the size of the DO can be adjusted appropriately, so that the three rounds of caving ensure enough rear coal, and the size of the DO position further ensures that the large gangue will not fall into the rear AFC, thus avoiding the damage to the equipment and the impact on the coal quality. According to the current test situation, 8102 panel have realized automatic caving [27].

5.5.3.2 Intelligent Coal Overload Alarm Device

Tongxin Coal Mine is equipped with an automatic coal quantity overload caving alarm device, and the coal quantity monitoring and alarm system is added to the belt-inclined roadway control center (Fig. 5.33). When the motor current of the current rear AFC exceeds 180 A and the transfer current exceeds 200 A, the alarm system at the distribution point of the inclined roadway will predict: the 'motor overheating' voice. The alarmed voice sends a voice alarm to each stopper in the panel through the power distribution point of the inclined roadway through the stopper, so that the panel personnel can realize a large amount of coal, and the post type of work will respond accordingly, such as the driver of the unit stops cutting coal, the support worker stops walking, and the caving worker stops coal caving, to ensure that the coal volume of each transport machine is not overloaded and avoid the occurrence of safety accidents.

(a) (b)

Fig. 5.33 Coal quantity monitoring alarm system: **a** voice alarm device; **b** panel carrier voice and alarm device

5.5.3.3 Upgrading of Hydraulic Support

To better prevent the influence of mine pressure control and coal flow system in the 8102 panel during the initial weighting period and the square pressure period, there are the following changes compared with the previous panel.

(1) Change support selection. The resistance of the middle support increases from the previous 15,000 kN to the current 21,000 kN and the resistance of the end support increases from the previous 21,000 kN to 30,000 kN. Up to now, the panel has been pressed 17 times, and the maximum pressure of the panel in each pressure process has reached more than 58 MPa. However, through these times of pressure, the blasting side of the panel is small in the process of pressure, and some are almost none, indicating that the 21,000 kN support can well cope with the pressure of the panel and support the roof of the panel well.
(2) Change the size of the support rib protection device. The rib protection device of the 21,000 kN support is greatly changed from the previous panel, which increases the length of the rib protection device and can support the roof of the coal wall and the roof crushing area in a limited way.
(3) Increase the front girder side protection. The support of the panel has increased the function of the front girder side protection. Since the mining of the panel, the roof of the first half is broken, and the coal leakage between the supports is serious. After increasing the function of the front girder side protection, the coal leakage between the supports can be effectively reduced.
(4) Increase the power of the front and rear flow motors. The motor of the front and rear AFC increases from the previous 1050 kW to the current 1200 kW. Up to now, there has been no phenomenon of sliding after pressure. Increasing the power of the motor mainly considers that the panel is the longest inclined panel in Tongxin Mine, and at the same time considers the reason for the thicker coal seam in the panel.

5.5.3.4 Application of Long-Distance Liquid Supply and Power Supply

For the first time, the 8102 panel uses a long-distance liquid supply for power supply, and at the same time moves the operator from the string car position to the belt head position, which truly realizes the liberation of personnel from the panel and roadway. Using a long-distance liquid supply power supply has the following advantages:

(1) Reduced the workload of mobile bunching. The long-distance liquid supply power supply shortens the length of the train, from the original 230 m to the current 88 m, which greatly facilitates the maintenance shift and reduces the workload.
(2) It is more adaptable to roadways. Before the change, when the undulating slope of the roadway is relatively large, it is necessary to set up two return vehicles, and furthermore, the number of stoppers must be increased to ensure that there is no risk of running a car. However, after the number of cars is greatly reduced, the adaptability to the roadway is stronger and the safety factor is higher.
(3) The ability to adapt to complex geological conditions is stronger. The 8102 panel is the first island panel in Tongxin Mine, and a small coal pillar of 6 m is set in the goaf on both sides. According to the mining situation of 8305 single-side small coal pillar, the pressure of 6 m small coal pillar is relieved compared with the previous 38 m large coal pillar mining, however, the pressure is still serious. After long-distance liquid supply and power supply, because there is no shift and liquid pump and other heavy vehicles when floor heave and side heave occur, the adaptability is stronger.
(4) The stability of the pumping station is increased. After the long-distance liquid supply is adopted, the three liquid pumps are stably located at the long-distance liquid supply station, and do not need to migrate with the car, which increases the stability of the pump station, facilitates installation, management, and maintenance, and is more convenient for disassembly. The maintenance space is increased. Furthermore, when the emulsified oil is transported, manual handling is no longer used, saving a lot of labor.
(5) Multiple fully mechanized mining panels can be repeatedly served. Selecting the appropriate location of the long-distance liquid supply station can be used for multiple panels, reducing the cost of equipment removal, installation, and repair, and reducing production costs.
(6) Avoid the difficulties caused by the string car when stopping mining. At present, to recover more coal resources and reduce the coal loss caused by the overlarge stop coal pillar, the coal industry group is required to shorten the stop coal pillar, by a minimum of 80 m. After the long-distance liquid supply power supply is adopted, the length of the car is reduced, which effectively avoids the difficulty caused by the long storage of the equipment car when the small coal pillar is stopped.

(a) (b)

Fig. 5.34 Remote-control center of intelligent centralized console: **a** Intelligent centralized control room; **b** remote control center

5.5.3.5 Remote-Control Integration

The remote-control integrated operation platform of the 8102 panel is located in the inclined roadway between the headgate and the panel area, and the personnel is liberated from the LTCC panel. The remote control and monitoring system (Fig. 5.34) mainly includes the remote-control system of the unit, the remote-control system of the hydraulic support, the remote-control system of the "three machines", the remote-control system of the belt, the Beck voice communication system, the video monitoring system and so on. Among them, the remote-control system of the unit can realize the memory coal cutting, intelligent height adjustment, and remote intervention control of the shearer; the remote-control system of hydraulic support can realize' visual remote control and automatic control', automatic decision-making and control of the middle of the hydraulic support, oblique cutting feed, end floating coal, automatic propulsion of the loader and automatic continuous production of the panel. The Beck voice communication system can realize the voice communication of the panel, the emergency stop locking unit, and the support function. The video monitoring system is equipped with three mine-used intrinsically safe cameras for every six supports, which are installed on the top beam of the support. It can track the shearer in real-time, automatically complete the functions of video follow-up push and video splicing, and realize the high-definition video monitoring of the production system without blind spots. The advantages of the device can be summarized as safety, a good working environment, fewer workers, and high production efficiency.

5.5.3.6 Intelligent Belt Conveyor Inspection Robot

The inspection robot is installed in the transportation belt roadway (Fig. 5.35), which realizes the environmental monitoring and equipment condition monitoring of the inspection area, and carries out reliable, stable, and real-time data interaction with the centralized control center. The data such as field temperature, sound, and image are transmitted to the monitoring center in real-time, which can realize the real-time

Fig. 5.35 Intelligent belt conveyor inspection robot

monitoring of the belt surface, roller, frame, and material of the belt conveyor. In case of abnormal conditions, it can intelligently identify and take photos to upload and alarm. Furthermore, it can intelligently identify and alarm the abnormal conditions of the regional environment, cables, and pipelines, and exchange data with the sensors and base stations passing through the region, to achieve the purpose of replacing manual inspection.

References

1. Wang JC, Wei WJ, Zhang GY et al (2022) Top coal drawing mechanism and intelligent drawing. Science Press, Beijing
2. Wang GF, Guo YC, Wang JC (2023) Introduction to intelligent mine. China University of Mining and Technology Press, Xuzhou
3. Tan J (2013) Research on the current situation and development trend of coal mine automation construction in China. Entrep World 513(7):47–48
4. Xu XL, Peng SP, Ma Z et al (2022) Principle and key technology of dynamic detection of coal-rock interface in coal mine based on air-coupled radar. J China Coal Soc 47(8):2961–2977
5. Xu XL, Ma Z, Chen LZ (2023) Progress and thinking of transparent detection technology for hidden geological hazards in coal mines. J Green Mine 1(1):56–69
6. Xu XL, Peng SP, Ma Z et al (2022) An intelligent detection and recognition system and method for mine coal-rock interface. China, ZL202010304390.7
7. Xu XL, Peng SP, Ma Z et al (2022) Intelligent detection and recognition system and method for coal-rock interface of mine. The United States of America, US11306586B2
8. Zheng M, Xu XL, Peng SP et al (2023) Development and application of air-coupled GPR antenna for coal-rock interface detection in mines. J Appl Geophys 219
9. Xu XL, Ma Z, Sun D et al (2022) A geological radar antenna support device for mine coal rock level identification. China, ZL202010004915.5
10. Zhu PQ, Xu XL, Ma Z et al (2022) Research on accuracy and error analysis of coal and rock strata detection based on air-coupled GPR. IEEE Geosci Remote Sens Lett 19:1–5
11. Zhu PQ, Xu XL, Peng SP et al (2024) Research on intelligent identification algorithm of coal and rock strata based on Hilbert transform and amplitude stacking. Geophys Prospect
12. Xu XL, Wang YD, Zhu PQ et al (2022) Research on coal and rock horizon identification and tracking method based on high frequency radar waves. Coal Sci Technol 50(7):50–58

13. Wang LJ, Wang HZ, Wu ZZ (2015) Fully mechanized coal mining face in fewer people "three machine" linkage control strategy. Coal Mine Mach 36(3):90–91
14. Wang GF, Du YB, Chen XJ et al (2023) Development and innovative practice from coal mine mechanization to automation and intelligence: Commemorating the 50th anniversary of the founding. J Mine Autom 49(6):1–18
15. Editorial Department of China Journal of Highway and Transport (2015) Review on China's tunnel engineering research: 2015. China J Highw Transp 28(5):1–65
16. Wang ZY, Wang JW, Qu L (2008) Discussion on optimization of drilling and blasting heading and roadheader heading. Coal Sci Technol 9:59–61
17. Li XL (2015) Study on the roadway drivage technology development. Energy Energy Conserv 11:187–188
18. Ren BR, Liu JP (2003) Application and development of quick heading equipment in coal entry. Colliery Mech Electr Technol 5:52–54
19. Wang BK (2020) Current status and trend analysis of roadway driving technology and equipment in coal mine. Coal Sci Technol 48(11):1–11
20. Wang GF (2022) New technological progress of coal mine intelligence and its problems. Coal Sci Technol 50(1):1–27
21. Wang QH (2023) Discussion on automation technology of mine heading face. Metall Mater 43(8):94–96
22. Zhang XG (2022) Application and development trend of intelligent technology in roadway roadheaders in coal mines. Energy Energy Conserv 4:197–199
23. Wang JC, Pan WD, Zhang GY et al (2022) Principles and applications of image-based recognition of withdrawn coal and intelligent control of drawing opening in longwall top coal caving face. J China Coal Soc 47(1):87–101
24. Wang JC, Li LH, Yang SL (2018) Experimental study on gray and texture features extraction of coal and gangue image under different illuminance. J China Coal Soc 43(11):3051–3061
25. Wang GF, Pang YH, Zhang CC et al (2016) Intelligent longwall mining technology and equipment and adaptability in super large mining height working face. Coal Eng 48(9):6–10
26. Wang GF (2014) Development orientation of complete fully-mechanized automation, intelligent and unmanned mining technology and equipment. Coal Sci Technol 42(9):30–34
27. Li Z, Zhao L, Li Y et al (2023) Intelligent technology for safe and efficient mining in Tongxin coal mine. J Shanxi Datong Univ (Nat Sci Ed) 39(1):88–93

Chapter 6
Surrounding Rock Control of Thick Coal Seams Mining

Determination of roof pressure is one of the core contents of rock control in comprehensive mechanized mining panel. Its goal is to make the panel roof control achieve great results under the premise of economic rationality to meet the safe production and efficient mining of the panel. The theory and method of determining the roof pressure are basically mature for the panel of ordinary mining height, however, the determination of the roof pressure for top coal mining panels and large mining height panels under thick and extra-thick coal seams is still in the exploratory stage, and it is often based on the on-site mining experience for analogical determination. In the process of exploration of roof control, some mines have paid a heavy price. For example, more than 30 supports in a LTCC panel were crushed to death at one time, the front pillar of the support penetrated the roof beam, the back pillar was pulled off, and the top coal and roof steps sank, etc. Also, the supports often appeared to be crushed during the mining process of some large mining height panels, which brought a new topic to the determination of the roof pressure.

6.1 Common Theories for Roof Pressure Calculation

6.1.1 Theory of "Masonry Beams"

6.1.1.1 Structural Model "Masonry Beam"

Through the observation of the internal movement of a large number of mining strata in the past, and on the basis of summarizing the articulated block hypothesis and the preformed fracture hypothesis, Academician Qian Minggao, Chinese Academy of Engineering, proposed the "masonry beam" structural model of the rock body in the late 1970's and early 1980's [1], seeing Fig. 6.1. Macroscopically, the "masonry beam" structural model divides the overlying rock formation into three zones from

J. Wang and W. Wei, *Thick Coal Seam Underground Mining*,
https://doi.org/10.1007/978-981-96-3874-1_6

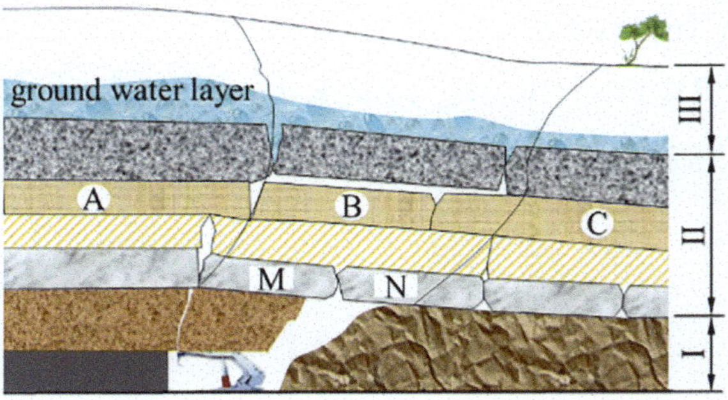

Fig. 6.1 Structural model of a "masonry beam" in the overlying strata

the bottom to top, i.e., collapse zone I, fracture zone II, bending subsidence belt III. Collapse zone I refers to the part of the collapsing immediate roof irregularly, and the fracture zone II belongs to the category of the main roof in the collapsed overlying strata. From the front to the back of the panel, it is divided into three zones, i.e., wall supporting zone *A*, separation zone *B* and recompacting zone *C*. From the point of view of the stress, the wall supporting zone is also known as the high stress Zone, separation zone is the unloading zone, and recompacting zone is the stress recovery zone.

6.1.1.2 Mechanical Analysis of Masonry Beam Structure

The masonry beam structure refers to the surface-like beam composed of broken rock blocks in the fracture zone of the separation zone. In essence, it is a kind of arch structure. Generally, the whole structure may be composed of broken rock layers of multiple groups, as shown in Fig. 6.2. The stress characteristics of masonry beam structure can be analyzed by selecting any group of structures. There are the following basic assumptions:

(1) The skeleton of overlying strata structure is the hard rock layer in overlying rock, and the overlying strata can be divided into several groups. Each group is based on the hard rock layer, and the upper soft rock layer can be regarded as

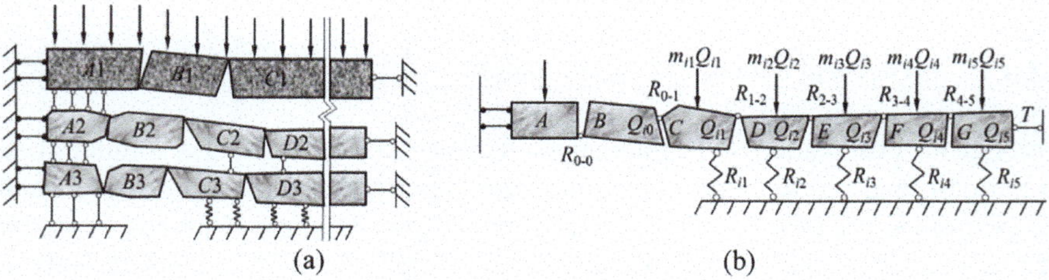

Fig. 6.2 Mechanical analysis of the masonry beam structure in the overlying strata

the load directly acting on the skeleton, and is also the cushion connecting the upper hard rock layer with the lower skeleton.

(2) With the advance of the panel, the hard rock layer above the goaf will break into neatly arranged rock blocks in the fracture zone, and the horizontal extrusion pressure caused by rotary deformation will form an articulated relationship between the rock blocks.

(3) Due to the weak ability of the soft rock layer to transmit shear force, the connection between the two layers of the skeleton can be replaced by a retractable support rod.

Select any group from the structure in Fig. 6.2a, as shown in Fig. 6.2b. There is a rock block *B* in the overhanging state, each symbol in the figure has two feet, the first foot indicates the layer of the rock, the second foot indicates the position of the rock block along the strike direction. For example, L_{i1} indicates that the length of block *C* in Group *i*. Taking Fig. 6.2b as an example, the moment balance equation is set out respectively for the front and back two articulated points of each rock block, and it can be obtained:

$$\{M_i\} = [F_i]\{R_i\} \tag{6.1}$$

where $\{M_i\}$ is the moment matrix; $[F_i]$ is the coefficient matrix; $\{R_i\}$ is the force matrix. To make a rough estimation of the force relationship between this structure, according to the characteristics of the rock movement curve, the slopes of the two neighboring blocks can be approximated as equal, then the horizontal thrust force T_i required to form the equilibrium of this structure can be found as:

$$T_i = \frac{L_{i0}Q_{i0}}{2(h_i - S_{i0})} \tag{6.2}$$

where L_{i0} is the broken length of the overhanging rock block, m; h_i is the thickness of the rock layer, m; Q_{i0} is the load on the overhanging rock block, kN; S_{i0} is the sinking amount of the overhanging rock block, m. Under the condition of structural equilibrium and assuming that the rotational sinking deformation of the rock block is a very small value compared to the length of the broken rock block, the plumbing force at each articulation point can be obtained as:

$$\begin{cases} (R_i)_{0-1} = 0 \\ (R_i)_{0-0} = Q_{i0} \\ R_{i1} = m_{i1}Q_{i1} \\ R_{i2} = m_{i2}Q_{i2} \\ \cdots \\ R_{in} = m_{in}Q_{in} \end{cases} \tag{6.3}$$

where Q_{ij} is the weight of the rock beam *j* block in layer *i*, KN; m_{ij} is the load factor of the rock beam *j* block in layer *i*. From the above analysis, this structure can be characterized as:

(1) The weight of the overhanging rock mass (*B*) in the separation zone and the overlying loads are borne almost entirely by the front support point;
(2) The shear between rock block *B* and *C is* close to zero, and this occlusion point corresponds to the vault of a half-arch;
(3) The maximum shear force of this structure occurs between rock block *A* and *B* and is equal to the weight of rock block *B* itself and its load;

The above analysis shows that the formation of the masonry beam structure must have a certain horizontal thrust T_i. As shown in Eq. (6.2), the thicker the main roof is (the larger h_i is), and the smaller the deformation of the rotary subsidence is (the smaller S_{i0} is), the smaller the horizontal thrust T_i required for the formation of the masonry beam structure is. When $S_{i0} = h_i$, T_i tends to infinity and the structure cannot be formed. Therefore, in the overlying strata, only rock layers with a certain thickness can form this structure. The small rotary subsidence deformation S_{i0} can be found in the following cases: the immediate roof is thicker, and can fill up the goaf area after bubbling down, and contact with the main roof, which can limit the rotary subsidence deformation of the broken rock block of the main roof; the goaf area is treated by filling method, which can also limit the rotary subsidence deformation of the broken rock block of the main roof; the coal seams with small mining height and the rock seams far away from the mined coal seams.

From the analysis of slip instability between rock blocks, seeing Eq. (6.4), the conditions that must be satisfied to require structural equilibrium are:

$$T_i \tan(\varphi - \theta) > (R_i)_{0-0} \tag{6.4}$$

where φ is the friction angle between the rock blocks, °; θ is the angle between the broken surface and the vertical plane, °. In the case that the rock block occlusion is not damaged, T_i in Eq. (6.2) can be brought into Eq. (6.4), thus obtaining:

$$\frac{L_{i0}Q_{i0}}{2(h_i - S_{i0})} \tan(\varphi - \theta) > (R_i)_{0-0} \tag{6.5}$$

6.1.1.3 Determination of Roof Pressure

Before the panel periodic weighting, the support is under the protection of the structure formed by masonry beam, as shown in Fig. 6.3a. The front arch foot of the structure is in the coal wall of the panel, with the advancement of the panel, the front arch foot of the masonry beam reaches the top of the support in the panel, currently, the support has to bear the force of the front arch foot of the masonry beam and the load of the upper part of the immediate roof. Therefore, according to the theory

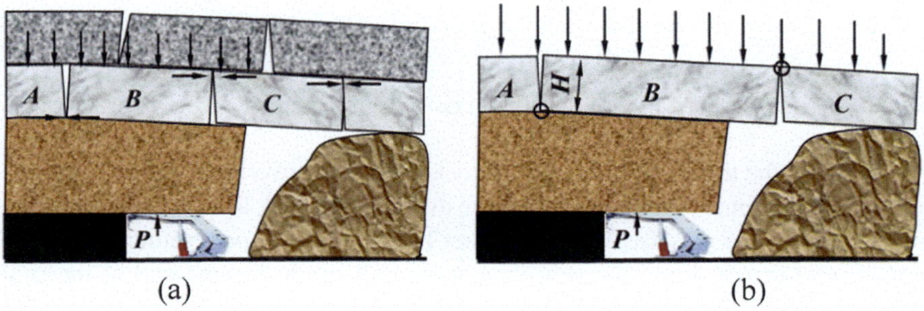

Fig. 6.3 **a** Support protected by masonry beam structure; **b** masonry beam model for work resistance calculation

of masonry beam, to prevent the rock block *B* and upper rock layer from cutting down along the coal wall, and to avoid the instability of the main roof sliding down (Fig. 6.3b), the working resistance of the support *P* is:

$$P = Q_{A+B} - \frac{L_B Q_B}{2(H - S_B)} \tan(\phi - \theta) + Q_D \tag{6.6}$$

where Q_{A+B} is the weight of rock block *A* and *B* and its upper load, kN; L_B is the length of rock block *B* (outcropping rock block), m; Q_B is the weight of rock block *B* and its upper load, kN; *H* is the thickness of the main roof, m; S_B is the sinking amount of rock block *B*, m; θ, φ is the angle of breakage and internal friction angle of the rock block, °; Q_D is the weight of the immediate roof, kN, which is related to the structure of overlying strata. As an estimation, it can be considered that the immediate roof height of the fall is mainly related to the mining height, and the rock block *B* touches the goaf in the process of slewing, otherwise, it is not easy to form the masonry beam structure. Q_D can be obtained by Eq. (6.7).

$$Q_D = L_D \cdot \sum h \cdot \gamma \tag{6.7}$$

where L_D is the length of the immediate roof rock beam above the support, m; $\sum h$ is the thickness of the immediate roof calculated according to the contact between the fallen immediate roof and the main roof, m; γ is the volumetric weight of the immediate roof, kN/m^3. The immediate roof height can be determined according to the actual situation when the boundary of coal seam and roof is obvious. When it employ the top coal caving method:

$$Q_D = Q_T + L_D \cdot \sum h \cdot \gamma \tag{6.8}$$

where Q_T is the weight of top coal, kN.

6.1.2 *Theory of "Transferring Rock Beam"*

6.1.2.1 Structural Model of "Transferring Rock Beam"

According to the theory of "transferring rock beam", the overlying strata overhangs with the advancement of the panel, and the overhanging rock layer bends and settles under the effect of gravity. With the further increase of the span, settlement development to a certain limit, the overlying strata will reach into the end of the coal wall cracking and cracking in the middle to form the "pseudoplastic rock beam". Its two ends support by the coal seam, or one end support by the front coal seam of the panel, one end support by the goaf area, to maintain the force transfer in the advancing direction. When the settlement value exceeds the allowable settlement value of "pseudoplastic rock beam", the overhanging rock layer will collapse by itself. Each group of rock strata moving simultaneously (or nearly simultaneously) is regarded as a moving whole, which is called "rock beam transferring force", or "transferring rock beam" for short. The structural model of "transferring rock beam" in the overlying strata is shown in Fig. 6.4, which was proposed by Academician Song Zhenqi [2].

Based on the Mohr's Coulomb criterion of the joint surface and the maximum curvature (ρ_{max}) and maximum deflection (W_{max}) of the rock settlement, it is determined whether each rock layer moves simultaneously or whether it is out of the layer. When $\rho_{\text{max upper}} \geq \rho_{\text{max lower}}$ or $W_{\text{max upper}} \geq W_{\text{max lower}}$, the two rock layers are combined to form a single transfer rock beam moving at the same time. On the contrary, the two rock layers will form two transfer rock beams moving separately. From the above criteria, the panel to the ground can be divided into fall zone, fracture zone and bending subsidence zone. Among them, it is the lower 1–2 transferring rock beams in the collapse zone and fracture zone that have obvious influence on the appearance of mine pressure in the panel. The collapse zone is called the immediate roof; the 1–2 lower transfer rock beams that have obvious influence on the mine pressure appearance of the panel are called the main roof, and all the rock layers

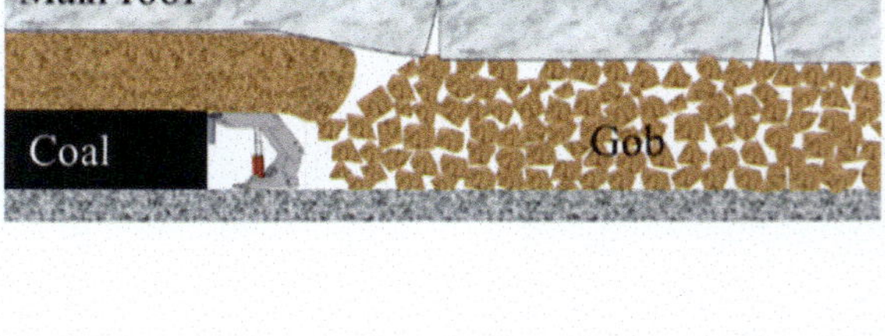

Fig. 6.4 Structural model of a "transferring rock beam" in the overlying strata

between the immediate roof and the main roof are the range of rock layers that need to be controlled in the panel.

The rock control theory of "transferring rock beam" emphasizes the control of rock movement. When determining the working resistance of support, firstly, researchers should find out the range of immediate roof and main roof and its movement development law, and then determine the support working resistance needed to achieve the degree of control requirements. The immediate roof refers to the cantilever beam that collapses in the goaf area as the panel advances and forms under the supporting effect of the hydraulic support, and the structure is characterized by the fact that the horizontal force cannot be continuously transmitted in the direction of the panel advancement, which requires the support to be able to bear all the weight of the immediate roof during its movement. The main roof is the sum of the transferring rock beams that have a significant effect on the manifestation of mine pressure in the panel during movement. After the initial incoming pressure, the main roof is a set of fracture beams that can transmit horizontal forces in the direction of panel advancement. Its mechanical characteristic is that no matter in the stable state or in the rotary sinking, it can always transfer the load to the coal wall, the support and the goaf in the mining area, and this process is realized by the deformation produced by the rock beam itself. Hence, the size of the pressure borne by the support is related to the controlled rock beam position. Therefore, the roof to support force comes from the immediate roof "given load" force and the main roof "given deformation" force.

6.1.2.2 Determination of Support Working Resistance

Whether it is immediate roof or main roof, when it is in a relatively stable state, the threat to the panel is very small, and there will be no obvious change in the force of the support, once the balance is destroyed and it enters into a significant movement, a series of obvious pressure manifestation will take place, thus when determining the working resistance of the hydraulic support, the mechanical relationship between the support resistance and the positional state of the main roof destruction should be taken into consideration. Support and immediate roof mechanical relationship model is shown in Fig. 6.5.

The immediate roof starts to settle gradually from *OA* position under the action of self-weight *G* and external load *Q* of the main roof beam, and comes to a standstill when the settlement reaches Δh_i, at which time the immediate roof is in *OB* position. The maximum force R_{zi} given to the support by the immediate roof satisfies the moment balance equation:

$$R_{zi} l_i = G \cdot \frac{l_z}{2} \cos \theta \tag{6.9}$$

where *G* is the immediate roof weight, l_z is the immediate roof hanging distance, and α is the maximum rotation angle of the immediate roof. From this we get:

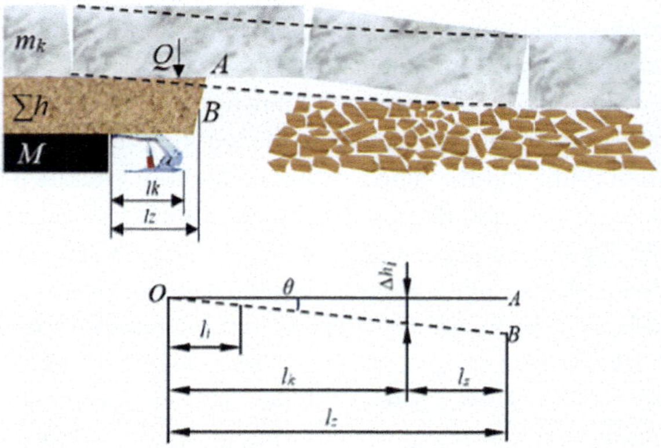

Fig. 6.5 Mechanical relationship between the support and the immediate roof

$$f_z = \frac{(l_k + l_s)^2}{2l_i l_k} = \frac{1}{2n_i}(1 + \frac{l_s}{l_k})^2 \tag{6.10}$$

where f_z is the immediate roof overhang coefficient, the value is related to the size of the controlled roof distance and the joint force point of the support; n_i is the ratio of the combined support reaction force in the roof control area at the location of l_i and the controlled roof distance of l_k, $n_i = l_i/l_k$. Obviously, the immediate roof force to the support can be approximated to be a constant that is independent of the immediate roof position, and the immediate roof overhang will have a significant effect on the magnitude of this value.

When the immediate roof is in the goaf with no overhanging roof conditions, i.e., when $l_s = 0$ and $l_i = l_k/2, f_z = 1$. Currently,

$$P_{zi} = \sum h\gamma = A \tag{6.11}$$

That is, the force of the immediate roof on the support depends only on the thickness of the immediate roof and the volumetric weight of the rock. When the immediate roof consists of thick sandy shale, calcareous sandstone, limestone and other hard rock layers, the overhanging distance l_s is more than zero, i.e., $l_s > 0$, at this time, the immediate roof overhanging coefficient $f_z > 1$.

6.1.2.3 Calculation of Main Roof Loads

In the main roof to support load calculation, the rock beam motion state and force analysis is shown in Fig. 6.6, Q is the rock beam load, is the sum of the self-weight of the rock beam fracture block and the force of the weak layer on it, that is:

$$Q = (m_1 + m_2)\gamma_E C \tag{6.12}$$

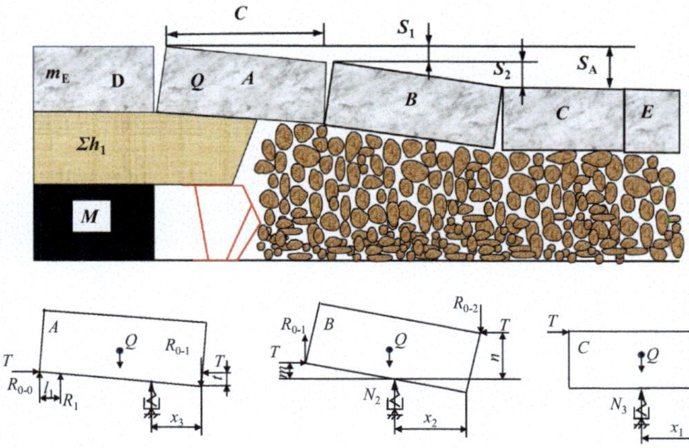

Fig. 6.6 Rock beam motion state and force analysis

where m_1 is the thickness of the supporting layer of the main roof, m; m_2 is the thickness of the weak layer on the main roof, m; C is the movement step of the rock beam, m.

In Fig. 6.6 R_1 is the combined force between the main roof and the immediate roof, which mainly depends on the support strength of the panel and the force of the immediate roof; l_1 is the distance of R_1 from the former occlusion point, which mainly depends on the distribution of the support reaction force in the panel and the stability of the immediate roof; t is the distance between the occlusion point at the occlusion point of the two rock blocks A and B from the lower edge of the block, m; m_E is the total thickness of main roof and weak layer on the main roof; $\sum h_1$ is the thickness of immediate roof; T is Horizontal thrust; R_{0-0} is the shear force at the left end of rock A; R_{0-1} is the shear force at the occlusion point of A and B rock blocks; R_{0-2} is the shear force at the occlusion point of B and C rock blocks; N_1, N_2, N_3 is the gangue counterforce, which depends on the range of the gangue support L_s, the stiffness of the gangue support K, and other factors. $N_1 \approx Q$, $R_{0-2} \approx 0$, and N_2, N_3 increase with the rock beam settlement.

The main roof breaks at the end and detaches from the whole. Acting on the coal wall, support and goaf in the form of "load". Due to the main roof rock beams are delaminated from the rock layer above, its movement shows independence in a certain degree. At this time, the role of the support is to increase the reverse bending moment, to prevent the main roof of the rotary sinking movement. Rock beam motion state linear simplified model, as shown in Fig. 6.7.

When the main roof sinks to the final position (without the action of resistance control force), the articulation point *a of A* and *B* two rock blocks has the maximum sinking value S_A, the rock beam hanging span is close to L_A, (approximated as the main roof movement step C), at this time, the roof sinking at the controlled roof distance is Δh_i, by the geometrical relationship:

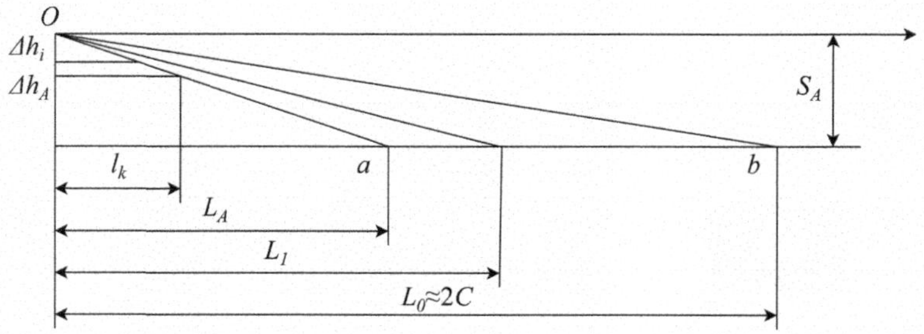

Fig. 6.7 Linear simplified model of rock beam motion state

$$\triangle\ h_i = \frac{S_A}{L_A} l_k \approx \frac{S_A}{C} l_k \tag{6.13}$$

In this position, the maximum load on the support from the rock beam (considered as anti-shear) is

$$P_{EA} = \frac{m_E \gamma_E L_i}{K_T l_k} \tag{6.14}$$

where K_{T} is the proportionality coefficient for the weight of the rock beam borne by the support, $K_{\mathrm{T}} \leq 2$. When the rock beam has not reached the final position, i.e., when it is in the II position, the span of the rock beam is L_i, and at this time, the amount of roof subsidence at the roof distance l_{k} is Δh_i. From the geometrical relationship,

$$\triangle\ h_i = \frac{S_A}{L_i} l_k \tag{6.15}$$

Maximum possible load on rock beam,

$$P_{Ei} = \frac{m_E \gamma_E C}{K_T l_k} \cdot \frac{\triangle\ h_A}{\triangle\ h_i} = K_A \frac{\triangle\ h_A}{\triangle\ h_i} \tag{6.16}$$

where K_{A} is the rock beam state constant, i.e., the force per unit area of rock beam when the amount of roof subsidence is Δh_{A}. Generally, in mining conditions, $\frac{m_E \gamma_E C}{K_T l_k} \triangle\ h_A$ is a constant, written as B, then Eq. (6.16) can be written as

$$P_{Ei} = \frac{B}{\triangle\ h_i} \tag{6.17}$$

Under the working condition of "limited deformation", there is a hyperbolic relationship between the load given to the support by the main roof beam and the amount of roof subsidence required to be controlled. Therefore, the relationship between the support and the surrounding rock in the working condition of "limited deformation" is considered. When the subsidence Δh_i is generated at the controlled roof distance l_k, the reasonable support strength is divided into two parts, i.e. immediate roof force and main roof beam force, which can be expressed as follows:

$$P_T = P_{zi} + P_{Ei} = A + K_A \frac{\Delta h_A}{\Delta h_i} \tag{6.18}$$

where P_T is the force exerted by the roof on the support at Δh_i to control the amount of roof subsidence; $P_{Ei} = A$ is a constant independent of the control position.

Equation (6.18) reflects the interaction between rock beam motion and support. Under the effect of support strength P_T, the significant movement of rock beam will develop to Δh_i to form a stable structure. As the rock beam is in a certain state, the amount of roof subsidence is different at different controlled roof distances in the panel. Therefore, Eq. (6.18) should be regarded as the equation of the interaction between the state and the support when the rock beam movement is stabilized, i.e., Eq. (6.18) is the "Rock Beam State Equation". The relationship between the two is a hyperbola with $\Delta h_i = 0$ and $P_T = A$ as asymptotes, as shown in Fig. 6.8.

In Eq. (6.18), the state constant is expressed differently depending on the state in which the rock beam is located. When the rock beam motion parameters m_E and C are known, the state constant K_A can be directly expressed by the rock beam motion parameters as:

$$K_A = \frac{m_E \gamma_E C}{K_T l_k} \tag{6.19}$$

At this point the equation of state is also called the limiting equation of state and K_A is called the limiting state constant.

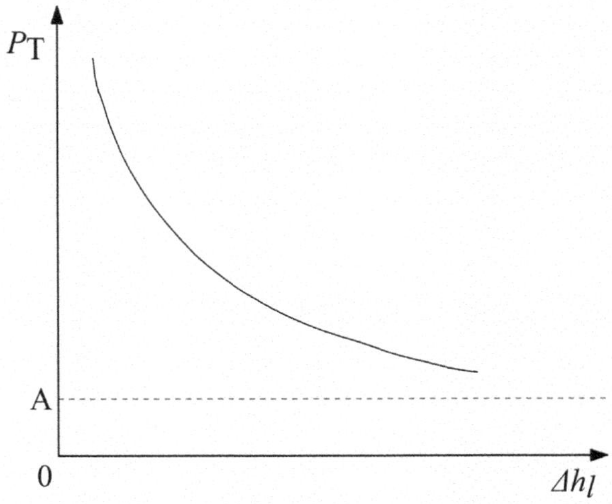

Fig. 6.8 Relationship between reasonable support strength and roof subsidence

6.1.3 Calculation of Roof Dynamic Load

The rapid innovation of mining techniques and supporting equipment has significantly intensified the exploitation of coal resources. High-intensity stopes, characterized by large-scale spatial dimensions, high-power equipment, and rapid advance velocity, have become widely adopted. The unique mine pressure behavior of high-intensity stopes, such as the overall cutting and breaking of the roof and even the entire bedrock along the rib, leading to press shelf phenomena, has led to the realization that the dynamic load impact caused by broken rock blocks after roof failure on the supports of the panel is a prerequisite for large-scale rock strata disasters. The dynamic load impact refers to the non-static contact between the broken rock blocks of the main roof and the immediate roof or the supports of the panel, resulting in collision phenomena. This can generate impact forces far greater than the static weight of the rock blocks in a short period of time, thereby triggering large-scale disasters in the rock strata.

6.1.3.1 Establishment of Mechanical Models

Following the breakage of the main roof, the fractured rock blocks can form a "three-hinged arch" structure through three hinge points, as depicted in Fig. 6.9. For the purpose of analysis, a unit width of 1 meter is considered. The length of the main roof rock block is denoted as L_B, the initial weighting interval as L, the overhanging length of the immediate roof as L_D, the controlled roof distance of the support as L_S, the thickness of the main roof strata as H, the cumulative thickness of the immediate roof strata as Σh, and the mining height of the panel as M. Prior to the collapse of the main roof, as the panel advances and the immediate roof undergoes fall, the main roof fractures into a "three-hinged arch" configuration. With continued panel advance, the main roof structure experiences rotational deformation, leading to the sinking of the central hinge point or further fracturing into a structure composed of multiple rock blocks. During the sinking and rotational processes, horizontal compressive forces (T) are generated at the three hinge points. According to the equilibrium principle of the "three-hinged arch," the horizontal force required to form an arch and maintain the stability of the rock block structure can be determined:

$$T = \frac{qL^2}{8H} \tag{6.20}$$

During the sinking and rotational processes of the main roof rock block, the load is transmitted to the support through the immediate roof. The support partially restricts the rotation of the main roof. However, fundamentally, the support is incapable of controlling the rotational deformation of the main roof. As the main roof rock block rotates and sinks, the support is compressed. Nevertheless, the primary load of the overlying strata is still borne by the structure formed by the main roof. Therefore,

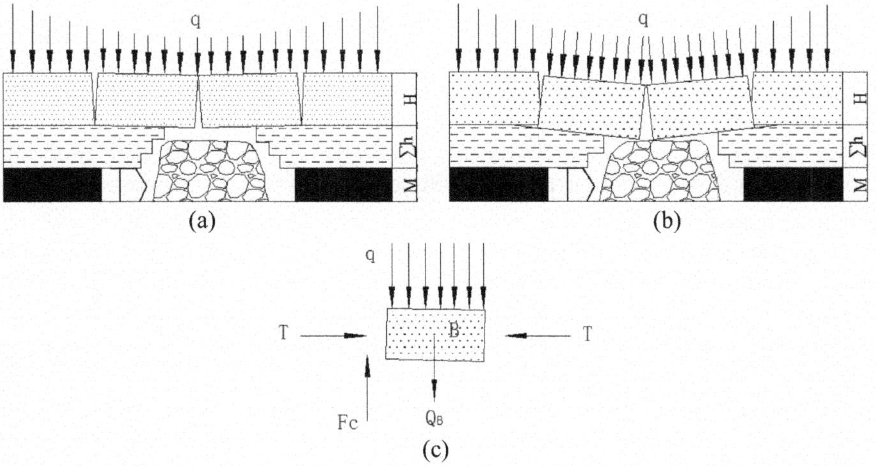

Fig. 6.9 Basic model for roof load calculation: **a** before the main roof rock block slides; **b** after the main roof rock block slides; **c** schematic diagram of forces acting on the sliding rock block

prior to the pressure, the load of the main roof rock block and the overlying strata is primarily supported by the two hinge points at the arch footings. When the structure of the main roof rock block becomes unstable, the load of the main roof rock block and the overlying strata suddenly acts on the immediate roof and is then transferred to the support. This is essentially an impact, or dynamic load, of the main roof rock block on the immediate roof. At this moment, considering that the two rock blocks forming the balanced structure slide downward simultaneously without any tendency for relative motion between them, the frictional force between the two sliding rock blocks is not considered. The mechanical model is as follows:

As depicted in Fig. 6.9, the pressure P generated following the instability of the main roof can be expressed as: $P = Q_D + (Q_B + qL_B - F_c)K_d$ K_d represents the dynamic loading coefficient; Q_D is the weight of the immediate roof; F_c denotes the frictional force experienced by Rock Block B during its sliding process, which can be calculated as $F_c = Ttan\varphi$, where T is the compressive force between rock blocks, and φ is the internal friction angle between the fractured rock blocks, typically ranging from 30° to 40°; Q_B is the weight of Rock Block B; and q represents the load intensity of the overlying strata acting on Rock Block B. The combined effect of the weight of Rock Block B, the load of the overlying strata qL_B, and the frictional force during sliding F_C is referred to as Q_H, expressed as $Q_H = Q_B + qL_B - F_c$. Consequently, the working resistance of the support can be determined based on these parameters:

$$P = Q_D + K_d Q_H \tag{6.21}$$

6.1.3.2 Calculation of Dynamic Load Coefficient

The magnitude of the impact force generated by the main roof rock block on the immediate roof at the instant of structural instability is related to Q_H, as well as to the off-layer quantity Δh between the main roof and the immediate roof and the combined elastic modulus E_d of the immediate roof and the support. The most unfavorable scenario, where the main roof fractures above the rib, is considered, and the immediate roof and support are treated as an integrated system. For the sake of simplifying the calculations, the weight of the main roof and the follow-up rock blocks is designated as Q_H, which falls from a minimal height Δh above the immediate roof and impacts it. Under the action of the dynamic load, the immediate roof undergoes deformation Δd, as illustrated in Fig. 6.10.

The following assumptions are made: the deformation of the main roof rock block is neglected, and no rebound occurs after contact with the immediate roof. The mass of the immediate roof is considered negligible compared to that of the main roof rock block and can be ignored. The impact stress is assumed to propagate instantaneously through the impacted object, and the material is assumed to obey Hooke's Law. Energy losses due to sound, heat, and other factors during the impact process are considered minimal and can be disregarded.

Based on the assumptions, during the impact process, when the main roof rock block meets the immediate roof and its velocity reduces to zero, the lower boundary of the immediate roof reaches its lowest position. At this moment, the deformation of the immediate roof is at its maximum, denoted as Δd, and the corresponding impact load is F_d. According to the law of conservation of mechanical energy, the kinetic energy E_k and potential energy E_P of the main roof rock block during the impact process should be converted into the energy of volume change U_V and distortional energy U_d of the immediate roof. The distortional energy is released in the form of heat,

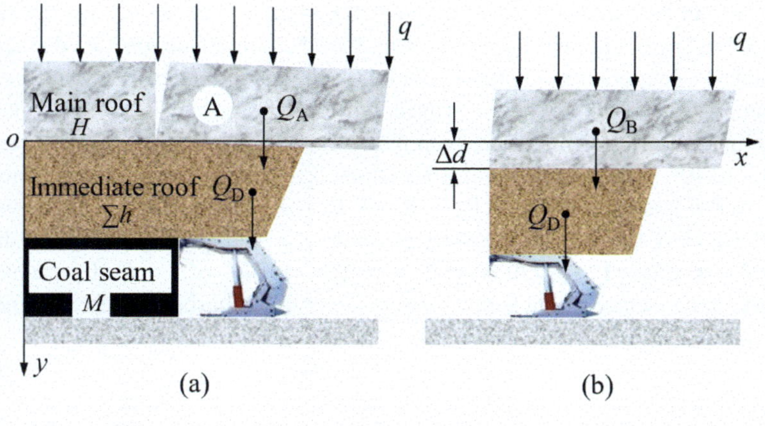

Fig. 6.10 Dynamic load calculation model for sudden instability of the main roof rock block: **a** the positional relationship prior to the pressure event of the main roof; **b** the positional relationship for dynamic load calculation during the pressure event of the main roof

sound, and other forms when the immediate roof undergoes plastic deformation. For the sake of simplifying the calculations, the mass of the immediate roof is neglected, and thus, the changes in kinetic and potential energy of the immediate roof are also disregarded.

$$E_k + E_p = U_v + U_d \tag{6.22}$$

At the instant of impact, it is assumed that the higher rock strata above the main roof do not move in conjunction with it. Therefore, only the weight of the main roof and the follow-up load Q_H are considered. When the lower boundary of the immediate roof reaches its lowest position, the reduction in potential energy of the main roof is given by:

$$E_p = Q_H(\Delta d + \Delta h) \tag{6.23}$$

Given that the initial and final velocities of the main roof rock block are both zero, the kinetic energy is therefore:

$$E_k = 0 \tag{6.24}$$

If the material obeys Hooke's Law, the increase in the energy of volume change and distortional energy of the immediate roof can be calculated through the work done by the impact load F_d over the displacement Δd. Therefore, the relationship can be expressed as:

$$U_v + U_d = \frac{1}{2} F_d^* \Delta d \tag{6.25}$$

For the immediate roof, the relationship between F_d and Δd is given by:

$$F_d = \frac{EL_S}{\Sigma h} \Delta d \tag{6.26}$$

Substituting the above expression into Eq. (6.25) yields:

$$U_v + U_d = \frac{1}{2} \frac{EL_S}{\Sigma h} \Delta d^2 \tag{6.27}$$

Substituting Eqs. (6.23), (6.24), and (6.27) into Eq. (6.22) yields:

$$Q_H(\Delta d + \Delta h) = \frac{1}{2} \frac{EL_S}{\Sigma h} \Delta d^2 \tag{6.28}$$

When the main roof rock block acts as a static load on the immediate roof, the static displacement of the immediate roof Δ_{st} is given by:

$$\Delta_{st} = \frac{Q_H \sum h}{EL_S} \tag{6.29}$$

Thus, Eq. (6.28) can be simplified to:

$$\Delta d^2 - 2\Delta_{st}\Delta d - 2\Delta_{st}\Delta h = 0 \tag{6.30}$$

Since the above equation yields two roots for Δd, and the root greater than Δ_{st} is selected, the result is obtained as follows:

$$\Delta d = \Delta_{st}(1 + \sqrt{1 + 2\frac{\Delta h}{\Delta_{st}}}) \tag{6.31}$$

Substituting the value of Δd obtained above into Eq. (6.26), the impact load F_d can be derived as follows:

$$F_d = \frac{EL_S}{\Sigma h}\Delta_{st}(1 + \sqrt{1 + 2\frac{\Delta h}{\Delta_{st}}}) \tag{6.32}$$

In the equation:

$$Q_H = \frac{EL_S}{\sum h}\Delta_{st}$$

The expression within the parentheses on the right-hand side of the equation is denoted as:

$$K_d = \left(1 + \sqrt{1 + 2\frac{\Delta h}{\Delta_{st}}}\right) \tag{6.33}$$

In the equation, K_d is referred to as the impact dynamic load coefficient. Thus, Eq. (6.32) can be rewritten as:

$$F_d = K_d Q_H \tag{6.34}$$

As can be seen from Eq. (6.33), reducing the free-fall height Δh of the main roof rock block will lower the impact dynamic load coefficient K_d. In the control of rock strata in the panel, Δh can be reduced by increasing the initial support force of the stand. When Δh approaches zero, that is, when no off-layering occurs between the main roof and the immediate roof, it is equivalent to the main roof rock block being suddenly added to the immediate roof. In this case, the impact dynamic load coefficient is given by:

$$K_d = \left(1 + \sqrt{1 + 2\frac{\Delta h}{\Delta_{st}}}\right) = 2 \quad (6.35)$$

Integrating the above analyses and considering the immediate roof as the object of study, the load on the hydraulic support can be expressed as:

$$P = Q_D + K_d Q_H$$

In the control of rock strata in the panel, increasing the initial support force of the stand can reduce the off-layer quantity between the immediate roof and the main roof. When no off-layering occurs, Δh equals zero, and the dynamic load coefficient K_d is equal to 2. However, due to the elastic cushioning effect of the immediate roof and the fact that part of the immediate roof (or top coal) is squeezed into the goaf during the impact of the main roof, absorbing part of the impact load of the main roof rock block and cushioning the impact of the main roof and follow-up load, the actual dynamic load coefficient K_d for support load calculation lies between 1 and 2.

For large mining height panels with hard main roofs and thin immediate roofs, where the main roof rock block is unlikely to come into contact with the caved immediate roof debris during its rotation before structural instability, the dynamic load coefficient K_d can be taken as 1.8 to 2. For conventional longwall panels with the same roof conditions, K_d can be taken as 1.6 to 1.8. For top coal caving panels with thin and hard immediate roofs, K_d can be taken as 1.5 to 1.6. For conventional longwall panels with medium stability main roofs and distinct pressure events in Class II, Type 2 roofs, or hard main roofs with thick immediate roofs (where the caved immediate roof can fill the goaf), K_d can be taken as 1.3 to 1.5. For conventional top coal caving panels with thick immediate roofs, K_d can be taken as 1.2 to 1.3. For panels with soft roofs and indistinct pressure events, K_d can be taken as 1.1 to 1.2.

6.1.4 Calculation Methods in Other Countries

6.1.4.1 Method of Calculating "S. S. Peng"

The external loads to which the support is subjected include those acting on the roof and the shield beams. The loads acting on the support include vertical loads caused by the weight or movement of the overlying rock, and lateral forces parallel and perpendicular to the panel caused by horizontal movement of the main roof. The loads acting on the shield beams are the weight of the rock debris accumulated on the shield beams [3]. Most researchers have concentrated their studies in the last few years on deriving the vertical loads due to the weight of the overlying strata. Although there are differences in the roof loads from one model to another, they can be summarized into the model shown in Fig. 6.11. The main roof is fractured

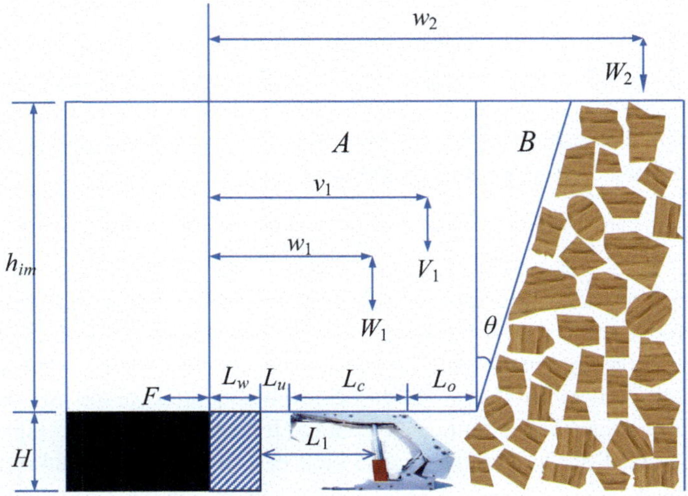

Fig. 6.11 Generalized main roof load model

along the demarcation line between the panel and the rib, and the edge of the goaf is fractured along the vertical or at a certain angle θ inclined to the goaf. Depending on the nature of rock and geological conditions of the roof, the roof sometimes hangs and protrudes, and sometimes breaks along or in front of the edge of the goaf. The force acting on the roof is not uniform because the suspended volume of the roof increases from the panel to the direction of the voiding zone, and because of the structural rigidity of the roof and the position of the support struts. Therefore, the analysis of the roof load should include two parts: the location and the magnitude of the combined force.

1. Location of combined roof loads

The weight on the immediate roof support W_1 is

$$W_1 = h_{im}\gamma_{im}B\left(L + \frac{h_{im}}{2}\tan\theta\right)\cos\alpha \tag{6.36}$$

where h_{im} is the thickness of immediate roof; γ_{im} is the volumetric weight of immediate roof; B is the center distance of the support; α is the dip angle of the coal seam. The position of W_1 acts on the main roof is follow:

$$w_1 = \frac{L + h_{im}\tan\theta}{2} \tag{6.37}$$

When the immediate roof is strong and overhanging, $w_1 > L_C$, L_C is the distance from the front to the back of the roof beam of the support, and the back of the support will be subjected to larger loads. On the contrary, when the immediate roof is weak, the immediate roof will break before exceeding the edge of the goaf, $w_1 < L_C$, which generally indicates that the front part of the support will bear larger load. When the periodic weighting, the support will bear additional weight W_2. The location of the action of W_2 on the immediate roof is very important. This is because

$$v_1 V_1 = w_1 W_1 + w_2 W_2 \tag{6.38}$$

besides

$$V_1 = W_1 + W_2 \tag{6.39}$$

where V_1 is the total combined force acting on the support, v_1 is the distance from the panel to the point of action of the combined force V_1; w_1 and w_2 are the distances from the panel to the points of action of the forces W_1 and W_2. V_1 is determined by W_1 and W_2. When the immediate roof is stronger and W_2 is larger, V_1 may also be larger. This means that the combined roof load acts near the edge of the goaf or even further away. On the contrary, when the immediate roof is weak and there is no overhang, the combined roof load acts at the leading edge of the support. Under ideal conditions, the combined roof load acts in the same line as the support of the support. In practice, however, they rarely act in the same line.

2. Determination of main roof loads

Calculation of roof loads using the detached roof block method has been commonly used by support manufacturers and coal operators. The method utilizes two factors, stratigraphic sequence and volume factor, to estimate the height of the separating top block. First, holes are punched to determine the contact surface between the immediate roof, which will span the fall immediately after the panel advances, and the main roof, which will overhang. (Note that the immediate roof is not necessarily a single layer; it may consist of several different rock types and thicknesses.) The location of the top of the seam above the contact surface is the thickness of the immediate roof, which determines whether it will fill the void zone (or mining height H) created by the complete mining of the seam. The height of the immediate roof fall required to fill the void created by a given mining height is determined by the coefficient of expansion. If the buildup height of the postmining straddle-fall rock equals or exceeds the determined straddle-fall height, then the straddle-fall rock mass can fill the void and providing support for the overlying rock formation. Therefore, the height of the rock mass is the height of the rock mass supported by the support. The weight of the separated rock mass is then:

$$W_f = \frac{HB\gamma_{im}L}{K_o - 1} \tag{6.40}$$

where K_0 is the crushing expansion for the immediate roof, 1.1 to 1.5. Thus, the weight of rock on the support is equivalent to the weight of rock at 2 to 10 times the mining height. In using this method, due to the increasing support capacity of the supports employed by the coal industry from the 1970s to the late 1990s, the required crushing expansion had to be correspondingly reduced from 1.5 to less than 1.1, a considerable range, to match the increasing carrying capacity. As noted earlier, in the 1860s and 1870s, the use of a maximum crushing expansion of 1.5 to determine support capacity resulted in an underestimation of the loads generated by the immediate roof, whereas in recent years the use of crushing expansion equal to or less than 1.1 has resulted in an overestimation of the loads generated by the immediate roof. If the volume factor method is used to determine the support capacity, the actual volume factor of the roof formation must be used, and the support capacity thus determined may be less than the actual support capacity used. Peng recommends a crushing expansion of 1.25.

6.1.4.2 WILSON Estimation Method in the UK

British WILSON (WILSON) takes the balance problem of the immediate roof as the basis for determining the work resistance research, considering the fracture angle of the rock layer and the coal dip angle, points out that the support should control the thickness of the immediate roof of the two times the mining height, i.e. the collapsed rock crushing expansion of 1.5. As shown in Fig. 6.12, the rock fracture angle α has different values: the main roof is very loose, at $\alpha = 0°$; main roof broken, at $\alpha = 15°$; main roof medium stable, at $\alpha = 30°$; main roof hard, at $\alpha = 45°$; main roof very hard, at $\alpha = 60°$. The support strength R is calculated according to the following equation:

$$R = \frac{W}{BD}\left(L + \sum h \tan\alpha\right) \tag{6.41}$$

where W is the weight of immediate roof rock block, kN, $W = \gamma LH$; γ is the volumetric weight, kN/m^3; B is the center distance of support, m; D is the distance from the coal to the point of supporting force, m; L is the distance from the coal wall to the back of roof beam, m; $\sum h$ is the height of the immediate roof, m. Generally, $\sum \mathrm{h} = 2\mathrm{M}$; M is the mining height, m.

For inclined coal seams, the slip along the inclined layer needs to be considered. Then the correction equation for the support strength R_K is:

$$R_K = \frac{W}{BD}\left(L + \sum h \tan\alpha\right)\left(\frac{\sin\alpha}{\sin\theta} + \cos\delta\right) \tag{6.42}$$

where δ is the dip angle of the coal seam; tanθ is the friction coefficient of the layer, generally, tan$\theta = 0.4$, $\theta = 21.8°$.

$$F_B = \frac{1}{f - b}\left[f(W + R_1) - (W\omega + R_1 r)\right] \tag{6.43}$$

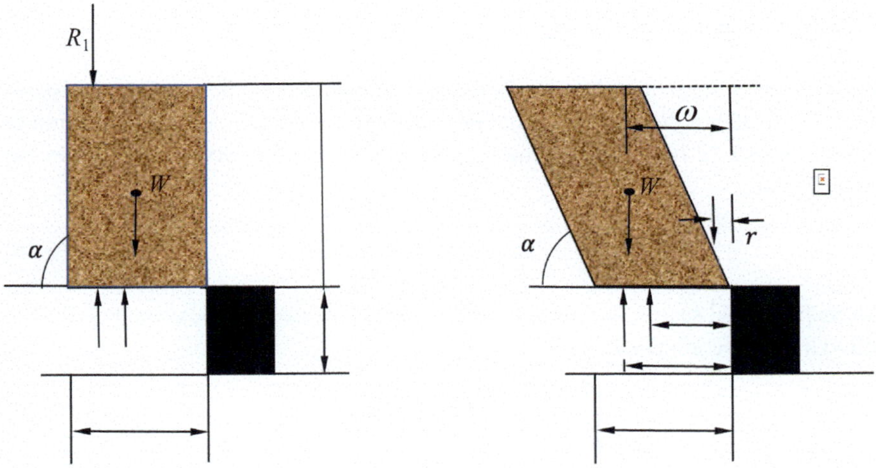

Fig. 6.12 WILSON model on front and rear strut resistance calculation

$$F_F = \frac{1}{f-b}(W + R_1) - \frac{(W\omega + R_1 r)}{f-b} \tag{6.44}$$

where F_B is the front column resistance, kN; F_F is the rear column resistance, kN; R_1 is the additional force acting on the main roof, kN; f, b, r, w are the size of the action points of each force indicated in Fig. 6.12.

The British Coal Board, based on WILSON's equation, specifies that the initial support for the strength of the support R_0:

$$R_0 \geq 75M \tag{6.45}$$

where M is the mining height. This is because when a support is lowered on a column, the neighboring frame is subjected to 1.5 times the additional load and, therefore,

$$R_0 = 1.5 \times 25 \times 2M = 75M \tag{6.46}$$

The rated working resistance of the support must take into account the worst case scenario. The support has not yet moved forward after the passage of the shearer and is at the maximum controlled roof distance, taking two times the safety coefficient:

$$R_H = 2 \times R_0 = 150M \tag{6.47}$$

6.1.4.3 German Methodology

The former Federal Republic of Germany stipulated that the minimum support strength R (kPa) of the support should be considered to bear the weight of the immediate roof rock (2 times the mining height) and to take into account 1.6 times the safety factor.

$$R = 1.6 \times 25 \times 2M = 80M \tag{6.48}$$

For inclined coal seams, the strength of the support R_{K} needs to be considered in the following correction equation:

$$R_K = (50 + 1.5E_\alpha)M \tag{6.49}$$

where E_α is the inclination angle of the coal seam, expressed in gon, $1\text{gon} = 0.9°$.

In the 1970s, the former Federal Republic of Germany specified a minimum rated working resistance R_{H} (kPa) of:

$$R_H = 80M$$

Calculated by the German EVANS method (seeing Fig. 6.13), the hydraulic support's support strength P is:

$$P = \frac{l_S\left(\gamma - \frac{2C}{l_S}\right)M}{2k\tan\varphi}\left(1 - \exp\left(\frac{-2k\tan\varphi}{l_S}\right)H_m\right) \tag{6.50}$$

where c is the rock bonding force; k is the ratio of horizontal stress to vertical stress; H_m is the mining depth, m; φ is the angle of internal friction of the rock, (°); γ is the rock volumetric weight, kN/m^3.

When $C = 0$ and H is infinity, then the support strength:

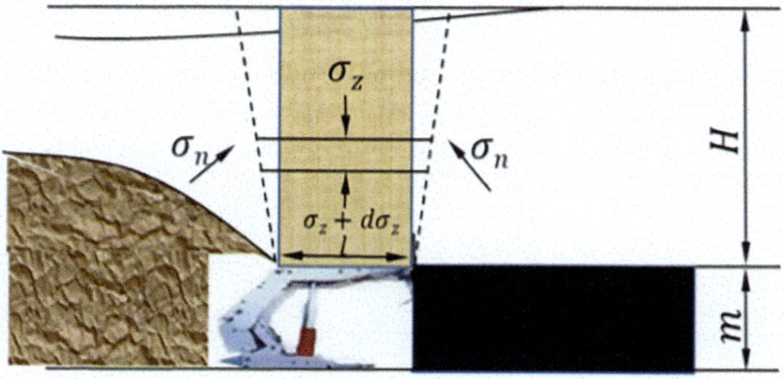

Fig. 6.13 EVANS computational mechanics system

$$P = \frac{\mathrm{l_s}\gamma}{2\mathrm{k}\tan\varphi} \tag{6.51}$$

6.1.4.4 Franch Methodology

France proposed an empirical equation for determining the rated support strength based on the amount of roof subsidence:

$$C_s = 200(\mathrm{q_c}M)^{\frac{3}{4}}H_m^{-\frac{1}{4}}\left(\frac{340}{P} + 0.33\right) \tag{6.52}$$

where C_s is the advancing subsidence per meter, mm/m; M is the mining height, m; H is the mining depth, m; q_c is the coefficient to consider the filling degree of the goaf area, collapse method $q_c = 1$; wind filling $q_c = 0.5$; water–sand filling $q_c = 0.2$; P is the strength of the support per meter, kN/m. When the critical subsidence C_k is determined, the necessary rated support strength P can be determined by substituting into Eq. (6.52).

6.2 Calculation of Roof Pressure in LTCC

The large mining thickness of LTCC causes significant disturbance to the overlying strata. The primary support object for the top coal caving method is the impending fall of the top coal, with the support situated within a system of "immediate-roof-support-floor". The stiffness characteristics of each part of the support system directly impact the working resistance. Therefore, the stiffness of the top coal directly effects the support resistance. Top coal caving results in loosening and deformation of the immediate roof, especially the top coal, leading to reduced stiffness and hindering transmission of force from overlying strata onto the support. As a result, part of this force of the main roof can be transferred to supporting by the coal wall, with transfer magnitude varying based on the stiffness of top coal and immediate roof. Obviously, higher stiffness leads to greater load on the support. Therefore, two scenarios arise during top coal caving: firstly, when the top coal is loose, the working resistance of the support may be lower than normal (sometimes as low as 70–80% at equivalent height), resulting in greater load on front columns compared to rear columns, typical for most operations. Secondly, when high stiffness in top coals turns them into pressure-transferring mediums causing significantly increased working resistance compared to general panels, leading to more load on rear columns than front columns.

6.2.1 *Working Resistance of LTCC Support Based on Top Coal Damage*

The top coal moves from the front of the coal wall to the top of the support as the panel advances, and the structure has been damaged, basically presenting a loose body. The deformation pressure of the roof strata causes the deformation of the top coal, and transfers the vertical deformation pressure to the support through the medium layer of the top coal. Therefore, the support load F consists of two parts: the gravity W of the top coal body above the support ($W = \delta_d Lb\rho$, δ_d is the thickness of the top coal, L is the roof-control distance, b is the width of the support, and ρ is the density of coal) and the pressure F_y of the roof strata to promote the vertical deformation of the top coal. The pressure of the roof strata to promote the vertical deformation of the top coal is calculated as follows [4].

6.2.1.1 Establishment of Physical Equation of Top Coal Motion Damage

Based on the field test of multiple top coal displacements and deformations, it can be inferred that the damage variable D is related to the combined displacement of top coal s and the fracture quantity N. Furthermore, s and N change exponentially with respect to the distance l from the panel at different positions, so it can be assumed that $D = AB^l$. According to the basic principles of damage mechanics, the basic elastic modulus E and the equivalent elastic modulus E_{ef} have the following relationship: $E_{ef} = E(1 - D) = E(1 - Ab^l)$. The damage property equation is as follows:

$$\sigma = E_{ef}\varepsilon = E(1 - D)\varepsilon \tag{6.53}$$

where A and B are constants, ε is the strain.

By using Eq. (6.53) and the displacement characteristics of the top coal in the horizontal and vertical directions, the damage physical property equations in the horizontal and vertical directions can be established respectively.

6.2.1.2 Deformation Pressure of the Top Coal

According to the principle of damage mechanics, damage is a property of object deterioration, and the process of deterioration is caused by changes in external conditions, such as external load and changes in the environment. As the panel advances, the roof interacts with the top coal. The roof bends and sinks, and the stress acting on the top coal changes, which appears in the deformation of the coal as well as the volume expansion of the coal. Due to the difference in the principal stress direction, horizontal and vertical deformation will inevitably occur, that is, the top coal

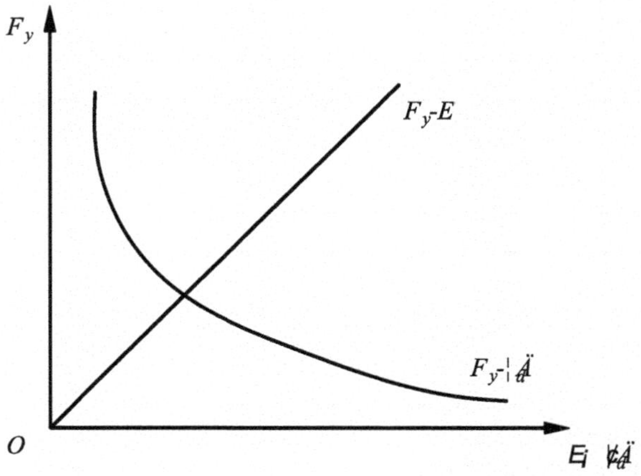

Fig. 6.14 The relationship between the vertical pressure of top coal by roof strata and its elastic modulus as well as the thickness of top coal

body is damaged in both horizontal and vertical directions furthermore, therefore, the vertical deformation pressure acting on the top coal body is as follow.

$$F_y = \frac{\eta_r EB}{\delta_d} \int_0^{L_k} D_h D_v \Delta dl \tag{6.54}$$

where η_r is the weakening coefficient of elastic modulus of coal, considering the mutual influence effect between rocks, $\eta_r = 1/8\text{–}1/25$; D_h and D_v are the damage variables in the horizontal and vertical directions of the top coal body, Δ is the vertical deformation of top coal, L_{k} is the length of the top beam of support.

From Eq. (6.54), the deformation pressure of the top coal is proportional to the top coal's elastic modulus (strength), and inversely proportional to the thickness of the top coal, as shown in Fig. 6.14.

The above-mentioned can effectively explain that under the same surrounding rock conditions, the softer the coal seam is, the smaller the support load is, and the deformation pressure of the support under the roof strata is not proportional to the mining thickness, that is, the roof activity space of the goaf. The overall change trend is shown in Fig. 6.15a. The relationship between the load of the support and the thickness of the top coal is superimposed by the linear increase of the top coal weight force and the deformation pressure of the inverse proportional function. This kind of "F-δ_d" type of top coal change relationship exists in the case of adjusting mining height to change the thickness of top coal in the same top coal caving panel, and it also exists in the working resistance of support designing and top coal caving panel with similar surrounding rock conditions and mining height. Figure 6.15b is the relationship between the measured load of the support and the mining thickness in Yangquan No. 4 Coal Mine, Xingtai Mine (high-DO caving), Wulan Mine, Lu'an Mine and Micun Mine.

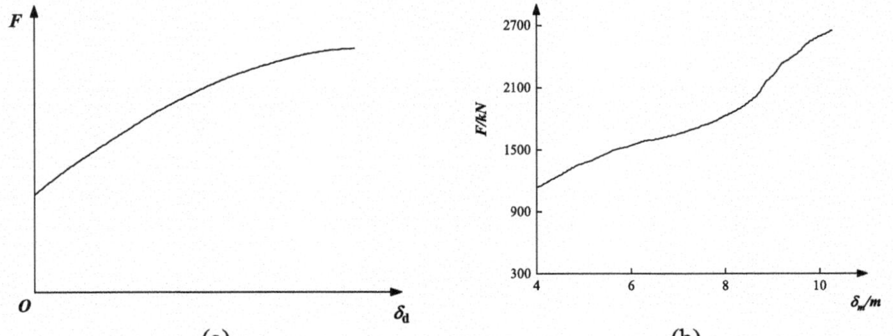

Fig. 6.15 Change curves of the support load with the seam thickness: **a** theoretical analysis; **b** measured results

6.2.2 Support Resistance of LTCC Panels in Extra-Thick Coal Seams

The movement space of strata in the goaf of the large mining height top coal caving panel is large and the activity is intense, and the layer of stable structure is higher. For the extra-thick coal seam with a thickness of 15 m and above, in the case of small mining height, the main roof can form a stable structure on the overlying and this rock strata transformed into the immediate roof in the case of large mining height top coal caving. The horizontal force can not be transmitted after the main roof is broken, however, the cantilever structure acting on the supports is easily formed in the main roof with high strength. Furthermore, the hard strarum at a high level as the main roof can also form the masonry beam structure, as shown in Fig. 6.16. When the working resistance of the support is small, it is difficult to prevent the overlying roof from sinking violently. The key blocks of the masonry beam structure will sink further, forcing the cantilever beam structure to rotate. The rotary sinking of the cantilever beam will lead to the decrease of the support resistance acting on the masonry beam, especially during cyclic weighting, which is easy to cause the slumping instability of the masonry beam, resulting in strong mine pressure and dynamic load effect of the support [5].

6.2.2.1 Normal Mining Stage

In the normal mining stage of the panel, the support only needs to endure the force acted by the cantilever beam structure. It can be calculated from the cantilever structure.

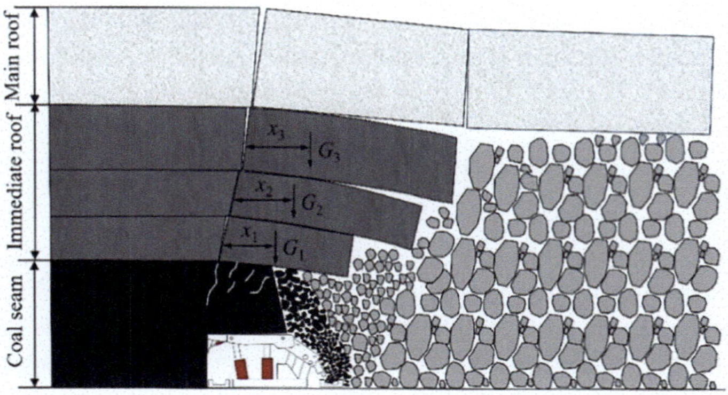

Fig. 6.16 Roof structure of large mining height top coal caving panel

$$\begin{cases} L_i = h_i\sqrt{R_{Ti}/3q_i} \\ G_i = L_i b h_i \\ x_i = \dfrac{L_{bi}}{2} \\ F_c = \gamma_m \delta_d b l_d + \dfrac{k}{l_r}\sum_{i=1}^{n} G_i x_i \end{cases} \tag{6.55}$$

where h_i is the thickness of one end fixed cantilever beam of layer i, (m), γ_m is the volumetric weight of the top coal (kN/m$^{3)}$, δ_d is the thickness of top coal (m), b is the width of the support (m), l_d is the length of the support top beam (m), L_i is the length of one end fixed cantilever beam of layer i (m), R_{Ti} is the tensile strength of rock, and it can be determined by experiments (MPa), q_i is the load on unit length of the cantilever beam (MPa), G_i is the weight of cantilever beam of layer i (kN), x_i is the horizontal distance from the coal wall to the center of gravity of the cantilever beam of layer i (m), l_r is the distance from the coal wall to the center line of the support column (m), k is the design coefficient considering the forward movement of adjacent supports (taking 1.10–1.25), and F_c is the support should bear the cantilever beam force (kN).

6.2.2.2 Weighting Stage

As the panel advances, the cantilever beam gradually extends and breaks with the drawing of top coal. In the stage of weighting, the key to controlling the stability of the overlying strata in the panel by the support is to control the masonry beam structure. In addition to providing the resistance to support the cantilever beam, the support should also provide a certain support force to the masonry beam structure to control the stability of the composite structure of the cantilever beam and the masonry beam. Therefore, the working resistance required to effectively control the roof is divided into two parts: one part is used to prevent the instability of the high masonry beam structure, the other part is used to support the force of the low cantilever beam structure. It can be obtained from masonry beam structure.

$$F_j = \left[2 - \frac{l\tan(\varphi - \theta)}{2(h - s)}\right] Q_0 b \tag{6.56}$$

where F_j is the force that the support needs to bear the roof (kN), l is the periodic weighting step (m), φ is the friction angle between rock blocks (°), θ is the rock breaking angle (°), h is the layer thickness of the key block (m), s is the subsidence of the key block (m), Q_0 is the load of the key stratum itself and its upper control strata (kN). Therefore, the calculation expression of the working resistance of the support in the top coal caving panel with large mining height is as follow.

$$F = \left\{\gamma h l_d + \left[2 - \frac{l\tan(\varphi - \theta)}{2(h - s)}\right] Q_0 \right\} b \tag{6.57}$$

where γ is volumetric weight of the rock. When the support resistance is not enough to prevent the instability of the masonry beam, the slumping instability of the masonry beam will inevitably lead to the shear failure of the cantilever beam. After the cantilever beam is broken and unstable, it will only act on the support within the length of the top beam of the support. At this time, the "cantilever beam-masonry beam structure" evolves into a "masonry beam structure".

6.2.2.3 Example Analysis

1. Project profile

Buliangou Coal Mine is a ten-million-ton mine located in Zhungeer Coalfield, Ordos. It has loess plateau landform and gully development. The F6202 LTCC panel is the second mining face in the second panel. The No. 6 coal seam is mined. The coal seam has a gentle attitude and relatively developed cracks. The coal seam is 11–21 m thick, with an average thickness of 15.2 m and an average dip angle of 4°. The LTCC with large mining height is adopted. The panel tends to be 240 m long, the strike is 1300 m long, the mining height is 3.8 m, the caving height is 11.2 m, and the mining and caving ratio is 1:2.95. ZF13800/27/42 four-column support produced by China Coal Beijing Coal Mine Machinery Co, Ltd. is adopted.

The conventional method is used to observe the six periodic weighting in front of the F6202 panel. From the statistical data, the periodic weighting step distance of the large mining height LTCC panel is not much different from that of the ordinary LTCC panel. The weighting step distance is 7.6–16.8 m, with an average of 11.5 m, in the normal stage, the average working resistance of the hydraulic support is 7586 kN, and the working resistance of the support increases sharply during the pressure period. The pressure generally lasts for 1–3 coal cutting cycles, the average working resistance is 14,998 kN, and the average dynamic load coefficient is 1.83. It shows that the roof activity is intense, the dynamic load phenomenon is obvious, the static pressure is small, the dynamic pressure is large, and the support safety valve is opened frequently. The field observation also shows that the roof sinking speed is fast and the sinking amount is large during the pressure period, showing the "moving column rapid contraction" phenomenon. Especially when the panel advances slowly, the roof sinking amount is larger, and there is a danger of crushing the frame, which is visible. The hydraulic support with rated working resistance of 13,800 kN cannot meet the needs of roof control.

2. Calculation of working resistance of support

Based on the geological conditions of F6202 panel, according to Eq. (6.55), the working resistance of the support to bear the top coal and the immediate roof cantilever beam can be calculated as follows.

$$F_c = \gamma_m h_m b l_d + k(\sum_{i=1}^{n} G_i x_i) l_r \tag{6.58}$$

The volume force of the top coal is 14 kN/mm^3, the thickness of the top coal is 11.2 m, the width of the support is 1.75 m, the length of the top beam of the support is 5.5 m, the distance between the coal wall and the center line of the action of the support column is 3.8 m, and the other parameters are substituted into Eq. (6.58) to obtain $F_c = 7403$ kN. This is also the reason why the support is in the normal stage and the working resistance of the support is concentrated in 6500–8500 kN.

In the weighting stage of the panel, to prevent the instability of the high masonry beam structure, the support resistance is calculated according to Eq. (6.57). The average periodic weighting step distance of the panel is 11.5 m, the friction angle between the rock blocks is 45°, the key rock block subsidence is 3.54 m, and the key rock block thickness is 14.3 m. The parameters are substituted into Eq. (6.56) to obtain $F_j = 10{,}576$ kN. To prevent the slumping instability of the masonry beam structure, the resistance required by the support is F_j, while maintaining the stability of the cantilever beam structure, and the resistance required by the support is F_c. At this time, the working resistance of the support is $F = F_j + F_c = 18{,}000$ kN. The working resistance of the effective control of the support to control the stability of the inverted step composite cantilever beam and the masonry beam structure is at least 18,000 kN. When the working resistance of the support is less than 18,000 kN, the masonry beam structure will slide and lose stability, which will cause the support to bear greater roof force and even impact the support. This is also the reason why the support of 13,800 kN working resistance in Buliangou Coal Mine is prone to crushing.

6.2.3 *Support Resistance of STCC Panel*

6.2.3.1 Roof Failure and Displacement Form

1. Project profile

Jiangcang No. 1 Coal Mine is in the northeast of Qinghai Province, on the south bank of Jiangcang River, with an altitude of about 3800 m. The mine area is 3.31 km^2, and the designed annual production capacity of the mine is 900,000 t/a. The regional structure of the mine field is relatively simple, which is an axial nearly east–west, two-wing asymmetric syncline structure. The dip angle of the south wing of the syncline is steep from east to west, about 50° in the east, and 60–70° in the west. The upper part of the field is the outcrop of coal measures, the shallow part is open-pit mining, and the deep part is transferred to underground mining. In the preliminary design, the vertical height of the district is 50 m, which is divided into three sublevel caving. To improve the efficiency, the plan is changed to two sublevel caving. The height of each sublevel is 25 m, of which the mining height is 2.8 m, the caving height is 22.2 m, the mining and caving ratio is close to 1:8, the average length of the panel is 17 m, and the dip angle of the coal seam is 55°. The cutting depth is 0.6 m, and the coal cutting cycle progress is 0.6 m. The panel adopts the pre-blasting method to loosen the top coal, and the top coal caving method adopts multi-round, interval and sequential caving, and the coal caving from the floor to the roof. The layout of the panel and the situation of the roof and floor are shown in Fig. 6.17.

2. Recovery of upper sublevel

During the upper sublevel mining, the top coal caving is carried out three rounds from the floor to the roof, and the displacement of the key points in the process of

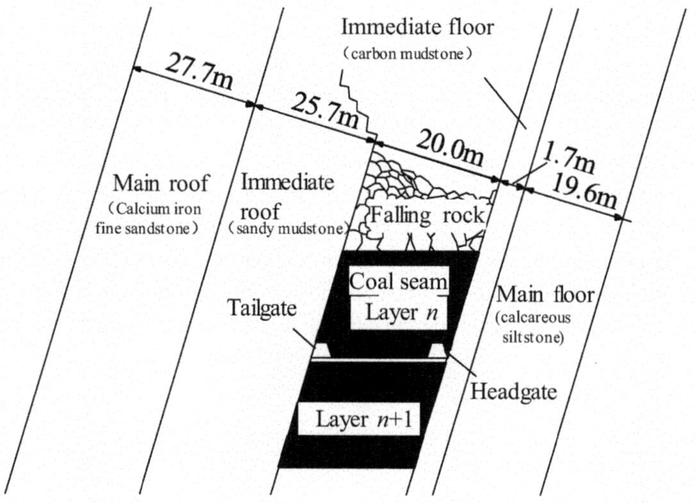

Fig. 6.17 Layout of the panel and conditions of roof and floor

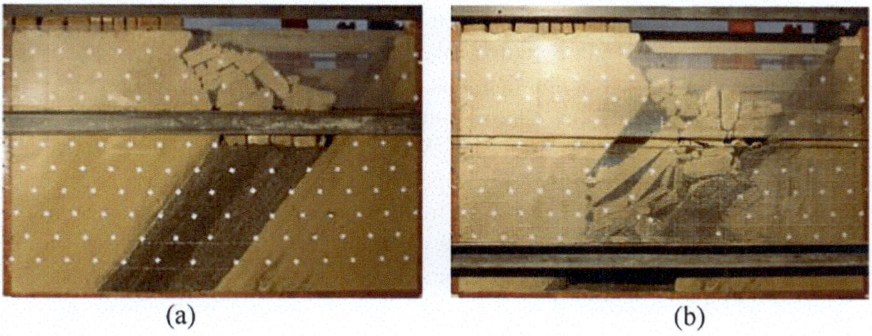

Fig. 6.18 Physical test of roof failure: **a** upper sublevel's "toppling" damage; **b** lower sublevel's "slumping" failure

immediate roof caving is observed. With the gradual release of top coal, the roof is gradually suspended. Under the action of gravity and overburden load, tension will occur at the upper position of the roof. When it exceeds its tensile strength, it will break and form rock blocks. The immediate roof near the coal seam is relatively broken, and the height of the formed rock block is small. As it extends deeper into the roof, the height of the rock block gradually increases to the maximum and then gradually decreases, showing a "short-long-short" distribution. Furthermore, at the moment of the roof breaking, the roof rock block suddenly rotates along the fulcrum at the bottom of the rock block, toppling and falling on the top coal, showing "toppling" damage, which is mainly determined by the dip angle of the coal seam, the thickness of the coal seam and the height of the horizontal section. The toppling failure is transmitted to the deep rock block of the roof in turn, and the toppled rock blocks are superimposed in turn, showing a "domino" chain failure, as shown in Fig. 6.18a.

For the rock block system, if the toppling rock block has enough movement space, the toppling rock block acts on the lower rock block, so that the lower rock block may slide along the bottom. With the gradual deepening of the mining level, this phenomenon is more obvious, and the roof is more likely to be broken and toppled with slumping failure, showing a "slumping" fracture form, as shown in Fig. 6.18b. The mechanism of collapse is also the fracture and toppling of rock strata, however, compared with a single toppling, the collapse has a greater impact on the goaf. In the upper sublevel mining, the rock mass is mainly toppling failure.

3. Recovery of lower sublevel

The rock of roof from the upper sublevel topples in the lower sublevel roof, which promotes the fracture and toppling of the lower sublevel roof. When the lower sublevel is mined, the roof is destroyed and then toppled or collapsed on the falling goaf formed by caving and the residual coal that is not released in the upper section. With the gradual drawing of the top coal in the lower sublevel, the goaf and residual coal in the upper sublevel will move downward, and the roof is equivalent to the free end at the upper end due to the collapse of the upper sublevel at this time. Therefore, the roof stores less strain energy before breaking the rock block than in the upper sublevel mining stage. Furthermore, when the lower sublevel is mined, the immediate roof near the coal seam is more broken under the influence of mining, so the height of the rock block is smaller than that of the upper section, and it is not easy to cause the toppling failure along the fulcrum rotation, and it is easier to slide along the bottom surface and fall into the goaf. With the collapse of the immediate roof near the coal seam filling the goaf, the upper immediate roof rock block away from the coal seam has better roof integrity due to the limited movement space. The height of the rock block formed after the fracture is larger, and the toppling failure is more likely to occur. The toppled rock block promotes the slumping of the lower height rock block, which is manifested as slumping failure. In general, in the lower sublevel mining, the collapse failure is the main failure form, as shown in Fig. 6.18b, when the goaf is fully filled, it may also be manifested as "stable" failure, which is often an intermediate process and will eventually evolve into "toppling" or "slumping" failure.

6.2.3.2 Mechanical Analysis of Toppling-Slumping Failure

To obtain the thickness and range of immediate roof caving, as well as the conditions of failure, the roof strata can be simplified into a discrete rectangular rock block system with a set of steeply inclined joints and cracks, and the failure surface is a step shape. The geometric model of roof toppling-collapse failure is established [6], as shown in Fig. 6.19. Where q is the overburden load, which is regarded as uniform load to simplify the model construction. Δx and y_n are the thickness and height of the rock respectively. Ψ_d is the dip angle of steep joint, Ψ_p ($\Psi_p = 90° - \Psi_d$) is the normal dip angle of steeply inclined joints, H is the vertical caving height of the immediate roof, Ψ_f is the lower heap panel angle, Ψ_s is the upper heap panel angle, Ψ_b is the dip angle of the step-like failure surface formed by the rock base. The number of rock blocks in the system is N, and the expression of N can be obtained according to the geometric relationship of Fig. 6.19.

$$N = \frac{H}{\Delta x}\left[\csc(\psi_b) + \frac{\cot\psi_b - \cot\psi_f}{\sin(\psi_b - \psi_f)}\sin\psi_s\right] \tag{6.59}$$

Starting from the smallest rock block at the bottom, the rock blocks are numbered in turn, that is, the smallest rock block at the bottom is recorded as rock block 1, the highest rock block is recorded as rock block N, and any upper rock block is recorded as rock block n. The location of the largest rock block is called the top of the heap, and the rock block is recorded as n_m. The height y_n of rock n sees Eq. (6.109).

$$y_n = \begin{cases} n(a_1 - b), & n \le n_m \\ y_{n-1} - a_2 - b, & n > n_m \end{cases} \tag{6.60}$$

Among them, Eq. (6.61) can be derived from the geometric relationship.

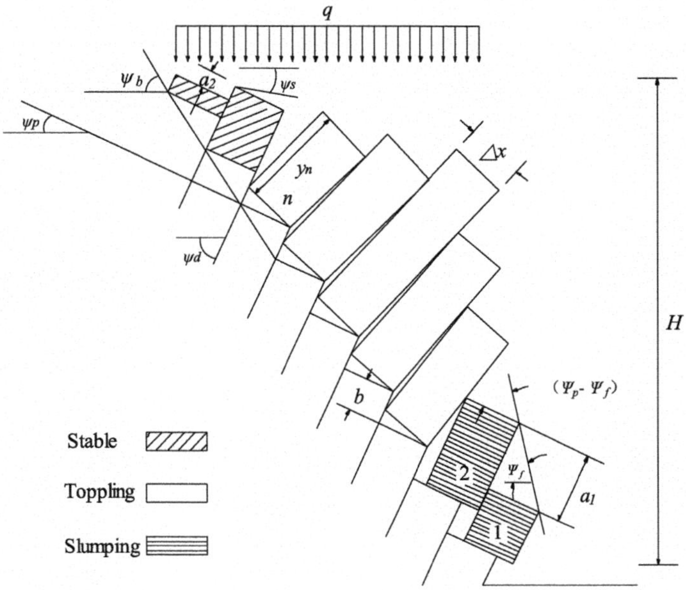

Fig. 6.19 Geometric model of toppling-slumping failure

$$\begin{cases} a_1 = \Delta x \tan(\psi_f - \psi_p) \\ a_2 = \Delta x \tan(\psi_p - \psi_s) \\ b = \Delta x \tan(\psi_b - \psi_p) \end{cases} \tag{6.61}$$

The premise of toppling stability analysis of rock mass is that the system does not slide as a whole, that is, it needs to meet,

$$\psi_\mathrm{p} < \Phi_\mathrm{p} \tag{6.62}$$

where Ψ_p is the normal dip angle of steep-dig joint (°), and Φ_p is the friction angle of rock undersurface (°).

1. Discrimination of roof rock failure form

There are three main states of rock blocks formed by roof strata: stability, toppling and slumping. Distinguishing the failure form is the basis for the analysis of roof caving form. Rock block n is taken as the research object and simplified into a rectangle for force analysis, as shown in Fig. 6.20. Where λ is the ratio of overburden load q to rock weight γ, P_n and P_{n-1} are the lateral force of rock block $(n + 1)$ and rock block $(n - 1)$ to rock block n. The distance from the action point to the bottom of rock block n is Mn and Ln.

$$M_n = \begin{cases} y_n, & n < n_m \\ y_n - a_2, & n \geq n_m \end{cases}$$

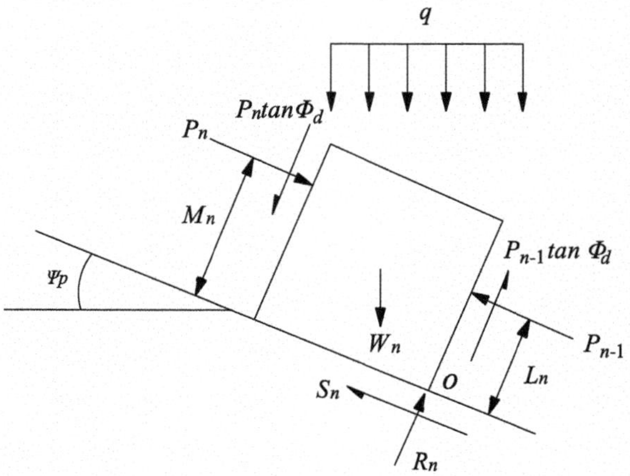

Fig. 6.20 Toppling analysis of the rock block *n* of the roof

$$L_n = \begin{cases} y_n - a_1, & n \le n_m \\ y_n, & n > n_m \end{cases} \tag{6.63}$$

Under the condition of thick immediate roof, the roof stability analysis of STCC panel in steeply inclined thick coal seam is based on the stress analysis of single rock block. At this time, the effect of upper and lower adjacent rock blocks can be ignored, and P_n, P_{n-1}, M_n and L_n are all 0. From Fig. 6.20, under the action of overburden load, the necessary condition for the toppling of a single rock mass is that the sum of the torque of the overburden load and the gravity of the rock mass to the rotation point O is greater than zero, as shown in Eq. (6.64).

$$\frac{y_n}{\Delta x} > \frac{\lambda + y_n}{y_n \tan \psi_p + 2\lambda \tan \psi_p} \tag{6.64}$$

From the rock block N downwards in turn, the rock block that Eq. (6.64) established is denoted by n_1, and the rock block n_1 and below is likely to be toppled, while the rock block above the rock block is in a stable state.

By this step, the number of rock blocks in a stable state and possible toppling failure can be determined, that is, the range of possible toppling failure can be determined, and then these rock blocks with potential toppling failure can be mechanically modeled for in-depth analysis.

2. Construction of toppling mechanics model

When the thick immediate roof is destroyed and the block toppling failure is satisfied, the roof block will be toppled and heaped up to form a heap shape. The immediate roof near the coal seam is more broken, and the height of the formed rock block is smaller because of the affected by mining. Due to the limitation of the movement space and the decrease of the mining influence, the degree of roof failure decreases and the height of the rock block increases gradually. However, due to the existence of upright texture·, the height of the rock block increases to the maximum value and then decreases gradually, as shown in Fig. 6.21. The stress and boundary conditions of rock blocks at different positions above the top of the heap ($n > n_m$), at the top of the heap ($n = n_m$) and below the top of the heap ($n < n_m$) are different, so the toppling mechanics models of rock blocks at different positions are constructed and discussed.

Assuming that the upright texture surface satisfies the limit equilibrium condition, the ultimate static friction forces on both sides of the rock block are shown as Eq. (6.65).

$$\begin{cases} F_n = P_n \tan \Phi_d \\ F_{n-1} = P_{n-1} \tan \Phi_d \end{cases} \tag{6.65}$$

where Φ_d is the side friction angle of rock block.

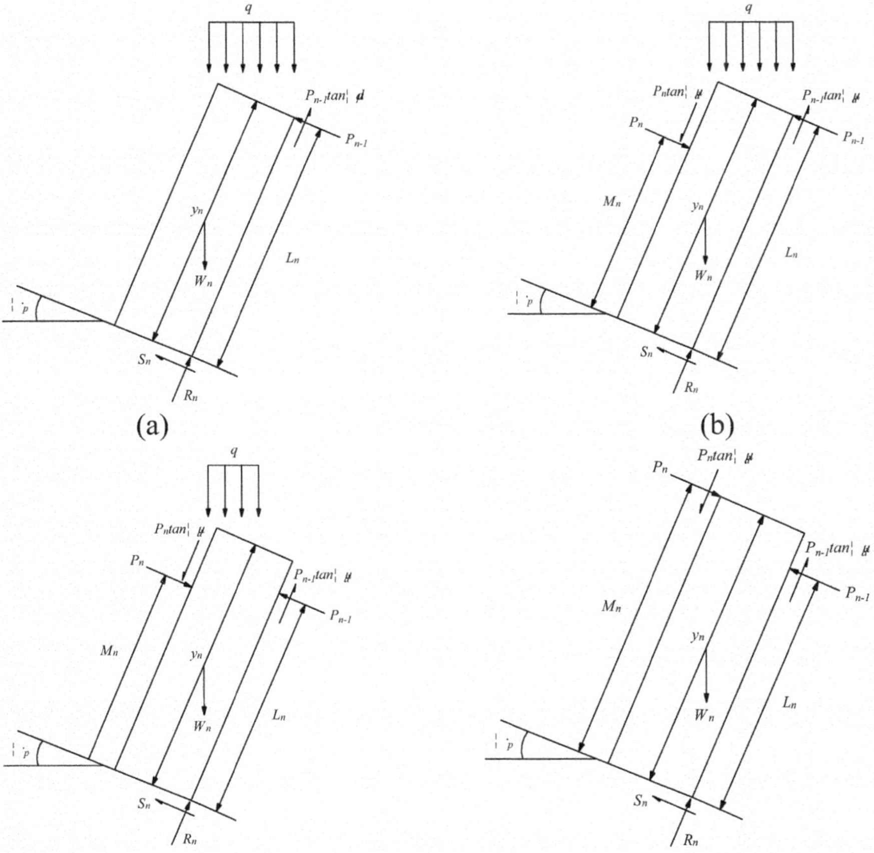

Fig. 6.21 Mechanical model of rock toppling at different positions: **a** $n = n_1$ **b** $n_m<n<n_1$; **c** $n = n_m$ **d** $n<n_m$

(1) When the rock block n is located above the top of the heap and is the first rock block to fail, that is, $n = n_1$, at this time, because the rock block $(n + 1)$ is in a stable state, there is a tension crack between the rock block n and the rock block $(n + 1)$ that gradually narrows from top to bottom, so the rock block n is only subjected to the lateral force P_{n-1} from the rock block $(n - 1)$ in the direction perpendicular to the side of the rock block. At this time, the rock n is affected by the overburden load, as shown in Fig. 6.21a.
(2) When $n_m < n < n_1$, the rock n is subjected to the lateral force P_{n-1} from the rock $(n - 1)$ and the lateral force Pn from the rock $(n + 1)$ in the direction perpendicular to the side of the rock, as shown in Fig. 6.21b.
(3) When $n = n_m$, due to the rotation of the rock block to the free-space, the top of the rock block is not fully subjected to the action of the overburden load. It is assumed that the range of the overburden load is only half of the thickness of the rock block, as shown in Fig. 6.21c.
(4) When $n < n_m$, the rock block is not affected by the overburden load, as shown in Fig. 6.21d.

The limit rotation equilibrium equation of rock n is,

$$P_{n-1,t} = \omega_n P_n + \zeta_n W_n + \xi_n Q_n \tag{6.66}$$

where $P_{n-1,\,t}$ is the critical value of lateral force required to prevent the toppling failure of rock block n, ω_n, ζ_n and ξ_n are the toppling transfer coefficient of rock block, the gravity transfer coefficient, and the overburden load transfer coefficient, meet,

$$\begin{aligned}
\omega_n &= \begin{cases} 0, & n = n_1 \\ (M_n - \Delta x \tan \Phi_d)/L_n, & n < n_1 \end{cases} \\
\zeta_n &= (y_n \sin \psi_p - \Delta x \cos \psi_p)/2L_n \\
\xi_n &= \begin{cases} (2y_n \sin \psi_p - \Delta x \cos \psi_p)/2L_n, & n_m < n \le n_1 \\ (2y_n \sin \psi_p - 3\Delta x \cos \psi_p)/4L_n, & n = n_m \\ 0, & n < n_m \end{cases}
\end{aligned} \tag{6.67}$$

Therefore, the normal force of rock block n acting on rock block $(n - 1)$ due to toppling failure is,

$$P_{n-1,t} = \sum_{j=n+1}^{N} \left[(\zeta_j W_j + \xi_j Q_j) \prod_{k=n}^{j-1} \omega_k \right] + \zeta_n W_n + \xi_n Q_n \tag{6.68}$$

3. Discrimination of transition point of failure form

In addition to the single toppling failure of the roof strata, the slumping failure accompanied by slumping and toppling may also occur. There is a change in the form of roof failure in the landslide failure, which means that the stability of the system has changed. The state of the rock block near the transition point of the failure form has a significant impact on the overall stability of the system. By analyzing the stress of the rock block along the normal and tangential directions, the slumping failure of the rock block can be judged. When the rock block satisfies Eq. (6.69), the rock block reaches the limit slumping state.

$$S_n = R_n \tan \Phi_\mathrm{p} \tag{6.69}$$

The limit rotation equilibrium equation of rock n is obtained by force analysis,

$$P_{n-1,s} = \omega_n' P_n + \zeta_n' W_n + \xi_n' Q_n \tag{6.70}$$

where $\omega_n{}'$, $\zeta_n{}'$ and $\xi_n{}'$ are the transfer coefficient of rock slumping, the transfer coefficient of gravity and the transfer coefficient of overburden load respectively, and satisfy,

$$\begin{aligned} &\omega_n' = \begin{cases} 0, \ n = n_1 \\ 1, \ n < n_1 \end{cases}, \\ &\zeta_n' = \frac{\sin \psi_p - \cos \psi_p \tan \Phi_\mathrm{p}}{1 - \tan \Phi_\mathrm{d} \tan \Phi_\mathrm{p}} \\ &\xi_n' = \begin{cases} \frac{\sin \psi_p - \cos \psi_p \tan \Phi_\mathrm{p}}{1 - \tan \Phi_\mathrm{d} \tan \Phi_\mathrm{p}}, & n_m < n \le n_1 \\ \frac{\sin \psi_p - \cos \psi_p \tan \Phi_\mathrm{p}}{2 - 2\tan \Phi_\mathrm{d} \tan \Phi_\mathrm{p}}, & n = n_m \\ 0, & n < n_m \end{cases} \end{aligned} \tag{6.71}$$

In general, slumping failure usually occurs below the top of the pile ($n < n_m$). The ratio of the number of rock blocks with slumping failure below the top of the pile S_s to the number of rock blocks with toppling failure S_b is recorded as the toppling-slumping coefficient m^*, to facilitate the discussion of the influence of overburden load q on the transition point of failure form.

$$m^* = S_s / S_b \tag{6.72}$$

4. Stability analysis steps

Based on the limit equilibrium analysis and the transfer coefficient method, the toppling stability of the roof rock block in STCC panel of steeply inclined thick coal seam is analyzed. The analysis process is as follows:

(1) Firstly, the failure mode of rock is analyzed. If the roof rock block system satisfies Eq. (6.61), it indicates that the overall slumping failure of the roof will not occur, and the stability analysis can be further carried out.
(2) Distinguish the stability state of rock. Starting from the rock block N, according to Eq. (6.64), whether the rock block will be toppled is judged in turn, and the first rock block toppled is recorded as rock block n_1. Starting from the rock block

n_1, according to the different positions of the rock block, the critical values of the lateral force $P_{n-1,\,t}$ and $P_{n-1,\,s}$ required to prevent the rock block from toppling and slumping failure are calculated according to Eq. (6.66) and Eq. (6.70). The larger values of the two are recorded as Eq. (6.73).

$$P_{n-1} = \begin{cases} P_{n-1,t}, \ P_{n-1,t} > P_{n-1,s} \\ P_{n-1,s}, \ P_{n-1,t} < P_{n-1,s} \end{cases}. \tag{6.73}$$

If $P_{n-1} = P_{n-1,\,t}$, it indicates that the rock block may be toppled. If $P_{n-1} = P_{n-1,\,s}$, it shows that the rock block may occur slumping failure. If all the rock blocks satisfy $P_{n-1} = P_{n-1,\,t}$, it shows that the toppling failure will extend to the lowest rock block without slumping failure. If a rock block satisfies $P_{n-1} = P_{n-1,\,s}$, then all the following rock blocks, including this rock block, are in a critical state of slumping failure.

According to the calculation results, the first rock block that satisfies $P_{n-1} < 0$ is called the key block. The existence of the key block prevents the continuous toppling of the roof rock block and makes the system stable. The roof cannot collapse in time and causes a large area of suspended roof, which will cause a large area of roof pressure and is not conducive to safe production. Loosening measures are taken for the location of the key block to destroy the stable state of the key block, which can minimize the overall stability of the system with the minimum energy and avoid large area suspension.

The filling degree of the goaf has an important influence on the roof activity. The influence of filling on the mechanical model is reflected in the force P_0 of the rock block 1 perpendicular to the side of the rock block. If the goaf accumulated in the goaf is more compact, the force is larger, which is helpful to the overall stability of the rock block system. On the other hand, if the goaf is fully filled and the goaf accumulation height is great, the rock block involved in the mechanical model will also be reduced accordingly, indicating that the roof damage is small and stable. With the continuous release of top coal and the influence of mining, the stability of the roof gradually decreases, and eventually toppling or slumping will occur.

(3) Determine the toppling-slumping coefficient. The instability failure of the roof rock block falls above the residual coal in the goaf, which will exert an impact load on the support of the panel. If the toppling-slumping coefficient is larger, the more slumping rock blocks, the more serious the overall damage of the roof, and the greater the impact on the residual coal in the goaf. Therefore, the study of the toppling-slumping coefficient can qualitatively predict the strength of the roof impact load to a certain extent and ensure the safety of recovery.

6.2.3.3 Example Analysis

According to the physical and mechanical parameters of the example given in Table 6.1, the stability analysis is carried out according to the above steps, and the changes of $P_{n,\,t}$, $P_{n,\,s}$, R_n and S_n are obtained, as well as the influence of overburden rock load on toppling-slumping coefficient under different rock thickness.

1. Changes in $P_{n,\,t}$, $P_{n,\,s}$, R_n and S_n

In this example, $N = 16$, $n_m = 10$ can be calculated by Excel. Assuming that the overburden load $q = 200$ kN/m, according to the steps of stability analysis, rock block 16 and rock block 15 are stable. Therefore, the mechanical modeling of rock block 14 and below is carried out. The calculation results of $P_{n,\,t}$ and $P_{n,\,s}$ is in Fig. 6.22.

The upper part of rock block 14 is a stable rock block, which has no lateral force on rock block 14, and the height of rock block 14 is also small. Therefore, the force required to prevent the toppling of rock block 13 is relatively low, which is 0.9 MN. After that, the toppling failure is transmitted to the lower rock block. This process is also accompanied by the gradual increase of the weight of the rock block and the superposition of the effect of the lateral force, which makes the toppling more likely to occur, and the force required to prevent the toppling also gradually increases. When the toppling failure is transferred below the top of the pile, because the rock block is no longer subjected to the overburden load, the weight of the rock block is gradually reduced, the trend of rock block toppling is slowed down, and the growth ratio of $P_{n,\,t}$ is reduced. When $n = 6$, $P_{n,\,t}$ reaches the maximum value, $P_{n,\,t,\,\max} = P_{6,\,t} = 5.8$ MN, after the peak value of, $P_{n,\,t}$ gradually decreases, that is, the closer to the foot of the pile, the lower the possibility of toppling damage.

In Fig. 6.22 that when $8 < n < 13$, $P_{n,\,s} < 0$, indicating that under the parameter conditions provided by this example, it is difficult to slide from rock block 14 to rock

Table 6.1 Examples of physical and mechanical param

H/m	Δx/m	γ/(KN/m^3)	Ψ_d/(°)	Ψ_p/(°)	Ψ_f/(°)	Ψ_s/(°)	Ψ_b/(°)	Φ_d/(°)	Φ_p/(°)
92.50	10.00	25.00	60.00	30.00	56.60	4.00	35.70	38.15	38.15

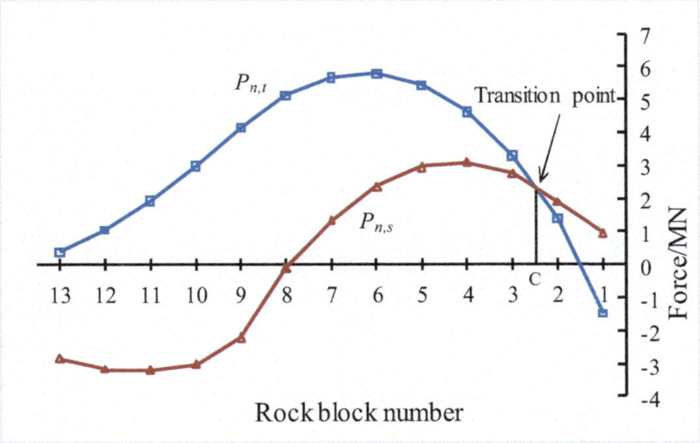

Fig. 6.22 Variation curves of $P_{n,\,t}$ and $P_{n,\,s}$

block 9. After that, $P_{n,\,s}$ increases to the peak value and then decreases gradually. The closer the rock block is to the pile foot, the lower the possibility of slumping failure is. There is an intersection point between the two curves in the graph, and the abscissa of the intersection point is marked as $c(c \in (2, 3))$, which is the transformation point of the failure form. When $n > c$, $P_{n,\,t} > P_{n,\,s}$, that is, the rock block 14 to the rock block 4 is toppling failure; when $n < c$, there is $P_{n,\,t} < P_{n,\,s}$, that is, slumping failure occurs from rock block 3 to rock block 1.

The rock block 16 to the rock block 1 and the calculation results are shown in Fig. 6.23. For all rocks, $R_n > S_n > 0$. The normal force and tangential force of block 16 are smaller, which are 2.6 MN and 1.5 MN respectively. After that, with the toppling to the lower rock block, R_n and S_n gradually increase in the trend of approximate linear function, and the ratio of R_n increase is greater than that of S_n. Both reached the maximum at rock block 10 (i.e. rock block n_m), which were 8.6 MN and 4.4 MN, respectively, and then gradually decreased to a lower level.

2. The influence of overburden load on toppling-clumping coefficient

Under the condition that other conditions remain unchanged, the thickness of rock block is taken as 2 m, 4 m, 6 m, 8 m and 10 m respectively, and the influence of overburden load on toppling-slumping coefficient is discussed. The calculation results are shown in Fig. 6.24. In general, under the condition of constant overburden load, the toppling-slumping coefficient increases with the increase of rock thickness.

When the thickness of rock block Δx is 4 m, 6 m and 8 m respectively, and the value of overburden load is small ($0 < q < 200$), the overburden load has little effect on the toppling-slumping coefficient, and the toppling-slumping coefficient of different rock block thickness has no obvious difference. As the overburden load increases ($200 < q < 400$), the toppling-slumping coefficient also increases, and the greater the thickness of the rock block, the greater the ratio of increase in the toppling-slumping coefficient. After increasing to a certain value, the toppling-slumping coefficient

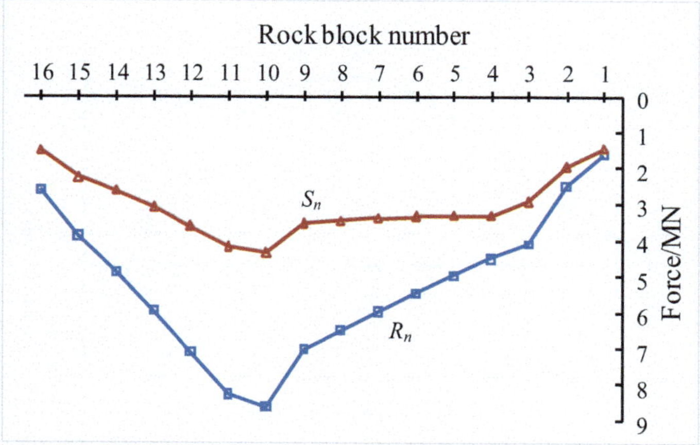

Fig. 6.23 Variation curves of R_n and S_n

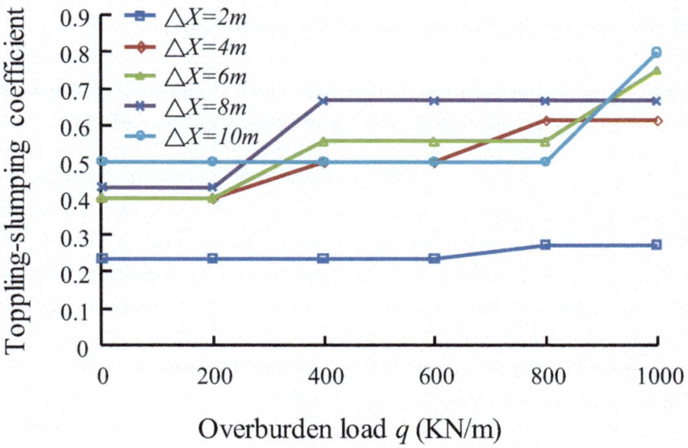

Fig. 6.24 Relationship between overburden load and toppling-clumping coefficient

remains stable with the increase of overburden load ($400 < q < 600$). After that, the overburden load continues to increase. When $q = 600$, the toppling-slumping coefficient of $\Delta x = 4$ m becomes larger again. When $q = 800$, the toppling-slumping coefficient of $\Delta x = 6$ m also increases, showing "hysteresis", while the toppling-slumping coefficient of $\Delta x = 8$ m does not change with the change of overburden load at this stage.

When the rock thickness $\Delta x = 10$ m, the toppling-slumping coefficient remains unchanged with the increase of overburden load within a certain range ($0 < q < 800$), showing obvious "passivation". Only when q reaches a large value ($q = 1000$), the toppling-slumping coefficient increases significantly. With the increase of rock thickness, the influence of overburden load on the transition point of failure mode decreases.

In this example, when $\Delta x = 2$ m, $N = 72$, $n_m = 43$, $y_{43} = 34.3$ m, the block height is large and the thickness is small, and it is easy to slump, which reduces the influence of overburden load on the transition point of failure form, so that the curve does not change significantly.

6.2.4 *Buffering Effect of Loose Top Coal on Roof Dynamic Load*

When the coal seam is soft and the fracture is developed, the top coal above the support often enters the granular state, and the dynamic load formed by the granular top coal on the overlying strata will form an obvious buffering effect. The degree of slowing down is related to the degree of fracture development of top coal, the size of broken block and the law of release.

6.2.4.1 Buffering Effect Calculation of Dynamic Load

Figure 6.25 is the calculation model of the dynamic load buffering effect of loose top coal. To simplify the calculation, the assumptions are as follows.

(1) Only consider the impact of a single roof rock block on the broken coal and gangue above the support, regardless of the interaction between the rock blocks.
(2) Crushing coal and gangue is an ideal elastic–plastic body
(3) The roof rock block is simplified as a cuboid with uniform mass distribution, the dynamic load is caused by the contact between the side of the cuboid and the coal and gangue.
(4) The stiffness of roof rock is much greater than the stiffness of coal and gangue, roof rock can be considered as a rigid body.

Suppose that a rock block with mass m collapses from height H to coal and gangue ($\sum$H is the thickness). The initial velocity is v_0, and the angle with the horizontal line is $90° - \theta$, where θ is the dip angle of the coal seam. The initial velocity $v_0 = 0$ indicates that the roof rock mass is a single toppling failure; when $v_0 \neq 0$, it means that the roof rock mass is toppling-slumping failure. When the roof rock block hits the broken coal and gangue, the broken coal and gangue is still in an elastic state. The maximum deformation after impact is Δ_d, and the kinetic energy becomes 0, so the change of kinetic energy is,

$$\Delta T = \frac{1}{2}m(v_0 \sin(90° - \theta))^2 \tag{6.74}$$

The change of potential energy is,

$$\Delta V = Q(H + \Delta_d) \tag{6.75}$$

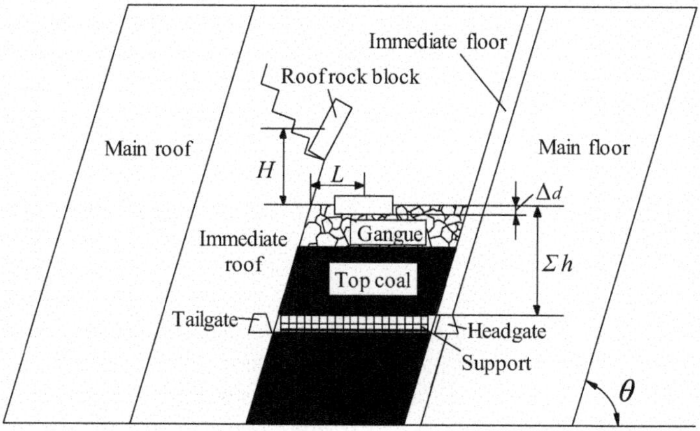

Fig. 6.25 Calculation model of dynamic load buffering effect of loose top coal

where Q is the weight of the cuboid rock (kg). If the energy dissipation and release during the impact process are ignored, according to the conservation of energy, the change of kinetic energy and potential energy of the system is equal to the strain energy of the broken goaf, that is,

$$\Delta T + \Delta V = V_{\varepsilon d} \tag{6.76}$$

where $V_{\varepsilon d}$ is the strain energy of broken coal and gangue.

When the system speed is 0, the dynamic load of broken top coal and goaf is F_d. Under the condition that the coal and gangue obeys Hooke's law, F_d is proportional to the deformation of coal and gangue. Therefore, the work done by the dynamic load during the impact process is $F_d\Delta_d/2$, that is,

$$V_{\varepsilon d} = \frac{1}{2}F_d\Delta_d \tag{6.77}$$

According to the characteristics of linear elastic deformation, the relationship between dynamic load, deformation and stress can be derived,

$$F_d = \frac{\Delta_d}{\Delta_{st}}Q, \sigma_d = \frac{\Delta_d}{\Delta_{st}}\sigma_{st} \tag{6.78}$$

where Δd and Δst are the deformation of broken coal and gangue under dynamic load and static load respectively.

$$\Delta_{st} = \frac{Q\sum h}{EA} \tag{6.79}$$

where E is the elastic modulus of broken coal and gangue, which can be measured by experiment. A is the contact area between the roof rock and the broken coal and gangue.

Substituting Eq. (6.78) into Eq. (6.77),

$$V_{\varepsilon d} = \frac{1}{2}\frac{\Delta_d^2}{\Delta_{st}}Q \tag{6.80}$$

Substituting Eqs. (6.74), (6.75) and (6.77) into Eq. (6.76),

$$\Delta_d = \left(1 + \sqrt{1 + \left(\frac{2HQ - m(v_0\sin(90^\circ - \theta))^2}{Q\Delta_{st}}\right)}\right)\Delta_{st} \tag{6.81}$$

Substituting Eqs. (6.79) and (6.81) into Eq. (6.78), the dynamic load of roof rock on loose coal and gangue is obtained,

$$F_d = K_d Q = \left(1 + \sqrt{1 + \left(\frac{2HQ - m(v_0 \sin(90^\circ - \theta))^2}{Q^2 \sum h}\right) EA}\right) Q \tag{6.82}$$

where K_d is the dynamic load coefficient. When $v_0 = 0$,

$$K_{d0} = K_d|_{v_0=0} = 1 + \sqrt{1 + \frac{2HEA}{Q \sum h}} \tag{6.83}$$

where K_{d0} is the dynamic load coefficient of single toppling failure of roof rock block. During the impact process, the broken coal and gangue will undergo plastic deformation, dissipate and release energy at the same time, showing a significant buffering effect. Therefore, it is necessary to modify the obtained dynamic load calculation equation, that is,

$$F_d^* = (1 - \alpha) F_d = (1 - \alpha)\left(1 + \sqrt{1 + \left(\frac{2HQ - m(v_0 \sin(90^\circ - \theta))^2}{Q^2 \sum h}\right) EA}\right) Q \tag{6.84}$$

where α is the buffer coefficient, $0 < \alpha < 1$. The larger the α, the greater the energy dissipation and release during the impact process.

6.2.4.2 Determination of Buffer Coefficient

As shown in Fig. 6.26, to further determine the dynamic load buffer coefficient, a roof dynamic load test platform was developed [7, 8]. The platform includes heavy hammer, test box, hydraulic cylinder, pressure sensor, data acquisition and analysis system. The dynamic load generated by the roof rock block is simulated by a heavy hammer with a diameter of 0.30 m, a height of 0.10 m, and a mass of 25 kg. The size of the test chamber is 0.4 m × 0.4 m × 0.6 m, and the mass is 55 kg, which is composed of transparent bulletproof glass. The bottom of the test chamber is a stainless-steel plate with a thickness of 0.03 m. Four CST-502-30 pressure sensors are installed under the steel plate to collect the dynamic load and static load of the system in real time. The pressure sensor is matched with the data acquisition and analysis system, and the sampling frequency is greater than 100 Hz.

As shown in Fig. 6.27, six coal samples with different particle sizes were selected as experimental materials, and the thickness of coal samples was 0.1 m and 0.2 m. The drop height of the hammer is 0.05 m, 0.10 m, 0.15 m and 0.20 m respectively, a total of 48 experimental schemes. To ensure the accuracy of the test, 25 groups of experiments were carried out for each scheme.

Figure 6.28 are the results of dynamic load test. With the increase of drop height, the dynamic load increases gradually, while with the increase of particle size and

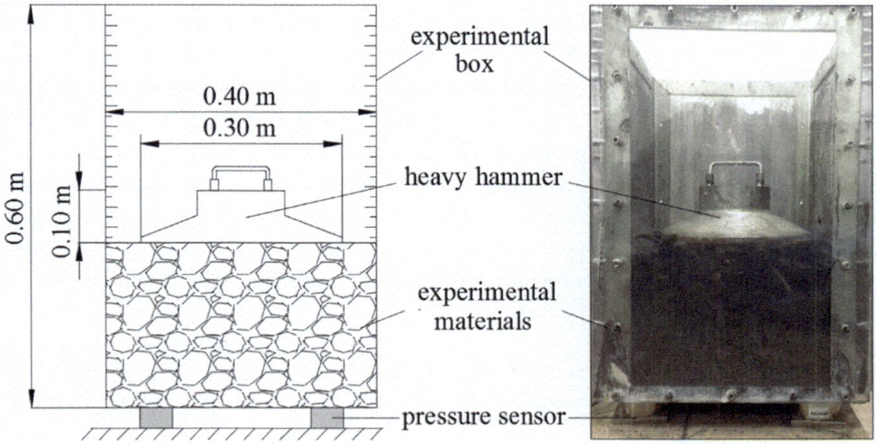

Fig. 6.26 Roof dynamic load test platform

Fig. 6.27 Simulation test materials of top coal with different particle sizes

laying thickness, the dynamic load decreases. That is, the larger the size of the broken top coal block and the thickness of the loose layer, the more energy dissipation during the impact process, and the higher the stability of the panel support.

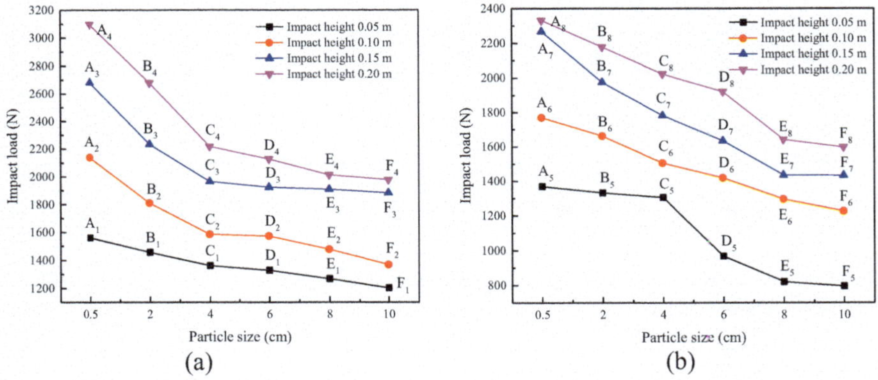

Fig. 6.28 Roof dynamic load test results under different conditions: **a** laying thickness is 0.1 m; **b** laying thickness is 0.2 m

The experimental parameters are brought into Eq. (6.82) to calculate the dynamic load, and the results are shown in Fig. 6.29. The theoretical calculation results are basically consistent with the dynamic load change trend of the indoor test results. The difference is that the former is larger than the latter. The main reason is that the granular top coal causes some energy dissipation during the impact process.

Comparing the theoretical calculation and laboratory test results, the buffer coefficient α can be obtained, as shown in Fig. 6.30. The dynamic load buffer coefficient α is mainly concentrated between 0.41 and 0.62. Considering the residual coefficient of 5%, the range of buffer coefficient α is finally determined to be (0.40, 0.65).

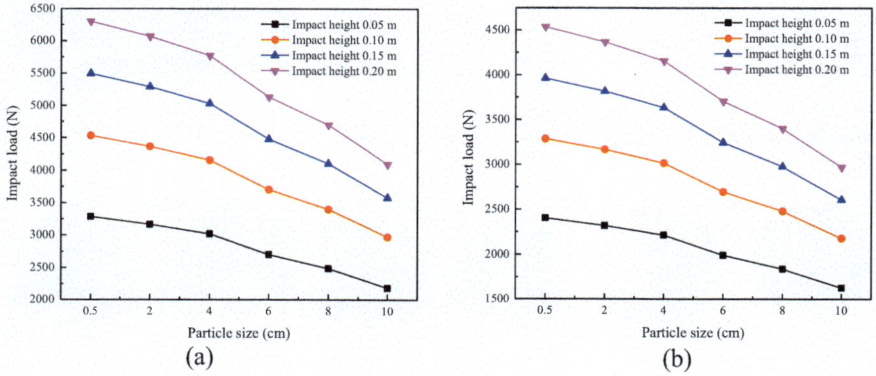

Fig. 6.29 Theoretical calculation results of roof dynamic load: **a** laying thickness is 0.1 m; **b** laying thickness is 0.2 m

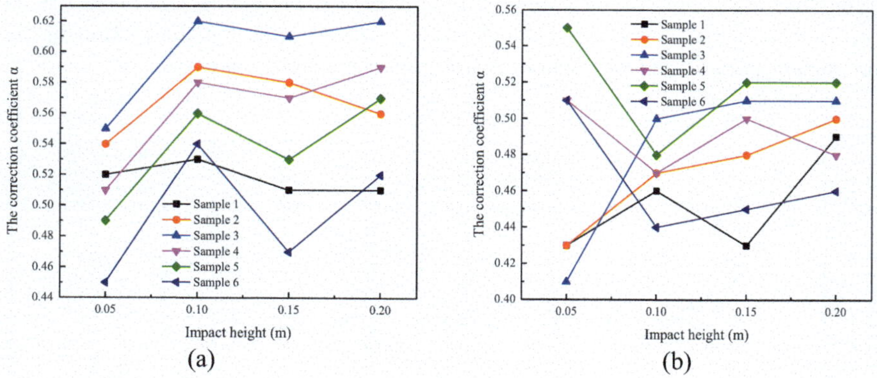

Fig. 6.30 The variation trend of dynamic load buffer coefficient: **a** laying thickness is 0.1 m; **b** laying thickness is 0.2 m

6.3 Calculation of Roof Pressure in Large Mining Height

The large mining height technology is distinguished mainly by its large volume of extracted space. The disturbance to the overburden rock by the panel is heightened, with an increased range, inevitably causing alterations to the movement law of extra-thick overlying rock broken, the roof structure shape, and the destruction characteristics of the coal wall. Thus, elucidating the patterns of roof fracturing and the features of roof pressurization in large mining height is essential for the computation of roof pressure. Moreover, when determining support resistance, it should be predicated on maintaining the stability of the surrounding rock in the panel. The manifestation and control of ground pressure in comprehensive panels with extra-large mining heights panel two principal issues: firstly, as the mining height augments, the problem of rib spalling at the panel worsens; secondly, due to the substantial mining height, the increased collapse height of the goaf roof broadens the area of overburden rock influencing the ground pressure behavior at the panel, resulting in pronounced ground pressure manifestations and a marked escalation in both roof pressure and support load. Consequently, it is imperative to grasp the structural shape of the overlying rock in the comprehensive panel with large mining height and its impact on the ground pressure behavior at the panel. Calculations based on the structural shape shaped by the key stratum are necessary to fulfill the roof control requirements in actual mining operations.

6.3.1 Interaction Between Support and Surrounding Rock

6.3.1.1 Interactive System of Support and Surrounding Rock

The hydraulic support, acting as a structural element for the support of the roof and the maintenance of safe panel production, is not an isolated entity but is instead part of a support system that includes both the supports and the surrounding rock. When viewed from the direction of the panel's advance, the structure created by the main roof layers in proximity to the coal seam within the overlying rock above the entire panel is sustained by the "coal wall-hydraulic support-goaf with collapsed goaf" support system. Given the starkly different properties of the coal wall and the fallen goaf, the functionality of the supports greatly influences their load-bearing status. Consequently, to enable the supports to uphold the roof with the minimal application of support force, the study of the supports' performance and load characteristics emerges as a significant area of inquiry [9].

As depicted in Fig. 6.31, from a vertical perspective, the main roof structure is upheld by the "immediate roof-support-floor' support system. The support system of the top coal caving panel should also incorporate the top coal. Given the existence of the coal wall support influence angle within the panel and the ongoing progression of the panel, the rotation of the main roof is uncontrollable, resulting in a predetermined

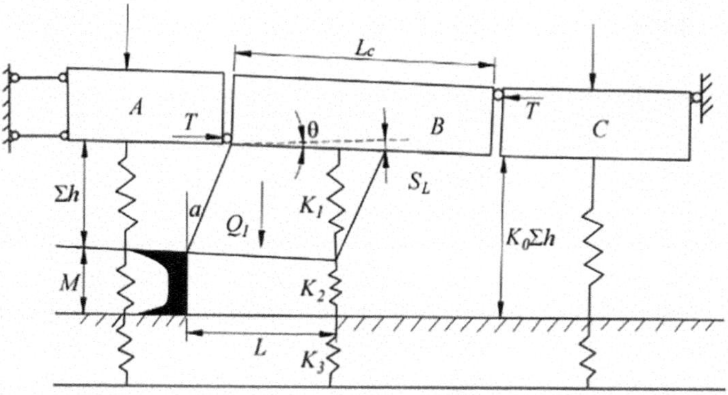

Fig. 6.31 Overall mechanical model of mine pressure in the panel

deformation. Consequently, within the interactive system of the support and the surrounding rock, the main roof's movement and influence are predominant, with mutual interaction and influence between the support and the surrounding rock. The state of motion of the surrounding rock impacts the operational status and load-bearing characteristics of the support, while the operational status of the support reciprocally affects the effectiveness of the roof maintenance.

Overall, the supports within the panel and the surrounding rock they support constitute a contradictory unity that interacts with each other. The design of the support structure and its performance must align with the movement law of the surrounding rock in the panel; it is only thus that the design of the support structure can be both cost-effective and sensible. Furthermore, only with a suitable distribution of support force and reliable control technology of roof can the roof be adequately maintained to ensure the safety of the operating personnel and the smooth progression of production. Consequently, examining the relationship between the support and the surrounding rock essentially entails analyzing how the support's performance and structure affect the support force and the movement of the surrounding rock, as well as the reactions displayed by the support under diverse conditions of the surrounding rock. This analysis aids in determining the most logical structure and parameters for the support.

6.3.1.2 Characteristics of the Interaction Between Support and Surrounding Rock

(1) The support and the surrounding rock interact as a pair of forces. In a localized area, the roof pressure exerted by the surrounding rock can be regarded as an acting force, and the support can be viewed as a reacting force; they should be compatible with each other, equal in magnitude, and ideally act at the same point of application.

(2) The magnitude of the support force and its distribution along the panel are related to the support's performance. The extent of the support force is associated with the properties of the surrounding rock, that is, the combined characteristics established by the support and the surrounding rock.
(3) The influence of support structure and dimensions on roof pressure has been demonstrated in actual production, where selecting the appropriate type of support allows for maintaining the roof with minimal working resistance. For instance, under certain conditions, shield support with short beams (with a working resistance of merely 800 KN) can yield superior maintenance outcomes compared to the stack type with four-leg (with a working resistance of 2400 KN).

It is evident from the aforementioned that the design of support structures must be adapted to the conditions of the surrounding rock, and the performance of the support should be designed as constant-resistance where possible, ensuring that during the application of support force, it remains as consistent as possible with the roof pressure. The most critical parameters in support design are determining the working resistance and the permissible amount of convergence. In the context of the surrounding rock, the primary consideration is the state of the roof under the action of various support reactions.

6.3.1.3 Relationship Between Support Working Resistance and Roof Subsidence

The relationship between support working resistance and roof subsidence reflects, to a certain degree, the interactive relationship between the support and the surrounding rock. Back in the 1960s, numerous experiments were conducted both domestically and internationally to study the relationship between support working resistance P and the ultimate roof subsidence ΔL. As depicted in Fig. 6.32, experiments in China were carried out in laboratories, whereas other countries relied on statistical analysis of field measurement data. The most complete study was the pressure regulation experiment of the support conducted by the Soviet Union on a comprehensive panel, yielding results that were largely consistent with those from laboratory findings. This substantiated that there is an approximate hyperbolic relationship between working resistance P and ultimate roof subsidence ΔL, known as the "P-ΔL' curve.

In the early 1950s, some scholars, based on actual measurements from individual support panels, concluded that it was possible to control the roof, that is, the amount of roof subsidence, under certain conditions. In practical applications, certain benefits to roof management have been observed after reducing the amount of roof subsidence in some panels. However, the "P-ΔL" curve theory indicates that there is a limit to controlling roof subsidence; beyond this limit, the support becomes ineffective. Therefore, increasing the working resistance significantly impacts roof subsidence only when it is initially low, such as under certain conditions with individual supports. Moreover, the "P-ΔL" curve does not provide a comprehensive relationship between

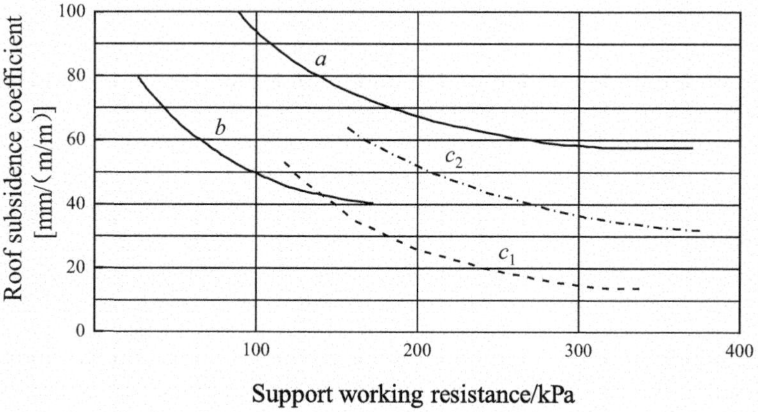

Fig. 6.32 "*P-ΔL*" curve (*a*-Soviet Union; *b*-United Kingdom; c_1, c_2-Ruhr Coalfield, Germany)

the amount of roof subsidence and the state of the roof. Indeed, the permissible subsidence varies among different rock formations. For instance, in hard rock formations, roof subsidence might cause overall fracturing, however, the local area could remain largely intact; whereas in formations with low strength and high brittleness, even minor subsidence could lead to severe roof damage. Consequently, in practical operations, measures to protect the roof are often employed. The advancement of shield support has further demonstrated that improving roof conditions by using protective measures under certain conditions is viable.

In panels where periodic weighting is intense and bench subsidence frequently occurs, the reasonable working resistance refers to the resistance needed to prevent such occurrences. This is the resistance required to balance the weight of the immediate roof and the main roof rock blocks that cannot achieve equilibrium on their own during either the first weighting or periodic weighting. The prerequisite is that the support must have a compressive capacity that corresponds to the subsidence of the roof resulting from the movement of the main roof rock layers.

6.3.2 *Stiffness Model of the Support and Surrounding Rock System*

6.3.2.1 Construction of the Support-Surrounding Rock System Stiffness

In the support-surrounding rock system, the main roof strata typically consist of thick and rigid rock layers. As the panel is extracted, the deformation, fracturing, and structural instability of the main roof are generally uncontrollable. The weaker overlying above the main roof can be managed as a load and the main roof strata can

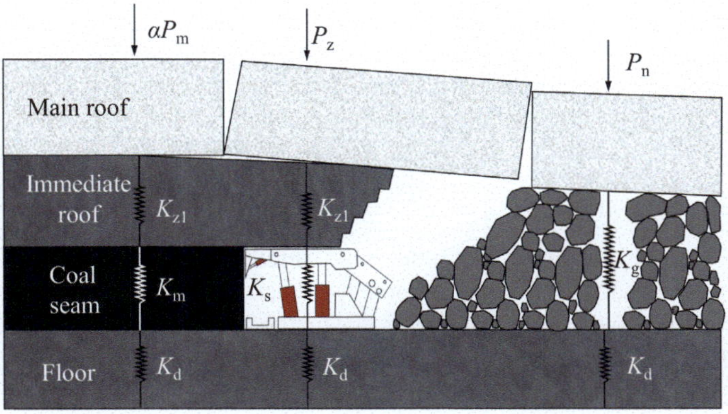

Fig. 6.33 System stiffness model of the coal seam, hydraulic support, and surrounding rock

be viewed as a rigid body, marking the upper boundary of the coal seam, support, and surrounding rock system. Vertically, the hydraulic support and surrounding rock system can be considered a series system made up of the immediate roof with certain stiffness, the coal seam, and the floor for System 1; or the immediate roof, support, and floor for System 2; or the goaf and floor for System 3. Horizontally, it can be seen as a parallel arrangement of Systems 1, 2, and 3. Within the entire interactive system of support and surrounding rock, the performance of the support can be optimized if the deformation and stiffness are in harmony. The stiffness model is depicted in Fig. 6.33.

ap_m represents the vertical stress on the main roof of System 1 ahead of the rib, and for a conservative calculation, this value can be equivalent to the peak of the abutment pressure in front of the rib. P_z is the vertical roof pressure experienced by System 2 in the roof-control area, which can be calculated based on the deflection of the rock layer under intact conditions; in panels with thin basic rock and thick loose layers, it can be computed based on the weight of the loose sand within the arch after the formation of the arch in the loose layers. P_n is the average vertical stress on System 3 in the goaf, usually less than the in-situ rock stress, and for conservative calculations, it can be assumed to be equivalent to the in-situ rock stress. K_{Z1} is the immediate roof stiffness in System 1; K_{Z2} is the immediate roof stiffness in System 2; K_g is the stiffness of the loose gangue; K_m is the stiffness of the coal seam; Ks is the stiffness of the support [10].

1. Support stiffness

Hydraulic supports are composed of a base, hydraulic cylinders, a canopy, and a rear four-bar linkage structure. Their mechanical characteristics are collectively affected by the hydraulic cylinders and the overall configuration. The stiffness of hydraulic supports typically refers to the stiffness during the stage where the hydraulic cylinders increase resistance. The higher the stiffness of the hydraulic support, the greater the resistance increase per unit of plunger compression, and the better the control effect on the roof, that is,

$$K_s = N \cdot k \cdot \cos\beta \tag{6.85}$$

where N represents the number of hydraulic support columns; k is the stiffness of the support columns; β is the angle between the support columns and the vertical direction.

2. Immediate roof stiffness

The stiffness of the immediate roof reflects its overall mechanical characteristics and its load-bearing capacity. Previous research has determined that the magnitude of the immediate roof stiffness is dictated by its elastic modulus E and the actual size ratio n, that is, $K_{\mathrm{Z1}} = E/\mathrm{n}$, where n is the ratio of the thickness of the immediate roof to its bearing width.

3. Coal seam stiffness

The degree of fracturing within the coal body, coal quality, the quantity of parting, and the strata can influence its stiffness K_{m}. Coal seams with varying stiffnesses exhibit differing levels of stability in the coal wall during mining; the lower the stiffness, the more susceptible it is to damage, and conversely, the coal wall remains more stable. In practical mining operations, aside from grouting to reinforce the coal wall in panels with severe rib spalling, it is typically difficult or unnecessary to alter the stiffness.

4. Goaf stiffness

Owing to the intricate nature of roof collapses in the goaf area, the goaf stiffness K_{m} refers to the ability of the post-collapse gangue structure to resist elastic deformation when subjected to force. Its magnitude is associated with factors such as the collapse height of the immediate roof, the particle size of the gangue, the thickness of the gangue accumulation, and the degree of compaction.

5. Floor stiffness

Low floor stiffness can readily cause the support base to sink, such as with the coal seam floor, which is akin to increasing the plunger subsidence when the column resistance is constant, thus reducing the stiffness of the support system. Floor stiffness K_d can be deduced through analysis of the specific pressure that reflects the floor's compressive properties. Floors of different rock types possess varying stiffnesses. To construct a coupled model of the support and surrounding rock stiffness, the following assumptions are established:

(1) The vertical stresses experienced by the coal seam, support, and surrounding rock system are assumed to be equal in magnitude prior to roof fracture or prior to structural instability caused by roof fracture, i.e., $\alpha p_m = P_z = P_n$.
(2) The floor stiffness is considered infinitely large, i.e., $K_d \to \infty$, and thus can be regarded as a rigid body.
(3) The impact of mining activities on the establishment of the system stiffness model is disregarded.

(4) The heights of the coal seam in systems 1 and 2 are equal to the height of the panel, with the immediate roof thickness being uniform.

For the coal seam, support, and surrounding rock system, the three subsystems can be individually analyzed and then combined for stiffness coupling to address various practical issues. Issue one: Discuss the relationship between the hydraulic support plunger subsidence and the deformation of the rib. Issue two: Discuss the coupling relationship between the hydraulic support plunger subsidence and the deformation of the gangue in the goaf.

6.3.2.2 Relationship Between Support Subsidence and Coal Wall Deformation

In the analysis of deformation issues, the following physical quantities are symbolically defined: K_1 is the stiffness of System 1, K_2 is the stiffness of System 2, K_3 is the stiffness of System 3; Δs_1 is the total deformation of System 1, Δs_2 is the total deformation of System 2, Δs_3 is the total deformation of System 3; Δs_{z1} is the deformation of the immediate roof in System 1, Δs_{z2} is the deformation of the immediate roof in System 2; Δs_g is the deformation of the goaf, Δs_m is the deformation of the coal seam, Δs_s is the deformation of the hydraulic support.

There are two assumptions in the discussion of Issue One. Assumption 1: The total deformation of Systems 1 and 2 is the same, i.e., $\Delta s_1 = \Delta s_2$; Assumption 2: The total deformation of the systems is not the same but the stiffness of the immediate roof is the same, i.e., $\Delta s_1 \neq \Delta s_2$, $K_1 = K_2$.

1. Discussion of Assumption 1

Since the premise of Assumption 1 is that the floor behaves as a quasi-rigid body, the stiffness of System 1 is determined by the combined stiffness of the immediate roof and the hydraulic support.

$$\frac{1}{K_1} = \frac{1}{K_{z_1}} + \frac{1}{K_{\mathrm{m}}} \tag{6.86}$$

According to Hooke's Law, it obtains,

$$\frac{K_{z_1} K_m}{K_{z_1} + K_m} \Delta s_1 = K_{z_1} \Delta s_{z_1} = K_m \Delta s_m = \alpha P_m \tag{6.87}$$

where α is the concentration coefficient for front abutment pressure. A substantial number of actual measurements have indicated that the stress concentration coefficient typically falls within the range of 1–2. In the stiffness coupling analysis, the position where System 1 is located adopts a stress coefficient α of 1.5 to deduce the corresponding approximate conclusion.

Similarly, for System 2, we have,

$$\frac{1}{K_2} = \frac{1}{K_{z_2}} + \frac{1}{K_s} \tag{6.88}$$

$$\frac{K_{z_2} K_s}{K_{z_2} + K_s} \Delta s_2 = K_{z_2} \Delta s_{z_2} = K_s \Delta s_s = P_z \tag{6.89}$$

Since under this assumption $\Delta s_1 = \Delta s_2$, then $\frac{\Delta s_m/\Delta s_1}{\Delta s_s/\Delta s_2} = \frac{\Delta s_m}{\Delta s_s}$, combining Eqs. (6.86) with (6.89) we get:

$$\frac{\Delta s_m}{\Delta s_s} = \frac{\alpha}{K_m/K_s} \tag{6.90}$$

Let $x = K_m/K_s$, $f_1(x) = 1/x$, then the function $f_1(x)$ has a domain of $(0, +\infty)$ and a range of $(0, +\infty)$. Construct the composite function $Z(x) = f_1(x)/(1 + f_1(x))$, which has a domain of $(0, +\infty)$ and a range of (0, 1). This maps the deformation ratio onto the composite function. According to probabilistic logic, a value typically below 20% suggests a lower likelihood of an event, whereas a value above 80% suggests a higher likelihood. Solving for the domain range of the function $Z(x)$ yields: when $x < \alpha/4$, $Z(x) > 80\%$; when $x > 4\alpha$, $Z(x) < 20\%$; and when $\alpha/4 \leq x \leq 4\alpha$, $20\% \leq Z(x) \leq 80\%$. Assuming $\alpha = 1.5$, the following conclusions can be drawn:

(1) When $K_s > 4km/\alpha$, that is, $K_s > 8km/3$, the deformation of the coal seam and the support is primarily determined by the coal seam.
(2) When $K_s < K_m/4\alpha$, that is, $K_s < K_m/6$, the deformation of the coal seam and the support is primarily determined by the support.
(3) When $K_m/4\alpha \leq K_s \leq 4km/\alpha$, that is, $K_m/6 \leq K_s \leq 8km/3$, the deformation of the coal seam and the support is shared by both.

2. Discussion under Assumption 2

Assumption 2 considers that the total deformation of the systems is different and the stiffness of the immediate roof is the same, i.e., $\Delta s_1 \neq \Delta s_2$, $K_{z1} = K_{z2}$. Hence, the ratio of deformation proportions is not the ratio of absolute deformations. Combining Eqs. (6.87) with (6.89) yields,

$$\frac{\frac{\Delta S_m}{\Delta S_1}}{\frac{\Delta S_s}{\Delta S_2}} = \frac{\frac{K_{z1}}{K_{z1}+K_m}}{\frac{K_{z2}}{K_{z2}+K_s}} = \frac{K_{z1}}{K_{z2}} \cdot \frac{K_{z2} + K_s}{K_{z1} + K_m} \tag{6.91}$$

Given the assumption that $K_{z1} = K_{z2}$, let's use K_{z1} to represent its value for now. In the analysis of actual problems, the stiffness of the support can typically be controlled by humans, while geological conditions such as the occurrence of coal seams are set. Thus, considering K_m, K_{z1} has known, K_s is the variable. Define $x = K_s/K_m$, $f_2(x) = \Delta s_m/\Delta s_s$, and by rearranging Eq. (6.91), $f_2(x) = (x/C_1)\text{-}(1/C_1) + 1$ can be got, where $C_1 = 1 + K_{z1}/K_m$, and $C_1 \in (1, +\infty)$.

Fig. 6.34 Function graph of $f_2(x)$

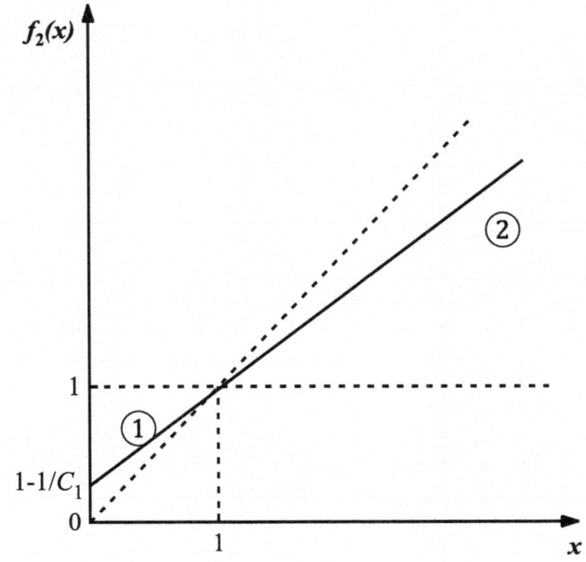

3. Graphical analysis

Since Systems 1 and 2 are identical when the stiffness of the coal seam and the support are the same, the relative deformation ratio of the coal seam and the support is 1 at this point, meaning the function $f_2(x)$ must pass through the point (1, 1). As the function is a linear univariate function in x, it is a straight line with a y-axis of $1–1/C_1$. Hence, the graph of the function (Fig. 6.34) must revolve around the point (1, 1) and will not extend beyond the two dashed lines that enclose areas ① and ②. Therefore, the following conclusions are obtained:

(1) When $K_s > K_m$, $1 < \frac{\Delta S_m/\Delta S_1}{\Delta S_s/\Delta S_2} < \frac{K_s}{K_m}$, the relative deformation ratio of the coal seam and the support is greater than 1 and will not surpass the inverse of their stiffness ratio.
(2) When $K_s < K_m$, $\frac{K_s}{K_m} < \frac{\Delta S_m/\Delta S_1}{\Delta S_s/\Delta S_2} < 1$, the relative deformation ratio of the coal seam and the support is less than 1 and will not fall below the inverse of their stiffness ratio.
(3) When $K_s = K_m$, $\frac{K_s}{K_m} < \frac{\Delta S_m}{\Delta S_s}$, the relative deformation ratio of the coal seam and the support equals 1 and is equivalent to the inverse of their stiffness ratio.

6.3.2.3 Relationship Between Plunger Subsidence and the Deformation of Goaf

In the discussion of this issue, it is assumed that the deformation of Systems 2 and 3 is equal, that is, $\Delta s_2 = \Delta s_3$. Then, by combining the equal deformation of Systems 2 and 3 and establishing Eq. (6.89), we obtain:

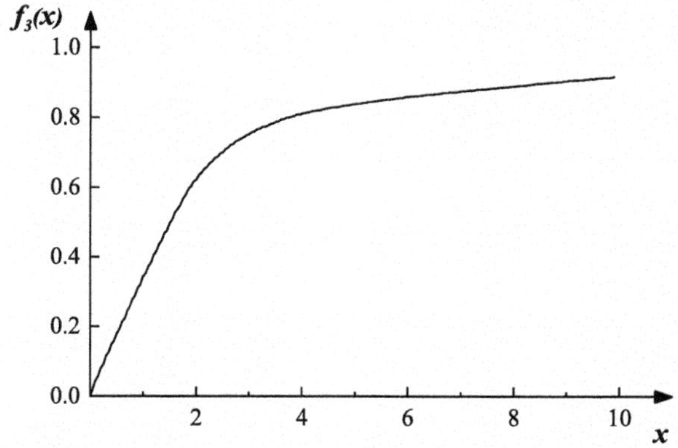

Fig. 6.35 Function graph of $f3(x)$

$$\frac{\Delta S_s}{\Delta S_3} = \frac{\Delta S_s}{\Delta S_2} = \frac{K_{z2}}{K_s + K_{z2}} = \frac{K_{z2}/K_s}{1 + K_{z2}/K_s} \tag{6.92}$$

Let $x = K_{z2}/K_s, f3(x) = x/1 + x$, the function graph is shown in Fig. 6.35. Graphical analysis leads to the following conclusions:

(1) When $\Delta s_s/\Delta s_3 > 80\%$, that is, $K_s > 0.25K_{z2}$, the deformation of the goaf and the support is primarily determined by the support.
(2) When $\Delta s_s/\Delta s_3 < 20\%$, that is, $K_s < 4K_{z2}$, the deformation of the goaf and the support is primarily determined by the goaf.
(3) When $20\% \le \Delta s_s/\Delta s_3 \le 80\%$, that is, $0.25K_{z2} \le K_s \le 4K_{z2}$, the deformation of the goaf and the support is jointly borne by both.

The analysis reveals that within the support and surrounding rock interaction system, the support significantly influences the overall system stiffness. It is possible to determine a suitable support stiffness through support design and selection, ensuring the support is neither crushed nor overwhelmed by the pressure of the rib, and possesses sufficient stiffness to control the dynamic load of a hard and thick roof's collapse. Therefore, under hard roof conditions, the selection of hydraulic supports should fully consider the impact of dynamic loads and also account for the support's compression deformation characteristics, that is, to comprehensively consider the coupling effect of support dynamic load and stiffness, which is a critical factor in the control of panel surrounding rock and support selection.

If reasonable support selection still cannot control the dynamic load formed by the collapse of a hard roof, it is necessary to consider treating the hard roof, such as hydraulic fracturing, pre-splitting blasting, etc., to reduce the step distance of the main roof's fracture, thereby decreasing the dynamic load, although this is relatively challenging to implement in engineering. When selecting supports, it is important to consider not only sufficient support strength but also stiffness that matches the surrounding rock. Hence, under hard roof conditions, the roof load should be calculated using the dynamic load method to determine the appropriate support strength; also, consider the subsidence when the hard main roof collapses to determine a reasonable support stiffness, ensuring the support is not crushed and does not cause damage due to excessive stiffness. Therefore, under hard roof conditions, hydraulic supports should consider the dynamic load method and stiffness coupling double factors in support selection.

In engineering, reasonable support selection is fundamental and should be complemented by enhanced panel management, such as timely support movement and sufficient initial support. If this still does not meet production needs, necessary presplitting measures for the roof can be taken to shorten the weighting interval distance. During the recovery of the panel, the advance speed of the panel can also be appropriately increased to mitigate the ground pressure behavior at the panel, however, excessively fast advance speeds may also exacerbate the instability of a hard roof's collapse.

6.3.3 *Determination of Working Resistance in Different Roof Structures*

6.3.3.1 Determination of the Working Resistance for the Masonry Beam Structure of Key Stratum

When key stratum is positioned at a considerable distance from the coal seam, a masonry beam structure can typically form. At this time, the key stratum has a relatively small space for rotation, and the ground pressure behavior at the panel is similar to that of a panel with a typical mining height. Hence, the working resistance of the supports can be approximated using the 4–8 times mining height rock weight method or the theoretical equation based on the equilibrium relationship of the masonry beam structure. However, due to the substantial mining height in extra-large mining height panels, the range of values calculated by the 4–8 times mining height rock weight method is quite broad, making it challenging to ascertain a reasonable value, however, it can serve as a reference for other computational methods. When performing calculations based on the equilibrium relationship of the masonry beam structure, the calculation equation is:

$$P = Bl_k \sum h_i \gamma + \left[2 - \frac{l \tan(\varphi - \theta)}{2(h - \delta)} \right] Q_0 B \tag{6.93}$$

where B represents the width of the support; l_k is the roof-control distance; $\sum h_i$ is the thickness of immediate roof below the key stratum; h is the thickness of key stratum; l is the fracture step of key stratum; φ and θ represent the friction angle and the breakage angle between rock blocks, respectively; δ represents the subsidence of the broken rock block; Q_0 is the load of the key stratum's broken rock block itself and the overlying controlled rock layer; P is the working resistance of the support; γ is the unit weight of the rock layer.

6.3.3.2 Determination of the Working Resistance for the Cantilever Beam Structure of Key Stratum

When key stratum is near the coal seam and situated within the caving zone of the overlying rock, they will fracture in the form of a cantilever beam structure. As the rotation of the overlying strata diminishes, the hinged rock layer of the fracture zone may emerge at a certain location between the two-sub key stratum. At this time, the calculation model for the support resistance is depicted in Fig. 6.36. Since the sub key stratum cannot form a stable structure during the process of fracture and rotation, the support resistance should ensure that it does not slide or become unstable, prevent delamination between the caving zone rock layers and the fracture zone rock layers, and also provide an acting force to the lower hinged rock layer of the fracture zone to balance part of its load and ensure the structural stability [11].

Therefore, when the key stratum forms a cantilever beam structure, the support resistance should be calculated in two parts: one part for the weight of the caving zone rock layers, and the other for the balancing force of the hinged rock layer below the fracture zone. The load of the caving zone rock layers should be calculated in two parts: the load Qz of the sub key stratum's immediate roof should be calculated based on the support's roof control distance, and the load Q_1 of the sub key stratum 1 and the overlying strata up to the top interface of the caving zone should be calculated based on the breakage length of the sub key stratum. The balancing force P_{H1} required by the hinged rock layer below the fracture zone from the support can be calculated using the masonry beam structural theory equation. The calculation equation for the support working resistance is as follows:

$$P = Q_z + Q_1 + P_{H1} \tag{6.94}$$

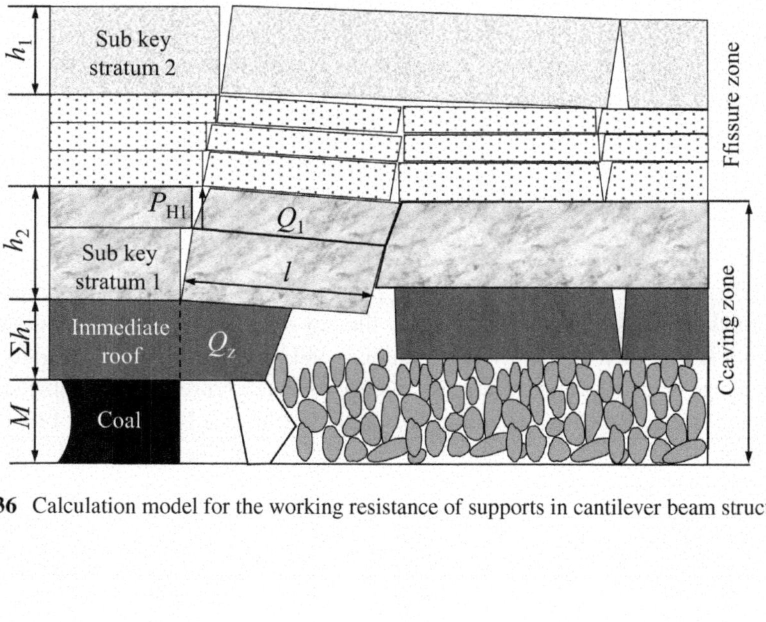

Fig. 6.36 Calculation model for the working resistance of supports in cantilever beam structure

Among them:

$$Q_z = Bl_k \sum h_i \gamma, Q_1 = Blh_2 \gamma, P_{H1} = \left[2 - \frac{l \tan(\varphi - \alpha)}{2(h - \delta_r)}\right] Q_r B$$

where h_2 represents the thickness of the sub key stratum and the rock layers within the caving zone above it, δ_r is the subsidence of the hinged rock blocks at the base of the fracture zone, Q_r is the load of the hinged rock blocks at the base of the fracture zone itself and the overlying controlled rock layers, and α is the break angle of immediate roof.

6.3.3.3 Determination of the Working Resistance When Sub Key Stratum Affecting the Strata Pressure

If the fracture of the sub key stratum 2 impacts the fracture of the lower sub key stratum 1 and the strata pressure in the panel, the calculation of the resistance of the hydraulic support must consider the role of the sub key stratum 2. When performing the calculation, it should be based on the critical scenario where the sub key stratum 1 is considered a cantilever beam structure, and its support resistance calculation model is depicted in Fig. 6.37.

When the fracture of the sub key stratum 2 impacts the fracture of the sub key stratum 1, it compels the load of the rock layers between the two key strata to be exerted onto the fractured block of the sub key stratum 1. Consequently, at this juncture, the load on the support will consist of three components: the load Q_z of the immediate roof beneath the sub key stratum 1, the rock weight Q_2 within the fracture

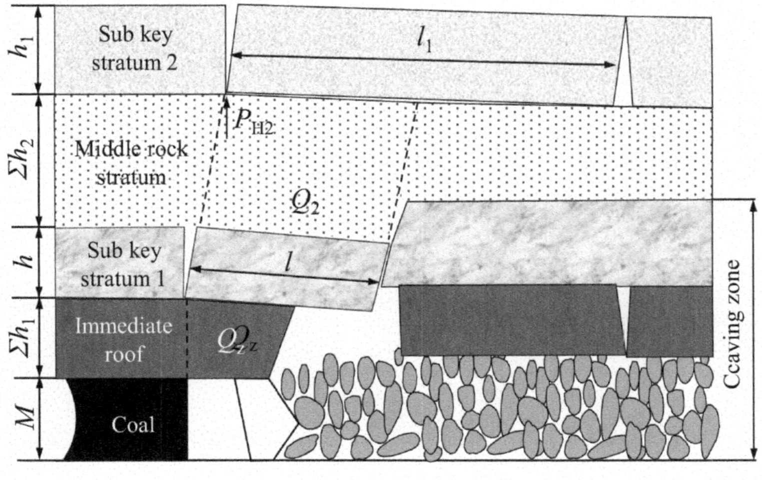

Fig. 6.37 Calculation model for the working resistance of supports when sub key stratum 2 affects strata pressure

interval between the sub key stratum 1 and the sub key stratum 2, and the balancing force P_{H2} necessary to balance the hinged structure of the sub key stratum 2.

Therefore, when the fracture of the sub key stratum 2, which is near the upper region, influences the strata pressure, the calculation equation for the working resistance of the support is as follows:

$$P = Q_z + Q_2 + P_{H2} \tag{6.95}$$

Among them:

$$Q_z = Bl_k \sum h_i \gamma, Q_2 = Bl\gamma(h + \sum h_2), P_{H2} = \left[2 - \frac{l_1 \tan(\varphi - \alpha)}{2(h_1 - \delta_1)}\right] Q_3 B$$

where $\sum h_2$ denotes the total thickness of the rock layers situated between the two sub key strata, h_1 is the thickness of the sub key stratum 2, δ_1 signifies the subsidence of the hinged rock blocks within the sub key stratum 2, and Q_3 accounts for the load of the hinged rock blocks of the sub key stratum 2 itself along with the superincumbent controlled rock layers.

6.4 Stress Distribution and Strata Control in Deep Buried Thin Bedrock Panel

Thick coal seams under thin bedrock in deep deposits are widely distributed in the Jiaozuo coal field in Henan, the Juye coal field in Shandong, and the Huainan coal field in Anhui. These seams are classified as high-quality coal resources in eastern China's mining areas. Safe and efficient development of such thick coal seams ensures the growing energy demand in the eastern region. However, the unique occurrence conditions significantly increase the mining difficulty of these thick coal seams. Thick alluvial layers have poor bearing capacity, acting on the thin bedrock roof as a dynamic load, leading to high support resistance at the panel and noticeable dynamic load effects. Mining-induced fractures originating from the high-level thick alluvial layers rapidly extend downward, causing the thin bedrock to break and form mining-induced fractures that penetrate the panel and the alluvial layers. These large opening mining-induced fractures create water and sand flow channels, triggering water and sand inrush disasters that threaten production safety.

6.4.1 *Project Background*

6.4.1.1 Coal Seam Occurrence and Mining Conditions

Zhaogu No. 2 Mine's 14,030 panel is buried at a depth of 750–800 m, with an inclination length of 200 m and a strike length of 2100 m. The upper side of the panel is the goaf area of 14,021, while the lower side is an unmined solid coal area. The 14,030 panel primarily extracts coal seam No. 2–1, with a coal seam thickness ranging from 3.9 to 6.5 m, averaging 5.9 m, and a seam dip angle of 3–5°. This is classified as a near horizontal thick coal seam, with a hardness coefficient of 1.25–1.62. The mining process employs a large mining height method and roof management through the caving method. The bedrock thickness at the panel varies from 0 to 50 m, with multiple regions exhibiting bedrock absence. The immediate roof consists of interbedded mudstone and sandy mudstone with a thickness of 3.6–7.3 m, featuring well developed horizontal bedding that can cave immediately upon mining, with a uniaxial compressive strength of 18 MPa. The main roof comprises thick sandstone of 6.2–13.5 m, with mud-siliceous cementation and numerous mud bands, weak water-bearing properties, and a uniaxial compressive strength of 36 MPa. Above the main roof, there are inter bedded mudstone and coarse sandstone layers with individual layer thicknesses of less than 5 m and noticeable pore structures. Above the bedrock is an alluvial layer, containing less than 2.0 m thick and discontinuously distributed mudstone, sandy mudstone, and coarse sandstone layers, with an overall thickness of up to 700 m. This panel is classified as a deep buried, thick alluvial layer with weakly cemented thin bedrock and thick coal seam. The geological conditions and mining characteristics are shown in Fig. 6.38. The panel uses ZY18000/30/65D hydraulic supports for roof support, with a support strength of up to 1.9 MPa.

6.4.1.2 Characteristics of Dynamic Pressure on the Roof

The mining practice of the 14,030 panel indicates strong manifestations of mining pressure under the influence of thick alluvial layers and deep-buried weakly cemented thin bedrock. Hydraulic supports frequently endure dynamic load effects caused by roof collapses, with a dynamic load coefficient reaching 2.1 and instances where safety valves on columns have been observed to open. During the advancement of the panel, the spatial distribution characteristics of support resistance are shown in Fig. 6.39a: there is a clear periodicity in roof pressure, with an initial interval of 42 m for the first pressure cycle and subsequent cycles occurring at intervals ranging from 8 to 15 m. The support resistance exhibits a pattern of higher resistance in the middle of the panel and lower at both ends. Influenced by the dynamic load effects of the main roof, local peaks in support resistance are noticeable, with the maximum column load reaching 50 MPa.

During three instances of roof pressure intervals, the distribution of column loads for the #66 hydraulic support is depicted in Fig. 6.39b. In normal advancing stages,

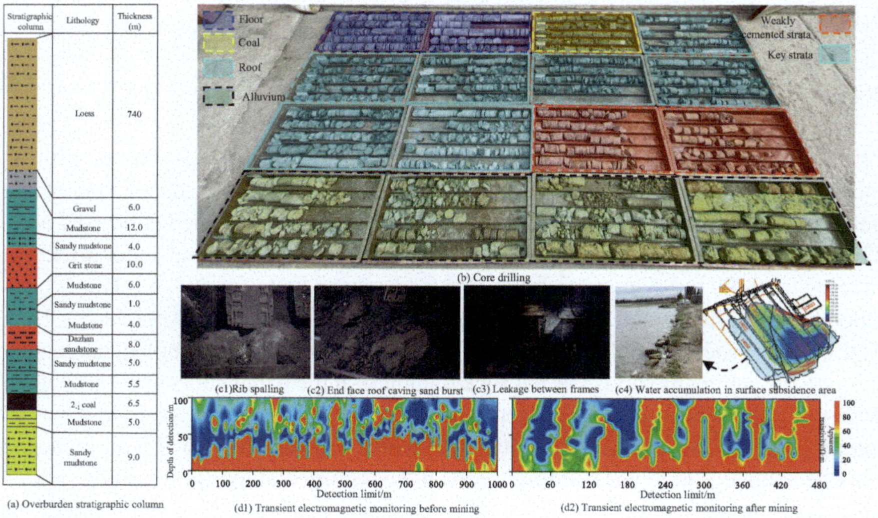

Fig. 6.38 Geological and mining conditions of 14,030 panel in Zhaogu No. 2 Coal Mine

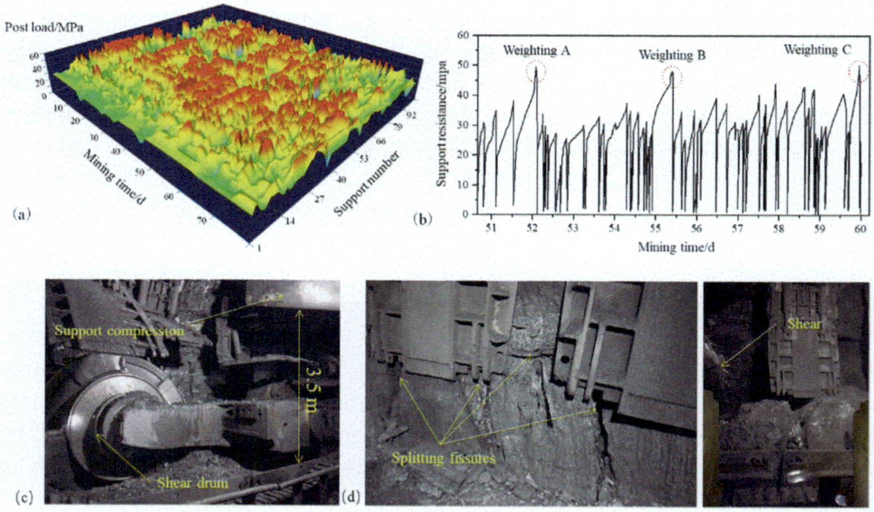

Fig. 6.39 Strong occurrence of mining pressure in the panel

the trend of roof pressure increase is relatively stable, however, the speed of pressure application is rapid, causing a sudden spike in support load. The peak curve is steep, indicating significant dynamic load effects on the roof. Under severe mining pressure, the 14,030 panel has experienced multiple incidents of frame pressure, reducing the effective support height to 3.5 m. With a shearer drum diameter of 3.0 m, shearer cannot pass under the compressed supports, disrupting normal production, as seen in Fig. 6.39c. The dynamic roof load is similarly transferred to the rib, leading to the development of longitudinal splitting cracks in the rib. Large coal blocks detach from the rib, obstructing the shearer's passage and reducing operational efficiency, as shown in Fig. 6.39d.

Field research indicates that during non-dynamic load stage in deep buried weakly cemented thin bedrock mining areas, column pressures on hydraulic supports range between 30 and 40 MPa with minimal column compression. However, under dynamic load conditions, columns rapidly compress, reaching pressures up to 50 MPa. This represents a low static load-high-dynamic load scenario. If safety valves do not open promptly, there is a risk of cylinder bursting, highlighting the significant challenges in roof control. The characteristics of mining pressure in deep-buried weakly cemented thin bedrock mining areas differ from those in shallow buried weakly cemented thin bedrock mining areas, which typically exhibit high-static load and low-dynamic load types. In these scenarios, roof pressure mainly results from static loads, and column compression rates are slow. Dynamic loads only cause intense vibrations in supports, allowing sufficient time for safety valves to open and ensuring that supports maintain adequate strength and compressibility to effectively control the roof [12].

6.4.2 *Characteristics of Roof Breakage Under Deep Buried Thin Bedrock*

6.4.2.1 Collapse Form of Overlying Strata Under Deep Buried

Figure 6.40 illustrates the characteristics of roof failure and movement in deep buried weakly cemented thin bedrock thick coal seam panels. The set-up room is set 30 m from the model boundary with a span of 10 m. Timely support with props is employed to prevent roof collapse in large-span set-up rooms, as shown in Fig. 6.40a. When the panel advances to 30 m, the immediate roof collapses with a height of about 7 m, forming a trapezoidal boundary shape. Supported by the panel props, there is a noticeable difference in the angles of roof breakage between the front and rear sides of the goaf, measuring 60° and 45° respectively, as seen in Fig. 6.40b. After the initial collapse of the immediate roof, the underlying main roof loses support. As the panel advances to 50 m, the main roof reaches its limit span, leading to tensile fracture. Rock blocks fracture and sink, generating high horizontal squeezing forces at the hinge points. Weakly cemented rocks undergo secondary damage without forming a stable structure, as depicted in Fig. 6.40c. Following the initial fracture of the main roof, the collapse of the overlying strata forms a trapezoidal shape, with a collapse height of 13.7 m. The fracture angles of the rock layers on both sides of the goaf increase to 70° and 50° respectively, indicating significant influence of hydraulic supports on the initial fracture of the main roof. At this point, the exposed span of the thick alluvial layer is small, with only a phenomenon of high-level mining-induced fractures, as shown in the magnified area of Fig. 6.40c. As the panel continues to advance by 10 m, the main roof undergoes a second fracture, and the alluvial layer collapses for the first time, forming a symmetrical arch of roof fall with a height of up to 24.6 m, as shown in Fig. 6.40d. The consistent fracture angles of the rock layers on both sides of the goaf indicate a weak control effect of supports on the main roof fracture. The expansion trajectory of the overlying strata mining-induced fractures indicates that the second fracture of the main roof is caused by the downward expansion of alluvial layer mining-induced fractures. Rapidly expanding fractures reach the interface between the alluvial layer and the main roof and cross this interface under inertia, forming inertial cracks. This deteriorates the bearing capacity of the main roof cantilever beams. The upward expansion of mining-induced fractures causes the collapse of the alluvial layer, rapidly transferring load to the main roof containing inertial cracks. The inertial cracks exhibit Type II expansion, resulting in shear fracture of the main roof and causing intense roof pressure in the stope.

As the mining range of the coal seam continued to increase to 75 and 90 m, the main roof experienced two cycles of periodic fractures, with a fracture spacing of

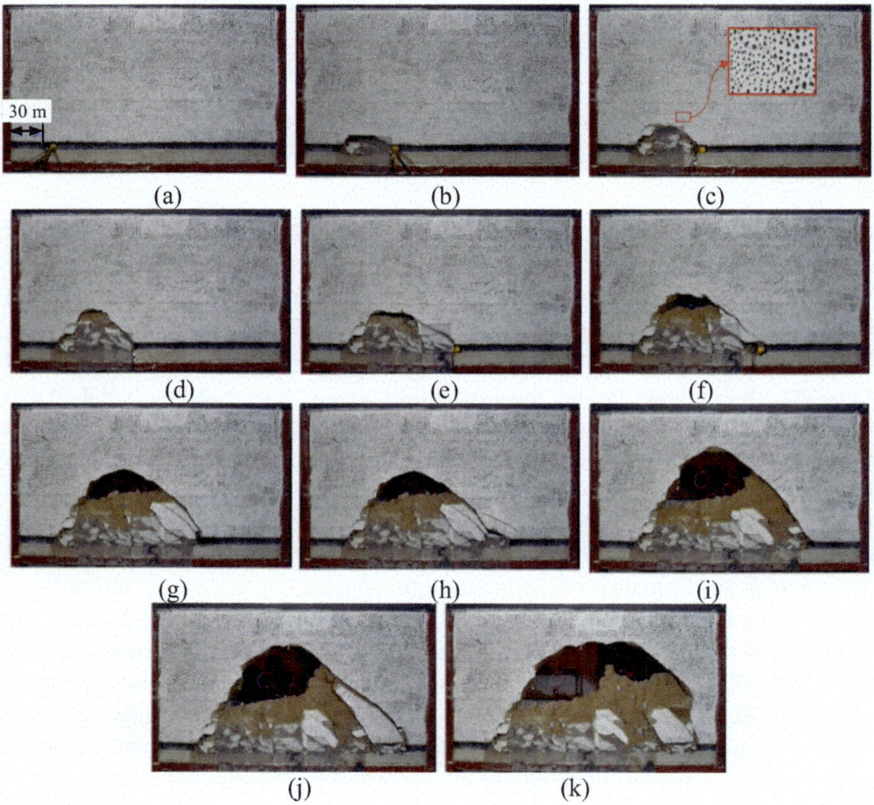

Fig. 6.40 Evolution of the boundary shape after caving of weakly consolidated overburden: **a** 10 m; **b** 30 m; **c** 50 m; **d** 60 m; **e** 75 m; **f** 90 m; **g** 100 m; **h** 110 m; **i** 125 m; **j** 140 m; **k** 145 m

15 m. High-level mining-induced fractures also originated in the alluvial layer, where upward expansion triggered localized collapses of the alluvial layer, maintaining the overall stability of the overlying strata collapse arch. Under the load of the alluvial layer, the inertial cracks in the main roof exhibited Type I expansion, with fracture angles smaller on the set-up room side. Newly formed collapse arches showed asymmetric shapes, with arch heights increasing to 28.5 m and 38.7 m respectively. These features indicate that at this stage, the collapse arches of the overlying strata have self-stabilizing capability. The load on the main roof from the alluvial layer is relatively small, and the influence of supports on the fracture angles of the roof is significant. The main roof undergoes tensile fractures, and the roof pressure on the panel is relatively moderate. As the panel advanced to 100 m, the collapse arches of the overlying strata reached their maximum span, and upward expansion of mining-induced fractures caused instability in the collapse arch structure. Extensive collapses occurred in the alluvial layer, rapidly transferring gravitational load to the main roof. The inertial cracks in the main roof exhibited Type II expansion, with fracture angles consistent with those on the set-up room side. This indicates a weak control effect of supports on rock layer fractures. Newly formed collapse arches showed symmetric shapes, with arch heights increasing to 58.1 m. The main roof underwent shear fractures, resulting in high intensity roof pressure on the panel. Subsequently, as the panel advanced to 110, 125, 140 m, and 145 m, the main roof experienced four instances of fracture. At positions 110 and 140 m, the collapse arches of the overlying strata had self-stabilizing capability, with localized collapses in the alluvial layer. Newly formed collapse arches showed asymmetric shapes, and the main roof underwent tensile fractures, with relatively moderate roof pressure on the panel. At positions 125 and 145 m, the collapse arches of the overlying strata reached their limit state and experienced overall instability. Extensive collapses occurred in the alluvial layer, and newly formed collapse arches showed symmetric shapes. The main roof underwent shear fractures, resulting in strong roof pressure on the panel.

As the panel advances, the shape of the collapse arch of overlying strata alternates between asymmetric and symmetric arches. During the evolution stages of asymmetric arches, localized collapses occur in the alluvial layer, with a small increase in arch height. During the evolution stages of symmetric arches, extensive collapses occur in the alluvial layer, resulting in a large increase in arch height. From Fig. 6.41, the characteristics of the span and arch height changes caused by nine instances of roof pressure on the panel are observed. As the panel advances to 145 m, the increase in arch height decreases significantly under the support of solid collapses in the goaf. Experimental results indicate that the initial fracture spacing of the main roof reached 50 m, with periodic fracture spacing ranging from 10 to 20 m. Influenced by the physical and mechanical parameters of the experimental materials and excavation spacing (5 m), the initial fracture spacing of the main roof in the experiments was slightly larger than the actual initial weighting interval distance of the 14,030 panel. However, the periodic weighting interval distances in the experiment match the actual data from the 14,030 panel, indicating that this physical simulation test effectively reproduced the characteristics of overlying strata fracture and movement during the advancement process of the 14,030 panel.

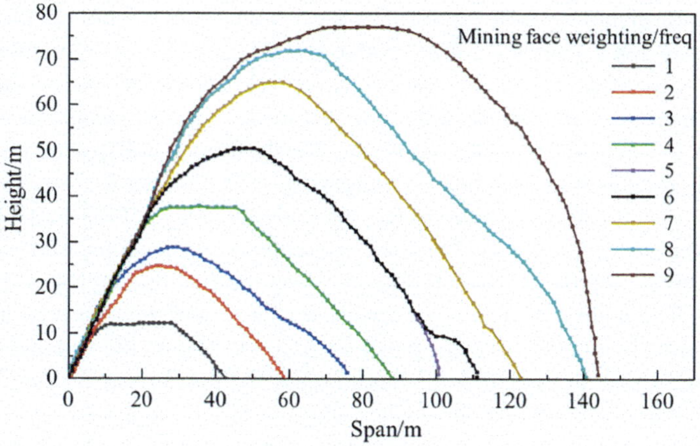

Fig. 6.41 Evolution of collapse arch of the overlying strata

6.4.2.2 Characteristics of Roof Pressure Variation in the Panel

The stability of the overlying strata collapse arch in thick alluvial layers results in two types of main roof fracture and rock movement trajectories, leading to different conditions for the hydraulic supports. The characteristics of roof pressure variation on the hydraulic supports during the 7th and 8th instances of roof pressure in the physical simulation test are shown in Fig. 6.42. During the experiment, the initial support load on the support columns was set to 2 MPa. During the 7th instance of roof pressure, the collapse arch of the overlying strata became unstable, causing shear fracture of the main roof. The fractured rock blocks directly impacted the lower immediate roof, resulting in a sharp increase in support resistance, with a peak value reaching 4.3 MPa. However, the duration of the peak load was very short, and afterwards, the support load dropped to 3.5 MPa, indicating that the supports were subjected to dynamic load effects. The instantaneous impact of dynamic loads on the roof could potentially lead to a risk of cylinder explosion if the hydraulic pressure of the support columns reaches the allowable value before the safety valve can open. During the 8th instance of roof pressure, partial instability occurred in the collapse arch of the overlying strata, leading to tensile fracture of the main roof. The support resistance increased slowly, bearing only the deformation pressure caused by the rotation and sinking of fractured rock blocks. The maximum value of the support column load was approximately 3.8 MPa, significantly lower than the dynamic load caused by shear fracture of the main roof.

6.4.2.3 Expansion Path of Overlying Strata Mining-Induced Fractures

Physical test results indicate that during the mining process of deeply buried weakly cemented thin bedrock thick coal seams, fractures in the overlying strata first originate

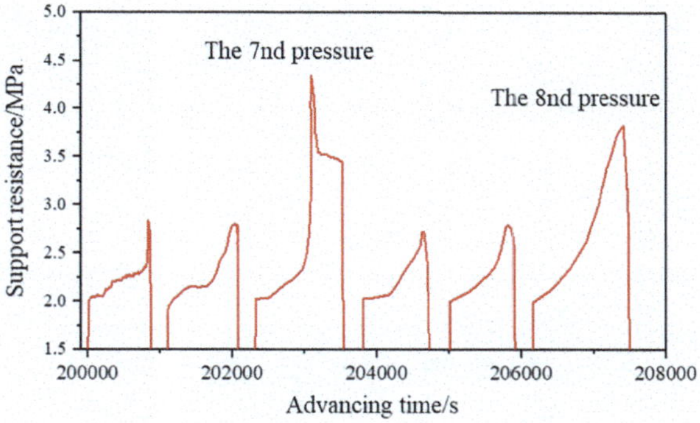

Fig. 6.42 Evolution of the support load

in the high-level thick alluvial layers. Under the influence of fracture propagation patterns, the main roof exhibits two forms of fracture: tension and shear. If the collapse arch of the overlying strata remains structurally stable, upward expansion of mining fractures causes local collapses in the alluvial layers. Downward expansion triggers tension fractures in the main roof, resulting in asymmetrical new collapse arches and mild pressure on the panel. Conversely, if the collapse arch structure of the overlying strata becomes unstable, upward expansion of mining fractures leads to extensive collapses in the alluvial layers. Downward expansion triggers shear fractures in the main roof, resulting in symmetrical new collapse arches and intense pressure on the panel. Anomalies in pressure on the panel are a focal point of this study. To understand the development of shear fractures in the main roof during basic shear fracture pressure, a Digital Image Correlation (DIC) monitoring system was used to obtain the strain field distribution in the overlying strata. The most pronounced precursor to rock failure is localized strain concentration, reaching critical values that initiate the formation and expansion of fractures. Based on the strain distribution characteristics, the expansion path of mining fractures in the overlying strata can be analyzed. The evolution of strain fields in the overlying strata during the 5th and 7th instances of pressure in the physical tests is shown in Fig. 6.43. Due to the presence of local block adhesion during the collapse process of the non-uniform simulated materials, some local blocks did not detach timely from the alluvial layer matrix. Between these local blocks and the alluvial layer matrix, discontinuous deformation already existed, causing concentrated strain phenomena around the boundaries of the collapse arch. The elastic strain values caused by stress concentration around the collapse arch differ significantly in magnitude from the discontinuous deformations. The DIC strain monitoring equipment used in this study struggled to capture these differences and was not the focus of this section's analysis.

Figure 6.43 indicates that prior to the 5th and 7th advance of the panel, fractures in the overlying strata first develop in the alluvial layer, originating asymmetrically above the impending collapse arch. Under mining-induced stress, these fractures propagate upward and downward. The time for upward propagation to the top of the collapse arch is roughly equivalent to that for down-ward propagation to its base. High-frequency DIC strain monitoring equipment did not capture

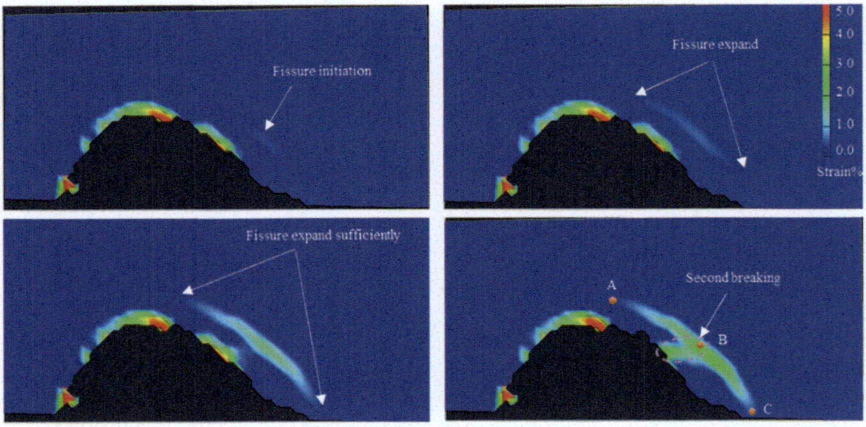

(a)

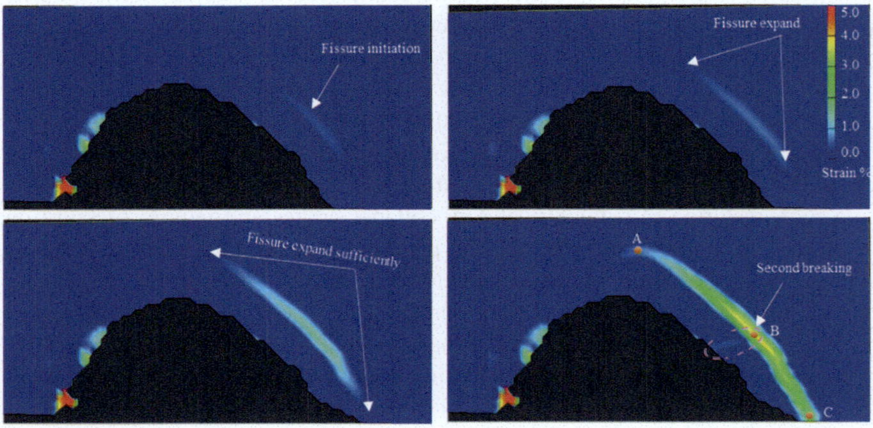

(b)

Fig. 6.43 Strain field evolution of thin bedrock: **a** the 5nd pressure; **b** the 7nd pressure

the expansion process of mining-induced fractures on the left side of the collapse arch, indicating that after ample development of fractures on the panel side, the collapse body collapses rapidly. This simultaneously triggers a sudden expansion of mining-induced fractures on the left side of the collapse arch, rapidly forming a new overlying collapse arch. During testing, secondary fracturing was observed in the alluvial layer during collapse. Figure 6.43 shows that after sufficient development of mining-induced fractures on the right side of the collapse arch, there is a noticeable concentration of strain in the fracture zone, roughly halfway up the height of the collapse arch. After the collapse arch structure becomes unstable, the load on the overlying strata above the main roof mainly consists of the gravity of the current collapse arch boundary, newly formed mining-induced fractures, and enveloped zones of secondary fractures.

According to the development characteristics of overlying strata mining-induced fractures, monitoring points *A*, *B*, and *C* are arranged at upper, middle, and lower positions on the mining-induced fractures. The evolution of maximum principal strain at these three monitoring points during the advancement of the panel is shown in Fig. 6.44. Under conditions not affected by mining activities, the initial value of maximum principal strain in the overlying strata is 0. As the mining area of the panel expands, monitoring point *B* is first affected by mining activities, with the maximum principal strain increasing first. Local strain concentration phenomena first appear at this point, indicating that mining-induced fractures in the overlying strata first originate at point *B*, specifically originating in the upper thick alluvial layer. The mining-induced fractures originating at point *B* propagate upward and downward, with the strain values at points *A* and *C* starting to in-crease after a delay of 3–5 s from point *B*. Moreover, the strain values at points *A* and *C* increase simultaneously, indicating that the mining-induced fractures originating at point B propagate to points *A* and *C* simultaneously after 3–5 s. By the 5th and 7th advance of the panel, the height of the overlying collapse arch has reached 50–60 m. The mining-induced fractures in the overlying strata originate in the middle of the height of the collapse arch. Assuming an expansion time of 4 s, the propagation speed of the mining-induced fractures reaches 7–8 m per second. Mining-induced fractures expanding at such high speed enter the main roof layer under inertial effects. Once the collapse body detaches from the alluvial matrix, the DIC monitoring system loses focus and cannot continue to capture changes in strain values at the monitoring points. Based on the coordinates of the end points of the strain evolution curves in Fig. 6.44, it is evident that the collapse body first detaches from the alluvial matrix at point B. The detachment times at points *A* and *C* are approximately the same. At the moment when the collapse body detaches from the alluvial matrix, the strain value is highest at point *B*, followed by point *C* at the arch foot, and the strain value is lowest at point *A* at the arch top.

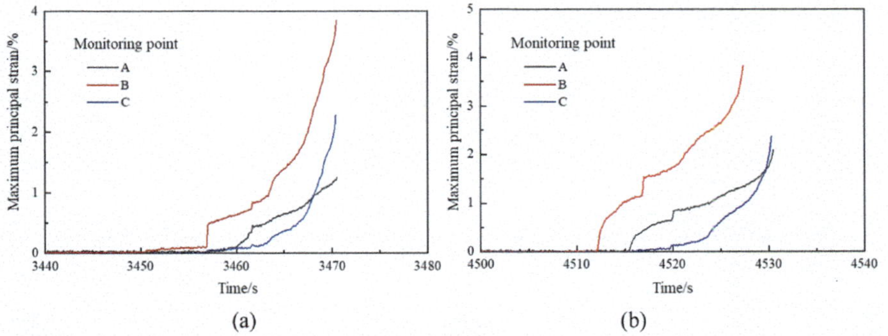

Fig. 6.44 Strain changes at the monitoring points in the panel weighting period: **a** the 5nd pressure; **b** the 7nd pressure

6.4.3 Mechanism of Dynamic Loading Effect on the Roof

The development pattern of mining-induced fractures in deeply buried, weakly cemented thin-bedded rock and thick coal seam mining areas is shown in Fig. 6.45. Mining-induced fractures first originate in the thick alluvial layer, with the initiation point located at point *O* in the right front of the overlying collapse arch.

Under external disturbances, mining-induced fractures rapidly expand upward and downward. The upward expansion fractures lead to the evolution of the overlying collapse arch shape, while the downward expansion fractures, under inertial effects, penetrate the interface between the alluvial layer and the bedrock, compromising the load bearing capacity of the main roof. The expansion trajectory of mining-induced fractures simultaneously reaches the top and foot of the collapse arch, forming new fracture lines in the collapse arch. Due to the higher strength of the main roof compared to the alluvial layer, mining-induced fractures stop expanding at point P, forming small scale inertial cracks in the main roof. After upward expansion mining-induced fractures connect with the left boundary of the collapse arch, extensive collapse of the thick alluvial layer occurs, followed by further expansion of inertial cracks, resulting in fracture of the main roof.

The mining-induced fractures in the overlying strata extend from the low strength alluvial layer to the main roof, with the expansion path crossing the interface between the alluvial layer and the main roof. The expansion mechanism at the rock layer interface is similar to the development process of fractures in weak to strong rock combinations. The upper part of the combination represents rocks with low strength, indicative of the thick alluvial layer, while the lower part represents rocks with high strength, indicative of thin-bedded rock. Under uniaxial compression conditions, the stress strain response of the two types of rocks in the combination is shown in Fig. 6.46a, with hard rock having higher strength than soft rock. When a uniaxial load is applied to the combination, the axial load first reaches the initial yield point *A* of the soft rock, where microcracks begin to form in the soft rock. The soft rock enters the nonlinear deformation stage, and with continued increase in axial load, reaches

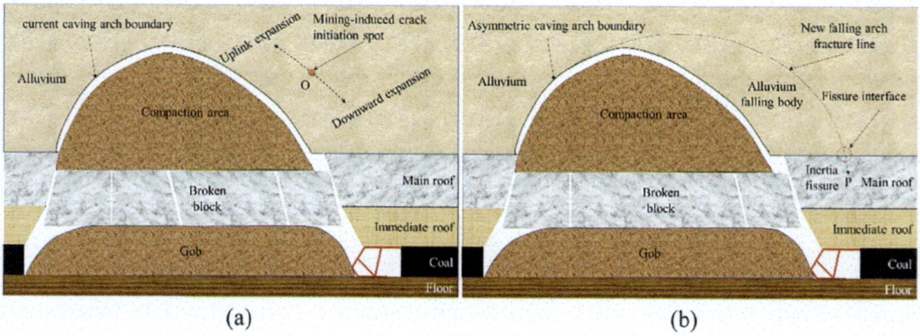

Fig. 6.45 Development mode of mining-induced fractures in deep buried thin bedrock: **a** Mining-induced crack initiation; **b** mining-induced crack propagation

point *B* where it reaches its ultimate bearing capacity. This indicates that microcracks initiated in the soft rock expand to form large-scale through type fractures and develop to the rock interface in the middle part of the combination. The peak strength *B* of the soft rock is much lower than the bearing capacity *C* of the hard rock, indicating that the hard rock at point *B'* is still in the elastic deformation stage. However, experimental results show that rapidly expanding fractures can, under inertial effects, cross the rock interface and enter the hard rock portion. This results in the sudden release of stored elastic strain energy in the hard rock, transforming it into surface energy of the fracture, driving the continued expansion of fractures in the hard rock. In the upper part of the combination, the soft rock exhibits two types of failure: axial splitting and wedge-shaped shear, as depicted in Fig. 6.46b. Both forms of failure lead to higher radial deformation in the upper soft rock compared to the lower hard rock, creating high horizontal radial shear stress at the interface between them. The appearance of radial shear stress further increases the tensile stress level at the tip of inertial cracks in the hard rock, causing Type I expansion of inertial cracks and ultimately resulting in axial splitting failure mode in the hard rock portion of the combination.

The inertial development mode of overlying strata mining-induced fractures in deeply buried thin bedded rock mining areas is similar to the fracture development mode in rock formations. However, after fractures cross the interface, the hard rock in the formation undergoes a post-peak unloading stage, while the load on the main roof tends to increase. Therefore, the subsequent expansion mode of inertial cracks in the main roof differs from the fracture mode in the formation. If the structure of the overlying collapse arch remains stable, with partial collapse of the alluvial layer, the newly formed collapse arch exhibits an asymmetric shape. During the descent of the collapse mass along line *MN*, a secondary fracture occurs, dividing the collapse mass into parts *A* and *B*. Part *A* is supported by goaf in the goaf, while the gravity in part *B* rapidly transfers downward to the main roof. Inertial cracks experience

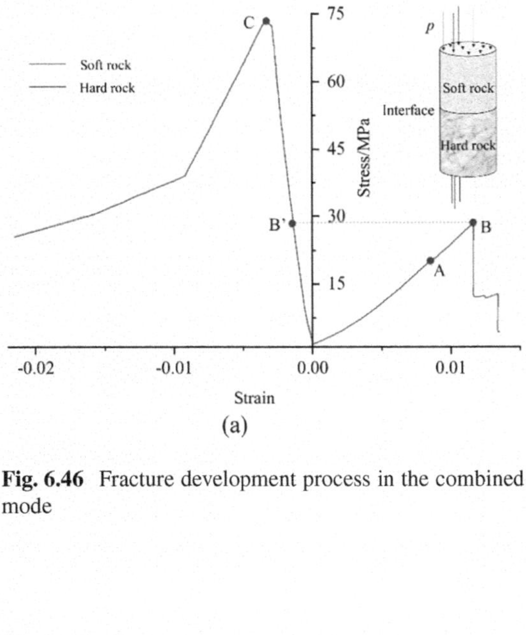

(a)

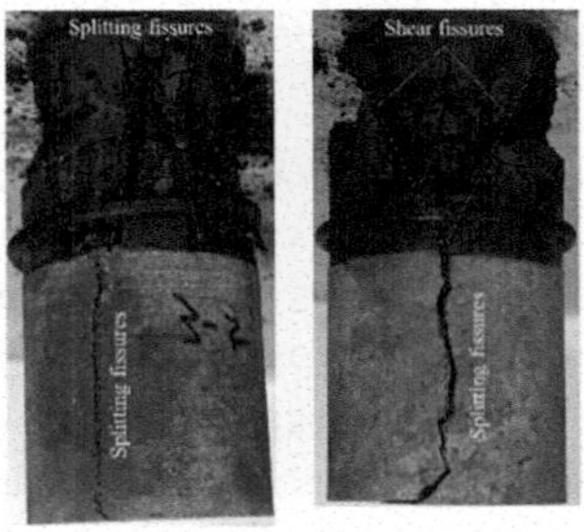

(b)

Fig. 6.46 Fracture development process in the combined sample: **a** stress–strain curve; **b** failure mode

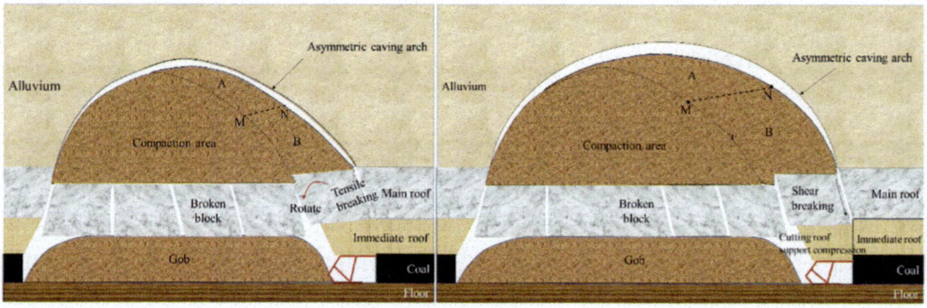

Fig. 6.47 Failure modes of main roof in deep buried thin bedrock thick coal seam: **a** tensile breaking; **b** shear breaking

Type I expansion at point *P*, causing the main roof to undergo tensile fracture. The fractured rock mass quickly rotates and forms an equilibrium structure with the rock mass in the goaf. The support structure only bears deformation pressure, as shown in Fig. 6.47a. If the structure of the overlying collapse arch becomes unstable, with extensive collapse of the alluvial layer, the newly formed collapse arch exhibits a symmetrical shape. After secondary fracture, the gravity of the collapse mass in part *B* acts on the main roof. Inertial cracks undergo Type II expansion at point *P*, causing shear fracture of the main roof. The elastic strain energy stored in it is rapidly released and transformed into initial kinetic energy of fractured rock masses. The fractured rock mass moves vertically along the shear fracture surface, unable to form an equilibrium structure with the rock mass in the goaf. This leads to dynamic contact with the lower immediate roof, generating dynamic loads far greater than the gravitational force of the fractured rock mass, transmitted to the hydraulic support at the panel, resulting in dynamic load effects on the roof, as depicted in Fig. 6.47b.

6.4.4 Methods for Determining Dynamic Loads on the Roof

6.4.4.1 Boundary Line of Collapse Arch Load of Thick Alluvial Layer

The collapse arch boundary line expression of the thick overburden layer overlying the thick coal seam stope with developed fracture characteristics of the overlying strata due to deep burial and weakly cemented thin bedrock indicates that the instability of the main roof shear rupture and the alluvial arch structure are closely related. To determine the dynamic load on the roof, it is necessary to first obtain the expression for the collapse arch boundary line. The upper bound theorem is used in this section to determine the boundary equation of the collapse arch. Parabolic and elliptical equations are commonly used to represent the collapse arch, with the latter adopted here for ease of calculation. In accordance with Prosser's arch theory results, the relationship between the arch span and arch height is $h = l/f$, where h and l are the height and half span of the collapse arch respectively, and f is the Prosser hardness coefficient of the alluvial layer. Due to the low strength of the alluvial layer, the Prosser hardness coefficient is less than 1, hence the arch height corresponds to the long axis of the elliptical collapse arch, and the span corresponds to the short axis of the elliptical collapse arch.

The equation of the initial collapse arch boundary in a thick alluvial layer:

$$\frac{1}{a_1^2}x^2 + \frac{f^2}{a_1^2}y^2 = 1 \tag{6.96}$$

In the thick alluvial layer, the development of fractures due to mining is caused by tensile stresses generated by the self-weight of the collapsing mass. According to the upper bound theorem, the ratio of energy dissipation due to the expansion of fractures in the alluvial layer is equal to the ratio of work done by the gravitational sinking of the collapsing mass. Based on these principles, the instantaneous conditions for the initial collapse in a thick alluvial layer satisfy the following:

$$\frac{1}{2}\gamma \pi a_1 h_1 = \sigma_t(\pi a_1 + 2h_1 - 2a_1) \tag{6.97}$$

where a_1, h_1 are the step length and height of the initial collapse arch in the alluvial layer, m; γ is the unit weight of the alluvial layer, N/m^3; π is the mathematical constant *pi*; σ_t is the tensile strength of the alluvial layer, MPa.

After the initial collapse, as the coal wall advances, the collapse arch in the alluvial layer undergoes periodic structural instability, leading to fundamental roof shear failure and dynamic pressure phenomena on the coal wall. The instantaneous instability conditions of the collapse arch satisfy the following:

$$\frac{1}{2}r\pi(a_i h_i - a_{i-1}h_{i-1}) = \sigma_t(\pi a_i + 2h_i - 2a_i) \tag{6.98}$$

where, a_i, h_i are respectively the half span and height of the collapse arch in the alluvial layer at the occurrence of structural instability, m.

In the thick alluvial layer, with a Prosser hardness coefficient of 0.8 and a tensile strength of 0.25 MPa, the evolution characteristics of the collapse arch can be obtained using the above method as shown in Fig. 6.48. The calculated initial collapse step length of the collapse arch is 38 m, slightly less than the initial advance step length of 14,030 panel. The calculated first periodic collapse step length of the collapse arch is 23 m, which gradually decreases to 20 m with increasing instability occurrences and stabilizes thereafter. Due to the support effect of the collapsed goaf, the height of the overlying collapse arch remains relatively unchanged after the coal wall advances 100 m. The periodic collapse step length of the collapse arch is significantly greater than the main roof fracture step length, which is caused by local collapse of the alluvial layer during the instability period of the overlying collapse arch, leading to stretching and fracture of the main roof.

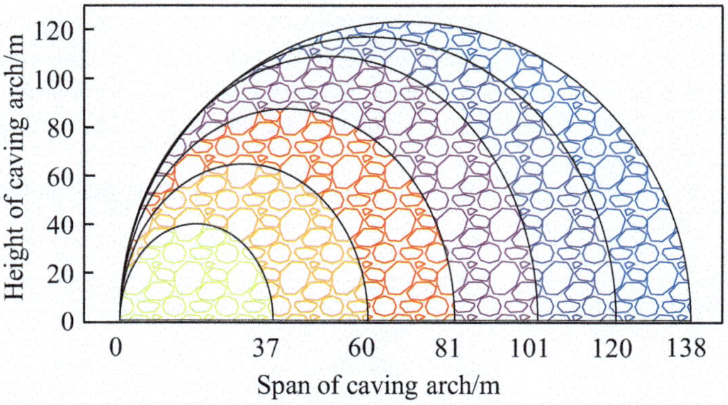

Fig. 6.48 Geometric evolution of the overburden caving arch

6.4.4.2 Conditions Leading to the Extension of Inertial Cracks in the Main Roof

After the collapse arch in the alluvial layer becomes unstable, secondary fracturing occurs during the descent of the collapsed mass, with the fracture line roughly located in the middle of the collapse arch height. The gravity of the collapsed body *B* can be determined by stepwise integration. However, through multiple analyses, it has been found that simplifying the shape of collapsed body *A* to a triangle and collapsed body *B* to a parallelogram, the simplified gravity of region *B* is 2/3 of the collapsed body's total gravity. The gravity determined by this method has an error of less than 5% compared to the integrated results. Therefore, using the simplified method, the gravity *Q* of collapsed body *B* can be determined as:

$$Q = \frac{\pi\gamma}{3f}(a_i^2 - a_{i-1}^2) \tag{6.99}$$

Under the self-weight load of the collapsed mass in the alluvial layer, the shear stress generated at the fixed support end of the main roof cantilever beam is:

$$\tau = \frac{\pi\gamma}{3fL_{\max}H}(a_i^2 - a_{i-1}^2) \tag{6.100}$$

where H is the thickness of the main roof, $L_{\max}$ is the limit span of main roof.

If we consider the shear stress determined by Eq. (6.100) as the far field load acting on the inertial crack at the surface of the main roof, then the stress intensity factor at the tip of the edge crack under far field shear stress is

$$K_{II} = Y\tau\sqrt{\pi s} \tag{6.101}$$

where K_{II} is the stress intensity factor for Mode II cracks, s is the length of inertial crack development on the main roof surface, and Y is the dimensionless geometric correction factor, whose expression is:

$$Y = 1.12 - 0.23\frac{s}{H} + 10.55\left(\frac{s}{H}\right)^2 - 21.72\left(\frac{s}{H}\right)^3 + 30.39\left(\frac{s}{H}\right)^4 \tag{6.102}$$

After instability of the collapsed mass in the alluvial layer, under the self-weight load, the conditions for Mode II extension of edge cracks on the surface of the weakly cemented thin bedrock of the main roof:

$$K_{II} \geq K_{\mathrm{IIc}} \tag{6.103}$$

where K_{IIc} is the fracture toughness of the main roof, MPa·m$^{1/2}$.

6.4.4.3 Initiation Speed of Broken Rock Blocks Ang Determination of Dynamic Load

Under the load of the alluvial layer, the cantilever beam of the main roof stores elastic strain energy. During the Mode II inertial crack extension leading to shear failure of the main roof, part of the elastic strain energy transforms into surface energy of the fracture gap, and part transforms into initial kinetic energy of fractured rock blocks. This transformation leads to non-static contact with the immediate overlying strata and hydraulic supports, triggering dynamic loading. Before the main roof fractures, it behaves as a cantilever beam under the load given by Eq. (6.98), and the stored elastic strain energy can be determined by the following equation [13]:

$$W_{\mathrm{e}} = (1 + \nu^2)\left(\frac{q^2L^5}{40EI} + \frac{q^2HL}{2E}\right) + (1 + \nu)\frac{q^2L^3}{3E} \tag{6.104}$$

where E is the elastic modulus of the main roof, GPa; v is the Poisson's ratio of the main roof; L is the span of the main roof, m; I is the moment of inertia of the cantilever beam section of the main roof per unit width, m^4; q is the load delivered to the key stratum.

The energy consumed during Mode II extension of edge cracks on the surface of the main roof can be determined by the following equation:

$$W_f = \frac{1}{2}H\tau_c\varepsilon_c \tag{6.105}$$

where τ_c is the shear strength of the main roof, MPa; ε_c is the corresponding ultimate strain value for shear strength of the main roof.

From Eqs. (6.104) and (6.105), the initial kinetic energy of the fractured rock block ($W_{\mathrm{k}} = W_{\mathrm{e}}$-$W_{\mathrm{f}}$) can be determined. Consequently, the initial initiation velocity of the fractured rock block is:

$$V = \sqrt{\frac{2W_{\mathrm{k}}}{m}} \tag{6.106}$$

where m is the mass of the fractured rock block of the main roof, kg; V is the initial initiation velocity of the fractured rock block of the main roof, m/s.

According to the law of conservation of momentum, the dynamic load exerted by the fractured rock block of the main roof on the immediate roof and support is obtained,

$$F = Q_m + \frac{mV}{t} \tag{6.107}$$

where Q_m is the gravity of the fractured rock block of the main roof, kN; t is the duration of the dynamic loading effect, s.

The immediate roof enters a yielding state under repeated support from hydraulic supports. Therefore, the dynamic loading effect between the main roof and the immediate roof can be regarded as a completely inelastic collision. In this dynamic loading effect, the initial initiation velocity of the main roof, determined by Eq. (6.106), remains unchanged. Thus, the duration of the dynamic loading effect on the main roof can be determined by the following equation:

$$t = \frac{S}{V} \tag{6.108}$$

where S is the amount of compression of the hydraulic support column, m.

6.4.5 Analysis of Dynamic Loading Effects of the Panel

6.4.5.1 Description of a Roof Collapse Accident at the Panel

In November 2020, when the 14,030 panel advanced to 200 m, the mining height reached 6.5 m, approximately 12 m from the previous load-bearing position. The mining pressure at the panel began to increase, as shown in Fig. 6.49. Serious coal wall damage occurred within the range of hydraulic supports 30# to 38#, resulting in roof leaks and deteriorating support conditions. The roof collapse reduced the coverage area, with some support columns experiencing minimal loads or even being unloaded. Subsequently, as the panel advanced, dynamic loading effects became pronounced. The sudden increase in roof subsidence led to significant compression of the active columns of hydraulic supports 32#, 33#, and 34#, reaching 0.8 m. The support canopy and forward beams contacted the floor, making it difficult to move the supports, resulting in them being crushed. To determine the fracture mode of the main roof, boreholes were drilled at the end panels of the roof with better

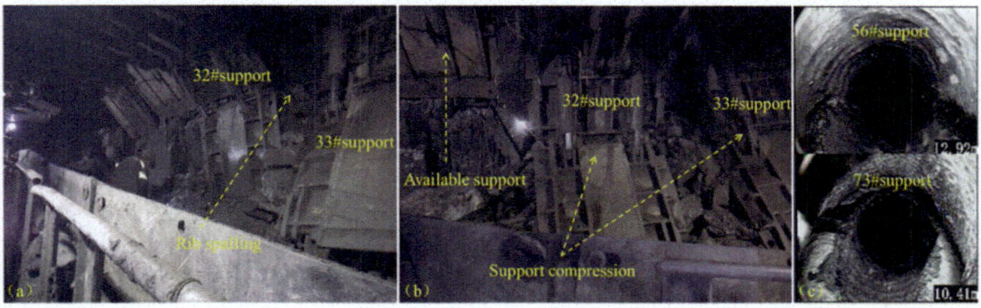

Fig. 6.49 Support compression accident induced by dynamic load

surrounding rock control, specifically at supports 55# and 73#. The borehole surveys detected the development of roof fracturing at depths of 12.9 and 10.4 m, where nearly vertical mining-induced fractures were observed. These fractures occurred within the sandstone of the main roof, with clear signs of shear displacement on both sides of the fracture surfaces, indicating shear failure of the main roof.

6.4.5.2 Analysis of Dynamic Loading Effects of the Roof

The onsite monitoring results indicate that the cause of the roof collapse accident at the 14,030 panel was as follows: the overall instability of the overlying strata collapse arch led to extensive collapse of the alluvial layer. This caused shear failure of the main roof and released strain energy. The fractured rock blocks rapidly initiated, resulting in dynamic loading effects on the immediate roof and hydraulic supports. Borehole survey results showed that the main roof thickness at the support location was 13 m, with a cantilever length of 12 m. Experimental results indicated an elastic modulus of 2.0 GPa, Poisson's ratio of 0.2, shear strength of 2.0 MPa, and ultimate shear strain of 0.01 for the main roof. The alluvial layer had a Proctor hardness coefficient of 0.8, tensile strength of 0.25 MPa, and rock density of 2500 kg/m^3. The hydraulic support center distance was 2.05 m, with a maximum active column compression of 1.0 m during support. By inputting these parameters into the equation for calculating roof dynamic load, the relationship curve between dynamic load, collapse arch span, height, and its instability frequency is obtained, as shown in Fig. 6.50. During the initial mining phase, the roof dynamic load was less than the rated working resistance of the supports at 18,000 kN, hence widespread support crushing did not occur. As the instability frequency of the collapse arch increased, its height and span increased, leading to greater collapse range of the alluvial layer. This increased the strain energy released from the shear failure of the main roof, and subsequently increased the initial initiation speed of fractured rock blocks, gradually raising the roof dynamic load. However, due to the rock's tendency to fragment after breaking, the parameters of the collapse arch do not increase indefinitely. Under conditions of full thickness mining of thick coal seams, the height of overlying strata damage H_f can be determined by empirical Eq. (6.109).

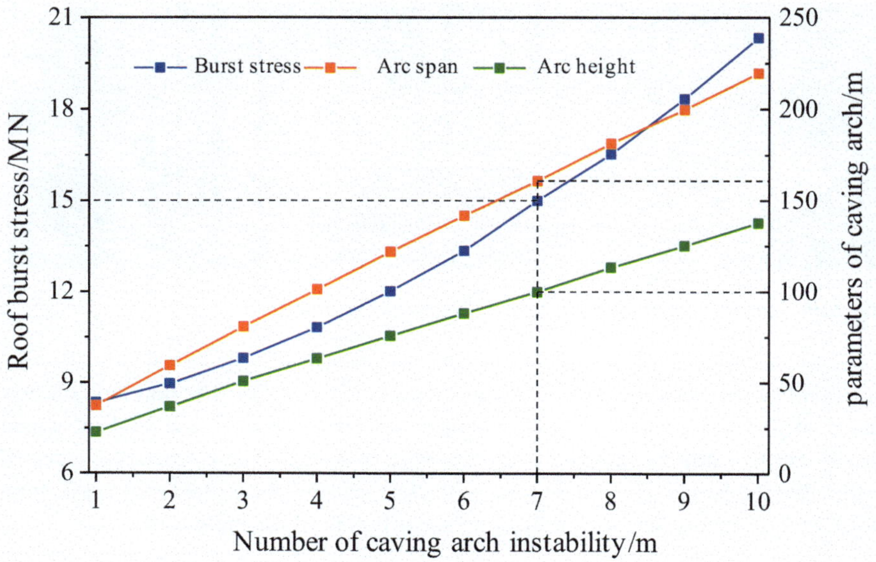

Fig. 6.50 Evolution of caving arch parameters and dynamic load

$$H_{\mathrm{f}} = \frac{100M}{0.2M + 6.0} + 10.5 \tag{6.109}$$

where M is the cutting height of the panel, m. The maximum mining height of 14,030 panel is 6.5 m, and the maximum developing height of overburden caving arch can be calculated by Eq. (6.109) to be about 100 m. The method for determining the caving arch boundary proposed in this section shows that when the caving arch of overlying rock on 14,030 face fails for the 7th time, the arch height increases to 101 m and the span is 162 m. According to the calculation results of the empirical formula, if the panel continues to be mined, the span of the caving arch of overlying rock will only increase, and the height will not show the increasing trend as shown in Fig. 6.50, however, will remain unchanged under the support of gangue in the goaf.

The position of the pressing frame in the 14,030 panel is 200 m away from the set-up room, indicating that the caving arch height of the overlying rock has reached the maximum. When the caving arch reaches 101 m, the dynamic load caused by the main roof breaking reaches 15,000 kN, and the self-weight of the main roof breaking rock block is 7900 kN. The ratio of the two is 1.9, which is close to the measured dynamic pressure coefficient 2.1. Under the condition that the roof-control distance is 6 m and the thickness of the immediate roof is 11 m, the immediate roof gravity acting on the support is 3400 kN, with the main roof dynamic load totally reaching 18,400 kN, which is greater than the rated working resistance of the hydraulic support, and eventually causes a large-scale pressure accident on the 14,030 panel.

6.4.5.3 Principles and Methods for Controlling Roof Dynamic Pressure

The mechanism of dynamic loading effects on the roof of mining sites with deep buried thin bedrock thick coal seams indicates that shear failure of the main roof and the non-static initiation of fractured rock blocks are the direct causes of roof collapse accidents. To prevent such accidents, it is necessary to either avoid shear failure of the main roof or reduce the initial initiation speed of the fractured rock blocks. The mining practice at the 14,030 panel has shown that the following two methods can significantly reduce the intensity of dynamic pressure on the panel: (1) When the panel advances to the stage of overall instability of the overlying strata collapse arch, reduce the coal cutting advance distance and increase the number of support movements. This increases the number of disturbances to the roof in the control area caused by the repeated support of the hydraulic supports, while lowering the support movement pressure and increasing the initial support force. This method can subject the overlying strata to high-frequency, high amplitude cyclic pressure, increasing the damage level of the overlying strata and enlarging the local collapse range of the alluvial layer during the stable stage of the collapse arch structure. This also increases the elastic strain energy consumption of the main roof, thereby reducing the dynamic load. (2) When the panel advances to the stage of overall instability of the overlying strata collapse arch, pre-split the main roof to release the stored elastic strain energy within it. This prevents the non-static contact between the fractured rock blocks and the immediate roof below, thereby preventing the occurrence of dynamic loads.

At the Zhaogu No. 2 Mine's 14,030 panel, pre-splitting blasting was used to pre-fracture the main roof in advance, releasing the strain energy within it. Deep hole pre-splitting blasting measures were implemented on the roof of the transport roadway, with boreholes arranged along the strike. The boreholes had a diameter of 50 mm, a depth of 16 m, a spacing of 1.8 m, and a dip angle of 90°. The boreholes were positioned 1.5 m from the outer edge of the roadway. Between two blasting boreholes, a guide hole was arranged, with the same parameters as the blasting holes. The blasting boreholes had a charge length of 9 m and a sealing length of 7 m, using 40 mm diameter double seam directional presplitting tubes as the charge carriers. The tube seams were aligned with the borehole arrangement direction. After blasting, large-scale cracks along the strike were formed in the roof, as shown in Fig. 6.51. After the presplitting blasting, the left, right, and rear sides of the main roof all became free or simply supported boundaries, significantly reducing the concentration of strain energy. Consequently, the amount of strain energy converted into the kinetic energy of fractured rock blocks upon roof failure decreased, preventing the non-static initiation of fractured rock blocks. After adopting pre splitting blasting measures, the dynamic load effects on the 14,030 panel's roof improved. The condition of the hydraulic supports improved, with the column load dropping to 40–45 MPa during pressure periods. The frequency of the safety valve activation on the supports decreased, and the roof control effect increased.

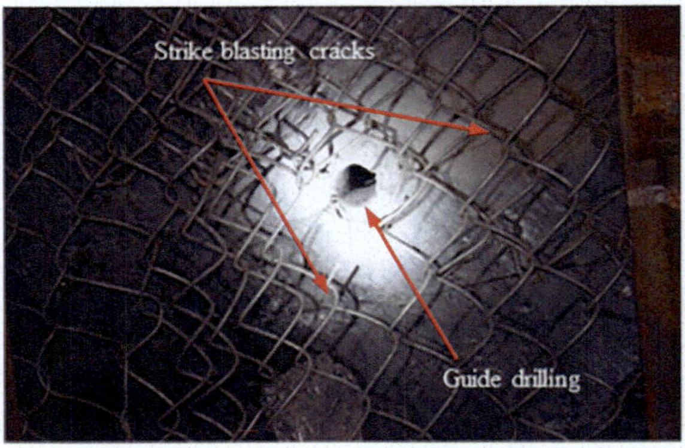

Fig. 6.51 Blasting-induced roof fractures in strike direction

6.4.6 Caving Arch and High Steep Rock Beam Composite Bearing Structure

A mechanical model was constructed for the composite structure of the thick alluvial layer caving arch and high steep rock beam at the arch foot in the overburden of a deep-buried thin bedrock panel. The stability of the caving arch and the high steep rock beam structure at the arch foot jointly ensures the safety of the mining space at the panel. Instability in either structure can cause severe mine pressure manifestations at the panel. To prevent impacts on mining activities, the bearing capacities of the high position caving arch and the low position high steep rock beam structure were analyzed separately [13].

6.4.6.1 Stability Analysis of the Thick Alluvial Layer Caving Arch

The thickness of caving arch in thick alluvium is consistent with the width of deformation concentration zone, and the ultimate bearing capacity of caving arch structure can be determined by Eq. (6.110) [14]. After the formation of the caving arch, the load of the thick alluvium in the mining range is quickly transferred to the lower bearing structure. If the actual load reaches the ultimate bearing capacity of the caving arch, the structural instability of the caving arch occurs.

$$q_{\max} = \frac{6E_s}{25}\frac{W^3}{L_s^3} \tag{6.110}$$

where E_s is the elastic modulus of the thick alluvial layer, W is the width of alluvial deformation concentration zone, and L_s is the limit span of caving arch.

The structural bearing mechanism of the caving arch in the thick alluvial layer involves the stress arch effect induced by uneven settlement. This effect grants the structure inherent self-supporting capacity, meaning not all loads from the affected alluvial layer due to mining activities are borne solely by the caving arch. As depicted in Fig. 6.52, above the caving arch, analysis of a thin layer shows that within the range of h_a influenced by mining activities, the maximum principal stress at the position of the thin layer is $\sigma_1 = \gamma_a h_a$. Applying principles from soil mechanics and neglecting cohesive forces within the alluvial layer, the minimum principal stress at the thin layer position, according to the ultimate balance principle, is $\sigma_3 = (1 - \sin\varphi)/(1 + \sin\varphi)\sigma_1$, where γ_a and φ represent the unit weight and internal friction angle of the alluvial layer, respectively. After uneven settlement, the deformation of the thin layer forms an arc-shaped band. Assuming the settlement range of the thick alluvial layer matches the span of the caving arch, shear effects occur at the boundaries, leading to shear stress development and rotation of mining-induced stresses. The trace of the minimum principal stress line after deformation matches the shape of the band, with the maximum principal stress distributed radially along the band.

The stress states at points *A*, *B*, and *C* of the thin layer at the ends and middle of the thick alluvial layer are shown relative to the Mohr stress circle in Fig. 6.52. The rotation angle of the principal stress on the shear planes on both sides is $\pi/2 - \theta$, where θ is the angle between the maximum principal stress and the horizontal plane. The value of θ_0 at the settlement boundary of the thick alluvial layer is generally taken as $\pi/4 + \varphi/2$, where φ is the internal friction angle of the alluvial layer. Based on the geometric relationships in the stress circle, the horizontal, vertical, and shear stress components at any position in the thin layer band can be determined.

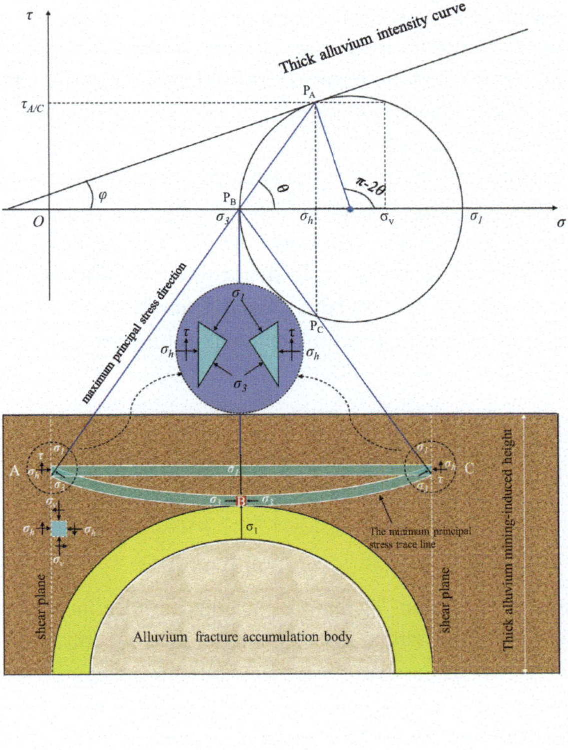

Fig. 6.52 Bearing mechanism of caving arch in thick alluvium

$$\begin{cases} \sigma_h = \left(\cos^2\theta + k\sin^2\theta\right)\sigma_1 \\ \sigma_v = \left(\sin^2\theta + k\cos^2\theta\right)\sigma_1 \\ \tau = \sigma_1(1-k)\sin\theta\cos\theta \end{cases} \tag{6.111}$$

where σ_h, σ_v and τ represent the horizontal, vertical, and shear stress components at any point in the thin layer respectively, and k denotes the ratio of the minimum to maximum principal stress, typically taken as $(1 - \sin\varphi)/(1 + \sin\varphi)$.

From Eq. (6.111), the trend of vertical and horizontal stress variations along the span direction of the caving arch can be observed, as shown in Fig. 6.53. The relative distance in the figure is defined as the ratio of the horizontal distance between a point on the thin layer and the center of the caving arch to half the span of the caving arch itself. The maximum principal stress at the thin layer position essentially represents the self-weight load of the alluvial layer within the mining range. Under the rotational effect of mining-induced stresses, the vertical stress in the upper thick alluvial layer tends to decrease, while the horizontal stress tends to increase. At point B, the center of the caving arch, where mining-induced stress rotation does not occur, the vertical stress equals the self-weight load of the alluvial layer affected by mining, and the horizontal stress remains unchanged. Moving from the center of the caving arch towards the arch foot, the horizontal stress increases, and the vertical stress decreases, demonstrating the stress arch effect. The stress arch effect induced by the rotational stress of mining operations causes the load of the alluvial layer to transfer towards both sides of the settlement boundary, thereby mitigating the load transmitted to the caving arch structure above. As the rotational angle of mining-induced stress increases, the load transferred to the alluvial layer above the caving arch decreases, indicating that the phenomenon of stress rotation benefits the stability of the thick alluvial layer caving arch.

Equation (6.110) represents the ultimate bearing capacity of the arch structure under uniformly distributed load, while Eq. (6.111) and Fig. 6.53 indicate that the actual overburden load borne by the thick alluvial layer caving arch exhibits non-uniform characteristics. To determine the stability of the caving arch structure, integrate Eqs. (6.110) and the second Equation of (6.111) along the span direction of the arch, to establish the allowable maximum concentrated force and actual concentrated force on the arch structure, respectively.

$$\begin{gathered} F_{\max} = \frac{12}{25}\frac{W^3}{L_s^2}E_s \\ F_v = \frac{1}{2}\sigma_1\left[(1+k)\left(\frac{\pi}{2} - \theta_0\right) + \frac{1}{2}(1-k)\sin 2\theta_0\right] \end{gathered} \tag{6.112}$$

where $F_{\max}$ and F_v denote the maximum concentrated force that the caving arch structure can withstand and the actual concentrated force transmitted to it respectively, kN. If the overburden load on the caving arch satisfies $F_{\max} = F_v$, then the arch structure reaches a state of ultimate equilibrium. If the overburden load continues to transfer downward, the caving arch may undergo structural instability, posing a threat

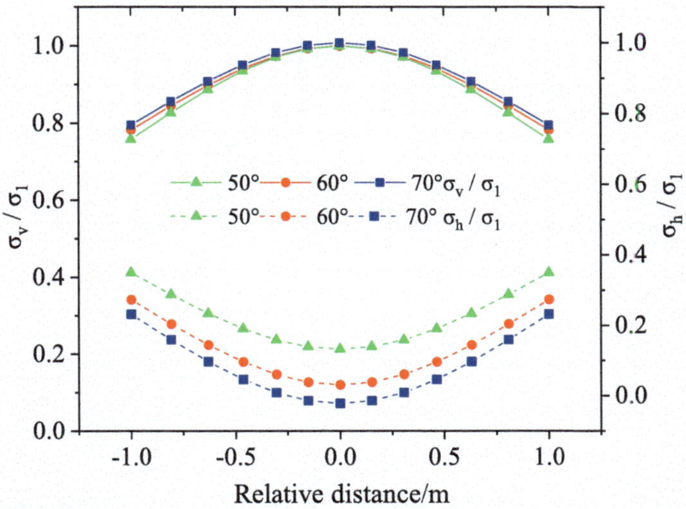

Fig. 6.53 Load distribution above the caving arch

to the safety of the coal wall. The step distance for instability in the thick alluvial layer caving arch structure is smaller than the span of a newly formed caving arch caused by material instability. The former leads to local instability of the caving arch, while the latter results in overall instability. These two instability modes alternate during the advancement of the coal wall.

6.4.6.2 Stability Analysis of Rock High Steep Rock Beams

After the formation of the thick alluvial layer caving arch, the underlying bedrock undergoes full depth shear fracture, forming a balanced high steep rock beam structure at the arch foot, as depicted in Fig. 6.54. The height of the rock beam exceeds its span, suppressing the rotational movement of fractured rock blocks, with sinking being the predominant mode of movement. During the sinking process, the shear fracture plane of the bedrock remains closed, and the fractured rock blocks maintain panel to panel contact with surrounding rock masses. The essence of the sinking process of the high steep rock beam involves shear sliding along the two successive fracture planes of the bedrock, constrained by unbroken rock layers ahead and recompacted rock blocks behind. The fractured rock blocks of the bedrock behind the coal wall maintain relatively intact states, with stiffness comparable to intact rock, and can be considered as elastic boundaries along with intact rock layers ahead of the coal wall. Therefore, the shear sliding process experienced by the front and rear fracture planes during the sinking of the high steep rock beam can be regarded as a constant normal stiffness (CNS) boundary condition. Additionally, the irregularity of the shear fracture planes in the bedrock contributes to their inherent roughness.

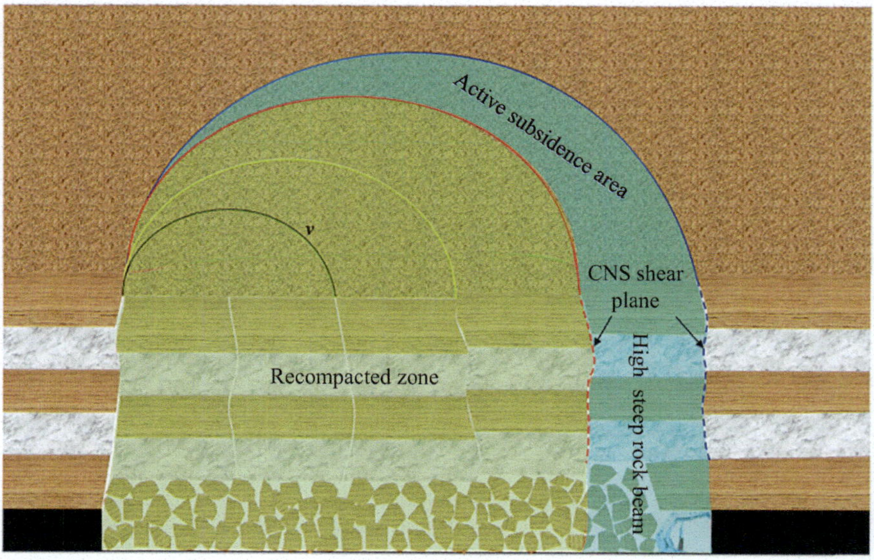

Fig. 6.54 Bearing mechanism of high steep rock beam in bedrock

The combination of rough fracture planes and CNS boundary conditions establishes the equilibrium state of the high steep rock beam structure.

During the shear sliding process under constant normal stiffness conditions in the high steep rock beam, the mechanical behavior on the shear plane differs significantly from that under constant normal stress conditions. Under constant normal stress conditions, the shear mechanical behavior on the failure plane is depicted in Fig. 6.55a. The shear displacement curve exhibits a peak point, and after reaching the peak, the shear strength of the failure plane decreases. This leads the high steep rock beam into an unstable state, causing continued increase in load transmitted to the supports above, thereby increasing the probability of roof instability and pillar collapse accidents. Under constant normal stiffness conditions, the shear mechanical behavior on the failure plane is shown in Fig. 6.55b. The shear dilation effect on the rough failure plane is suppressed by boundary conditions. The shear stress increases monotonically with the roughness of the failure plane and the displacement, enhancing the load bearing capacity of the high steep rock beam as the sinking of the roof increases. This results in improved roof stability.

Based on the above, it is evident that, under conditions that do not compromise mining safety, deep buried thin bedrock mining fields with thick overburden layers should allow for a certain amount of roof subsidence. This allows the high steep rock beam to exert its load bearing capacity, thereby enhancing the stability of the coal wall support-rock mass system. To determine the load bearing capacity of the high steep rock beam, it is necessary to analyze the variation characteristics of shear stress on the bedrock fracture plane with respect to roof subsidence. The shear mechanical model [15] of rock fractures under constant normal stiffness conditions is as follows:

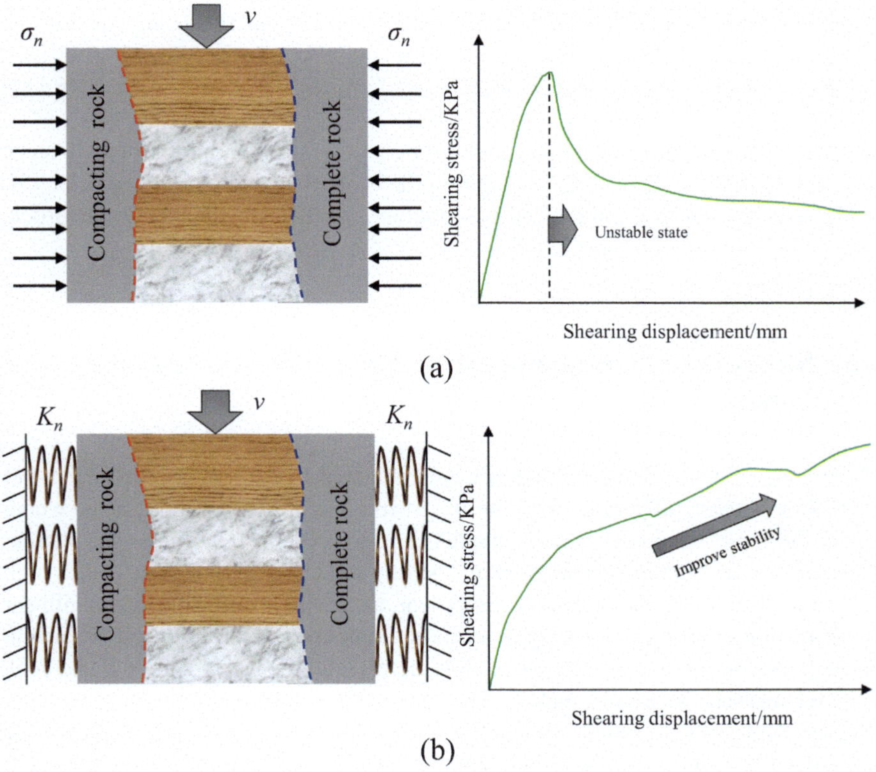

Fig. 6.55 Mechanical model for towering roof beam with different boundary conditions: **a** constant normal stress; **b** constant normal stiffness

$$
\begin{aligned}
\sigma_{\rm n} &= \sigma_{\rm ni} + 4.1 \times 10^{-3}\left(\frac{\mathrm{UCS}^{0.44}\mathrm{JRC}^{0.94}}{\sigma_{\rm nr}^{0.06}}\right)(K_s u_s)^{0.62} \\
\tau_{\rm s} &= \sigma_{\rm n} \tan\left[\varphi_{\rm r} + \mathrm{JRC}\log_{10}\left(\frac{\mathrm{UCS}}{\sigma_{\rm n}}\right)\right]
\end{aligned}
\tag{6.113}
$$

where σ_{ni} and σ_n represent the initial normal stress and normal stress on the bedrock fracture plane respectively, MPa; τ_s denotes the shear stress, MPa; JRC stands for the roughness of the bed-rock fracture plane; φ_r represents the friction angle of the bedrock, °; UCS indicates UCS of the bedrock, MPa; Ks is the shear stiffness of the slip plane; u_s is the slip amount.

To obtain the self-supporting capacity of the high steep rock beam, integrate the second equation of Eq. (6.113) over the bedrock shear fracture plane. The integrated result indicates that as the roughness of the bedrock fracture plane increases, the shear stress, and the area of the shear plane increase during the sinking process of the high steep rock beam. This enhances the load bearing capacity of the high steep rock beam. The support force required for the support to stop shear sliding of the high steep rock beam and enter a stable state is determined by the following equation:

$$F_s = G - \lambda H_r L_r \tau_s \tag{6.114}$$

where F_s is the support capacity required for ensuring the stability of the high steep rock beam structure, kN; G is the weight of the bedrock and overburden blocks, kN; L_r is for the step distance of the bedrock fracture, m; H_r is the thickness of bedrock, m; λ is the side pressure coefficient.

6.5 Stress Distribution and Strata Control in Extra-Long Panel

The depth of coal resource mining in China is increasing at a ratio of 10–25 m per year. This is especially evident in the long-exploited eastern mining areas, where nearly 50 mines have now reached depths of over 1000 m. Deep mine has characteristics such as high in-situ stress levels, strong mining disturbance effects, a high degree of coal and rock fracture development, a transition from linear to nonlinear mechanical properties, and strong deformation timeliness. During the advancement of the panel, the frequency of abnormal mine pressure behavior and coal and rock dynamic disasters increases, making the control of the surrounding rock more challenging. Moreover, to reduce coal pillar losses, lower support costs, and increase single-face production capacity, the length of panels has been increasing annually, with the current maximum panel length reaching 460 m. The disturbance effects of deep extra-long panel mining are further intensified, resulting in a more complex mining stress environment.

6.5.1 Engineering Background

The Kouzidong Coal Mine, part of the Huainan Coalfield, features the 140,502 panel as the initial coal panel of the 14th mining area. The panel extends 270 m in length and advances 820 m, primarily extracting the No. 5 coal seam at a burial depth ranging between 900 and 1000 m. The thickness of the coal seam varies from 4.0 to 8.0 m, averaging 6.6 m, with a dip angle ranging from 3° to 15°, averaging 10°, categorizing it as a gently inclined thick coal seam. The plan layout of the 140,502 panel is shown in Fig. 6.56a, utilizing large mining height extraction technology. Near the headgate side of the panel, there are two normal faults with displacement ranging from 0 to 5 m. Core samples obtained from drillhole reveal the comprehensive stratigraphic column of the 140,502 rock layers as depicted in Fig. 6.56b. The immediate roof is a composite roof consisting of sandy mudstone and coal lines, with a thickness ranging from 0.98 to 11.71 m and a hardness coefficient between 3.0 and 3.9. The main roof is fine sandstone, with a thickness ranging from 2.35 to 9.90 m and a hardness coefficient between 6.2 and 7.3. The thickness of the bedrock above the main coal seam ranges from 210 to 350 m, with unstable sandstone layers less than 10 m thick present within the overburden. Above the bedrock lies a thick alluvial layer of very low strength.

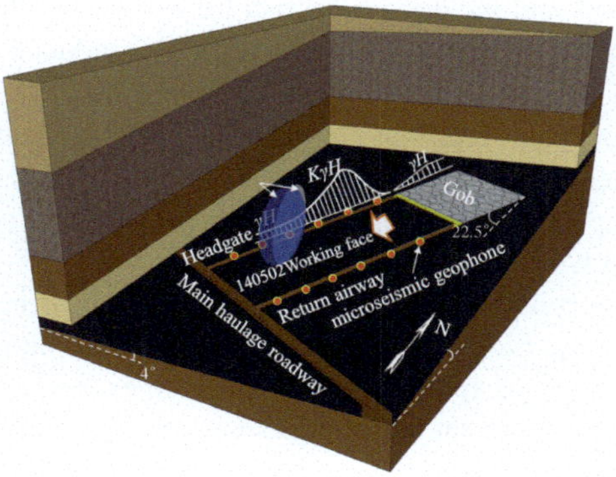

(a)

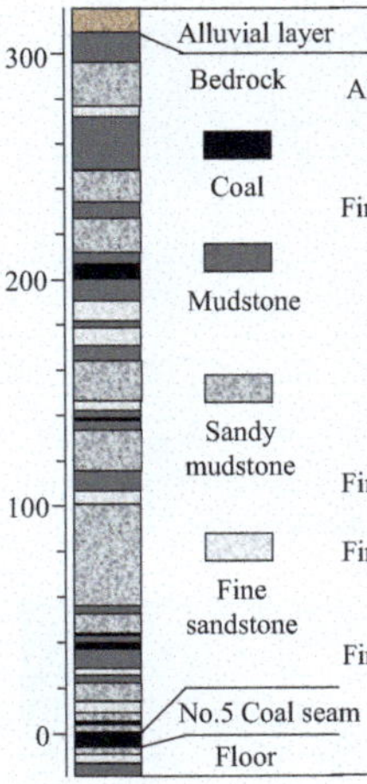

Lithology	Thickness/m	Depth/m
Alluvial layer	680	
Mudstone	13.21	-648.84
Silt-clay	19.38	-668.22
Fine sandstone	5.01	-673.23
Mudstone	24.04	-697.27
Siltstone	14.04	-711.31
Mudstone	6.92	-718.23
Siltstone	15.04	-733.27
Mudstone	5.28	-738.55
13-1 Coal	6.19	-744.74
Mudstone	9.61	-754.35
Fine sandstone	9.04	-763.39
Mudstone	3.01	-766.40
Fine sandstone	8.09	-774.49
Mudstone	6.99	-781.48
Siltstone	17.32	-798.80
Fine sandstone	4.50	-803.30
Mudstone	3.35	-806.65
11-2 Coal	1.23	-807.88
Mudstone	3.90	-811.78

Lithology	Thickness/m	Depth/m
Siltstone	18.06	-829.84
Mudstone	9.08	-838.92
Fine sandstone	6.05	-844.97
Siltstone	44.59	-889.56
Mudstone	3.41	-892.97
Siltstone	8.89	-901.86
9 Coal	1.07	-902.93
Mudstone	3.00	-905.93
8 Coal	2.70	-908.63
Mudstone	8.51	-917.14
Fine sandstone	3.07	-920.21
Mudstone	3.34	-923.55
Siltstone	7.92	-931.47
Fine sandstone	5.18	-936.65
Siltstone	3.82	-940.47
Coal lines	1.48	-941.95
Siltstone	2.65	-944.60
5 Coal	6.53	-951.13
Siltstone	3.13	-954.26
Fine sandstone	3.93	-958.19
Mudstone	5.63	-963.82

(b)

Fig. 6.56 General situation of 140,502 panel: **a** layout of the panel; **b** stratigraphic column of the rock layers

6.5.2 *Characteristics of Mine Pressure Appearance in the Panel*

6.5.2.1 Issues in Surrounding Rock Control in the Panel

The burial depth of the 140,502 panel reaches 1000 m, leading to an increase in both the level of in-situ stress and the degree of fracture development. After mining the thick coal seam, the peak value of the advanced mining-induced stress approaches UCS of the main roof. The primary fractures in the coal and rock extend ahead of the panel, accompanied by the initiation of new mining-induced fractures. As the panel advances, a high degree of spalling occurs after the coal wall is exposed, with the spalling depth reaching 1.0 m. The cumulative range of spalling covers one-third of the panel length, as shown in Fig. 6.57a.

The exposure of the roof reveals that it is fractured into irregular blocks, with the shape and size of these blocks closely related to the distribution of fractures. If the support posture is improperly controlled, leading to excessive gaps between the supports, inter-support roof fall may occur, as shown in Fig. 6.57b. When fractured

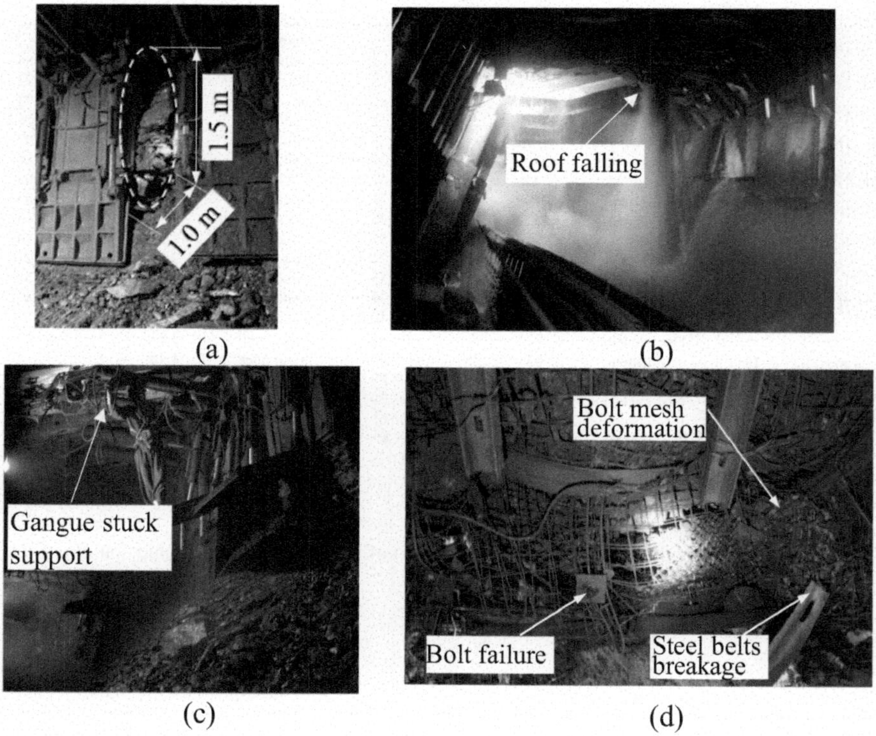

Fig. 6.57 Characteristics of mine pressure appearance in the extra-long panel: **a** rib spalling; **b** roof falling; **c** gangue stuck support **d** local large deformation and failure in roadways

roof blocks are located above the support beams, they may get stuck during the shifting of the frame, causing difficulty in moving the frame or even damaging components. This requires repairs by workers before the frame can be shifted again, as shown in Fig. 6.57c. If the range and depth of rib spalling increase, it can lead to the development of fractures, increasing the exposed area of the roof and further exacerbating roof broken degree. This can cause roof fall in front of the supports, burying mining equipment and threatening personnel safety. Additionally, under the influence of high peak mining-induced stress, the coal on both sides of the roadway can become severely fractured and extrude into the roadway space. Constrained by the bolting with wire mesh, this can form mesh bags. Under the pressure of the fractured coal, steel strips may break, and anchor rods can be pulled out from the fractured coal, resulting in the loss of anchoring effect. Ultimately, this leads to significant deformation and failure of the roadway in the advanced influence section, as shown in Fig. 6.57d. These surrounding rock control issues are driven by mining-induced stress. To improve the control of surrounding rock in extra-long panel, it is imperative to study the mechanisms of surrounding rock failure and the distribution characteristics of mining-induced stress.

6.5.2.2 Advanced Mining Influence Range

A measurement station was set up in the return airway to determine the range of mining influence. Anchor rods and anchor cables were installed 300 m ahead of the panel, with 7 sets of each, to monitor the stress conditions. Due to the large extent of the fractured zone in the roadway surrounding rock, some anchor rods (and cables) could not reach the stable zone of the surrounding rock to exert their anchoring effect, resulting in abnormal sensor data. Only effective data were analyzed. The ratio of the actual stress on the anchor rods (and cables) to the initial prestressing force is defined as the stress concentration factor. The dynamic changes in the effective stress concentration factor of the anchor rods (and cables) as the panel advances are shown in Fig. 6.58.

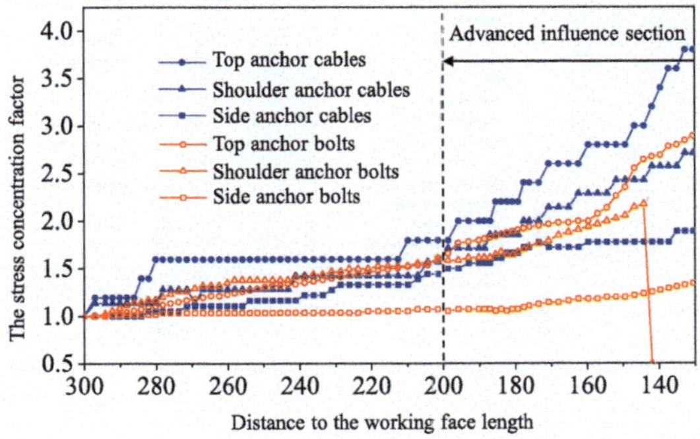

Fig. 6.58 The anchor rods (and cables) stress concentration coefficient as the panel advances

At the initial stage of monitoring, the surrounding rock of the roadway was not affected by the mining, and the force on the anchor rods (cables) showed a slow increase due to the roadway heading. When the distance between the measurement points and the panel was reduced to 200 m, the ratio of increase in the stress concentration factor of the anchor rods (cables) accelerated, indicating that the measurement point had entered the advanced mining influence zone. This shows that the advanced mining influence range of the 140,502 panel extends to 200 m. During the monitoring period, the maximum stress concentration factor reached 3.8, indicating strong disturbance characteristics of deep mining. Due to severe break of the return airway shoulder, the anchor rods in the shoulder area fell off 140 m ahead of the panel, causing the stress concentration factor to drop rapidly to 0. To avoid affecting the secondary repair work of the roadway, the actual measurement work ended at a position 130 m ahead of the panel.

6.5.2.3 Characteristics of Roof Fractures Development

To investigate the development of fractures in the roof under different degrees of mining influence, three measurement stations were arranged at positions 50, 100, and 200 m ahead of the panel in the return airway. A drillhole imaging device was used to observe the distribution characteristics of the roof fractures. The drillholes had a diameter of 42 mm and a depth of 15 m, oriented perpendicular to the roof strata. The fracture observation results are shown in Fig. 6.59.

The roof rock strata are influenced by geological structures, in-situ stress, and mining-induced stress, resulting in various forms of fractures within the drillhole walls, such as transverse, longitudinal, and inclined fractures. At a position 50 m ahead of the panel, the mining-induced stress has the greatest impact, causing severe roof damage in this area. Particularly within the 0–5 m range from the drillhole opening in the sandy mudstone, the drillhole wall is affected by intersecting fractures, leading to debris falling off. In the deeper fine sandstone layers, delamination and

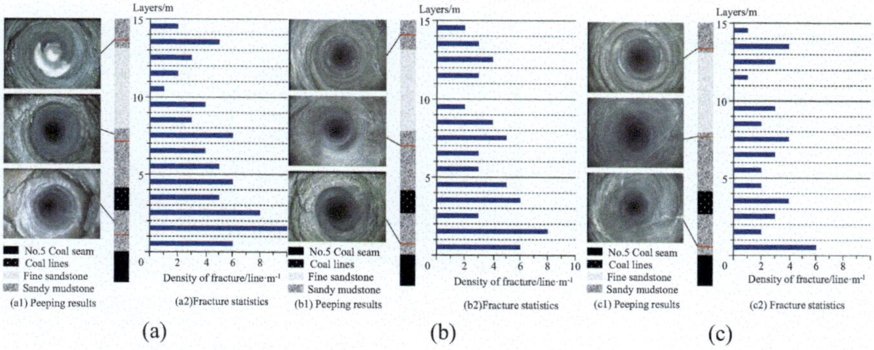

Fig. 6.59 Roof fractures drillhole imaging results: **a** 50 m ahead of the panel **b** 100 m ahead of the panel **c** 200 m ahead of the panel

dislocation phenomena occur, with an average fracture spacing of 0.21 m, as shown in Fig. 6.59a. At a position 100 m ahead of the panel, the impact of mining-induced stress decreases, and the number of roof fractures reduces accordingly. Transverse delamination fractures dominate, while longitudinal opening fractures result from the expansion of primary fractures, with the average fracture spacing increasing to 0.30 m, as shown in Fig. 6.59b. At a position 200 m ahead of the panel, the impact of mining-induced stress is minimal. The roof is primarily characterized by longitudinal primary fractures, with an average fracture spacing of 0.37 m, as shown in Fig. 6.59c, indicating an increased development of primary fractures in the deeper strata.

6.5.2.4 Characteristics of Roof Pressure Distribution

A total of 190 hydraulic supports were installed in the 121,304 panel. During the advancement of the panel, the spatial distribution of support resistance was monitored, as shown in Fig. 6.60a. The support resistance along the length of the panel exhibits a clear non-uniform distribution, forming a "valley shape" with lower resistance in the middle and higher resistance at both ends. The weighting interval varies significantly in different areas: the middle of the panel has a shorter weighting interval with a shorter duration, while the sides have a longer weighting interval with a longer duration. The resistance variation curves for hydraulic supports 20#, 100#, and 170# are shown in Fig. 6.60b. In the middle area of the panel, the weighting interval ranges between 8 and 12 m, with the maximum support resistance reaching approximately 38 MPa during pressure events. In the two side roof areas, the weighting interval ranges between 10 and 15 m, with the support resistance reaching up to 42 MPa during pressure events. On-site observations indicate that some supports exhibit safety valve activation. Additionally, during the normal advancement stage of the panel, the distribution curve of support resistance along the inclined direction is shown in Fig. 6.60c. It generally presents a valley-shaped distribution, with peaks in the two side areas and a trough in the middle area. Near the ends, due to the support provided by solid coal and coal pillars, the support resistance shows a decreasing trend. These differences in roof pressure characteristics cause issues in controlling the surrounding rock.

Based on the actual measurements of the support resistance in the 121,304 panel, it is shown that the overall roof pressure distribution in the extra-long panel exhibits a pattern of "high at both ends, low in the middle", which is different from the conventional panel distribution of "high in the middle, low at both ends." The masonry beam hypothesis suggests that the primary sources of roof pressure are the rotational deformation pressure of the main roof's failure rock blocks or the gravitational force of the cut-down rock blocks themselves. Therefore, the distribution of support resistance is closely related to the main roof's failure movement pattern. It can be inferred that the increase in depth and panel length has changed the main roof's failure movement form, thereby affecting the response characteristics of the panel hydraulic supports. According to the different states of the main roof, there are two working conditions for the hydraulic supports in the panel: given deformation and given pressure. The

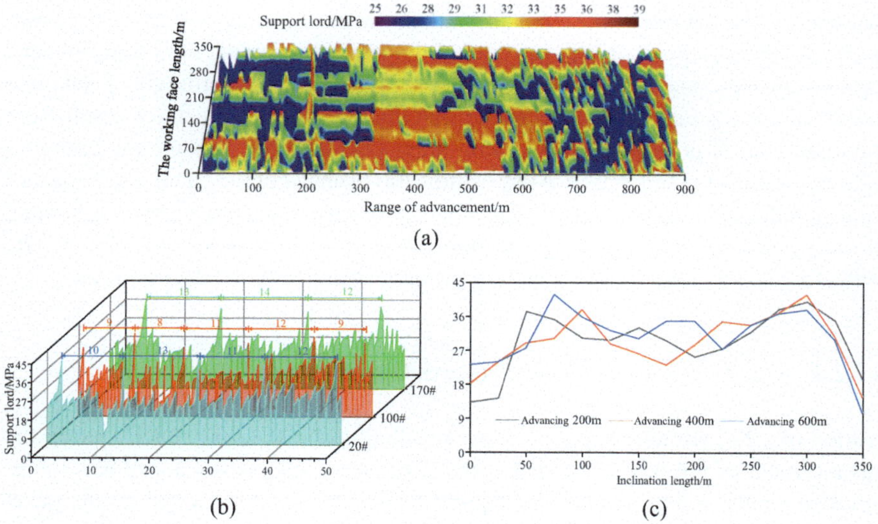

Fig. 6.60 Support resistance distribution characteristics of extra-long panel: **a** the spatial distribution of support resistance; **b** the strike evolution curve of support resistance **c** the inclined curve of support resistance

former refers to the condition where the main roof is unfractured or the masonry beam structure of the main roof remains stable. Under this condition, the hydraulic supports of the panel only bear the deformation pressure caused by the subsidence of the main roof, which corresponds to the support resistance during the normal advancement stage. The latter refers to the condition where the main roof structure becomes unstable, and the failure rock blocks' gravity is primarily borne by the hydraulic supports, corresponding to the support resistance during the pressure arrival stage. Under the given pressure condition, the support resistance is mainly influenced by the size of the failure blocks of the main roof. The larger the failure blocks, the greater the load on the supports from the roof, and vice versa. With the increase in the panel depth, the mining-induced stress increases, and under the drive of advanced mining-induced stress, advanced fractures develop in the main roof. Under the induction of advanced mining-induced fractures, the main roof experiences local failure, which affects the distribution characteristics of roof pressure along the length of the panel. Under the given deformation condition, the load on the hydraulic supports from the roof is mainly controlled by the subsidence of the roof. The subsidence of the exposed roof is primarily influenced by the roof load and the supporting effect of the goaf area. Under the condition of a 350-m extra-long panel, the compaction characteristics of the goaf area along the length of the panel inevitably exhibit more distinct zoning phenomena. The varying support capacities of the goaf for the roof lead to differences in the subsidence characteristics of the exposed roof, thereby influencing the distribution characteristics of roof pressure along the length of the panel.

6.5.3 *Main Roof Failure Characteristics of Extra-Long Panel*

6.5.3.1 Characteristics Analysis of Main Roof Failure in the Extra-Long Panel

Both empirical measurements and theoretical research results indicate that as mining depth increases, the degree of fracture development in the rock mass also increases, which in turn affects the type of rock mass failure. When studying the failure and movement patterns of the roof in deep mining panels, it is necessary to consider the variations caused by fractures [16].

1. **3D reconstruction of fractured main roof**

Field measurements show that fracture length follows a log-normal distribution, fracture dip and incline follow a normal distribution, and the spatial position of fractures follows a Poisson process. Based on the distribution forms of these fracture parameters, the Monte-Carlo simulation method is used to generate random numbers corresponding to different fracture occurrence parameters. Combining these random numbers with the Discrete Fracture Network (DFN) simulation method, randomly distributed fractures are generated in the main roof coverage area. These randomly distributed fractures are used to cut the intact main roof, thereby achieving a 3D reconstruction of the fractured main roof, as shown in Fig. 6.61. The random distribution of fractures in the main roof cuts it into blocks of greatly varying sizes and shapes, increasing the uncertainty of the main roof's failure patterns. Due to the random spatial distribution of fractures, the degree of cutting in the main roof varies significantly at different locations. The central area of the panel, enclosed by the white curve, has a relatively smaller main roof block size, while the block size at both ends is relatively larger.

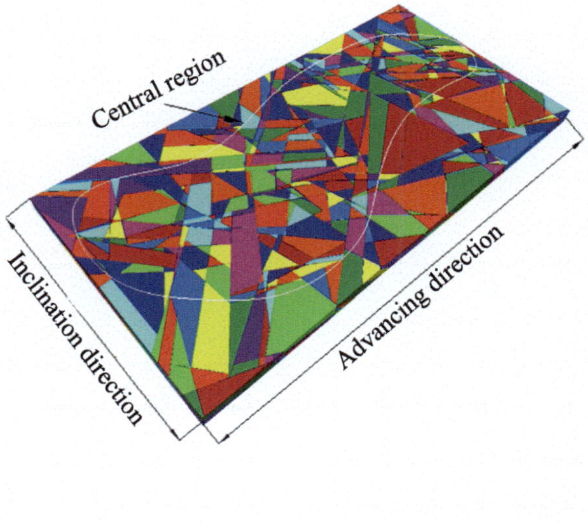

Fig. 6.61 Fractured main roof reconstruction

2. **Effect of panel length on main roof failure characteristics**

Under the condition of not considering fractures, the masonry beam theory states that the main roof undergoes an O-X type failure, as shown in Fig. 6.62. As the panel advances along the strike, after the main roof fractures, a trapezoidal plate forms in the middle of the panel, while arch-shaped triangular plates form at both ends. Both the triangular plates and the trapezoidal plate constitute a masonry beam structure that appears to be a beam externally but is essentially an arch in both the advancing direction and the panel direction.

From Fig. 6.62, under the condition of a fixed panel length b, the influence range of the arc-shaped triangular plates at both ends is m, while the influence range of the middle trapezoidal plate is $b - 2m$. The primary fracture in the roof follows a Poisson process in space, meaning that the probability of a fracture appearing at any point in the main roof is the same. As the panel advances, the ratio of the probability of primary fractures appearing in the coverage area of the middle trapezoidal plate to that in the coverage area of the arc-shaped triangular plates on both sides can be expressed by Eq. (6.115).

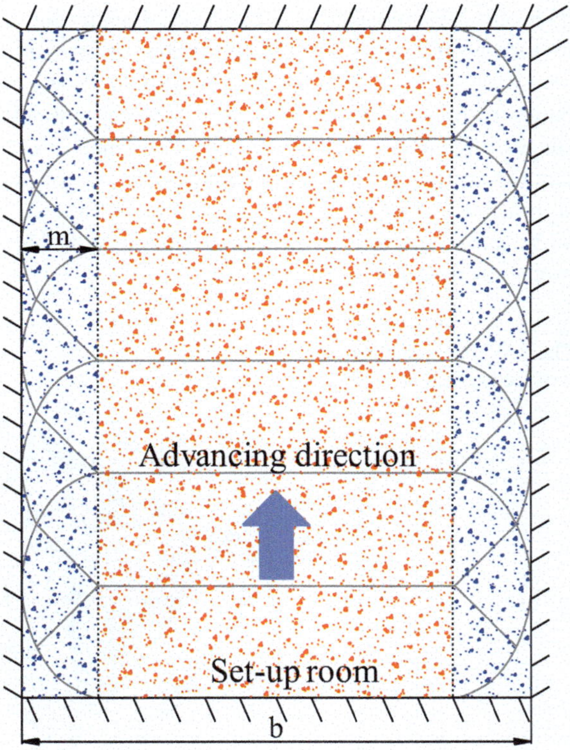

Fig. 6.62 Main roof O-X failure model

$$\frac{P_b}{P_m} = \frac{b}{2m} - 1 \tag{6.115}$$

where P_b represents the probability of fractures in the middle of the panel in the main roof, %; P_m represents the probability of fractures at the ends of the panel in the main roof, %; m represents the influence range of the arc-shaped triangular plate, m; b represents the length of the panel, m.

The influence range of the triangular plates at both ends of the panel is mainly affected by the width of the sectional coal pillar and the width of the mining roadway, and is less affected by the length of the panel. Therefore, as the length of the panel increases, the value of Eq. (6.115) shows an increasing trend. This means that the ratio of the probability of fractures appearing in the middle of the main roof to the probability of fractures appearing at the ends increases linearly. Compared to the strength of the rock, the strength of the fractures is very low, and they almost entirely lose their tensile capacity. As the panel advances, when fractures appear in the exposed main roof, the local bearing capacity of the fracture-affected area significantly decreases, making the main roof prone to local failure. The above analysis indicates that in extra-long panel, the main roof in the middle of the panel is likely to break first and then migrate to both sides. That is, the main roof of the panel is prone to forming zonal failure and motion phenomena.

6.5.3.2 Regional Failure Model of Main Roof in the Extra-Long Panel

1. Complete main roof of O-X type failure condition

Before the first weighting, the main roof can be regarded as a thin plate with four sides fixed. If the influence of fractures is not considered, O-X type failure occurs in the main roof, and the break lines are shown in Fig. 6.63. A total of 9 break lines are generated in the complete main roof of O-X type failure, triangular plates ADE and BCF are formed on both sides of the panel, and trapezoid plates CDEF and ABFE are formed in the middle of the panel.

The upper bound theory is used to determine the upper bound condition of occurring O-X type failure in the main roof, and the relationship between the limit load q_{si} that the main roof can withstand before the first weighting and the length of the panel and the advancing distance is obtained:

$$q_{si} = 8M_s \frac{2bm + a^2}{ma^2(3b - 2m)} \tag{6.116}$$

where M_s is the ultimate bending moment; a is the advancing distance of the panel. When the weight of the main roof and following load reach its ultimate bearing capacity and Eq. (6.117) is satisfied, the main roof will occur first failure.

$$q_r \geq q_{si} \tag{6.117}$$

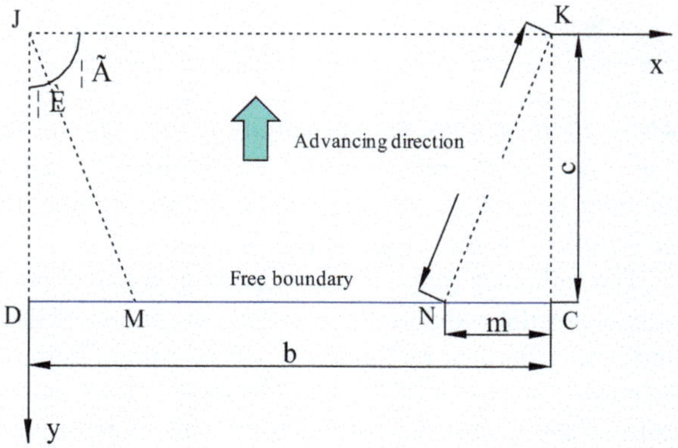

Fig. 6.63 First failure of the main roof

After first failure of the main roof, the panel enters the stage of periodic weighting, and the long side CD near set-up room becomes free boundary. At present, the main roof can be regarded as a thin plate of three edges fixed and one edge free. Because of the change of the boundary condition, the failure form of the main roof is changed, as shown in Fig. 6.64. When the roof occurs periodic failure, 5 break lines are generated in the main roof, and trapezoid plates are still formed in the middle of panel, and triangular plates are formed at both ends.

The relationship between the limit load q_{sp} that the main roof can withstand before the periodic failure and the length of the panel and the advancing distance is obtained:

$$q_{sp} = 6M_s \frac{2m^2 + bm + 4c^2}{mc^2(3b - 2m)} \tag{6.118}$$

where c is the advancing distance of the main roof during periodic failure. When Eq. (6.119) is satisfied, the main roof will occur periodic failure.

$$q_r \geq q_{sp} \tag{6.119}$$

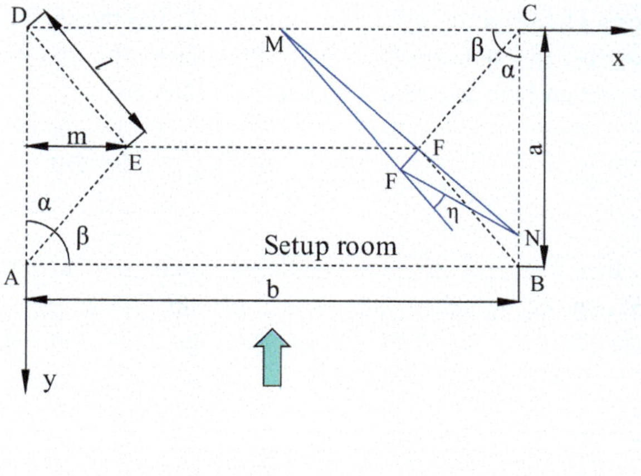

Fig. 6.64 Periodic failure of the main roof

When the O-X type failure occurs in the complete main roof, the triangular plates at both ends along the direction of the panel is significantly smaller than the trapezoid plates in the middle of the panel. Thus, support resistance distribution is usually characterized by large in the middle and small at both ends in the conventional panel.

2. Partition failure conditions of the main roof with fractures during the first weighting

The increase of fractures development degree increases the uncertainty of the failure form of the main roof in deep panel. Under the condition of random distribution of fractures, it is very difficult to determine the main roof type of failure, and only 1 and 2 fractures are considered in the main roof. In the first weighting stage, the main roof with 1 fracture is shown in Fig. 6.65 (the red solid line is a fracture): the fracture has no tensile strength, the crack has no tensile capacity, and the main roof is first destroyed along the fracture, then the fracture develops in the middle of the long edge. Furthermore, the increase of fractures development degree increases the uncertainty of the failure form of the main roof, which leads to the dynamic non-uniform distribution of microseismic events. The mechanical model of roof partition failure of fractures development is established, and the fracture has no tensile strength. The main roof failure is induced by the primary fractures at first, and the local failure occurs. After the failure, the main roof turns to tensile stress leading under the new boundary conditions, local failure phenomenon is driven to migrate to both sides of panel under the effect of load transfer of overburden rock, the evolution process of the failure form of the main roof with fractures is shown in Fig. 6.65.

3. Effect of partition failure of the main roof on support resistance

The roof pressure withstood by the support of panel is mainly provided by the main roof failure rock and the overlying following load, the decrease of lumpiness of the broken rock blocks will inevitably slow down the roof pressure on support. Thus, compared with support resistance in the conventional panel, the support resistance in the extra-long panel (especially the roof pressure withstood by the support in the middle of the panel) will not increase sharply. In addition, the increasing trend of lumpiness of the broken rock blocks from the middle to the two ends of the main roof in extra-long panel will lead to the distribution characteristics of the support resistance exhibiting small center and large ends in the panel or there is a lack of regularity in the distribution of support resistance along the direction of panel, which is different from the distribution characteristics of large center and small ends of support resistance in the conventional panel.

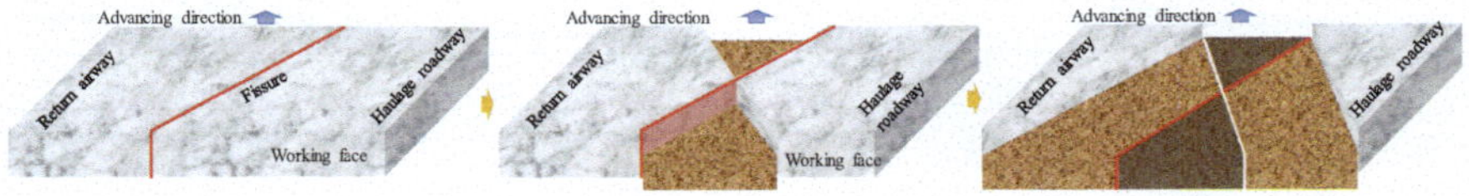

Fig. 6.65 Failure evolution process of the roof with fractures development

6.5.4 Characteristics of Roof Microseismic Activity in the Extra-Long Panel

6.5.4.1 Characteristics of Spatiotemporal Distribution of Microseismic Events

Between March 19, 2021, and November 13, 2021, the microseismic monitoring system recorded a total of 6553 microseismic events induced by mining activity at the panel. The 3D spatial distribution of these microseismic events, based on the location data, is illustrated in Fig. 6.66a. The mining area of the 140,502 panel is delineated by a straight line, with the blue region indicating the advance range during the monitoring period. Microseismic events are represented by small spheres, where the size of the sphere corresponds to the energy of the event, and the color represents the magnitude, which is calculated using Eq. (6.120) [17].

$$\lg E = 4.8 + 1.5M \tag{6.120}$$

where E and M represent the energy and magnitude of microseismic events, respectively. Figure 6.66a shows that microseismic events predominantly occur in regions with strong mining-induced impacts near the 140,502 panel. Due to the low strength

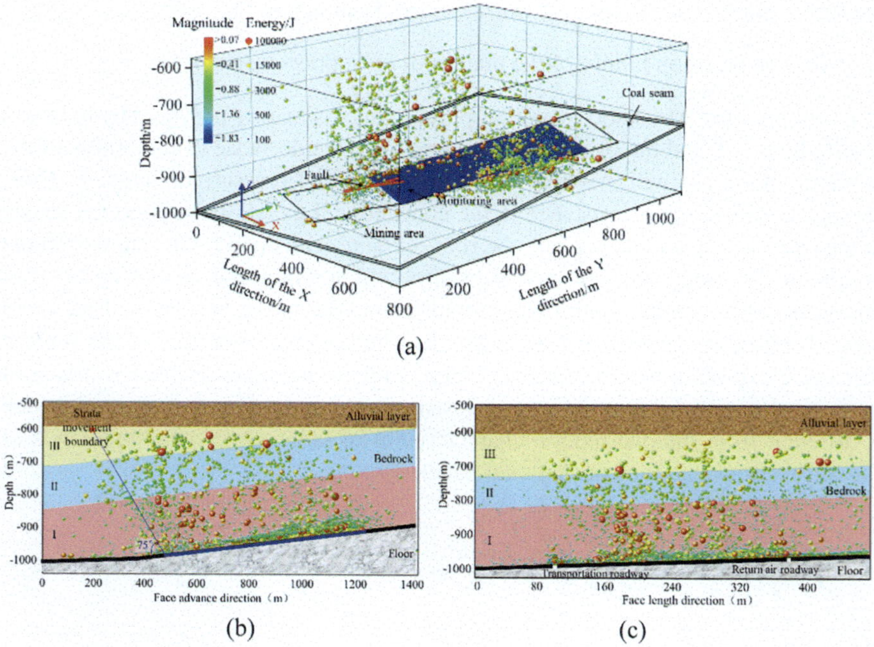

Fig. 6.66 Distribution characteristics of microseismic events: **a** 3D spatial distribution; **b** advance direction profile; **c** panel profile

of the coal seam and immediate roof, and high break degree under the influence of the shearer, a large number of low-energy microseismic events are distributed near the coal seam. As the roof layer elevation increases, the number of microseismic events gradually decreases, however, high-magnitude events (magnitude ≥ 0.5) become more frequent. This indicates an increase in rock strength, a reduction in break degree, larger fragment sizes, and higher energy release during the failure process. Additionally, microseismic events exhibit a clear spatial heterogeneity, with increased activity in fault-affected zones and on the roof rock strata of the return airway side.

When projecting the microseismic events onto a vertical section parallel to the advance direction, as shown in Fig. 6.66b, it is observed that microseismic events in the roof rock strata begin to occur approximately 180 m ahead of the monitoring stopping line, corresponding to the pre-mining influence range determined by anchor rod (cable) pulling force gauges. The vertical development height of microseismic events is located 300–350 m above the coal seam, equivalent to the thickness of the bedrock. Above the bedrock lies a low-strength alluvial layer, which does not release energy upon failure and thus cannot be detected by the sensors. This indicates that the entire thickness of the bedrock above the 140,502 panel falls within the strong mining influence zone. Connecting the position of the panel at the monitoring termination time with the initiation position of microseismic events on the upper surface of the bedrock gives a rock movement angle of approximately 75°. Vertically from the coal seam level, the mining influence range in the bedrock can be divided into three zones: Zone I is located 0–150 m above the coal seam, characterized by intense roof movements and a high number of high-energy microseismic events, representing the strong mining influence zone; Zone II is located 150–270 m above the coal seam, where the number of microseismic events decreases and high-energy events disappear, indicating an increased distance between the rock layer and the coal seam, entering a weak mining influence zone; the strata above Zone II, nearing the upper surface of the bedrock, experience renewed microseismic activity with scattered high-energy events, marking the active zone of the bedrock surface.

The microseismic events were projected onto a vertical section parallel to the long axis of the panel, as shown in Fig. 6.66c. The distribution of microseismic events along the long axis of the panel is asymmetrical. High-energy events within the intense mining-induced zone predominantly occur in the middle of the panel and the roof above the haulage roadway, with a few high-energy microseismic events observed in the active zone of the rock surface on the return airway side. This distribution is closely related to the development of roof fractures, which will be further explained later. High-energy events in the active zone of the rock surface are mainly distributed on the return airway side. This distribution is influenced by the inclination of the panel, causing the upper surface of the rock on the return airway side to enter a state of tensile stress, thereby promoting the occurrence of high-energy microseismic events through tensile failure.

The dynamic and cumulative distribution characteristics of microseismic events with respect to the advance time (distance) of the panel are illustrated in Fig. 6.67a. Microseismic monitoring began on March 19 when the panel had advanced 90 m past the set-up room, and it concluded on November 16, by which time the panel had advanced a total of 768 m. Most of the microseismic events captured by the monitoring system were less than 30,000 J in energy. These low-energy events were triggered by local damage in the roof and were approximately uniformly distributed within the same stratigraphic level of the roof strata, having minimal impact on the underlying panel operations and not causing widespread instability of the surrounding rock. High-energy events carrying more than 30,000 J were significantly fewer. These events are closely related to large-scale fracturing of the roof and are dispersed and unevenly distributed within the roof strata. However, they have a considerable impact on the underlying panel operations, potentially triggering dynamic loading effects in the roof.

When the panel advance distance was less than 500 m, the number of microseismic events increased rapidly, however, the proportion of high-energy events was small. After reaching 500 m, the ratio of increase in the number of microseismic events slowed, however, the proportion of high-energy events increased. This difference is attributed to changes in the advance speed of the panel. When the advance distance reached 500 m, the advance speed increased from 3 to 5 m/day. Essentially, an increase in advance speed elevates the loading ratio on the roof strata, leading to an increase in their load-bearing capacity, a reduction in the frequency of failures, an enhancement in energy storage capacity, a higher degree of strain energy accumulation, and a greater ratio of energy release upon failure. Consequently, there is an overall decrease in the number of microseismic events but an increase in the number of high-energy events. During the rapid advance stage of the panel, three extra-high-energy microseismic signals, each carrying nearly 150 kJ of energy, were recorded. Analysis of the hydraulic support working resistance at the corresponding positions of the panel, based on the location information of these three microseismic events, indicates that extra-high-energy microseismic events signify extensive roof activity. The hydraulic supports of the underlying panel experience significant dynamic loading effects from the roof, with the maximum instantaneous working resistance of the supports approaching their rated working resistance. Safety valves on the supports may activate, necessitating caution against potential roof collapse accidents induced by extra-high-energy microseismic events, which could

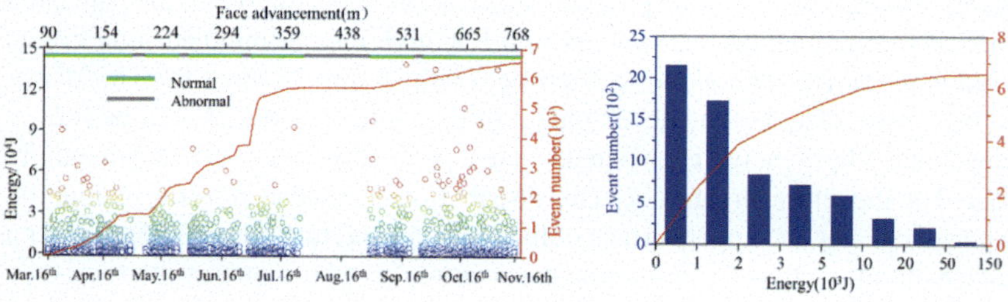

(a) (b)

Fig. 6.67 Distribution characteristics of microseismic events energy: **a** temporal distribution of microseismic events; **b** energy-frequency distribution even trigger extensive dynamic failures of the underlying coal and rock strata and induce dynamic disaster effects.

The bar and cumulative distribution characteristics of microseismic events concerning energy magnitude are shown in Fig. 6.67b. Over 90% of microseismic events carry less than 10,000 J of energy and have no significant impact on the stability of the surrounding rock. However, 8.36% of the events carry more than 10,000 J, warranting caution for strong mine pressure phenomena associated with these high-energy events during mining operations. Particularly notable are 34 instances of microseismic events carrying over 50,000 J of energy, which pose significant risks of dynamic loading effects and widespread roof collapse accidents.

Statistical analysis of the magnitude of microseismic events monitored during three advance stages of the panel (0–200, 200–400, and 500–700 m) yielded histograms as shown in Fig. 6.68. The magnitudes of roof microseismic events in all three stages follow a normal distribution, with mean magnitudes of – 1.06, – 1.04, and – 0.88 respectively, and variance values of 0.08, 0.10, and 0.18. This characteristic aligns with the pattern of microseismic event energy level distribution. Due to the slower advance speed of the panel in the initial two stages, the mean magnitudes are relatively small and approximately equal. However, in the final stage, as the advance speed increases to 5 m/day, there is a noticeable upward trend in mean magnitudes. The increasing trend in magnitudes also results in a corresponding increase in their dispersion, evidenced by a significantly higher variance in magnitude during the third advance stage compared to the preceding two stages. The observed increase in both mean magnitudes and variance indicates intensified disturbance effects on the roof strata during the rapid advance phase of the panel, leading to more active roof microseismic activity. This heightened activity complicates predictions of roof activity intensity and poses increased challenges for controlling the roof strata of the panel.

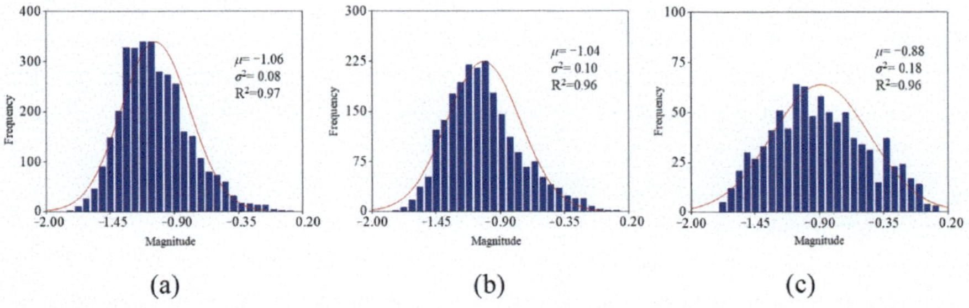

Fig. 6.68 Magnitude distribution characteristics of microseisms under different pushing progress: **a** 0–200 m; **b** 200–400 m; **c** 500–700 m

6.5.4.2 Methods for Hazard Zone Prediction Based on Microseismic Information

To develop a method for evaluating the stability of the surrounding rock in the panel based on microseismic information, an analysis was conducted on lower strata. Measurement points N_{11} to N_{ij} were arranged on the iOj plane. The residual energy of stress waves from microseismic sources with energy greater than 10,000 J, propagating to the target points, was superimposed, as illustrated in Fig. 6.69. During spherical propagation, the stress wave experiences both geometric attenuation and intrinsic attenuation due to internal friction and heat conduction within the propagation medium. The attenuation behavior of source energy during propagation is governed by Eq. (6.121).

$$E' = ER^{-2}e^{-\eta R} \tag{6.121}$$

where E' is the residual energy of stress waves reaching each monitoring point on the prediction plane, J; E is the energy of the seismic source, J; R stands for the distance from the seismic source to the monitoring point, m; η represents the attenuation coefficient, with a value of 0.00094.

The cumulative results of residual energy at the target point are normalized using Eq. (6.121), yielding a range of [0, 1] for the roof damage index. This allows for a quantitative classification of roof stability.

$$n = (E' - E'_{min})/(E'_{max} - E'_{min}) \tag{6.122}$$

where n denotes the damage index, E'_{min} represents the minimum energy value at each monitoring point, at J, E'_{max} represents the maximum energy value at each monitoring, at J.

Vibration energy ranging from 1×10^2 J to 1×10^{10} J exerts varying degrees of influence on the movement of rock strata. The instability of rock structure is closely related to the joint structure between rock blocks, fracture shape, and the physical–mechanical properties of rock blocks. By substituting the gradient values of

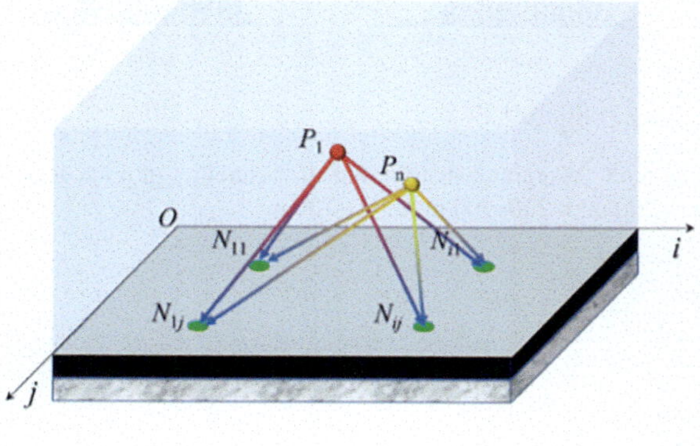

Fig. 6.69 The attenuation of source energy in rock strata

accumulated residual energy into Eq. (6.122), the damage indices corresponding to 1 $\times 10^2$ J and 1 $\times 10^3$ J are found to be 0.10 and 0.75, respectively. Therefore, a damage index of 0.10 is established as the threshold for the affected area, while a damage index of 0.75 serves as the danger zone threshold. The calculated results of the main roof damage index when the panel advances to 120, 260, and 660 m are shown in Fig. 6.70. When the panel reaches 120 m, microseismic activity primarily concentrates the damage to the front area of the panel, with the damage index evenly distributed along the length of the panel. However, the level of roof damage is relatively higher on the headgate side. The roof damage area is approximately 180 m ahead of the panel, and the calculated disturbance range is consistent with the measured results. At this point, the extent of roof damage is relatively low, at around 0.50. As the panel advances to 260 m, the damage area of the tailgate side extends 220 m ahead of the panel, with highly damaged areas shifting from the front of the panel to the vicinity of the panel. When the panel advances to 660 m, microseismic activity in the tailgate and central roof tends to stabilize. Due to the cumulative effects of fault activation, the roof degradation near the headgate within a range of 100 m is severe. At this stage, the fault has been activated by mining disturbances, however, high-energy events are generally distributed in high-level rock strata, resulting in weaker damage to lower-level rock strata. Therefore, no high-level roof damage appears in the fault area, and on-site investigations indicate that staff did not notice any abnormal roof phenomena during this period. During this stage, the highly damaged area of the rock strata shifts to the goaf behind the panel, indicating continued movement of overlying rock strata due to the action of thick alluvial layers. The main roof undergoes secondary fracturing, leading to a further increase in the damage index.

The distribution characteristics of the support resistance of the panel during the extraction period are shown in Fig. 6.71. The calculation results of the main roof damage index for the three advance stages correspond well with the high-resistance zones in the support resistance distribution. As the advance extent increases, the

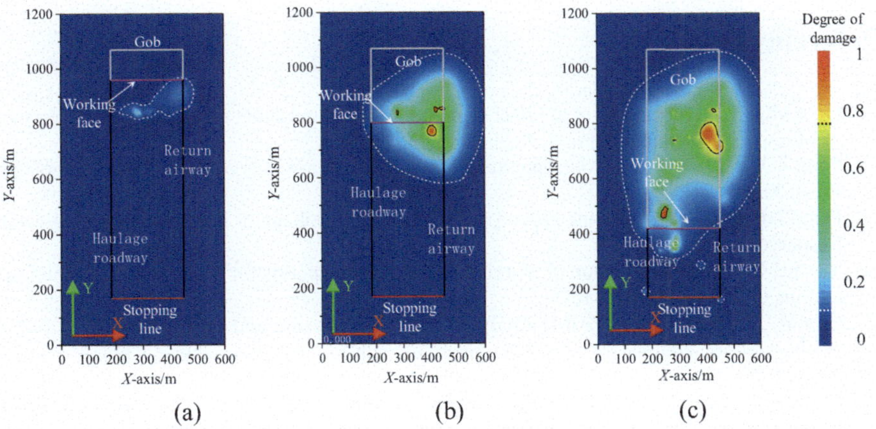

Fig. 6.70 Roof damage in different mining stage: **a** 120 m; **b** 260 m; **c** 660 m

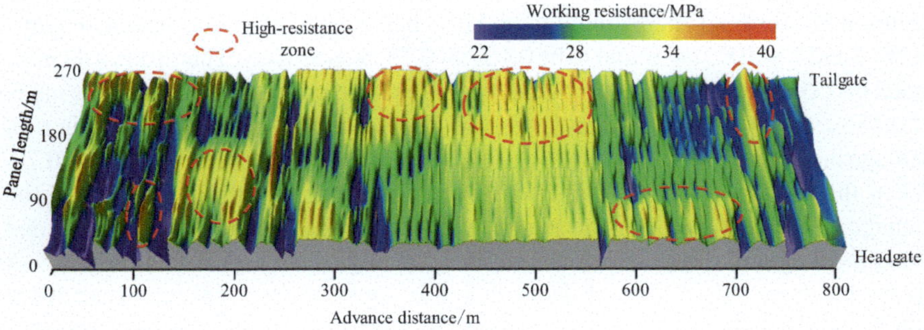

Fig. 6.71 Distribution diagram of support pressurein 140,502 extra-long panel

main roof damage index gradually rises, and the support resistance increases accordingly. Areas where the roof damage continuously escalates exhibit widespread and sustained increases in hydraulic support resistance at the panel. Initially, the roof transitions from being intact to undergoing localized damage, resulting in short-term, intense increases in support resistance over small areas, demonstrating significant dynamic load characteristics. This is evident in the affected zone near the tailgate ahead of the panel in Fig. 6.70c and is consistent with the high-resistance zone measured at an advance distance of 750 m in Fig. 6.71. The consistency between the measured and predicted results verifies the accuracy of the quantitative evaluation method for roof stability proposed in this study. Particularly in the later extraction stages of the panel, the increased extraction speed leads to a reduction in near-field roof damage, and the overall support resistance decreases. However, the randomness of the high-resistance zones increases. Predicting sudden increases in support resistance using the roof damage index impact area is reliable, helping to prevent the panel from being affected by dynamic load effects.

References

1. Qian MG, Xu JL, Wang JC et al (2021) Ground pressure and strata control. China University of Mining and Technology Press, Xuzhou
2. Song ZQ (1998) Practical ground pressure control. China University of Mining and Technology Press, Xuzhou
3. Peng Syd S (2008) Coal mine ground control, 3rd edn. College of Engineering and Mineral Resources
4. Yan SH (1997) Determination of working resistance of supports for sub-level caving mining. J China Coal Soc 22(1):15–19
5. Li HM, Jiang DJ, Li DY (2014) Analysis of ground pressure and roof movement in fully-mechanized top coal caving with large mining height in ultra-thick seam. J China Coal Soc 39(10):1956–1960
6. Wang JC, Yang SL, Li LH (2018) Toppling-slumping failure mode in horizontal sublevel top-coal caving face in steeply-inclined seam. J China Univ Min Technol 47(6):1175–1184
7. Yang SL, Li ZX, Wei WJ et al (2017) Influence of loose top-coal on support and surrounding rock relationship in longwall top-coal caving mining. J China Coal Soc 42(10):2511–2517

8. Yang SL, Li LH, Deng XJ (2020) Disaster-causing mechanism of roof "toppling-slumping" failure in a horizontal sublevel top-coal caving face. Nat Hazards 100(2):757–780
9. Qian MG, He FL, Wang ZT et al (1994) A further discussion on the theory of the strata behaviors in longwall mining. J China Univ Min Technol 23(3):1–9
10. Yang SL (2019) Study on the disaster-causing mechanism and control criteria of the hard and thick roof strata based on medium thick plate theory. China University of Mining & Technology
11. Xu JL, Ju JF (2011) Structural morphology of key stratum and its influence on strata behaviors in fully-mechanized face with super-large mining height. Chin J Rock Mech Eng 30(8):1547–1556
12. Wang JC, Wang ZH, Tang YS et al (2021) Experimental study on mining-induced dynamic impact effect of main roofs in deeply buried thick coal seams with weakly consolidated thin bed rock. Chin J Rock Mech Eng 40(12):2377–2391
13. Wang ZH, Tang YS, Li M et al (2023) Development and application of overburden structure composed of caving arch and towering roof beam in deep longwall panel with thin bedrock. J China Coal Soc 48(2):563–575
14. Xiang HF, Liu GD (1991) Stability and vibration of arch structure. China Communications Press, Beijing
15. Lee YK, Park JW, Song JJ et al (2014) Model for the shear behavior of rock joints under CNL and CNS conditions. International J Rock Mech Min Sci 70:252–263
16. Wang JC, Yang SL, Yang BG et al (2019) Roof sub-regional fracturing and support resistance distribution in deep longwall face with ultra-large length. J China Coal Soc 44(1):54–63
17. Wang JC, Tang YS, Wang ZH et al (2023) Characteristics of microseismic events and damage degree calculation method in kilometer deep fully mechanical longwall panel. J China Univ Min Technol 52(3):417–431

Chapter 7
Typical Cases of Top Coal Caving Mining

7.1 LTCC in Thick Coal Seams Under Favorable Conditions

7.1.1 LTCC Technology in Lu'an Mining Area

7.1.1.1 Historical Development of LTCC in the Lu'an Mining Area

The Lu'an mining area is in the southeastern part of Shanxi Province, at the western foot of the Taihang Mountains in the Changzhi Basin. The Lu'an Mining Bureau's affiliated mines primarily extract the No. 3 coal seam in the eastern part of the Qinshui coalfield. The seam is stable with a thickness ranging from 5.61 to 7.23 m, averaging 6.42 m. It dips at an angle of 3–5°, locally up to 10°, and includes 1–3 layers of parting. The coal is medium-hard with a hardness coefficient of 1.0–3.0. The roof is stable, composed of sandstone, while the floor consists of argillaceous shale.

Since 1987, the Lu'an mining area has been researching and designing LTCC technology. The first on-site industrial test began in 1988 at the 4309 panel of Wangzhuang Coal Mine, successfully initiating LTCC operations. Over the years, significant achievements have been made in equipment, caving method, and coordinated control of mining and caving operations [1].

In March 1989, Wangzhuang Coal Mine conducted an experimental trial of LTCC with a high-level window at the 4309 panel. The panel was 125 m long, with a cutting height of 2.8 m and a caving height of 4.06 m. They employed a caving interval known as two-cutting and one-caving. The monthly production from the panel reached 92,000 tons, with a top coal recovery of 78.97%. This successful trial enabled an annual production capacity of 1 million tons [2]. At the 4309 panel, the ZFD4000-17/23 high-level caving supports were used alongside the SGZC-730/320 AFC. Due to the medium-hard coal seam with underdeveloped fractures, water was

J. Wang and W. Wei, *Thick Coal Seam Underground Mining*,
https://doi.org/10.1007/978-981-96-3874-1_7

injected into boreholes in the mining roadway to soften the top coal, resulting in excellent effectiveness [3, 4].

In June 1991, at Wuyang Coal Mine, a trial of low-level LTCC was conducted using ZZPF4000-1.7/3.5 type supports. The coal seam was 6.42 m thick, with a cutting height of 2.6 m and a panel length of 165 m. They employed a staggered sequential multi-round caving technique with the caving interval of one-cutting and one-caving, operating in parallel. The maximum monthly production from the panel reached 200,000 tons, with an average monthly production of 120,000 tons and a coal recovery ratio of 91.1%. This fully demonstrated the high productivity and efficiency potential of LTCC mining.

In 1993, using the ZZP4800-17/33F low-level LTCC supports and associated equipment developed internally, the Lu'an mining administration achieved a national record at the Wangzhuang Coal Mine's 6111 panel with an annual production of 2.53 million tons. This accomplishment not only realized the goal of fully extracting thick coal seams in one pass but also enabled highly intensive coal production, achieving high productivity and efficiency. During that year, all mines under the Lu'an mining administration adopted LTCC technology, playing a crucial role in advancing China's LTCC techniques.

In July 1999, the Wangzhuang Coal Mine conducted a trial mining operation on the 4326 panel, which involved a 270-m-long super-long LTCC panel. They appropriately increased the power of the shearer and the front and rear AFCs. Simultaneously, they reformed roadway support techniques, adopting the small coal pillar bolting-wire mesh support technology. The annual production from the panel reached 4 million tons with a recovery ratio of 88.4% [5].

In 2005, the Lu'an mining area pioneered research on comprehensive supporting technologies for large mining height automated LTCC operations at the Wangzhuang Coal Mine's 6203 panel. The hydraulic support model used was ZF7000/20/40 with a support strength of 0.88 MPa. The mining machine employed was the MGTY400/930–3.3D electric traction shearer, and the AFC used was the SGZ-960/1400 type. The panel achieved a peak daily production of 33,186 tons and an average daily production of 26,068 tons. The highest monthly production reached 633,168 tons, with a peak labor productivity of 502 tons per worker and an average of 394 tons per worker. The average recovery ratio of the panel was 91.6%, reaching a maximum of 92.1%. The technology of large mining height automated LTCC increases the ventilation cross-section of the panel, facilitating methane emissions while improving mining efficiency. In recent years, guided by policies, intelligent mining has rapidly developed, with beneficial experiments in intelligent LTCC technologies at the Changcun Coal Mine's 2115 panel in the Lu'an mining area.

7.1.1.2 High-Yield and Efficient LTCC Panel at Wangzhuang Coal Mine

1. Basic conditions

The 91–208 LTCC panel at Wangzhuang Coal Mine is designed northward to interface with the 91–207 designed panel, southward adjoining the 91–209 already mined panel, with the 540 belt roadway extension to the east and the mine boundary to the west. The panel mines the No. 3 coal seam, situated in the lower part of the Permian Shanxi Formation. The seam exhibits stable thickness, with a maximum parting of 0.45 m below and a coal thickness of 7.28 m. The haulage gateway of this panel shows a monoclinic structure sloping from east to west, with local undulations and a coal seam dip angle ranging from 1 to 5°. The mine is characterized as a high gas mine, with coal seams resistant to spontaneous combustion and coal dust exhibiting explosive characteristics.

2. Arrangement of the panel and main equipment

The main equipment at the 91–208 LTCC panel includes a shearer, hydraulic supports, front and rear AFCs, transfer machine, crushers, belt conveyors, emulsion pumps, spray pumps, and others as depicted in Fig. 7.1.

The shearer model WG400/930-WD is suitable for mining heights of 2800–3500 mm, with a traction power of 110 kW and a total power of 930 kW. It is designed to operate on panels with inclinations ranging from 0 to 25°.

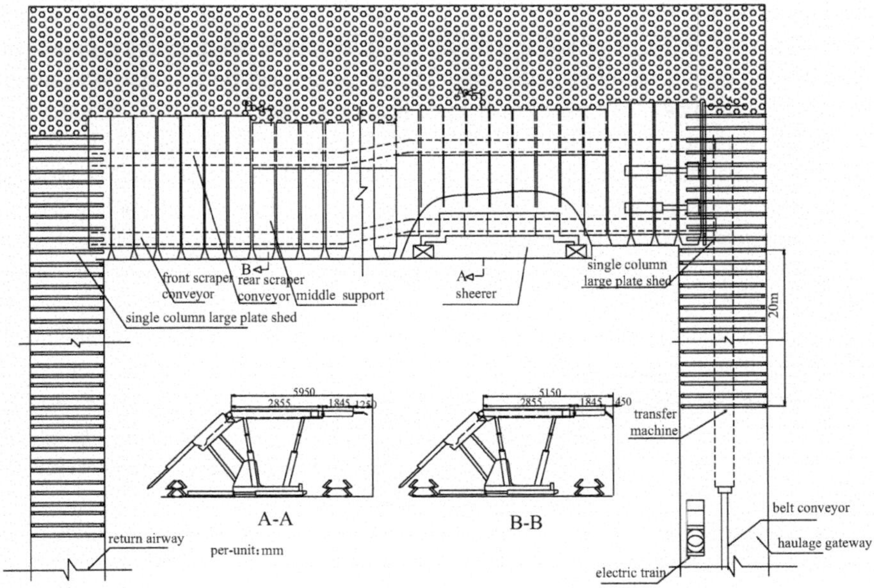

Fig. 7.1 Layout and main equipment of Panel 91-208

The panel is equipped with a total of 193 supports, including 182 middle supports of type ZF8000/20/38. These middle supports have a working height ranging from 2000 to 3800 mm and a support width of 1430–1600 mm. They provide an initial support force of 6972 kN and a working resistance of 8000 kN, operating under a rated hydraulic pressure of 31.5 MPa. The average support strength is 0.85 MPa, suitable for coal seam angles up to 20°. Additionally, there are 11 head (tail) supports of type ZFG9600/23/38, with 4 located at the head and 7 at the tail of the equipment area. These supports operate at heights from 2300 to 3800 mm and provide a working resistance of 9600 kN. Furthermore, positioned at the head is an end support frame of type ZT2200/25/40, with a working height from 2500 to 4000 mm and a support width of 320 mm.

The front and rear AFCs are both of SGZ1000/2 × 1000 type, totaling 2 units. Each SGZ1000/2 × 1000 AFC has a conveying capacity of 2000 t/h, operates at a chain speed of 1.35 m/s, and is powered by a motor of 1000 kW.

The transfer machine model is SZZ1200/525, with a conveying capacity of 3000 t/h and a length of 50 m. It interfaces with the conveyor belt over a length of 12 m at the tail end, and operates at a slope angle of 10° with a slope height of 1.45 m.

The crusher model is PLM-3500, with a crushing capacity of 3500 t/h. It has a maximum inlet cross-section of 1200 × 1000 mm and produces an output particle size smaller than 300 mm. The installed power is 250 kW.

In roadway 91–208, 1st belt conveyor model DSJ-140/230/3*400, transport capacity 2300 t/h. 2nd belt conveyor model SSJ-120/150/*315, transport capacity 2000 t/h.

The emulsion pump model is BRW-400/31.5, with 3 units installed, along with 1 reservoir tank. It has a nominal flow ratio of 400 L/min and operates at a nominal pressure of 31.5 MPa. The voltage rating is 1140/660 V, with a motor power of 250 kW. The pumps are designed to handle a working medium of neutral aqueous solution containing 4–5% emulsion oil, and they are installed in the haulage gateway.

The spray pump model is BPW-400/16, with 2 units installed, along with 1 reservoir tank. It operates at a nominal pressure of 16 MPa and has a nominal flow ratio of 400 L/min. The motor power is 110 kW, with a voltage rating of 1140/660 V. These pumps are installed in the haulage gateway.

3. Layout and support system of roadway

During the installation of equipment at the 91–208 panel, the roadway layout included four types: haulage gateway (headgate), ventilation roadway (tailgate), dedicated return airway, and high drainage roadway, as depicted in Fig. 7.2. According to the design of the panel, the haulage gateway at the 91–208 panel has a total length of 2462 m. Additionally, it utilizes the return airway and dedicated return air roadway from 91 to 209, with a net cutting length of 290.5 m.

The haulage gateway at 91–208 has a rectangular cross-section, with an average net width of 6 m and an average net height of 3.6 m. It spans a total length of 2462 m, with an average net cross-sectional area of 21.6 m^2. The cutting section is also rectangular, with a net width of 7.8 m, a net height of 3.6 m, and a net cross-sectional area of 28.08 m^2. For support, the roadway roof employs resin capsule extended

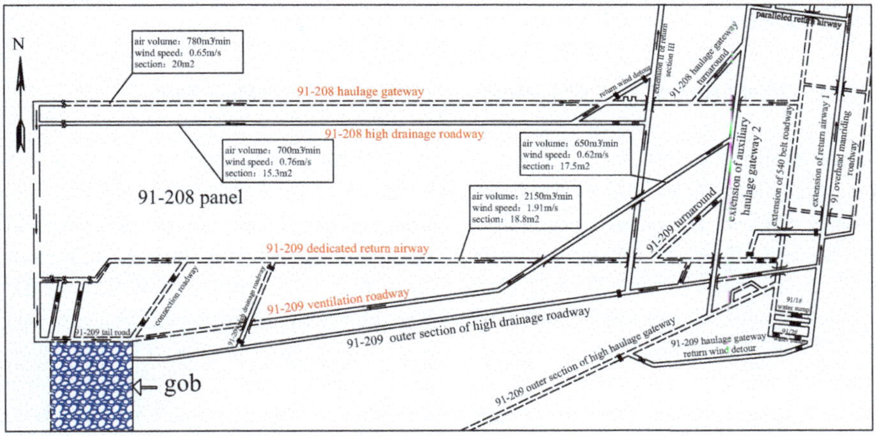

Fig. 7.2 Roadway layout of 91–208 panel

anchor LWGSSZMG500-Φ22L2400 mm high-strength threaded steel anchor rods and steel band chain anchor cables for reinforcement. It includes the installation of metal mesh, double-resistant mesh, and steel ladder beams, as shown in Fig. 7.3.

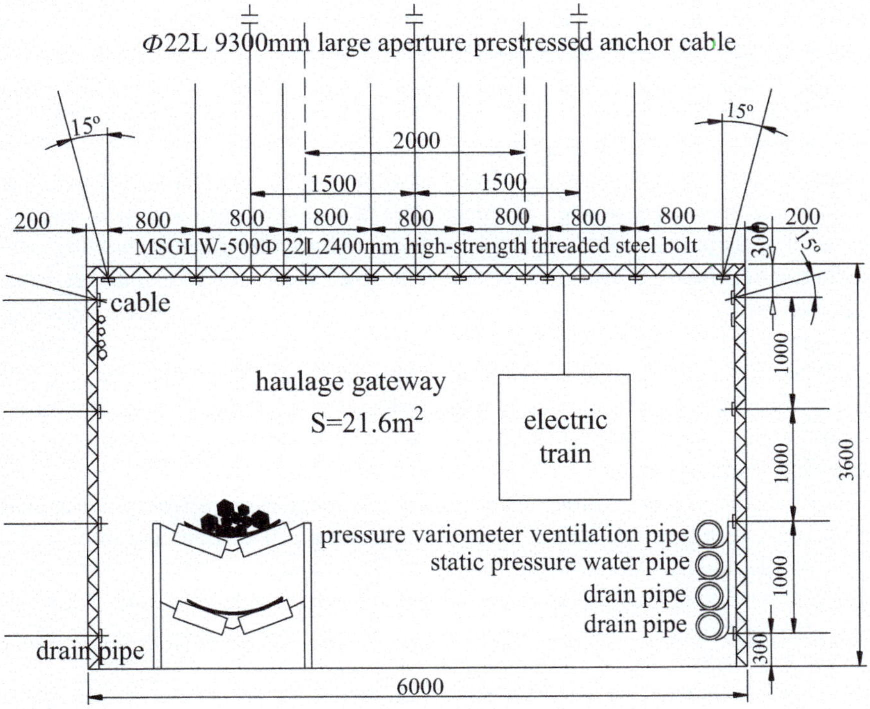

Fig. 7.3 Roof support for haulage gateway of 91–208 panel

7.1.2 LTCC Technology in Yanzhou Mining Area

7.1.2.1 Evolution of LTCC Technology in Yanzhou Mining Area

Yanzhou mining area includes the majority of the Yanzhou Coalfield and the southern part of the Jining Coalfield, both concealed coalfields of the Carboniferous Permian System covered by Quaternary alluvial deposits. The basement of the coalfield consists of Ordovician limestone, overlain by remnants of Upper Jurassic red sandstone. Coal seams suitable for LTCC mining include the 3rd upper coal seam of the Yanzhou Coalfield, ranging from 3.6 to 10.0 m thick with an average of 5.23 m, and the 3rd lower coal seam of the Jining Coalfield, ranging from 0 to 17.96 m thick with an average of 4.68–5.26 m. Gas emission levels in the mines of the area are low, with historical gas detection results consistently indicating low gas emissions. Each mineable coal seam presents a risk of coal dust explosions. All seams exhibit a natural tendency for spontaneous combustion, with thick coal seams having a natural combustion period of 3–6 months [6].

In 1992, LTCC was first implemented at Xinglongzhuang Coal Mine of Yanzhou mining group. This involved the use of equipment including ZFS-5200/17/35 supports, ZFG-5200/18/32 transitional supports, AM-500 shearer, SGB-764/264W front AFC, SGZ-764/320D rear AFC, SZB-764/132 transfer machine, SDJ-150 belt conveyor, and MRB-125/3.5 pump station. Operations commenced in June at the 5306 panel with trial production and full production began on July 1st. By the end of the year, the mine had produced 644,000 tons of raw coal, achieving a recovery ratio of 81%. The average monthly output was 99,750 tons, with a mining efficiency of 32.803 tons per man-shift, achieving safety and high productivity goals. This laid the foundation for LTCC in the Yanzhou mining area.

In March 1993, Baodian Coal Mine began employing the LTCC method. The coal seam at the panel was 8.5 m thick and 125 m long. The equipment used at the panel was similar to that at Xinglongzhuang Mine. The average monthly coal production reached 171,700 tons, with a mining efficiency of 74.8 tons per man-shift, demonstrating significant economic benefits. In 1994, the LTCC operations at Xinglongzhuang Mine's Mining Team 1 yielded 2.3 million tons, while Baodian mine's Mining Team 2 produced 2.2 million tons. By 1995, Xinglongzhuang Mine's Mining Team 1 achieved an actual coal production of 3,006,036 tons, becoming the first team in the country to exceed 3 million tons annually in comprehensive mechanized mining output.

In 1997, the second mining team of Dongtan Coal Mine achieved an annual production exceeding 4.1 million tons. The coal seam at Dongtan coal mine had a thickness ranging from 5.6 to 6.5 m, with an inclination angle of 3–8°. By 1998, four mining teams in the area had reached an annual production level of over 3 million tons, with Dongtan Coal Mine's mining team being the first to surpass the 5 million tons per year mark.

In 1999, Xinglongzhuang Coal Mine adopted the ZF6200-16/35 type high-resistance caving supports, mining a 8.2-m-thick coal seam with an inclination of

4–12°. The panel was 162 m long and employed a one cutting and one caving sequential caving method, with parallel cutting and coal shearing operations. The average monthly production at the panel was 350,000 tons, achieving a recovery ratio of 85.17%. The average efficiency of the mining personnel was 177.98 tons per worker.

In 2002, Dongtan Coal Mine mining team achieved an annual production of 6.07 million tons using domestically produced equipment. Xinglongzhuang Coal Mine mining team achieved an annual production of 6.4 million tons from the caving panel with electric-hydraulic automatic control developed during the "Tenth Five-Year" tackling project, both surpassing 6 million tons.

Since the beginning of the twenty-first century, the Yanzhou Mining Group in the Yanzhou mining area embarked on a path of international expansion. In 2004, Yancoal Australia Ltd. was established, and that same year, it acquired the Austar coal mine in New South Wales, Australia, marking the first use of LTCC technology in Australia. In 2019, the Jinjitan Coal Mine successfully applied the world's first set of 7-m extra-large mining height LTCC equipment, achieving successful extraction with an annual production capacity reaching tens of millions of tons.

7.1.2.2 LTCC Equipment for Yanzhou Mining Area

After adhering to domestically produced equipment, Yanzhou mining area has developed a process of "industrial testing, promotion and application, improvement and enhancement, and innovative development." Based on varying coal seam extraction conditions, it has preliminarily established a comprehensive set of high-yield and efficient LTCC mining equipment models: low (annual production 1–2 million tons), medium (annual production 2–3 million tons), high (annual production 4–5 million tons), and extra-high grade (annual production 6–7 million tons), tailored to different coal seam occurrences. Details of production targets, process parameters, and equipment characteristics for medium, high, and extra-high grade equipment models are outlined in Table 7.1.

7.1.2.3 Innovative Application of LTCC in Australia [7]

The Austar Coal Mine is located in the Hunter Valley region of New South Wales, Australia, approximately 160 km north of Sydney, 65 km west of Newcastle, and 8 km south of Cessnock. The mining lease covers an area of approximately 63 km^2, with geological reserves close to 140 million tons, and a recoverable coal volume of 50 million tons. The coal seams at the mine are stable, with a thickness ranging from 4.5 to 7.5 m and averaging 6.0 m, inclined at around 4°, and buried at depths of 450–700 m. The immediate roof consists of interbedded sandstone, conglomerate, and sandstone/siltstone layers, reaching up to 20 m in thickness. The immediate floor is composed of thinly bedded mudstone/claystone/siltstone/sandstone, underlain by a thick conglomerate layer. The coal seams exhibit cleat and joint development. The mine is classified as a low gas emission mine with explosive coal dust and a risk

Table 7.1 Characteristics of equipment at various production levels in Yanzhou mining area

Production levels		Medium grade	High grade	Extra-high grade	
Annual production/ten thousand tons		200–300	400–500	600–700	
Technological parameters	Mining height/m	2.8	2.8	3.0	3.2
	Caving height/m	2.8–5.6	2.8–5.6	5.4	5–6
	Cutting depth/m	0.6	0.8	0.8–1.0	
	Drawing interval/m	0.6–1.2	0.8–1.6	0.8–1.0	
	Face length/m	150–200	200–250	250–300	
Matching equipment	Shearer	AM500 three tractions	MGTY400/900–3.3D	SL300	SL750
	Hydraulic support	ZFS5600/17/32 ZTF5400/22/32	ZFS6200/18/35 ZTF6500/19/32	ZFS6800/18/35 ZTF7000/19/32	ZFY8500/21/40 ZFG10800/22/38
	AFC	SGZ-764/400 SGZ-830/500 SGZ-830/630H	SGZ-960/750 SGZ-900/750H	SGZ-1000/1200 SGZ-1200/1400H	SGZ-1000/1400 SGZ-1000/1400H
	Transfer machine	SZZ-1000/375 SZZ-1000/400S	SZZ-1000/375(400)	SZZ-1200/525	SZZ-1200/700
	Crusher	PLM-3000	PLM200	PLM-3500	PCM250
	Belt conveyer	SSJ-1200/3 × 200	SSJ-1200/3 × 200	SSJ-1400/3 × 400	SSJ-1400/6 × 400
	Pump station	GRB-315/31.5	GRB-315/31.5	EHP-3KE	GRB-315/31.5
Remark	"Eighth Five-Year" equipment		"Ninth Five-Year" equipment	"Tenth Five-Year" equipment	6 million tons of automated and informatized equipment

of spontaneous combustion, with a latent period of approximately 3–6 months. The total inflow of water into the mine averages around 2.0 ML/d. The mine adopts a combined development method of inclined and vertical shafts. The main equipment and performance parameters of the panel are depicted in Fig. 7.4.

The hydraulic support model is 2-LEG-SHIELD-1800/3500-971T, with a working resistance of 9710 kN, a center distance between supports of 1.75 m, a support range of 1.9–3.5 m, a moving step distance of 800 mm, weighing 31 tons, operated via PMCR support controller.

The shearer has an installed power of 1250 kW, a cutting depth of 800 mm, a cutting height range of 2.2–3.5 m, and uses VVVF variable frequency traction. The drum diameter is 2.2 m, operating at a voltage level of 3.3 kV.

Fig. 7.4 LTCC complete mining equipment in Austrada Coal Mine

The front AFC model PF6 is of the bottom-sealing type, with a power of 2 × 540 kW, chain speed of 1.12 m/s, throughput of 1800 t/h, trough width of 1000 mm, and soft start CST. The rear AFC model PF5 is of the open-bottom type, with a power of 2 × 540 kW, chain speed of 1.12 m/s, throughput of 1500 t/h, trough width of 1000 mm, and soft start CST.

The transfer machine model PF6/1342 has a power of 400 kW, throughput of 3000 t/h, trough width of 1200 mm, and length of 37 m. The crusher model SK11/18 has a power of 400 kW and a crushing capacity of 3000 t/h. The mobile substation has a capacity of 6250 KVA, with an input voltage of 11 kV and an output voltage of 3.3 kV. It moves on tracks for self-propulsion.

The pump station is configured as a crawler-mounted trailer with four pumps and one tank, consisting of two emulsion pumps (35 MPa, 306 L/min each), one high-pressure emulsion pump (42 MPa, 250 L/min), one spray pump (13 MPa, 630 L/min), and an emulsion tank capacity of 10,000 L. Monorail hoist with a storage length of 200 m, driven by pneumatic motors, used for auxiliary transportation of materials and other goods.

In the LTCC panel at Austrada Coal Mine, several advanced coal mining technologies have been integrated, including automatic memory cutting technology for the shearer, automatic support-following and machine-moving technology, automatic coal flow balancing technology, CST soft-start technology, panel condition monitoring technology, and remote data communication technology.

The panel is equipped with two-column shield-type hydraulic supports. Compared to traditional four-column top coal supports, this type offers greater effective support force and better adaptability. The front-end force of the support beam is beneficial for preventing roof collapse in the front area of the panel. Additionally, the design of the piping is simple, facilitating easier automation implementation.

At the end of the panel, a paired set of left–right separated stepping-type roadway ends supports have been implemented. This support system consists of independent left and right sections, each equipped with roof fall prevention jacks on the top beams. Between the supports and the roadway transfer machine, there are stepping jacks for incremental movement, acting as pivots for their mobility. The support adjacent to the panel moves one step at a time, while the outer support moves in two steps, first the front and then the rear parts, thereby reducing the unsupported area and duration of the roof.

The pump station provides high fluid pressure and the safety valve opening pressure is high, ensuring the supports are automatically replenished to maintain high support force. Since the late 1970s, the pump station pressure at the mining panel in China has consistently been maintained at 31.5 MPa. Recently, adjustments were made by setting the pressures of two emulsion pumps and the safety valve opening pressure of the supports to 35 MPa and 45 MPa, respectively. This was achieved without increasing costs and resulted in an 8.55% increase in initial support force and a 6.4% increase in working resistance. To further enhance the initial support force, modifications were made to the third emulsion pump, enabling it to achieve an outlet pressure of 41 MPa. An independent high-pressure fluid supply line was also installed specifically to supply fluid to the supports. This high-pressure supply system is controlled by a hydraulic valve pilot circuit, which activates a fluid-controlled check valve when the system pressure reaches 35 MPa in the lower chamber of the supports. This action automatically opens the high-pressure replenishment circuit, allowing 41 MPa high-pressure emulsion fluid to enter the lower chamber of the supports. These enhancements resulted in a 27% increase in initial support force.

The transportation equipment at the panel is advanced and highly reliable in operation. By adjusting the centerline distance of the front and rear AFCs, redesigning the front supports, and successfully resolving spatial constraints for the head supports, the use of CST soft-start devices in both front and rear AFCs effectively prevents chain breakage accidents. The optimization of transportation equipment speed reduced the chain speed of the AFC to 1.12 m/s and correspondingly decreased the chain speed of the transfer machine to 1.5 m/s. This adjustment decreased the AFC's transportation capacity from 2806 to 2395 t/h, and the transfer machine's hourly transportation capacity from 3945 to 3402 t, fully meeting the annual production requirements of the panel. Introducing a cross-side discharge arrangement at the rear transport machine head position after coal cutting resolved the issue of floating at the front and rear ends of the transport machine head. This implementation enabled unmanned operation, ensuring normal and efficient production at the mining panel. The coordinated use of roadway support shields and roadway supports ensured reliable and safe support at the ends, effectively achieving the displacement of the transfer machine and cross-side discharge machine heads, as depicted in Fig. 7.5.

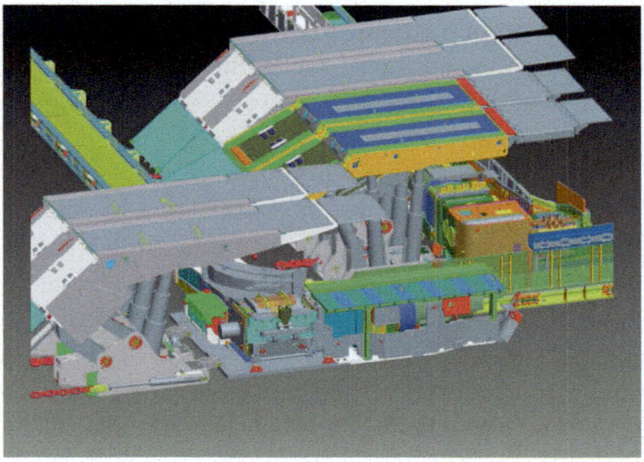

Fig. 7.5 The rear conveyor with cross-side discharge head

7.2 LTCC Technology with Two-Hard Thick Coal Seams

7.2.1 Geological Overview of Datong Mining Area

Datong Coalfield is located southwest of Datong City in Shanxi Province. The coalfield roughly forms a rectangular basin, trending northeast-southwest with a length of 85 km, dipping northwest-southeast with a width of 30 km, and covering an area of 1872 km^2. The coalfield contains two coal series: the Upper Jurassic coal series and the Lower Permian coal series. The Upper Jurassic coal series covers an area of 772 km^2 with a dip angle of 3–5°. The main coal-bearing section is the Datong Formation, which is 230 m thick with coal seams totaling 21 m in thickness and a coal quality grade of No. 2 weakly bonded coal ($f = 3$–4). There are 15 mineable coal seams with hard sandstone as the main roof and floor material. The Lower Permian coal series covers 1739 km^2, with geological reserves of 30.83 billion tons, an average thickness of 83 m, and a total coal seam thickness of 26.3 m, with a coal content ratio of 31.6% [8].

The Jurassic coal seams are hard and intact, with a hardness coefficient of $f =$ 3.0–4.5. The top coal does not self-break and is difficult to extract. The roof consists mainly of strong, thick sandstone and conglomerate, with undeveloped bedding planes and joints, characterized by its hardness. The hardness coefficient of the roof is $f \geq 8$, making it difficult for the roof to collapse. However, during collapses, sudden impact loads and intense pressure are common.

The coal series includes thick and extra-thick coal seams primarily comprising the combined seams 3–5 and seam 8. Seam 3–5 is mostly stable for mining in most areas, with a mineable thickness ranging from 1.63 to 29.21 m at depths of 300–500 m. The coal is of moderate to high hardness with developed fractures. The immediate roof consists mainly of kaolinite mudstone, carbonaceous mudstone, and typically 20 m of medium to coarse-grained quartz sandstone, gritstone, and conglomerate. Seam 8 is approximately 35 m thick on average and located about 35 m below seam

3–5. Its thickness ranges from 0.60 to 14.59 m, averaging 6.12 m. The immediate roof comprises mudstone, sandy mudstone, and siltstone, with a moderate hardness and thickness varying from 3.59 to 16.89 m. The main roof is stable coarse-grained quartz sandstone.

Datong Coal Mining Group Co., Ltd. (hereinafter referred to as Datong Coal Group), based on the maximal utilization of limited resources and the needs for sustainable enterprise development, has collaborated closely with universities and research institutions for over 20 years on LTCC technology. Starting from the late 1980s, they began with automatic network laying for Upper Jurassic integrated mining with LTCC, trials of upper-level LTCC in lower strata, and progressed to simultaneous mining of high-level, middle-level, and low-level with LTCC. They also researched low-level LTCC in lower strata and efficient low-level LTCC for extra-thick and complex coal seams of the Carboniferous period. Eventually, they developed a comprehensive set of LTCC technologies suitable for Datong's Jurassic and Carboniferous coal seam conditions, achieving safe mining with high productivity, efficiency, and extraction rates for "two-hard" thick coal seams.

7.2.2 Evolution of LTCC Technology in Datong Mining Area

Since 1991, Datong Coal Group has conducted experiments with high, middle, and low-level LTCC supports in nearly 30 panels at mines such as Meiyukou and Xinzhouyao. These experiments included lower layer LTCC under netting and full-thickness single-pass LTCC. At the 8809 panel of the 11–12 merged layer in Meiyukou mine, lower layer LTCC (upper layer adopting integrated mining with network laying) was carried out using 5600 kN high-level caving supports under a metal mesh false roof. The shearer had a cutting height of 2.8 m and a coal thickness of 2.0 m. Successively, ZFD5600, ZFD4500, and ZFD4400 types of LTCC supports were developed. Due to the high cost of network laying and the complexity of upper and lower layer mining processes, the monthly production at the panel was only about 60,000 tons.

In 1992, Xinzhouyao mine conducted trial mining using ZFS6000 mid-level LTCC supports for full-seam LTCC in hard coal conditions at a panel with a coal thickness of 10 m and near-horizontal coal seams, achieving great results. In 1993, at the 8920 panel of the 11–12 merged layer in Xinzhouyao mine, mid-level and low-level LTCC supports were successively used for LTCC trials, however, encountered issues such as difficult roof breaking and significant top coal loss. In 1995, the ZFSG6000 low-level LTCC supports were developed, and simultaneous trials of top coal blasting weakening technology achieved preliminary success. From 1996 to 1998, Datong Coal Group, China University of Mining and Technology (Beijing), Taiyuan Branch of China Coal Research Institute, and Taiyuan University of Technology formed a joint project team. Under the Ministry of Coal Industry's "Ninth Five-Year Plan" scientific and technological tackling project, they developed the new ZFS6000 low-level LTCC supports, studied the movement laws of top coal and roof, implemented roof stepping blasting for LTCC, and established a fractal energy release model for coal blasting. By conducting deep-hole pre-blasting of top coal in construction headings, they significantly improved the weakening effect of top coal blasting. During the industrial trial at the 8911 panel, monthly production reached 121,400 tons, with a mining recovery ratio of 80.3%. Annual production from LTCC panels under hard coal conditions exceeded one million tons, subsequently expanded to Meiyukou mine, Yungang mine, and others [9].

The Xinzhouyao and Meyukou mines mentioned above are mining Jurassic coal seams, which have hard coal seams and roofs. There are significant differences in the occurrence conditions of Carboniferous coal seams and Jurassic coal seams. Taking the Tashan Mine as an example, the current main coal seam is the merged coal seam 3–5 of Carboniferous, with an average thickness of 18.44 m and a dip angle of 1–3°, belonging to a relatively stable type of extra-thick coal seam. In response to the unique conditions of Carboniferous coal seams in Datong, Shanxi Coking Coal Group has researched proprietary mining technologies for extra-thick Carboniferous coal seams in Datong, achieving safe, efficient, and high resource recovery mining operations. In 2008, under the National "Eleventh Five-Year" Science and Technology Support Program, Datong Coal Group developed the ZF15000/28/52 Four-Column High-Cutting Top Coal Hydraulic Support and its ancillary equipment. They established a binary criterion based on roof pressure and coal wall stability to determine support resistance, along with a 3D top coal caving theory. They proposed a "cantilever beam-articulated rock beam" structural model post-roof breaking, and developed technologies to prevent and control high gas outbursts in low-gas coal seams. The annual production at the panel reached 10.85 million tons, with a top coal extraction ratio of 86.7%. In 2018, under the National "Thirteenth Five-Year" Key Research and Development Program, they further developed the ZF21000/27.5/42D Four-Column High-Resistance Support. Research and technological development focused on the intelligent aspects of large mining height LTCC techniques.

7.2.3 Top Coal Weakening Technology in Datong Mining Area

The Jurassic coal seams in the Datong mining area are exceptionally hard, and relying solely on mining-induced stress to fracture the top coal does not achieve adequate fragmentation necessary for efficient extraction. This limitation constrains the top coal recovery and poses significant safety risks to the panel. Basic methods for pre-weakening the hard top coal include water infusion weakening, blasting weakening, and combined water infusion and blasting weakening. These methods involve pre-breaking the top coal in the area of abutment pressure ahead of the advancing panel.

7.2.3.1 Water Infusion Weakening Technology

1. Coal seam water infusion process [10, 11]

Through trials conducted at the LTCC panels of Yungang Coal Mine in Datong mining area, effective water infusion parameters and techniques were obtained, such as Panels 8826, 8828, 8824, 8822, and 8818. The 8820 panel includes five roadway layouts: 2820 headgate, 5820 tailgate along the floor, and two roadways 2820^{-1} and 2820^{-2} along the roof, as well as one dedicated methane drainage roadway 5820^{-1}. Water infusion weakening of the top coal seam is concentrated within the two roadways (Fig. 7.6).

(1) Determining the water injection volume. The coal seam drillhole construction is advanced 100 m ahead of the panel, using a KHYD75DIA drilling rig with a diameter of 62 mm. It employs stacked three-wing drill bits and core drill bits to achieve smooth drillhole walls, which facilitates subsequent hole sealing. The drillholes are 30 m and 27 m deep, arranged as shown in Fig. 7.6(b). The injection pressure is 2.46 MPa, with an approximate single drillhole water injection volume Q of 32 tons.
(2) Holes sealing. The key to water injection in coal seams lies in hole sealing. Initially, the use of drillhole grouting packers is due to the presence of joints and fractures within the coal seam itself, which cause poor sealing of the drillhole openings, water seepage, and stress relief, thereby disrupting the later water injection effectiveness. A high-pressure rubber hose packer has been designed. Once the water control gate is opened, the high-pressure rubber hose rapidly expands and squeezes the drillhole walls, ensuring tight and seamless contact, thus ensuring the water injection pressure.

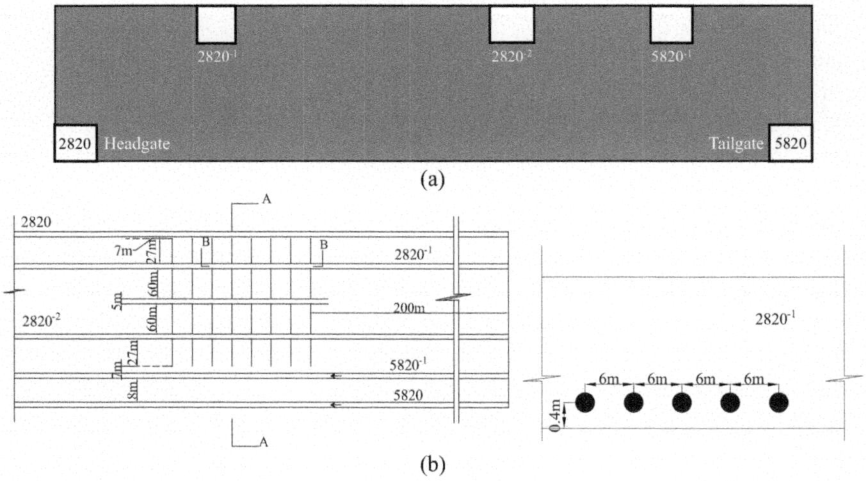

Fig. 7.6 Drilling layout diagram: **a** roadway layout; **b** drilling hole spacing arrangement

(3) Equipment and materials required for hydrostatic water injection are listed in Table 7.2. Each water injection cycle covers 48 m, with 32 drillholes simultaneously injecting water into two galleries, totaling an injection volume Q of 1054 tons. After a week of water injection, the coal body will be thoroughly moistened, causing the pre-development of fracture zones in the intact top coal and inducing structural changes. This preprocessing and inducing factors will expand and connect the primary fractures and geological weak planes in the coal seam, thereby altering the strength characteristics of the coal and facilitating further expansion and connectivity of the primary fractures thereafter.

2. Ground water hydraulic fracturing technology [12]

The Jinzhuang Coal Mine of Tongmei Group employed ground hydraulic fracturing technology to weaken and treat the overlying roof and top coal of the 8106 panel. This was done to enhance top coal recovery and reduce the impact of roof collapse. The center of the fracturing well was located 849 m from the set-up room of the 8106 panel and approximately 62 m horizontally from the panel. The fracturing layer selected was a composite sandstone layer 20.72 m thick, overlying the top coal and coal seam.

During the fracturing process, the fracturing time was 40 min, with a pump station flow ratio of 3 m^3/min and pump pressure of 20 MPa. A total of 200 m^3 of fracturing fluid was injected. Cable transmission perforation was used during fracturing, with guar gum fracturing fluid (0.2% thickener and organic boron crosslinker) selected. The actual fracturing time was 38 min, with a pump station flow ratio ranging from 2 to 4.5 m^3/min and a total volume of 240 m^3 injected. Pump pressure varied from 16.9 to 21.3 MPa, and the pressure change curve showed frequent jagged spikes, indicating the expansion of new fractures within the rock strata. This achieved the objective of fracturing to disrupt the integrity of the rock strata. The fractured area exhibited a 6.1% increase in top coal recovery compared to the unfractured area (Table 7.3).

Table 7.2 Coal seam water injection equipment configuration

Serial number	Name	Quantity	Specifications	Remarks
1	Drilling rig	2	KHYD75DIA	
2	High-pressure glue pipe hole packer	32	1.5m	Sealing water injection hole
3	Pressure gauge	2		
4	Flowmeter	2	DC-4.5/20	
5	Three-way	100		
6	Hose	100 m		Connect with main pipe

Table 7.3 Various indicators inside and outside the weakened zone of the top coal in 8106 panel

Indicators	Daily production/t	Daily cutting amount/t	Daily caving amount/t	Daily advance length/m	Panel recovery ratio/%	Top coal recovery/%
Weakened zone	28,767	6604	22,163	4.8	85.4	82.2
Non-weakened zone	30,481	7430	23,051	5.4	80.4	76.1

7.2.3.2 Pre-blasting Weakening Technology

At the Yungang Coal Mine, apart from injecting water into the coal body, pre-blasting is also employed to loosen the top coal. The pre-blasting technique for coal loosening primarily focuses on two pre-blasting roadways (Fig. 7.7), utilizing water injection drillholes directly as blasting holes for coal loosening. Due to factors such as the weight of the drill rod during construction, the inclination angle of the bottom segment of the original drillhole often falls below 0°, resulting in accumulation of coal dust and water in the hole. This leads to situations where the explosives, when loaded prematurely into the drillholes, are soaked for too long to detonate properly or result in misfires. Through analysis, adjusting the inclination angle of the drillholes to 2° in both the head and tail directions ensures no water accumulation post-charging, enabling the explosives to detonate effectively as intended. The construction of coal loosening pre-blasting holes is 60 m ahead of the advancing panel, with a minimum distance of 20 m from the panel for initiating pre-blasting. Each blasting operation covers 10 m, including simultaneous blasting of the roof holes within this range. By increasing coal water injection and modifying the coal loosening blasting technique, the recovery ratio of the panel has significantly increased to 83%.

Additionally, at the Meiyukou Coal Mine, the 81,020 panel employs a coal loosening blasting technique to weaken the overlying coal [13]. In the pre-blasting roadway, holes are arranged in a single-row pattern vertically along both sides of the roadway, spaced 1 m apart. The blasting holes are positioned 1.2 m from the roadway roof, with construction extending from the full-section to the stopping line at 20 m. Drilling proceeds from the pre-blasting roadway towards both ends, leaving 10 m of coal pillar in the tailgate and 10 m in the headgate. The stemming length for the charged holes is 10 m. Parameters for the coal blasting holes used are detailed in Table 7.4.

The specific blasting process includes charging, priming, sealing, and detonation:

(1) Charging: Employ a forward loading structure, as shown in Fig. 7.8.
(2) Priming: Utilize a single instantaneous detonator, connecting the holes in series.
(3) Sealing: Use detonating fuse for stemming, with water-soaked mud and clay as stemming material. Seal initially up to 0.8 m from the borehole mouth. For the second sealing, bundle the detonator with the detonating fuse, insert it into the void section, then seal with mud until fully packed and compacted, short-circuiting the exposed detonator leads.

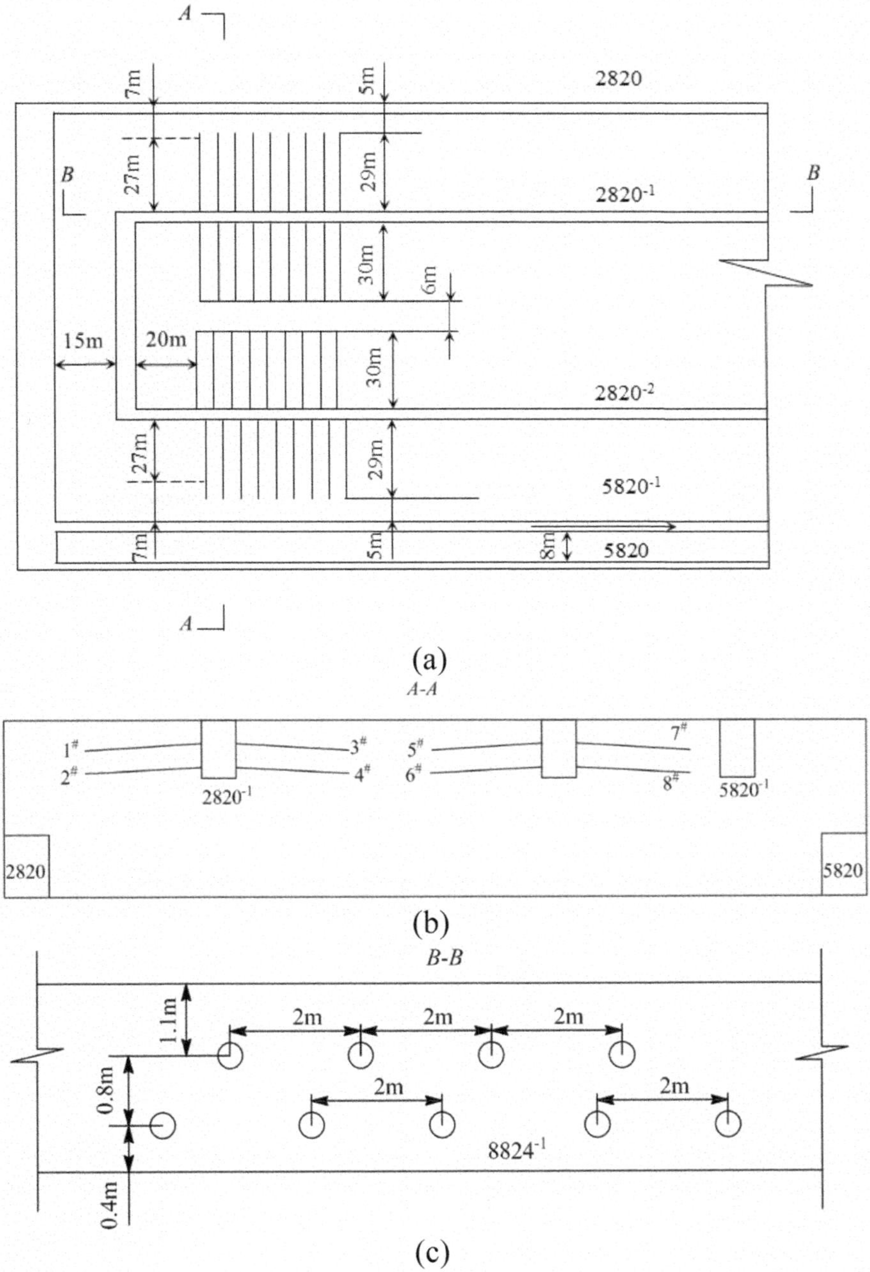

Fig. 7.7 Layout diagram of pre-blasting holes for top coal: **a** top view; **b** A-A cross-sectional; **c** B-B cross-sectional

Table 7.4 Blasting hole parameters for top coal in Meiyukou Coal Mine

Hole number	Hole depth/m	Angle of elevation/(°)	Aperture/mm	Sealing length/m	Loaded length/m	Explosive cartridge diameter/mm	Powder charge/kg	Explosive category
No. 1	57	3	63	10–12	47	50	94	Permissible emulsion explosive for Class III coal mines
No. 2	67	3	63	10–12	57	50	114	

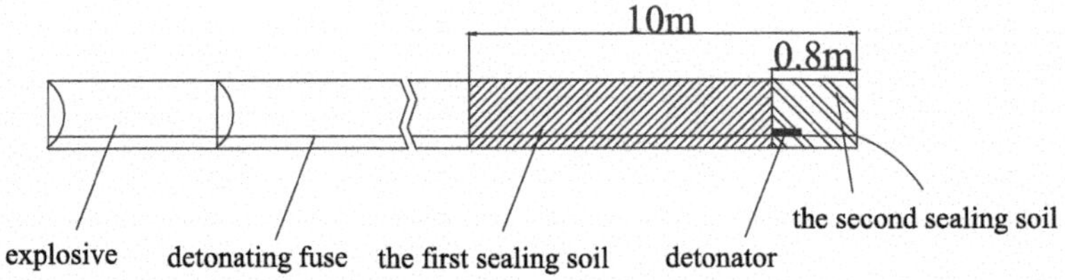

Fig. 7.8 Diagram of forward loading structure

7.2.3.3 Pre-blasting and Water Injection Combined Weakening Technology

Due to varying conditions of coal seam occurrence, sometimes a single weakening method alone may not effectively weaken the top coal. In such cases, multiple approaches can be employed. Specifically, begin with pre-blasting of the top coal and roof. When pre-blasting fails to meet on-site production requirements, proceed with water injection weakening of the coal seam to continuously expand fractures in the top coal, achieving the desired weakening effect on the coal seam. The temporal and spatial relationships between coal seam water injection, loose coal blasting, and roof pre-blasting can be summarized as follows:

(1) Water injection drill holes are advanced 200 m ahead of the panel to ensure one month of injection time.
(2) Drilling for loose coal blasting and roof pre-blasting holes is conducted at least 40 m ahead of the panel.
(3) Loose coal blasting and roof pre-blasting holes are horizontally spaced 20 m ahead of the panel for simultaneous detonation.
(4) When loose coal blasting holes and roof pre-blasting holes are located at the same position, they are detonated simultaneously; however, the number of detonations per heading must not exceed 8 at a time.
(5) Explosive charging maintains a safe distance of 10 m from construction drill holes.

These temporal and spatial arrangements are aimed at ensuring safety and efficiency during the construction process, avoiding accidents, and effectively controlling the impact range of blasting operations.

7.2.4 Intelligent LTCC Technology for Million-Ton-Level Coal Mine

7.2.4.1 Geological Conditions

The 3–5# coal seam at the Tashan coal mine of 8222 panel, has a thickness ranging from 8.17 to 29.21 m, with an average thickness of 15.76 m and a dip angle varying between 1 and 4°, averaging 2°. The coal seam exhibits significant natural fractures and complex structure. It contains 2–17 layers of parting, with a total thickness ranging from 0.26 to 5.20 m and averaging 1.33 m. The gangue primarily consists of carbonaceous mudstone and shale, occasionally interspersed with dark gray siltstone, spotted rock, and natural coke interlayers. During the mining of the 3–5# coal seam, the top coal can self-collapse into large blocks. The immediate roof predominantly comprises sandy mudstone and shale, with a thickness ranging from 1.51 to 14.52 m and averaging 8.22 m, including an interlayer of the 2# coal seam averaging 1.57 m thick. The main roof consists mainly of coarse sandstone and conglomerate, with a thickness ranging from 1.45 to 20.31 m and averaging 8.16 m, predominantly composed of hard quartz. The floor mainly consists of sandy mudstone and fine sandstone, with a thickness ranging from 1 to 10.42 m and averaging 5.32 m, including an interlayer of the 6# coal averaging 0.25 m thick. The base floor is primarily composed of coarse sandstone and conglomerate, with a thickness ranging from 0.85 to 18 m and averaging 8.9 m.

7.2.4.2 Arrangement of the Panel and Main Equipment [14]

The 8222 panel has a length of 230.5 m, and the advancing length is 2471 m. The cutting height of the panel is 3.8 m, with an average mining height of 11.96 m. The cutting-caving ratio is 1:3.14, utilizing one-cutting and one-caving method with a caving interval of 1.0 m. Equipment used at the 8222 panel includes the SL-500 shearer, PF6/1142 front AFC, PF6/1542 rear AFC, ZF17000/27.5/42D caving support, and ZFG14000/29/42D transition hydraulic support. Equipment selection parameters for the 8222 panel are based on production capacity, equipment performance, geometric considerations, and equipment compatibility, as detailed in Table 7.5. The layout of equipment at the panel is illustrated in Fig. 7.9.

Table 7.5 Detailed list of complete set of equipment for intelligent top coal extraction at Tashan coal mine

Equipment name	Specifications and models	Technical parameters	Quantity
Shearer	SL500AC	1715 KW, 3.3 kV, 2700 t/h	1
Middle support	ZF17000/27.5/42D	Support strength 1.45 MPa, 50 t/frame	125
Transition support	ZFG14000/29/42DA(B)	Support strength 1.2 MPa, 47.5 t/frame	8
End support	ZTZ30000/25/50D	Support strength 0.52 MPa, 100 t/frame	1
Front AFC	PF6/1142	2 × 1050 kW, 3.3 kV, 3591–5002 t/h	1
Rear AFC	PF6/1542	2 × 1600 kW, 3.3 kV, 5444–7582 t/h	1
ST transfer machine	PF6/1742	800/400 kW, 3.3 kV, 6236 t/h	1
Crusher	SK1422	700 kW, 3.3 kV, 6000 t/h	1

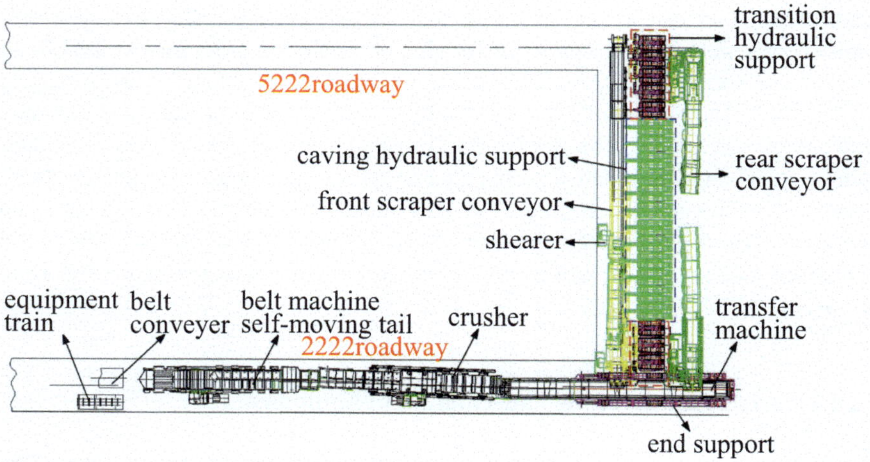

Fig. 7.9 Layout of mining equipment for the 8222 panel

7.3 LTCC Technology in High Gas Soft Coal

7.3.1 LTCC with High Gas Soft Coal in Huaibei Mining Area

Huaibei mining area is one of China's major coal production bases, established in 1958 with a mining history of over 60 years. The LTCC in Huaibei mining area mainly focuses on some thick coal seams around 10 m and extra-close coal seams with an interval within 2 m and a combined thickness of less than 14 m. Mining these coal seams has always been a technical challenge. Additionally, the complex geological structure of the mining area, with small blocks suitable for continuous mining, poses difficulties for large-scale mechanized production.

The coal seams are extremely unstable, with significant variations in thickness and complex structures containing 1–2 layers of parting. The coal is very soft ($f = 0.5$), making the face difficult to control and roadway support challenging. The gas content is high, posing a risk of outbursts, and the permeability of the coal seams is poor, with a risk of spontaneous combustion. For example, the gas content of the No. 8 coal seam at Luling Coal Mine is 16 m^3/t, with a maximum gas pressure of 2.83 MPa and a permeability coefficient ranging from 0.004843 to 0.098038 $m^2/(MPa^2d)$. Since the establishment of the mine, there have been over 20 coal and gas outburst incidents, making it a typical high gas outburst mine in China. As the mining depth increases and the scale of mining expands, the threat of water hazards in the mining area has become increasingly severe. The types of water hazards are comprehensive, including roof sandstone water, floor limestone water, and aquifers within the coal measures. The Permian sandstone fracture aquifers and the Taiyuan Formation limestone karst aquifers pose significant threats. The maximum water inflow in the mine area reaches 8900 m^3/h, with a minimum of 2890 m^3/h [15].

Due to the complex geological conditions, Huaibei mining area once primarily used blast-mining technology. This led to low output levels, high labor intensity, and poor safety. To solve these issues, Huaibei Mining Group reformed its mining technology. They adopted various methods, including grid-type support LTCC, Π-type beam simple LTCC, and light support LTCC. Eventually, they developed a system centered on fully mechanized mining and LTCC. This brought fundamental changes to production. From 2000 to 2012, Huaibei Mining Group's raw coal production increased from 16.93 to 40.19 Mt. The average monthly output per panel exceeded 70,000 tons, with a mechanization level of over 95% [16]. Since 2015, coal production in Anhui Province has generally declined, bottoming out in 2019, and then stabilizing with slight growth from 2020 to 2021. In 2022, Huaibei Mining's commodity coal production was 22.90 Mt, accounting for 20% of Anhui Province's total raw coal production. To achieve efficient LTCC in Huaibei mining area, the main technical challenges are: (1) preventing spalling and panel leakage in extra-soft coal seams; (2) gas management in high gas coal seams; and (3) efficient LTCC and increased recovery ratio in thick coal seams with partings.

7.3.2 LTCC with Soft Coal in Zhuxianzhuang Coal Mine

7.3.2.1 Geological Conditions

The 8# coal seam of Zhuxianzhuang Coal Mine is an extra-thick and extra-soft coal seam. The seam thickness ranges from 8 to 12 m, with an average of 9.98 m. The roof and floor are mainly composed of mudstone or mud-cemented rock layers, which are soft, making it a typical "three-soft" coal seam. The coal seam has a Protodyakonov hardness coefficient of $f = 0.30–0.57$. Since 1999, this coal seam has been mined using light support LTCC. The panels mined include 8413-2, 8415, 853, 874, 855, 877, II867, and 857, with an average single output of 50,000 tons. The primary reason for the low output is that during the retreat period, the coal wall is prone to spalling and roof leakage.

7.3.2.2 Measures for Preventing Spalling and Roof Falls

1. Local reinforcement of anchorage support

At the II867 light LTCC panel of Zhuxianzhuang Mine, severe spalling and roof leakage occurred when passing through faults and syncline axes. Based on the site conditions, the coal wall and roof were in a mudstone section. To stabilize the roof, fiberglass anchor rods were used. The rods were 1.8–2 m in length and 20 mm in diameter. The rods were arranged with a spacing of 750 mm, placed close to the lower end of the top beam, and angled upwards at 70–80° relative to the rib, with full-length anchorage. This technique effectively reinforced the coal wall and roof, preventing spalling and roof leakage. It is suitable for panels passing through faults or during uphill retreat mining.

2. Coal wall grouting

At a position 0.5 m below the top of the support, grouting holes are drilled (hole depth 6–7 m, drill diameter 42 mm, upward angle 40°, hole spacing 5–6 m). Under normal grouting conditions, the amount of grout per hole is 6–8 barrels, with a grouting time of 20 min per hole and grouting pressure less than 10 MPa. During grouting, two suction tubes are inserted into the barrels of Malisan resin and catalyst, respectively. The piston moves under the action of an air motor. Due to the pressure, the raw materials pass through the piston into the delivery tube and are transported to the injection gun. The raw materials are then injected into the formation through the injection gun, permeating the fractures and quickly reacting to reinforce the rib.

3. Coal wall water infusion

After multiple trials, the selected water infusion pressure in the Huaibei mining area is between 5 and 8 MPa. Two sets of drilling arrangements are used: one set of drill holes at a certain upward angle (30° for water infusion into the top coal), and the other set of drill holes for horizontal water infusion in front of the panel. The drilling sets alternate every other day. The spacing between the infusion holes is set at 5 m, and the hole length is set at 6 m, which is more suitable. During water infusion, water is supplied through the emulsion pump station, with the water pressure not exceeding 8 MPa. When using parallel multi-hole water infusion, 2–3 holes are infused at a time, using a hole sealer for the infusion. The operation is controlled using the hydraulic support valve handle to regulate the water flow. If a large amount of water leakage occurs from the drill hole or the surrounding rib, the infusion should be stopped immediately and handled. During water infusion, a small amount of spalling at the panel top is normal, however, if large-scale spalling occurs, the infusion should be stopped, and the coal wall should be managed in advance. If water flows from adjacent holes or the top of adjacent supports, the infusion should be stopped. Generally, the infusion time for each hole should be between 10 and 20 min.

4. Increase support setting load

Increasing the setting load of the support actively supports the top coal and roof above the support. This reduces the pressure from the roof on the rib, thereby decreasing the occurrences of spalling and roof leakage.

5. Strengthen the management of the panel

Keeping the panel straight and level can prevent spalling and roof leakage caused by stress concentration. When cutting coal, advancing the conveyor, and moving the support, it is essential to use alignment lines to ensure the panel remains straight. Additionally, the panel should be kept as level as possible to prevent spalling and roof leakage. For proactive management of areas prone to spalling and roof leakage, the following method is used: Use 1.4 m × 0.2 m × 0.15 m telescopic beams as the support beam, with individual props as legs. One beam is supported by two props, forming two sets per support. Place metal mesh over the top. The distance between the props and the AFC should be 0.9 m, based on the coal cutting machine's drum. This method effectively controls spalling and roof leakage but requires considerable labor and time to manage the roof.

6. Reduce the mining height

For example, the height of the ZF2400/24/16/BF support ranges from 1.6 to 2.4 m. While considering the height of the shearer and pedestrian clearance, appropriately reducing the mining height can help reduce rib spalling and roof leakage.

7. Increase the advance speed of the panel

Observations of the 8415 panel showed that with a faster advance speed, there were fewer instances of roof flaking ratio, spalling, and roof leakage. Conversely, a slower advance speed resulted in more occurrences of these issues. Therefore, reasonably increasing the advance speed can help prevent rib spalling and roof leakage.

7.3.2.3 Intelligent LTCC in Zhuxianzhuang Mine

The 8105 panel of Zhuxianzhuang Mine primarily extracts the No. 8 coal seam. The mineable strike length is 600 m, the panel length is 170 m, and the average coal thickness is 8 m, with a mineable reserve of 1.19 million tons. The coal seam dip angle ranges from 5° to 15°. The immediate roof consists of medium to fine-grained sandstone with an average thickness of 9.0 m, while the immediate roof is mainly mudstone with an average thickness of 6.5 m. The floor is also mudstone, with an average thickness of 6.8 m. This panel uses a new intelligent LTCC technology. It features intelligent control and automated monitoring in video, communication, information collection, electric control, and hydraulic control. This increases the degree of automation and safety production efficiency while significantly reducing labor intensity for workers. The main equipment layout and information for the panel are shown in Fig. 7.10 and Table 7.6.

Fig. 7.10 8105 intelligent LTCC panel equipment

Table 7.6 Main equipment configuration of the 8105 panel

Serial number	Equipment	Model	Quantity	Using sites	Voltage level (V)	Motor power (Kw)
1	Shearer	MG500/1180-AWD	1	Panel	3300	1180
2	Fornt AFC	SGZ-764/630	1	Panel	3300	630
3	Rear AFC	SGZ-764/630	1	Panel	3300	630
4	Hydraulic support	ZF7000/18/28	110	Panel	3300	630
5	Emulsion pump	BRW-400/31.5	2	Shaft bottom	3300	2 × 250
6	Crusher	PLM-2000	1	Headgate	1140	200
7	Reversed loader	SZZ-830/200	1	Headgate	1140	200
8	Belt conveyor	DSJ-100/100/2 × 125	1	Headgate	1140	2 × 125
9	Spray pump	BRW200/31.5	2	Shaft bottom	1140	125
10	Spray pump	PBW320/6.3	2	Shaft bottom	1140	2 × 45

Additionally, an intelligent caving control system was installed at the 8105 LTCC panel (see Sect. 5.5.1) to conduct gangue mixed ratio detection tests. Using image recognition intelligent caving online detection software, multiple on-site image analyses were performed. The test results showed that the projection area data of gangue detected by the images were reliable. The volume data of gangue obtained from the image-based volume model were also reliable. The ratio of the projection area to the modeled volume of gangue on the AFC fluctuated within a range of less than 10%. This indicates that the detected gangue area from the images can replace the volume and mass of the gangue.

7.3.3 LTCC with High Gas Coal Seam in Luling Coal Mine

The III811 LTCC panel at Luling coal mine is the first coal mining panel for the No. 8 coal seam in this mining area. The panel has an average strike length of 717.0 m and a dip width of 175.9 m, covering a mining area of 126,120.3 m^2. The surface elevation ranges from +22.0 to +23.0 m, while the panel elevation ranges from − 587.7 to − 701.4 m. The III811 LTCC panel coal-bearing strata belong to the Permian period. The No. 8 coal seam is relatively stable but structurally complex, with a thickness ranging from 3.33 to 14.75 m and an average thickness of 8.25 m. The No. 9 coal seam is unstable, structurally simple, with a thickness ranging from 0 to 4.25 m and an average thickness of 1.21 m. The distance between the No. 8 and No. 7 coal seams above the III811 panel is 8.72–26.08 m, with an average of 19.38 m. The immediate roof is sandstone, with a thickness of 0–10.78 m. The interlayer between the No. 8 and No. 9 coal seams consists of mudstone and sandstone, with an average thickness of 0.59 m. The immediate floor of the panel is mudstone, with an average thickness of 3.48 m [17].

7.3.3.1 Panel Mining Technique

Mining Position: In the area with the 9# coal seam, mining follows the floor of the 9# coal seam. In the area without the 9# coal seam, mining follows the floor of the 8# coal seam.

Mining process flow: Supporting rib protection device, cutting coal, extending the front beam, pushing the front AFC, moving the support, top coal caving, pulling the rear AFC.

Coal falling method: The two-cutting and one-caving method is used. The shearer (MG500/1080-WD) cuts two slices of coal (advancing 1.2 m), then top coal is released from the low-level DO at the rear of the support with a caving interval of 1.2 m. Multiple rounds of interval caving are employed, with the caving step and number of caving rounds adjusted based on on-site conditions. The amount of caved top coal is controlled to prevent gas overrun, and the spraying system is used to suppress dust on the current and adjacent supports. Due to the significant panel inclination, the lower part of the coal panel is compressed by the upper coal body, making it denser. Upward caving is more beneficial for top coal recovery.

Loading and transporting coal: The shearer's drum rotates to cut coal, which falls onto the front AFC. Top coal falls through the caving opening onto the rear AFC. The AFC transports the coal to the crusher. After crushing, the coal is transferred to the belt conveyor by the transfer conveyor. Small amounts of loose coal between supports and the upper section of the shearer tail are manually loaded into the conveyor.

Advancing the conveyor: During normal production, supports are advanced unidirectionally in sequence, lagging the shearer by more than 15 m. Supports are advanced sequentially along a unidirectional direction, ensuring the bending length is not less than 15 m.

Moving the support: Sequential and segmented support movement is used. The support must lag the shearer by 10–12 supports. The step distance for moving the support is 630 mm.

7.3.3.2 Roadway Layout and Support System

1. Headgate

The headgate is constructed along the floor of the 9# coal seam. The eastern section of the headgate uses a horseshoe-shaped U29 steel arch support with specifications: base width of 4900 mm, net height of 3400 mm, and spacing of 600 mm. The western section uses a flat-top U29 steel arch support with specifications: base width of 5409 mm, net height of 3300 mm, and spacing of 600 mm. The support system includes iron backboards and double-resistant mesh for sidewall and roof support, as shown in Fig. 7.11.

This roadway is primarily used for ventilation, pedestrian passage, power supply, installation of water spray systems, drainage, hydraulic pipelines, explosion-proof facilities, transportation equipment (coal transport), installation of endless rope haulage, and various safety facilities and accessories.

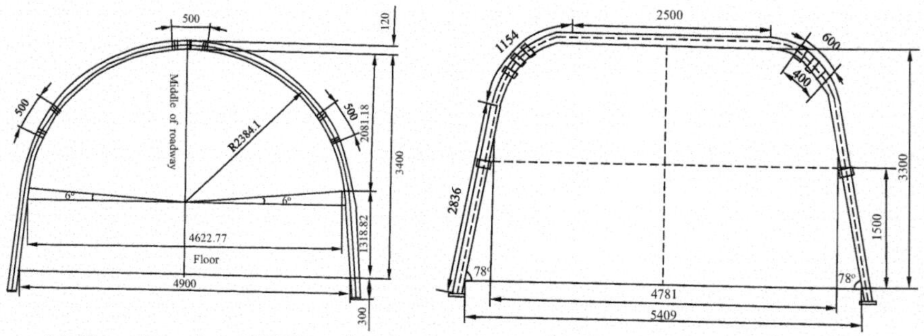

(a) Horseshoe-shaped U29 steel arch support (b) Flat-top U29 steel arch support

Fig. 7.11 Cross-section of support in the headgate

2. Tailgate

The tailgate of the panel is constructed along the floor of the 9# coal seam. The eastern section uses a horseshoe-shaped U29 steel arch support with specifications: base width of 4900 mm, net height of 3400 mm, and spacing of 600 mm. The support system includes iron backboards and double-resistant mesh for sidewall and roof support. This section is mainly used for the return air of the panel, pedestrian passage, material transport, water supply installation, drainage pipelines, and safety facilities for ventilation and fire prevention. The western section of the tailgate uses a flat-top U29 steel arch support. The set-up room is mainly used for installing hydraulic supports, coal transportation, and ventilation. The roadway is constructed along the floor of the 9# coal seam, using I-beam support for the initial excavation, followed by single props and π-type steel for inclined shelf support. The final dimensions after excavation are a net width of 7920 mm and a net height of 2800 mm. Other roadways in rock sections use bolting-wire mesh with additional steel bands for support, while the coal roadway sections and headgates use horseshoe-shaped U29 steel arch support.

7.3.3.3 Gas Control in the Panel

The 8# coal seam is prone to gas outbursts and has a relatively stable thickness, averaging 10 m. Due to the lack of protective seam mining conditions in this area, eliminating the outburst risk of the 8# coal seam in advance is critical. Based on actual conditions, large-scale pre-drainage of gas is the only feasible method. However, due to the soft coal seam, drilling is challenging, often resulting in blowouts, collapses, and stuck drills. The coal seam has poor permeability, and pre-drainage is generally ineffective, taking 8–12 months, which fails to meet the mine's production needs. Therefore, Luling Coal Mine cannot eliminate the outburst risk of the 8# coal seam through large-scale pre-drainage using cross-seam drilling.

To address the gas issues during the top-slice mining process, ventilation and gas drainage methods are used. Comprehensive gas drainage methods are adopted because most gas emissions from the panel come from the lower sections and adjacent seams' stress relief gas. High-level drilling in the roof is mainly used to drain the mining-induced stress relief gas, while embedded pipes are used to drain gas from the goaf and the upper corner of the panel. After top-slice mining, the gas outburst risk in the lower slice is assessed. Once the coal and gas outburst risk are eliminated in the lower slice, LTCC is carried out.

During LTCC in the lower slice, high-level roadways and drilling, combined with high-level inclined drilling in the return airway, are used to drain the mining-induced stress relief gas. Concurrently, in-seam drilling is arranged in the return airway and headgate to pre-drain and continuously drain coal seam gas during mining. Embedded pipes are also used to drain gas from the goaf and the upper corner of the panel, as shown in Fig. 7.12.

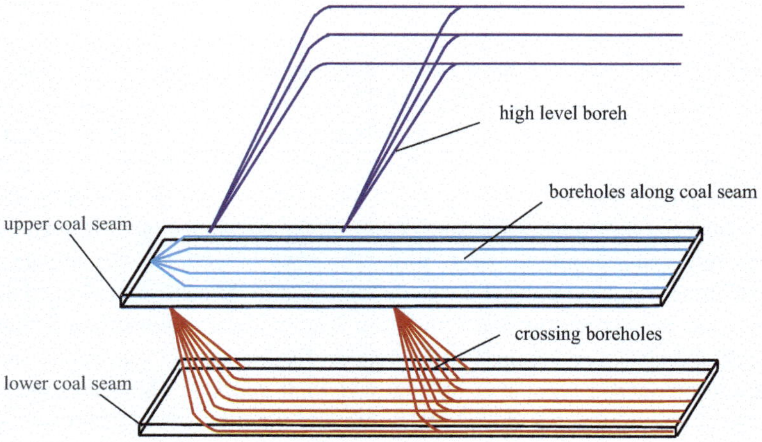

Fig. 7.12 Common techniques for gas drainage

7.4 LTCC Technology in Large Dip Angle Thick Coal Seam

7.4.1 Characteristics of LTCC in Large Dip Angle Thick Coal Seam

One of the significant characteristics of LTCC mining in large dip angle thick coal seams (coal seam dip angle greater than 35°) is that the collapsed roof will slide down along the coal seam floor. This results in uneven filling in the goaf, with dense filling at the bottom of the panel, uneven filling in the middle, and no filling at the top, forming voids, as shown in Fig. 7.13.

This outcome leads to uneven stress on the supports in the panel, with different interactions between the supports and the roof in different areas. For LTCC in thick coal seams, the extraction space is larger, making the unevenness of roof collapse and filling more pronounced. An unreasonable mining and caving process can cause the top coal to slide down along the top beam of the support, the shield beam, and the coal seam floor, leading to the tilting of the support and worsening the relationship between the support and the surrounding rock. According to the characteristics of roof collapse and movement in LTCC mining of large dip angle thick coal seams, the upper supports of the panel mainly bear the dynamic load impact from the main roof collapse, while the lower supports primarily bear the downward sliding compression of the top coal and fractured roof, as well as the static load from the upper supports. When selecting and calculating the supports, the working resistance of the supports can be calculated based on the dynamic load impact on the upper supports, while the anti-compression capacity of the side guard plates can be calculated based on the static load compression on the lower supports.

Recovery ratio of top coal in the panel area: The recovery ratio of top coal at the upper end is relatively low. Numerical simulations, 3D experiments, and field

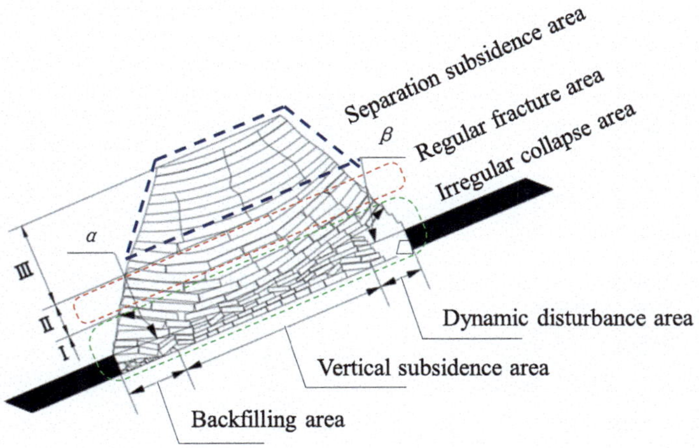

Fig. 7.13 Double-direction distribution characteristics of overburden fracturing in LTCC panel of large dip angle thick coal seams

measurements all show that the recovery ratio of top coal along the panel direction presents a "Π" shape distribution [18]. The recovery ratio of top coal in the middle of the panel can reach up to 90%, while at both ends it is around 75%.

Caving method: For top coal caving in large dip angle thick coal seams, when the panel angle exceeds the natural repose angle of the broken top coal or the friction angle between the broken top coal and the shield's canopy, the bottom-to-top caving method is no longer applicable. When caving starts at the bottom of the panel, it causes the top coal above the upper shields to flow downward, reducing the interaction between the upper shields and the top coal and roof, thus weakening the stability of the shields.

7.4.2 LTCC with Large Dip Angle Seam in Beixinyao Coal Mine

7.4.2.1 Geological Overview

Beixinyao mining field, located in the Xuangang mining area, spans across Ningwu, Yuanping, and Shenchi counties in Xinzhou city. The resources are primarily situated in the Ningwu compound syncline saddle and its southern side, containing two coal-bearing series from the Jurassic and Carboniferous-Permian periods. The coal seams in the mining field are relatively deep. The minable seams are the 2#, 5#, and 6# coal seams of the Taiyuan Formation. The 8103 panel is the first mining panel in the Beixinyao Coal Mine, located in the 11th mining district of Beixinyao Mine, adjacent to the + 1040 tailgate in the west. The ground elevation ranges from 1275.5 to 1352.7 m, and the panel elevation ranges from 918.7 to 1063.8 m. The two roadways

Table 7.7 8103 Panel roof and floor conditions

Strata	Rock name	Thickness/m	Lithological characteristics
Main roof	Medium to fine sandstone, medium to coarse sandstone	1.1–22.81 11.96	Grayish-white, semi-hard to hard, well-developed fractures, mainly composed of quartz and feldspar, part of the K3 sandstone aquifer at the interface of the Shanxi and Taiyuan formations
Immediate roof	Mudstone, sandy mudstone	0–17.25 8.63	Dark gray to black, relatively soft
Immediate floor	Mudstone, sandy mudstone	3.11–9.02 4.18	Gray to dark gray, soft
Main floor	Medium to fine sandstone, medium to coarse sandstone	2.81–5.25 3.83	Grayish-white, weak to hard rock, well-developed fractures, mainly composed of quartz and feldspar, part of the aquifer

are arranged in an approximately east–west direction, with the set-up room arranged in a near north–south direction. The average strike length of the panel is 1868.8 m, and the dip length is 165.9 m. The recoverable reserves amount to 1.2773 million tons. The coal seam thickness ranges from 4.3 to 7.0 m, with an average thickness of 5.6 m. The coal seam dip ranges from 19° to 37°, with an average of 22°. The true density is 1.59 t/m^3, and the apparent density is 1.48 t/m^3. The coal seam thickness varies little, mainly consisting of dull coal, followed by bright coal, with bands of vitrain and asphaltic luster. The coal seam structure is complex, containing 2–7 partings, with parting thickness ranging from 0.1 to 1.0 m. Partings are primarily composed of mudstone and dark gray carbonaceous mudstone. The roof and floor conditions of the 8103 panel are shown in Table 7.7.

7.4.2.2 Hydraulic Support Anti-toppling and Anti-sliding Techniques

1. Hydraulic supports for large dip angle seams

As shown in Fig. 7.14, the 8103 panel is equipped with 97 ZFQ13000/25/38 low-level caving hydraulic supports, 8 ZFQG13000/27.5/42 transition hydraulic supports (3 at the head and 5 at the tail), and a set of ZTZ20000/27.5/4 end hydraulic supports (one set consists of two supports). To better adapt to the actual conditions of the Beixinyao Mine and optimize the performance of the supports, a series of designs and modifications were made to the hydraulic supports used on the 8103 panel. These modifications included the hydraulic system, rib protection device, tail beam jack, side guard plate, base adjusting beam, and spray system.

Fig. 7.14 Hydraulic supports in the 8103 panel of beixinyao Mine: **a** central support; **b** transition support; **c** end support

2. "Double adjustment and one coordination" management

Through on-site investigation of the 8103 panel, it is believed that the four-link mechanism is an important part of the LTCC support on the 8103 panel. The existence of the four-link mechanism allows the support to withstand a large horizontal force, effectively improving the stability of the support. Furthermore, the coordination between the front and rear posts of the support during operation is also a prerequisite for ensuring the stability of the support, the supporting effect, and the smooth advancement of the panel. In actual production, a "double adjustment and one coordination" anti-tipping and anti-sliding technology for large dip angle panel supports has been summarized. This technology involves timely adjusting the position state of the supports during the normal advancement of the panel. When a support tipping occurs, the supports are adjusted sequentially from top to bottom. When jamming occurs, single hydraulic props are used as struts to coordinate the adjustment of the support position state.

7.4.2.3 Structure of Shearer and Anti-tipping/Anti-sliding Management Technology

1. Structure of the shearer

The model of the shearer used in the 8103 panel is SL500AC, as shown in Fig. 7.15. The cutting height is about 4995 mm, the drum diameter is 2300 mm, the cutting depth is 800 mm, the maximum traction force is 771 kN, and the maximum traction speed is 30.2 m/min. The shearer is equipped with a reliable brake, making it suitable for large dip angle coal seams without the need for additional anti-sliding devices. During coal cutting, the shearer operates from top to bottom, with the shearer ascending without cutting coal. The front AFC of the panel moves with the shearer to the coal wall and maintains forward hydraulic pressure, which prevents the shearer from sliding down during coal cutting. When the shearer stops operating, both drums of the shearer are inserted into the coal wall and dropped to the floor simultaneously. Additionally, anti-sliding pins are installed on the front conveyor to ensure the shearer does not slide down after stopping.

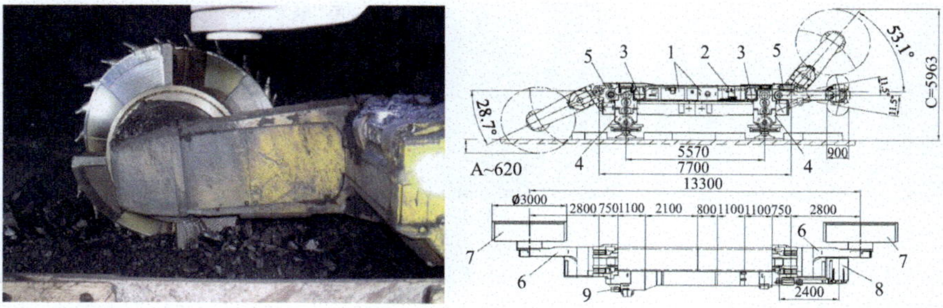

Fig. 7.15 The shearer in the 8103 panel. (1-Electrical unit with transformer box; 2-Hydraulic unit; 3-Traction section; 4-Sprocket box; 5-Arm bracket; 6-Cutting unit; 7-Drum; 8-External spray device; 9-Cable dragging device)

2. Shearer "forward push continuous hydraulic supply" anti-sliding technology

In coal mining technology and panel management, the "forward push continuous hydraulic supply" anti-sliding technology has been developed to effectively control the downward sliding of the shearer. The main points include: (1) The downward coal cutting speed of the shearer must not exceed 6 m/min; (2) The shearer's braking device must be tested every shift, with the downward speed not exceeding 2 m/min during the test; (3) In case of an emergency where the shearer slides down, the shearer operator should insert the locking pin into the rack rail or push the front AFC forward to stop the shearer against the wall, and force the lower drum to the floor as an emergency measure; (4) To enhance the braking force of the shearer, an additional brake is installed on the third shaft of the left arm; (5) During coal cutting, the shearer operates from top to bottom, with the shearer ascending without cutting coal. The front conveyor of the panel moves with the shearer to the coal wall and maintains forward hydraulic pressure to prevent the shearer from sliding down during coal cutting; (6) When the shearer stops operating, both drums of the shearer are inserted into the coal wall and dropped to the floor simultaneously. Anti-sliding pins are installed on the front conveyor to ensure the shearer does not slide down after stopping; (7) Each shift, a detailed inspection is conducted on the gaps and integrity of the shearer's upper and lower guide shoes. If the gaps are too large, they must be replaced; (8) The upper 60 m of the shearer's cable drag bracket use a Φ10 mm steel wire rope for auxiliary traction, fixed to the cable clamp every 5 m with small U-clamps; (9) When the shearer cuts coal downward, the lower arm should be raised once every 30 m to ensure good lubrication of the first and second shafts; (10) The M36 bolts on the upper and lower travel boxes are replaced with M30 hydraulic bolts to ensure the tightness of the shearer's travel boxes. Limit blocks are also added to the upper and lower travel boxes.

7.4.2.4 AFC "Pressure Prop and Anti-toppling Pick" Anti-sliding Technology

The 8103 panel is equipped with one SGZ-1000/2100 front AFC and one SGZ-1250/2100 rear AFC, as shown in Fig. 7.16. The AFC, as the main equipment for coal transportation in the panel, also serves as the track for the shearer. It is connected through push cylinders and hydraulic supports. During the normal coal mining cycle, hydraulic supports tend to slide down the coal seam floor under gravity, exacerbating the occurrence of toppling and jamming of the supports. To mitigate the sliding of the AFC, the lower roadway is typically advanced a certain distance in the roadway layout. This ensures that during each cycle of moving the supports, the AFC has an upward movement to balance the downward sliding. In the cutting direction, downward coal cutting can be adopted, which not only slows down the sliding of the AFC but also reduces the probability of chain breakage. Additionally, limiting the swing angle of the hydraulic cylinders connecting the hydraulic supports to the AFC can also help achieve the goal of restricting downward sliding.

In addition, other auxiliary equipment is used for anti-sliding purposes, such as the "pressure prop and anti-toppling pick" anti-sliding technology. The key techniques include: when the panel conveyor moves up or down, the support status must be immediately adjusted so that the support base is perpendicular to the conveyor; at least two pressure props must be set at the appropriate position of the conveyor head to firmly fix the head; an anti-sliding top pick must be installed on the rear transporter of the panel. If the conveyor is found to be sliding down, the anti-sliding top pick must be used to pull the rear transporter back to its original correct position. The technical parameters of the SGZ-1000/2100 front AFC are shown in Table 7.8.

Fig. 7.16 Front AFC in 8103 panel

Table 7.8 Parameters of the SGZ-1000/2100 front AFC

Item	Parameter
Design length	300m
Conveying capacity	3000 t/h
Installed power	3 * 700 kW
Motor	
Model	YBSD-700/350-4/8
Rated power	700/350 kW
Rated voltage	3300 V
Cooling method	Water cooling
Middle trough	
Dimensions (L*W*H)	1750 mm * 1000 mm * 372 mm
Structural form	Cast-welded sealed bottom
Connection method	Dumbbell pin
Connection strength	4000 kN
Scraper chain	
Chain specification	48 * 152 flat chain
Type	Double middle chain
Chain center distance	280 mm
Scraper pitch	6 * 152 mm
Unloading method	Cross-side unloading
Tensioning method	Hydraulic motor tensioning, hydraulic telescopic tail auxiliary tensioning
Total weight of standard section	781.53 t (300 m)

7.4.3 LTCC with Steeply Inclined Thick Seam in Dayuan Coal Mine

7.4.3.1 Geological Conditions

The coal seam mined by Shanxi Dayuan Coal Industry of the Fengfeng Group belongs to a typical steeply inclined (seam dip greater than 45°, included in the category of large dip angle seams) extra-soft thick coal seam. The mine field extends 2.33 km from north to south and 15 km from east to west, covering an area of 3.0801 km^2. The average seam dip is 53°, with a maximum of 62°. The hardness coefficients of the coal seam, roof, and floor are low, classifying it as a typical "three-soft" coal seam. The 1201 panel is buried at a depth of 195.6–242.6 m, with a coal seam thickness of 6–8 m and an average of 6.8 m. The coal seam hardness coefficient is $f = 0.3$. The immediate roof is gray, well-laminated siltstone with a thickness of 8 m and a hardness coefficient of $f = 4$–6.14. The main roof is medium sandstone, predominantly light gray to grayish-white medium-grained sandstone, locally fine sandstone, cemented by calcareous or argillaceous material, with a thickness of 5 m and a hardness coefficient of $f = 6.31$. The immediate floor consists of interbedded siltstone, coal, and carbonaceous mudstone. The 1201 panel is 80 m long, with a minable strike length of 680 m, and is arranged with a pseudo-dip of 13°.

7.4.3.2 Panel Layout and Mining Technology

1. Panel roadway layout

The 1201 LTCC panel's tailgate is arranged along the coal seam floor. The cross-section of the roadway is arched with dimensions: a bottom width of 3.80 m, wall height of 2.00 m, and arch height of 1.15 m. It employs a combined support system of anchor net, anchor cable, and W steel strips. This roadway is primarily used for material transport and return air. Dust control pipes (2-inch), compressed air pipes (4-inch), and drainage pipes (2-inch) are installed along the upper side of the roadway.

The 1201 headgate is arranged along the coal seam roof. The cross-section of the roadway is arched with dimensions: a bottom width of 4.30 m, wall height of 2.00 m, and arch height of 1.15 m. It employs a combined support system of anchor net, anchor cable, and steel strips. The headgate is mainly used for the arrangement of the panel transfer machine and belt conveyor for coal transportation and intake air; a drainage ditch is arranged along the lower side of the roadway for panel drainage.

2. Coal mining technology

The panel employs an MG200/500QWD shearer with chainless AC variable frequency electric traction for coal cutting. The panel conveyor uses an SGZ730/400 double-chain AFC, while the headgate uses SGW-620/40T and SGB-630/150 scraper transloader for coal transportation. The headgate belt uses a DSJ-1006.32 × 75 belt conveyor for coal transportation. The hydraulic support uses ZYF4800/17/28 type caving supports for roof support, with a mining height of 2.3 m and a cutting step of 0.6 m. This support is a specialized hydraulic support with an articulated front beam, bottom-sealed type, double-side movable wide side shields, two-post frame-type four-link mechanism. It has strong support and roof control capabilities, and good stability, effectively solving issues such as support toppling, sliding, base pressing, and coal and rock leakage between supports. The side shields of the support play an excellent role. The retreating longwall method is adopted for mining.

3. Coal caving technology

The caving method involves drawing coal with each cut, with a step distance of 0.6 m. A single-wheel staggered sequential drawing method is used, meaning that after the supports are moved, top coal is drawn sequentially from bottom to top. First, the insert plates are retracted and the tail beam is lowered, then the tail beam is raised repeatedly until about two-thirds of the coal is drawn out. The insert plates are then closed to block DO and move to the next support for drawing until the first round of drawing in this section is completed. When gangue is released, the insert plates are opened to block the gangue, preventing it from entering the rear AFC, and then the next support's drawing operation is conducted. Similarly, the other group draws coal following the same requirements. The drawing height is 4.5 m, the mining height is 2.3 m, and the mining-drawing ratio is 1:1.96.

7.4.3.3 Measures to Improve the Top Coal Recovery and Panel Management

1. High resource recovery technology

To further improve the recovery ratio of top coal and the stability of supports in steeply inclined panels, the "downward dynamic segmentation and upward drawing within segments" technology was developed for LTCC in steeply inclined thick coal seams [19]. This technology involves cutting coal from top to bottom with the shearer, moving supports from top to bottom, dynamically segmenting from top to bottom, drawing coal from bottom to top within each segment, and moving the front and rear AFCs as a whole from bottom to top. Dynamic segmentation from top to bottom refers to determining the number of supports in each segment based on the shape of top coal boundary and the drawing body. According to the actual conditions of the panel, the first segment at the upper part consists of 3 supports, the second segment consists of 4 supports, and subsequent segments consist of 5 supports each. This mining and drawing technology achieve a top coal recovery of 85% and a maximum monthly output of 86,000 tons at the panel.

2. Local reinforcement measures for surrounding rock

Considering the steep inclination of the panel, the soft coal seam floor, the softness of the coal seam, and the ease of top coal size and flow, certain local reinforcement measures were implemented during the initial stage of mining.

(1) Reinforcement of the set-up room floor

Before the installation of the panel, the floor and sleepers of the set-up room were reinforced with anchor net. The floor reinforcement used Φ18 mm × 1800 mm full-thread high-strength rods, in combination with 150 mm × 150 mm × 10 mm bearing plates and 350 mm × 280 mm × 3 mm steel strap shields, arranged in a staggered pattern with a spacing of 1300 mm × 1000 mm. A diamond-shaped metal mesh was laid across the entire floor and connected every 200 mm with wire. To ensure track stability, the sleepers were spaced at 500 mm × 500 mm, with each sleeper fixed by one bolt. The floor reinforcement is shown in Fig. 7.17.

(2) Local reinforcement of top coal at the upper end

Before the initial coal drawing, due to the softness and high flowability of the top coal, the top coal above the upper 5 supports was temporarily not drawn to prevent damage to the return airway during the drawing process. This top coal was reinforced with anchor cables. The anchor cables were Φ17.8 mm × 7 m, with a spacing of

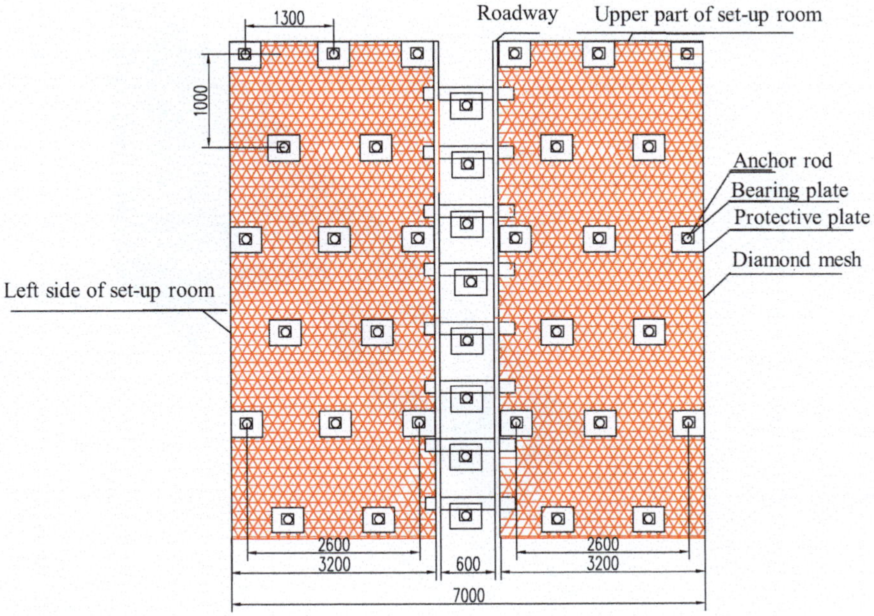

Fig. 7.17 Reinforcement of the 1201 panel set-up room floor

3 m, and each group consisted of 3 anchor cables. Inside the panel, the anchor cables were installed at an upward angle towards the coal wall (at an angle of 60° to the horizontal), spaced at 2.5 m along the panel, with the bolt length extending 1.5 m into the coal seam roof. This reinforcement is shown in Fig. 7.18.

Through practical exploration and the adoption of reasonable mining and drawing technique, it was found that after normal mining operations commenced, the reinforcement measures for the top coal at the upper end were no longer necessary. The top coal in the uppermost two supports was not drawn, effectively preventing any impact on the return airway from the drawing process.

7.5 STCC Technology in the Steeply Inclined Thick Coal Seam

7.5.1 Roadway Layout and Support

Wudong Coal Mine is situated in the northeastern part of Urumqi City and is under the jurisdiction of the Xinjiang Branch of the State Energy Group. This coal mine's mining field is in the southeastern section of the Zhunnan coalfield in Xinjiang, on the southern and northern wings of the Badaowan syncline. The +443m$B_{3\text{-}6}$ STCC

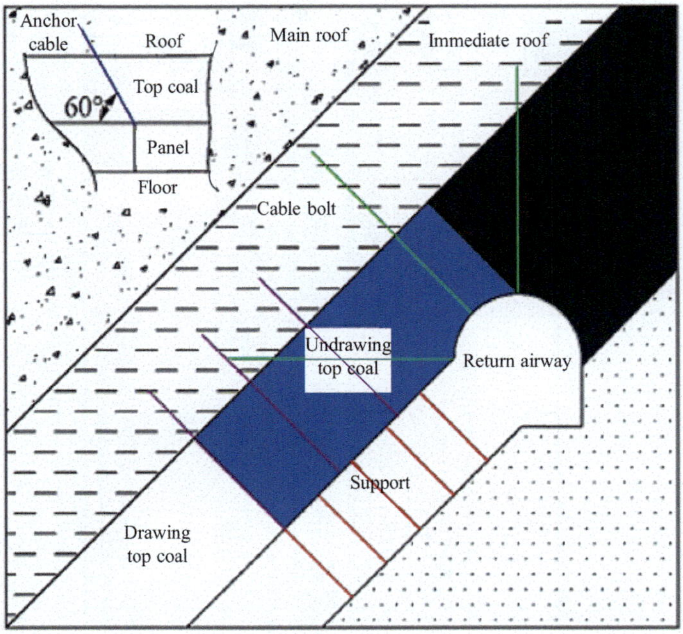

Fig. 7.18 Local reinforcement for top coal at the upper end

panel is positioned on the east side of the west district's auxiliary inclined shaft at the Wudong Coal Mine (46 m apart from the main and auxiliary inclined shafts), within a range of 84–1892 m from the centerline of the auxiliary inclined shaft. The surface corresponding to the $B_{3\text{-}6}$ coal seam is characterized by a strip of collapse, with the southern side being a strip of subsidence formed after the mining of the B1+2 coal seam, and the northern side consisting of barren hills. The ground elevation of the panel varies from + 765 to + 815 m, while the underground elevation ranges from + 443 to + 449 m. The + 443m$B_{3\text{-}6}$ panel is located on the east side of the + 443 m cross-cut, with the upper part being the goaf area resulting from the mining completion of the + 469m$B_{3\text{-}6}$ panel; the lower part consists of the + 417 m cross-cut and solid coal seam; to the north lies the siltstone layer between the B_6 and B_7 coal seams; to the south is the siltstone layer between the B_2 and B_3 coal seams. The panel measures 53.6 m in width and 1780 m in strike length, as depicted in Fig. 7.19.

The coal seam dip angle is between 83 and 87°, the structure is relatively straightforward. B_5 and B_6 coal seams within the $B_{3\text{-}6}$ coal seam are stable and relatively rigid, with well-developed stratification and joint systems. There are no normal or reverse faults or other geological structures within the extraction area of the panel. However, the phenomenon of variable thickness of coal seam can be observed locally, with the presence of false roofs and floors that are susceptible to collapse. Although the B_3 coal seam exhibits well-developed stratification and jointing, and lacks normal or reverse faults and other geological structures, it is revealed as loose and fragmented

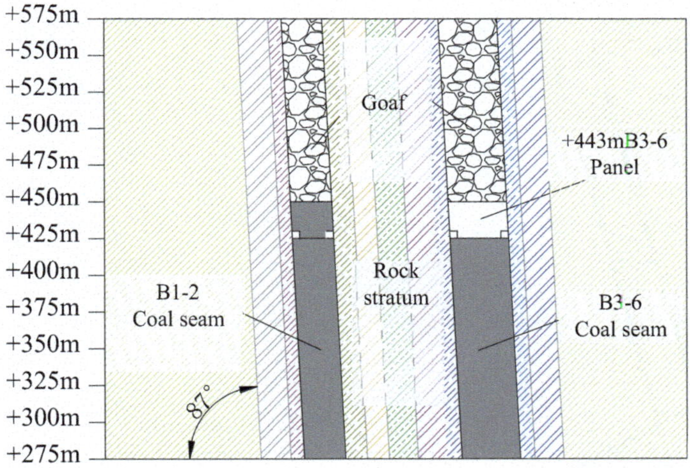

Fig. 7.19 Coal seam distribution and layout of the + 443mB$_{3\text{-}6}$ panel

during coal seam uncovering. The excavation and extraction of the panel roadway are prone to rib spalling and roof collapse. The phenomenon of variable thickness of the coal seam can be encountered locally, with the presence of false roofs and floors that are prone to collapse. Based on the excavation conditions, no faults or folds are present within the range of excavation. Details regarding the roof and floor conditions of the coal seam are presented in Table 7.9.

Wudong Coal Mine is characterized by steeply inclined coal seams, with the ground pressure behavior in the roadways marked by significant top pressure and minimal side pressure. The roadway roofs are susceptible to fragmentation and collapse, creating a natural equilibrium arch; hence, the cross-sectional design of the roadways features vertical walls and an arch shape. The B_3 and B_6 roadways are designed with a circular arc arch cross-section; the set-up rooms are designed with a

Table 7.9 Coal seam roof and floor conditions table

Stata	Rock name	Thickness/m	Lithological characteristics
Main roof	Sandstone	56	Gray sandstone, gray fine sandstone, sandstone with grayish-white color
Immediate roof	Sandstone	8	Gray contains thin layers of fine sandstone, layering and joint development, calcium silicate cementitious binder
False roof	Coal shale	0.6	Gray-black in color, thin and fragile
False floor	Coal shale	1	Gray-black, clayey cemented, containing thin coal
Immediate floor	Sandstone	49	Relatively hard, grayish-white in color, layered, with well-developed joints

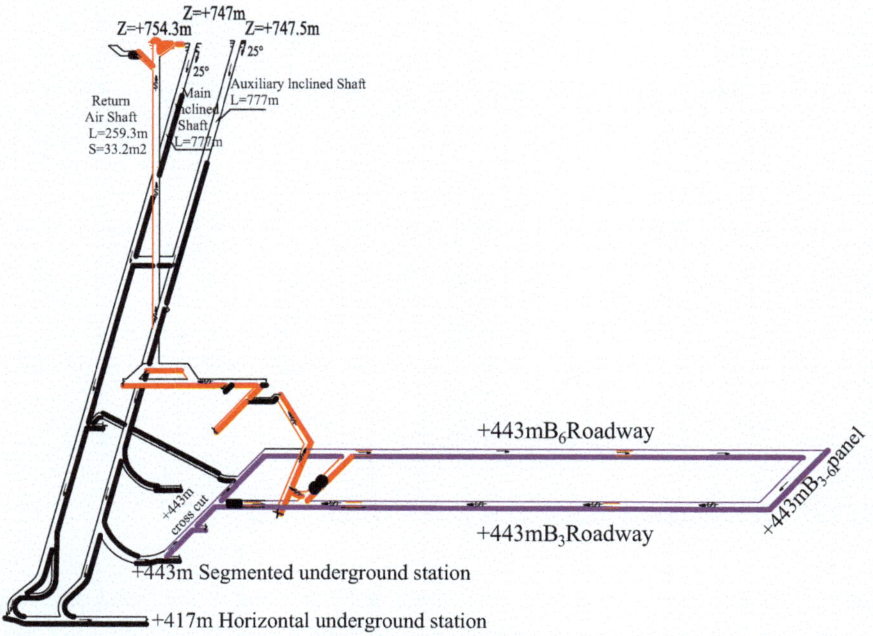

Fig. 7.20 Layout of roadways for the + 443mB$_{3\text{-}6}$ Panel

rectangular cross-section to accommodate the installation requirements of the panel equipment. The layout of the + 443mB$_{3\text{-}6}$ panel roadways is depicted in Fig. 7.20.

7.5.1.1 +443m B$_3$ and B$_6$ Roadways

The + 443m B$_6$ tailgate measures 4.6 m in width and 3.45 m in height, with a clear cross-sectional area of 14.62 m^2, and utilizes bolting-wire mesh shotcrete for support. The + 443m B$_3$ headgate is 4.9 m wide and 3.45 m high, with a clear cross-sectional area of 15.24 m^2, also employing bolting-wire mesh shotcrete for support. The shotcrete specification is C20, with an application thickness of 70 mm.

7.5.1.2 Winding Chamber

The winding chamber utilizes anchor net support, with a clear cross-sectional area of 10.38 m^2. The anchor rods are equal strength rods, 20 mm in diameter and 2500 mm in length. The rod spacing on the roof is arranged at 800 mm × 800 mm, and along the two ribs at 1000 × 800 mm. The anchor net employs a double-antibody net for support.

7.5.1.3 Set-Up Room

The set-up room is supported by a combination of anchor net, anchor cables, and individual hydraulic props. The west rib of the set-up room is secured with resin rods, 20 mm in diameter and 2000 mm in length. The roof rods in the set-up room are equal strength bolts, 20 in diameter and 2500 mm in length (fully threaded). The east rib of the set-up room uses ordinary round steel rods, 20 in diameter and 2000 mm in length. The rod spacing on the roof and west rib is 800 × 800 mm, while on the east rib it is also 800 × 800 mm. The anchor net uses a double-antibody net, the anchor cable are 18.9 mm in diameter and 9000 mm in length, spaced at 2000 × 2400 mm. The individual hydraulic props are spaced at 1400 × 3500 mm.

7.5.2 Panel Equipment Configuration

The comprehensive equipment set for STCC primarily consists of a shearer, hydraulic supports, AFCs, reversed loaders, crushing machines, and belt conveyors. These components are not standalone single machines but rather form an integrated whole that structurally requires mutual coordination and functionally demands harmonization, with stringent matching requirements and a high need for reliability. Each mechanical component of the top coal caving equipment set has specific application conditions, and improper selection can lead to misaligned equipment, reduced production efficiency, and diminished economic performance. Thus, the accurate selection and design of equipment are essential for maximizing their capabilities and achieving high-yield, high-efficiency, economical, and safe operation of the top coal caving panel. The main equipment for the + 443m$B_{3\text{-}6}$ panel is detailed in Table 7.10.

7.5.3 Mining and Caving Technology in +443m$B_{3\text{-}6}$ Panel

7.5.3.1 Mining Method

The + 443m$B_{3\text{-}6}$ panel utilizes STCC, with a sectional height of 26 m. The cutting height of coal is 3.2 m, the top coal caving height is 22.8 m, the cutting depth is 0.8 m, and the coal cutting cycle advance is 0.8 m. Following the beginning of top coal caving, the roof is treated with advanced pre-blasting. The distance between the last row of blasting holes in the initial caving and the advanced pre-blasting holes is 4 m. The top coal caving method employs a multi-round, interval, sequential, and equal quantity caving method, mined coal while simultaneously carrying out top coal caving. The caving interval is 1.6 m, referred to as the two-cutting and one-caving technique.

Table 7.10 +443m$B_{3\text{-}6}$ panel equipment installation schedule

Serial number	Equipment	Model	Unit	Device occupancy rate			Location of use
				Using	Backup	Total	
1	End hydraulic supports	ZFT44000/20/38	Set	1	0	1	B_3 end of the panel
2	Transition hydraulic supports	ZFG11000/25/38D	Shelf	3	0	3	Panel
3	Hydraulic support	ZFY10000/22/40D	Shelf	25	1	26	Panel
4	Advanced powered support	ZCH18000/20/38	shelf	3	0	3	B_6 roadway
5	Short-wall shearer	MG380/435-NWD	Set	1	0	1	Panel
6	Front AFC	SGZ–800/315	Set	1	0	1	Panel
7	Rear AFC	SGZ–1000/315	Set	1	0	1	Panel
8	Continuous crushing machine	PLM-2000	Set	1	0	1	B_3 roadway
9	Reversed loader	SZZ-1000/525	Set	1	0	1	B_3 roadway
10	Mobile substation	KBSGZY-1600/10/3.3 kV	Set	2	0	2	B_6 roadway
11	Mobile substation	KBSGZY-1600/10/1.14 kV	Set	1	0	1	B_6 roadway
12	Emulsion pump	BRW400/31.5	Set	1	1	2	B_6 roadway
13	Mine belt conveyor	DSJ-120/160/3 × 200	Set	1	0	1	B_3 roadway
14	Spray pump	BPW315/16(10)	Set	1	0	1	Cross-cut
15	Individual hydraulic prop	DW40-300/110X	Piece	80	10	90	B_6 and B_3 roadway
16	Advanced support hydraulic support	ZQ4000/19/40	Shelf	50			B_6 roadway

7.5.3.2 Top Coal Weakening Technology

The + 443m$B_{3\text{-}6}$ panel conducts advanced pre-blasting in the roadway to fulfill two goals: firstly, diminishing the impact force of extensive roof collapse; secondly, ensuring that the collapsed roof thoroughly fills the goaf areas, thereby securing the

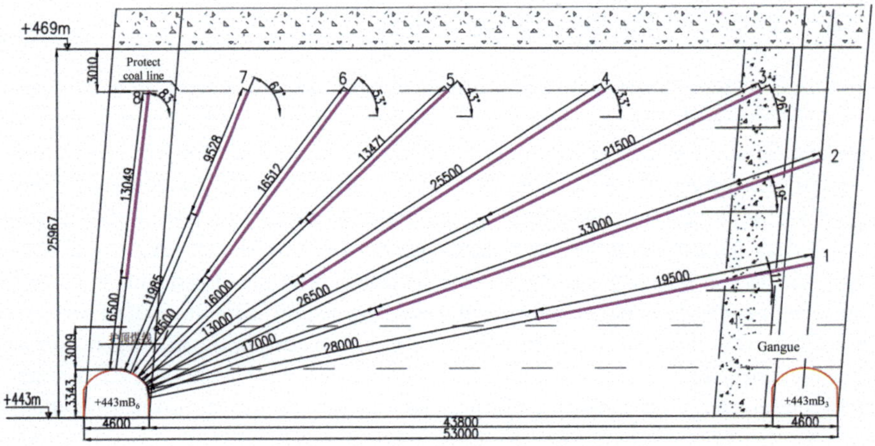

Fig. 7.21 Arrangement and parameters of blasting holes for top coal weakening

safe retreat mining of the panel. The advanced pre-blasting holes are created using a CMS1-4200/55 model track-mounted deep hole drill rig on the rib side of the B6 roadway, with a drill bit diameter of Φ113 mm. The layout for the panel's advanced pre-blasting holes consists of a 4-m row spacing, with 8 holes drilled per row, as depicted in Fig. 7.21.

Advanced blasting holes are charged with emulsified explosives through manual loading. To ensure the fluent initiation of the explosives, each primer is equipped with two detonators and two primer cords connected in series. The blasting hole is located more than 30 m away from the coal wall of the panel for charging, and the blasting is immediately initiated after the charging is completed. The sealing of the hole is manually performed using yellow soil, and it is required that the sealing length of the blasting hole is not less than the designed value. No fewer than three water cannon muds are used for sealing, arranged alternately with the terminal of the yellow soil cannon mud inside the hole. After each group of explosives is loaded, the foot line is twisted into a short circuit and wrapped with black tape at the end.

The blasting is carried out immediately following the charging of explosives. Before blasting, the advanced support section of the B_3 and B_6 roadways is inspected, and all personnel are withdrawn to a safe place more than 300 m away from the incoming airway blasting point where the roof support is complete. A special person is arranged for vigilance at the first intersection of + 443 m, and after the shift leader checks the personnel, a report is made to the production command center. Only after the command center issues the blasting command can the blasting proceed, and the warning distance for blasting must reach 300 m. The blasting location is no less than 30 m from the panel, and after blasting, a safety inspection is carried out by the gas inspector, safety officer, team leader, and shift leader at the blasting site after the blasting smoke has cleared or after waiting for 30 min.

7.5.3.3 Intelligent Top Coal Caving Mining

The STCC panel is short, with a fast advance speed and good working conditions, which facilitates the implementation of intelligent mining and aids in improving mining efficiency. When performing intelligent mining, there is no need to identify coal and rock when cutting the bottom coal; the shearer can utilize memory cutting, bypassing the current intelligent mining challenges where the precision and reliability of coal-rock identification by the shearer are not high. However, due to the short panel, the shearer must repeatedly perform inclined cutting feed end and cut triangle coal at both ends of the panel, along with releasing a significant amount of top coal. Therefore, in intelligent mining, the primary focus is on developing intelligent control for rapid cutting feed at the end and intelligent top coal caving technology. Currently, intelligent cutting feed technology at the end can basically meet production needs, however, there is still much room for improvement in terms of reliability, precision control, collision avoidance between equipment, and early warning technology for personnel approaching.

Since the thickness of the top coal is considerable, enhancing the caving efficiency and the recovery ratio of the top coal is an essential aspect. Utilizing multiple rounds of top coal caving from the bottom of the coal seam towards the roof helps maintain an even subsidence of the top coal boundary, reducing the contact between the top coal drawing body and the top coal boundary. This is an effective caving method to improve caving efficiency and recovery ratio while lowering the gangue mixed ratio. The third-generation top coal path tracker, developed independently, can accurately record the recovery ratio and drawing time of the top coal at various levels, providing a basis for determining the parameters for multiple rounds of caving. The use of intelligent top coal caving technology with image recognition before the conclusion of caving can achieve precise and intelligent control of caving [20].

References

1. Cui YP, Pan LQ et al (2012) Leading the way with a strategy of trial and error, achieving new heights beyond imagination-from "China Lu'an" to "internationalized new Lu'an." J Brand Res 2:28–37
2. Wang JC (2023) 40 years development and prospect of longwall top coal caving in China. J China Coal Soc 48(1):83–99
3. Su S (1997) Application of the fully-mechanized coal mining with top coal caving in Shanxi. Shanxi Coal 1:14–20+26
4. Wang AM (1998) Research and application of full-mechanized top coal caving technology in the Lu'an mining area. Coal 1:4–7
5. Cao SG, Zhang DS, Du WX (2003) Key mining technology of extra-long fully-mechanized caving face. J Min Saf Eng 4:69–71
6. Huang FC (2010) Development course and outlook of fully mechanized top coal caving mining technology in Yanzhou mining area. J China Coal Soc 35(11):1778–1782
7. Lai CL (2012) Innovative application and promotion of Yankuang fully mechanized top coal caving mining technology in Australia-a case study on the implementation of China fully

mechanized top coal caving mining technology at Austar coal mine. In: China coal society. Proceedings of the high-level academic forum on the 50th anniversary of the establishment of china coal society 18
8. Wang YL (2004) Selection and evaluation for the mining methods of high and extra-high seam of carboniferous period in Datong coal field. Jinneng Hold Sci Technol 1:1–2+5
9. Wu YP (2010) Fully mechanized top coal caving mining technology for ultra thick seam in Datong mining area. Coal Sci Technol 38(11):28–31
10. Yu B (2015) Coal seam mining in Datong mining area. Science Press, Beijing
11. Huang HJ, Li SJ, Guo J et al (2018) Experiments of top coal weaken with hydraulic fracturing in fully mechanized top coal caving working face. Coal Min Technol 23(S1):67–69
12. Xue JS (2023) Weakening technology for full-area hard roof and hard top coal in fully mechanized caving mining. Coal Eng 55(12):5–10
13. Wang ZW (2021) Research and application of top coal blasting weakening technology in 81020 fully mechanized caving face of Yukou coal mine. Coal 30(12):63–65
14. Wang ZG (2022) Research on top-coal group caving theory and intelligent top-coal caving control method in extra-thick coal seam. Henan Polytechnic University
15. Wang JC, Li W (2012) Theory and practice of fully mechanized mining in extremely soft and thick coal seam in Huaibei mining area. In: Mining Design Division of Tiandi Technology Co., LTD. (Mining Design Research Branch of Coal Research Institute) Mining Technology Research Institute. Innovative Development of Theory and Practice of Fully Mechanized Top Coal Caving—Collection of Scientific Papers on the 30th Anniversary of Fully Mechanized Mining. China Coal Industry Publishing House, p 7
16. Li W, Zhan ZJ (2013) Fully-mechanized coal mining technology in seam under complicated condition in Huaibei mining area. Coal Sci Technol 41(9):79–82
17. Chen TY (2021) Luling mine III 811 full-mechanized caving mining technology and mining pressure patterns appeared. Anhui University of Science and Technology
18. Wang JC, Zhang JW (2015) Study on top-coal recovery ratio distribution law of longwall top-coal caving mining in steeply thick coal seam. Coal Sci Technol 43(12):1–7
19. Wang JC, Zhao BW, Zhao PF et al (2017) Research on the longwall top-coal caving mining technique in extremely inclined and soft thick coal seam. J China Coal Soc 42(2):286–292
20. Wang JC, Pan WD, Zhang GY et al (2022) Principles and applications of image-based recognition of withdrawn coal and intelligent control of draw opening in longwall top coal caving face. J China Coal Soc 47(1):87–101

Chapter 8
Typical Cases of Large Mining Height Practice

8.1 10 m Extra-Large Mining Height Technology

8.1.1 Panel Conditions

Caojiatan Coal Mine is in Machang Village, Mengjiawan Township, Yuyang District, Yulin City, Shaanxi Province, China. The mining area belongs to the first phase of the Yushen mining area planning district and is a key development project in the national coal industry "12th Five-Year Plan" as a large-scale coal mine demonstration construction project. The mine field is approximately 10 km wide and 12.5 km long, covering an area of 108.49 km^2. The main coal seams mined are four layers, with recoverable reserves of 1.511 billion tons and an approved production capacity of 17.0 Mt/a.

The initial mining area, Sect. 8.12, has an east–west inclination width of 2.8 km and a north–south length of 12.5 km. The structure within the area is simple, with the coal seams generally flat and dipping less than 1°. The coal seam structure is simple, with good integrity and a Protodyakonov hardness coefficient of 2–3. The coal seam is buried at depths of 25–338 m. The thickness of the coal seam ranges from 8.1 to 12.7 m, with an average thickness of 11.2 m, varying slightly in different areas. The main roof consists primarily of massive, homogeneous siltstone and medium-grained sandstone with poorly developed joints, with thicknesses ranging from 3.94 to 20.83 m. The immediate roof mainly comprises siltstone, sandy mudstone, and medium-grained sandstone, generally 2 m thick, with a maximum thickness of 8.33 m. There is little development of the false roof, which is composed of sandy mudstone and mudstone interspersed with thin coal lines.

J. Wang and W. Wei, *Thick Coal Seam Underground Mining*,
https://doi.org/10.1007/978-981-96-3874-1_8

8.1.1.1 Panel Situation

122,104 panel is in the east wing of the 12th district. To the west of the panel are four main roadways serving the entire mine, while the eastern side borders the coal pillar at the edge of the mine field. The northern side is adjacent to the 122,102 panel (design), and the southern side to the 122,106 panel (closed). There are no surface water bodies, railways (or major highways), etc., above the 122,104 panel. The set-up room is situated at the eastern boundary of the mine field, with the roadway opening located within the protective coal pillar of the main roadway in the 12th district.

8.1.1.2 Coal Seam Conditions

Mining panel 122,104 has been mined the No. 2–2 coal seam, which has an average thickness of 10.5 m. The coal seam occurrence is relatively stable with a simple structure and an average dip angle of 0–6°. It appears as black blocky coal with a weak to moderate luster of asphalt and bitumen, and locally exhibits a greasy luster. Bright coal and dark coal are predominant with streaks of vitrinite. The coal quality is relatively hard, with a Protodyakonov hardness coefficient ranging between 2 and 3.

8.1.1.3 Surrounding Rock Conditions

The main roof consists of medium-grained sandstone, with an average thickness of 16.9 m. It appears grayish-white in color, predominantly composed of quartz with lesser amounts of feldspar. The sorting is moderate, with sub-angular grains and mudstone cementation, exhibiting blocky bedding. The immediate roof is composed of siltstone, with an average thickness of 21.7 m. It appears light gray in color, mudstone cemented, and shows blocky bedding with plant fossil inclusions. The immediate floor is siltstone, with an average thickness of 6.31 m. It appears dark gray in color, containing carbonized plant stem fossils, mudstone cemented, and exhibiting blocky bedding. The main floor is fine-grained sandstone, with an average thickness of 8.0 m. It appears grayish-white in color, primarily composed of quartz with lesser amounts of feldspar. The sorting is moderate, with sub-rounded grains and mudstone cementation, showing blocky bedding.

8.1.1.4 Underground Gas and Coal Dust

The mine has an absolute methane emission volume of 3.76 m^3/min and a relative methane emission volume of 0.10 m^3/t, indicating it is a low-gas mine. The maximum absolute methane emission volume at the panel is 0.84 m^3/min, with a relative emission volume of 0.04 m^3/t. The coal dust is explosive, and there is a natural tendency for coal spontaneous combustion.

8.1.1.5 Mining Method

The 122,104 panel adopts the retreating longwall mining method with roof management by full caving. The end oblique cutting method is used for the panel. The panel is arranged along the coal seam inclination and advances along the coal seam strike, with a strike length of 5977 m and an inclination length of 300 m. The mining height is 10 m in the panel. A total of four roadways are arranged in the panel: auxiliary transportation roadway, main transportation roadway, inner return airway, and outer return airway. All mining roadways have a rectangular cross-section.

8.1.2 Equipment and Technical Parameters of the Panel

As shown in Fig. 8.1, the 122,104 extra-large mining height panels equipped with major supporting equipment, including the MG1200/3350-GWD shearer (Fig. 8.2), ZYA29000/45/100D hydraulic support (Fig. 8.3), ZYG29000/42/90D (A/B) transitional support, SGZ1600/3 × 2000 AFC, PLM7500 crusher, BRW1250/40K emulsion pump, and BPW1000/16 spray pump. Detailed technical parameters of each piece of equipment can be found in Tables 8.1, 8.2, 8.3, 8.4, 8.5, 8.6, 8.7, 8.8, 8.9, 8.10 and 8.11.

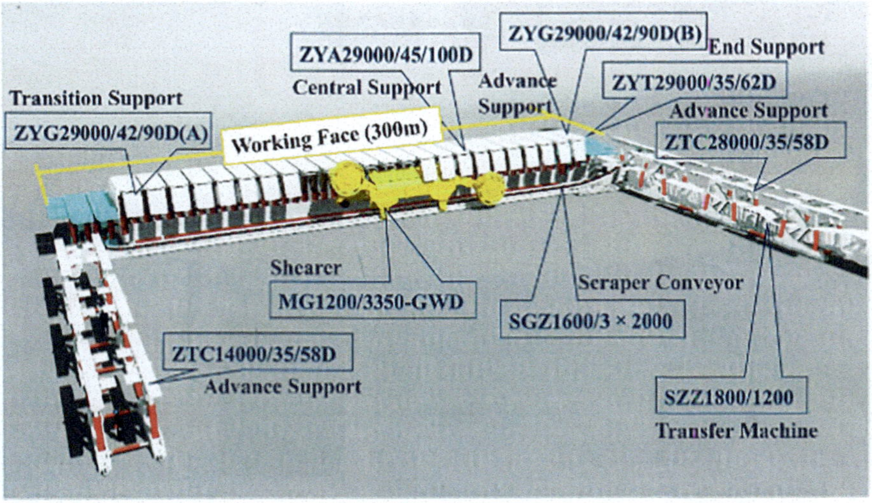

Fig. 8.1 Equipment layout of the 122,104 panel

Fig. 8.2 MG1200/3350-GWD shearer

Fig. 8.3 Hydraulic support for 10 m extra-large mining height panel: **a** Installation of hydraulic supports; **b** double-layer telescopic beam and three-level rib protection device

Table 8.1 Main technical parameters of MG1200/3350-GWD shearer

Item	Parameter	Item	Parameter
Cutting height range (m)	6.0–10.0	Drum diameter (mm)	4800
Production capacity (t/h)	5000	Power supply voltage (V)	3300
Cutting panel height (mm)	2250	Drum cutting depth (mm)	865
Traction drive method	AC VFD	Maximum bottom hole depth (mm)	572
Installed power (kW)	3350	Machine weight (t)	262t
Cutting motor power (kW)	2 × 1200	Traction motor power (kW)	2 × 250
Pump motor power (kW)	2 × 75	Crusher motor power (kW)	300

Table 8.2 Main technical characteristics of ZYA29000/45/100D type middle hydraulic support

Item	Parameter	Item	Parameter
Support type	Two-column shielded type	Support height (mm)	4500–10,000
Support center distance (mm)	2400	Support width (mm)	2260–2610
Support strength (MPa)	1.88–1.95	Advancing step (mm)	865
Initial support force (kN)	22,368	Working resistance (KN)	29,000
Operation model	Electro-hydraulic control	Total weight of support (t)	127

Table 8.3 Main technical characteristics of ZYT29000/35/62D end support

Item	Parameter	Item	Parameter
Support type	Two-column shielded type	Support height (mm)	3500–6200
Support center distance (mm)	2400	Support width (mm)	2260–2510
Support strength (MPa)	1.42–1.47	Advancing step (mm)	865
Initial support force (kN)	23,368	Working resistance (KN)	29,000
Operation model	Electro-hydraulic control	Total weight of support (t)	101

Table 8.4 Main technical characteristics of ZYG29000/42/90D (A/B) transition support

Item	Parameter	Item	Parameter
Support type	Two-column shielded type	Support height (mm)	4200–9000
Support center distance (mm)	2700	Support width (mm)	2560–2740
Support strength (MPa)	1.50–1.57	Advancing step (mm)	865
Initial support force (kN)	23,368	Working resistance (KN)	29,000
Operation model	Electro-hydraulic control	Total weight of support (t)	125

Table 8.5 Characteristics of ZTC28000/35/58D advanced support for main transport roadway

Item	Parameter	Item	Parameter
Support type	Four-support group, stepping split type	Support length (m)	24
Support height (mm)	3500–5800	Roof beam width (mm)	5050
Working resistance (kN)	28,000	Support strength (MPa)	0.16
Roadway width (mm)	7500	Advancing step (mm)	865
Operation model	Electro-hydraulic control	Total weight of support (t)	140

Table 8.6 Characteristics of ZTC14000/35/58D advanced support for return airway (12 groups)

Item	Parameter	Item	Parameter
Support type	Step-by-step sectional type	Support length (m)	45
Support height (mm)	3500–5800	Roof beam width (mm)	5912
Working resistance (kN)	14,000	Support strength (MPa)	0.34
Roadway width (mm)	6500	Advancing step (mm)	865
Operation model	Electro-hydraulic control	Total weight of support (t)	Approx. 678

Table 8.7 Main technical characteristics of SGZ1600/3 × 2000 AFC

Item	Parameter	Item	Parameter
Manufacturer	CCTEG	Capacity (t/h)	8000
Power supply voltage (V) and frequency (Hz)	3300, 50	Motor power (kW)	3 × 2000
Scraper chain type	Double middle chain	Scraper chain speed (m/s)	0–1.83
Chain specifications	60/135 × 181/197	Chute specifications (mm)	2400 × 1600 × 540
Horizontal deviation	± 1°	Vertical deviation	± 3°
Chain tightening method	Hydraulic tightening	Tightening stroke (mm)	800

Table 8.8 Main technical characteristics of SZZ1800/1200 transfer machine

Item	Parameter	Item	Parameter
Manufacturer	CCTEG	Capacity (t/h)	8500
Power supply voltage (V) and frequency (Hz)	3300, 50	Motor power (kW)	1200
Scraper chain type	Central double chain	Scraper chain speed (m/s)	2.2
Chain specifications	48 × 152-C	Chute specifications (mm)	Unsupported Section:3000 × 1600 × 1100 Ground Section:1500 × 1800 × 2275
Chain pitch	700 mm	Scraper pitch	912 (6 × 152) mm
Chain tightening method	Hydraulic tightening	Tightening stroke (mm)	500

Table 8.9 Main technical characteristics of PLM7500 crusher

Item	Parameter	Item	Parameter
Manufacturer	CCTEG	Capacity (t/h)	7500
Power supply voltage (V) and frequency (Hz)	3300, 50	Motor power (kW)	1200
Maximum feed size (mm)	2600 × 2100	Discharge size (mm)	≤ 425
Middle plate thickness (mm)	80	Inner width of tough (mm)	1800

Table 8.10 Main technical characteristics of BRW1250/40K emulsion pump

Item	Parameter	Item	Parameter
Model	BRW1250/40 K	Rated working pressure (MPa)	40
Power supply	3300 V, 50 Hz	Motor power (kW)	1000
Nominal flow rate (L/min)	1250	Total volume of emulsion tank (L)	10,000 (2 tanks)

Table 8.11 Main technical characteristics of BPW1000/16 emulsion pump

Item	Parameter	Item	Parameter
Model	BPW1000/16	Rated working pressure (MPa)	16
Power supply	3300 V, 50 Hz	Motor power (kW)	315
Nominal flow rate (L/min)	1000	Total volume of emulsion tank (L)	14,000 (2 tanks)

8.1.3 Cycle Work Mode of the Panel

The shearer cuts two slices of coal in each round trip, with each slice of coal constituting one operational cycle. The cycle progresses by 0.865 m, and the daily production organization completes 17 regular cycles. The cycle operation includes: the upper (lower) end head cuts the coal obliquely → moves the head (tail) and advances the support → the shearer cuts the coal → follows the machine and moves the frame → pushes the AFC → the lower (upper) end head cuts the coal obliquely → moves the tail (head) and advances the support.

8.2 Large Mining Height Technology in Shallow Thick Coal Seams

Shendong Mining Area is a major concentration area of thick coal seams in China. As China's coal mining focus has shifted westward over the years, the coal supply from the Shendong Mining Area holds a pivotal position in China's energy reserves and supply. Compared to the eastern coal resources, the coal seams in the Shendong Mining Area are characterized by shallow burial depth, thickness, and stable occurrence. At the beginning of mining in this region, the direction was set for large-scale, modern, specialized, and informatized development. The success of the Shendong Mining Area is attributed to innovative coal mining methods and the revolutionary development of supporting equipment. Initially, the main coal seams in various mines in the Shendong Coalfield ranged in thickness from 3.5 to 7.0 m, typical of thick coal seams. To enhance the efficiency and recovery ratio of thick coal seam extraction, Shendong Coal Company conducted pioneering practices in large mining height extraction technologies in 2007, 2009, and 2018, achieving unprecedented mining heights of 6.3 m, 7.0 m, and 8.8 m respectively, both domestically and internationally. This effort set records with a maximum daily production of 65,500 tons and a peak monthly production of 1.5 million tons from a single panel. It served as a model for comprehensive extraction of extra-thick coal seams (7–9 m) in China and demonstrated the feasibility of large mining height technology for thick and extra-thick coal seams in regions such as Shanxi, Shaanxi, Inner Mongolia, Ningxia, Gansu, and Xinjiang.

As shown in Fig. 8.4, the traditional movement laws of the overlying strata in mining areas can be categorized into "blocky beam" structure and "transmitting rock beam" structure. In the Shendong Mining Area, influenced by factors such as shallow burial depth, thin bedrock, thick loose layers, large mining heights, and extra-large panels, when the depth of the panel is within 150 m, the overlying strata often exhibit a "two-zone" characteristic. This characteristic refers to the periodic and cohesive breaking of the overlying strata in the mining area, where the broken range extends to the surface. This movement structure of the overlying strata is known as the "drawing body" structure [1].

In response to the "drawing body" structure of the overlying strata in the Shendong Mining Area, there is a need to strengthen research on the selection of hydraulic support types, the determination of support parameters, and the determination of guard parameters. Each increase in mining height in the fully mechanized mining panel represents a revolution in coal mining equipment and extraction technology, and witnesses the advancement of China's coal mining equipment manufacturing capabilities. To gradually improve the mining height of large mining height panels in the Shendong Mining Area, extensive research, exploration, and innovation have been conducted in areas such as strata control, development of mining equipment, and innovation of mining processes.

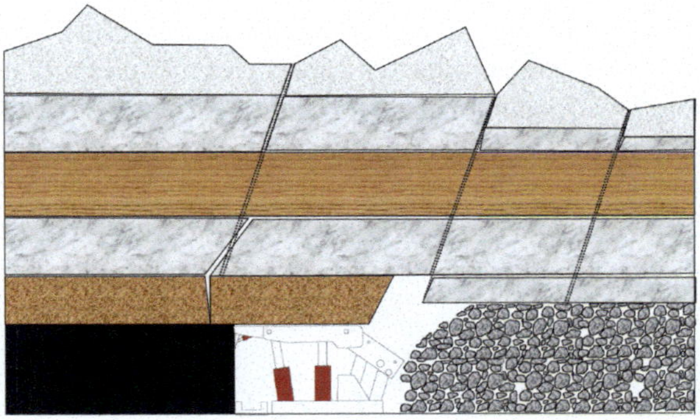

Fig. 8.4 'Drawing body' structure of shallow buried coal seam panel

8.2.1 6.3 m Large Mining Height Technology

Shendong Coal Branch Company has substantial reserves of coal seams with thicknesses exceeding 6.0 m. The recoverable reserves of coal seams thicker than 6 m in Shangwan Mine and Bulianta Mine amount to approximately 800 million tons. Before 2011, mining activities were conducted in the No. 1–2 coal seam of the second panel area of Shangwan Mine, where the coal seam thickness ranged from 6.11 to 6.69 m, with an average mining height of 6.3 m. Similarly, in the No. 2–2 coal seam of the third panel area of Bulianta Mine, the coal seam thickness ranged from 6.75 to 7.5 m, with a mining height between 6.5 and 7.2 m. The 51,202 large mining height panel at Shangwan Mine represents China's first 6.3-m large mining height heavy-duty fully mechanized panel. During the extraction process, hydraulic supports with a working resistance of 10,800 kN effectively controlled the roof. This operation also achieved significant technological innovations and breakthroughs in integrated mining, electromechanical equipment, and the automation and informatization of the coal mine. The highest daily production reached 40,214 tons, and the highest monthly production was 1.095 million tons. The annual production capacity increased to 12 million tons per year, resulting in an additional annual economic benefit of approximately 100 million yuan. The resource recovery ratio reached 94.2%, and the mine's service life was extended by 4.5 years.

The successful extraction of the 51,202 panel is mainly attributed to innovations in hydraulic support selection and associated equipment under the guidance of the "key strata" structure theory for overlying rock strata in the mining field. This success can be primarily attributed to two key points: (1) According to traditional mine pressure theory, increasing the mining height from 4.3 to 6.3 m requires only a 2000 kN increase in the working resistance of hydraulic supports to accommodate mine pressure. However, considering the "key strata" structure movement in the overlying rock strata of the 51,202 heavy-duty panel, it was necessary to estimate the residual load of the key strata. Consequently, the working resistance of the hydraulic supports was increased by 4085 kN, enhancing the support strength by 0.38 MPa. This led to the development of the ZY10800/28/63D type hydraulic support, which at the time had the highest support height among domestic two-leg hydraulic supports, and featured a three-stage side shield structure with a maximum side shield height of 3.2 m. (2) To improve the production capacity of the fully mechanized panel, a new AFC was developed and implemented, featuring a motor power of 3×1000 kW, a chain diameter of 48 mm, and a transportation capacity of 4200 tons per hour. Additionally, a coal shearer with a then-record power of 2390 kW was developed, equipped with a 3.2-m diameter drum. Furthermore, an emulsion pump with a maximum flow ratio of 430 L per minute was developed to support the increased operational demands.

8.2.2 7.0 m Large Mining Height Technology

The Bulianta Coal Mine of Shendong Coal Group has actively implemented the fully mechanized mining technology for single-pass full-height extraction. In 2009, the 22,303 fully mechanized panel was equipped with a heavy-duty panel with a mining height of 7.0 m. The working resistance of the hydraulic supports was increased to 18,000 kN, and the center distance was extended to 2.05 m. Since the commencement of extraction at the 7.0-m large mining height fully mechanized panel, similar operations have been conducted in other mines under the group, including Daliuta, Shangwan, and Sandaogou Mines, resulting in the recovery of 17 panels and the production of 150 million tons of raw coal, with a zero-fatality ratio per million tons, demonstrating excellent mining outcomes.

The 22,303 panel employed a fully mechanized mining method, utilizing the longwall retreating technique with complete caving for full-seam mining. This involved using a double-drum shearer for cutting and loading coal, hydraulic supports for roof control, and a combination of AFCs, transfer conveyors, and belt conveyors for coal transportation. The goaf was managed using the natural caving method. The shearer's cutting approach involved an oblique cut at the end, with the front drum cutting the top coal and the rear drum cutting the bottom coal. The sequence of operations included cutting coal, advancing supports, and pushing the AFC. Coal was loaded onto the conveyor by the rear drum and AFC's shovel plate, and transported via the crusher and transfer conveyor to the belt conveyor in the transport roadway. Advanced hydraulic supports were used for pre-supporting the two roadways of the panel.

Engineering practice indicated that increasing the mining height from 6.3 to 7.0 m required an increase in the maximum working resistance of the hydraulic supports by 7200 kN and an enhancement in support strength by 0.575 MPa. The shearer incorporated innovations such as fully straight-tooth transmission, cast-weld hybrid casing, time-sharing forced lubrication, and composite multi-path cooling, achieving a 7.0-m large mining height with domestically maintainable fully straight-tooth ranging arm five-stage straight-tooth transmission, improving the stability of the reduction mechanism. To address the severe coal loss at the ends of the large mining height panel, vertical transition supports were developed, increasing the coal recovery ratio. Auxiliary roadway multi-channel technology and dedicated withdrawal supports for the 7.0-m large mining height equipment were adopted, systematically resolving issues such as difficulty in final panel-through, high mine pressure, and equipment withdrawal challenges in large mining height fully mechanized panels.

During the mining of the 22,303 panel, the roof was generally controllable, with noticeable periodic roof movement. In the advancement direction, the mine pressure exhibited a "large-small" alternating pattern, with periodic weighting intervals ranging from 9.0 to 16.8 m, averaging 13 m. In the inclined direction, the periodic weighting intervals were symmetrically distributed around the central support, showing a "middle small, ends large" movement characteristic. The first key stratum of the overlying strata of the 22,303 panel typically exhibited a "cantilever beam" structure. Due to the different fragmentation coefficients of the caving blocks and the immediate roof caving rock behind the goaf, various movement forms were observed. Under the cantilever beam structure of the large mining height, the fracture span of the key stratum was larger than that of the ordinary mining height "masonry beam" structure, and the length of the panel under pressure was also greater than that of the ordinary mining height "masonry beam" structure. The mining pressure manifestation intensity was significantly influenced by the burial depth, with greater depth resulting in greater load transfer to the supports due to synchronous fracturing of multiple key strata, leading to more intense mining pressure manifestations in large mining height panels [2].

8.2.3 8.0 m Large Mining Height Technology

8.2.3.1 Geological Conditions

The 12,511 large mining height panel of Bulianta Coal Mine is in the fifth panel area of the No. 1–2 coal seam. The panel has a maximum mining height of 8.0 m, a panel length of 319.1 m, and an advance length of 3439.3 m. The coal seam structure is simple, with a thickness ranging from 3.95 to 9.50 m and an average thickness of 7.44 m. The seam has a dip angle of 1–3°, a Protodyakonov coefficient (f) of 2.5, a density of 1.33 g/cm^3, and a burial depth of 233–301 m. The mining area covers 1,002,452.7 m^2, with geological reserves of 10.133 million tons and recoverable reserves of 9.4235 million tons. The fully mechanized panel is arranged along the dip and mainly advances with a positive slope, using the longwall retreating method for single-pass full-height extraction with full caving for roof management. The 12,511 fully mechanized panel is adjacent to solid coal on the transport roadway side and the goaf of the 12,510 panel on the return airway side. A 20-m coal barrier separates the 12,511 transport roadway and the 12,512 return airway. The roadway height is 4.1 m, with a transport roadway width of 6.0 m and a return airway width of 5.4 m, both excavated along the floor.

8.2.3.2 Panel Layout and Major Equipment

The 12,511 panel employs a JOY07LS08 shearer with a drum diameter of 4.5 m and a single mass of 16 tons, making it the largest shearer drum in the world. The shearer has an installed total power of 2925 kW and produces 2864 tons of raw coal per cutting cycle. With 20 cutting cycles organized per day, the daily raw coal production reaches 57,271 t.

Double-shield hydraulic supports are used, including 149 ZY21000/36.5/80D supports for the main panel, 6 ZYT18000/28/55D supports for the end shields, 2 ZYG21000/36.5/80D short transition supports, and 2 ZYG21000/36.5/80DA/B long transition supports. The initial support force is 15,443 kN (35 MPa), the working resistance is 21,000 kN, and the support strength is 1.67 MPa, with a safety valve unloading pressure of 47.6 MPa. The support height ranges from 3.65 to 8.0 m, and the long and short transition supports enable two-stage vertical transitions, with wireless remote control functionality significantly improving the adaptability of the supports to the roof conditions.

An SGZ1400/4800 AFC with a transport capacity of 6000 tons per hour and a total power of 4800 kW is used. The conveyor features enhanced intelligence, variable frequency speed regulation, and a hydraulic automatic tensioning device in addition to traditional mechanical tensioning. The BRW630/37.5 (BRW200/45) pump station combination includes 8 pumps and 3 tanks, with a total power of 2994 kW, providing high flow and extra-high-pressure dual-circuit liquid supply, achieving digital pressure control adjustment, remote start/stop, and automatic oil injection. The transport roadway belt conveyor has a belt width of 1.6 m, a belt speed of 4.0 m/s, a length of 3200 m, and a power of 4×500 kW. Advanced technologies such as high-power variable frequency soft start, automatic tensioning, high-speed wear-resistant rollers, and fast self-moving tail enhance equipment uptime.

8.2.3.3 Panel Pressure Behavior

To avoid large-scale roof pressure during the initial mining period of the 12,511 panel, hydraulic fracturing technology was used to weaken the roof, reducing its strength and integrity, allowing the goaf roof to collapse in layers and stages, thereby shortening the initial and periodic weighting intervals.

1. First weighting of the panel

When the 12,511 panel advanced to approximately 16 m, the immediate roof fully collapsed and filled the goaf. When the panel advanced to 36 m at the headgate and 40.6 m at the tailgate, the pressure on the supports began to increase significantly, with some areas experiencing coal wall spalling but no safety valve activation. As advancing continued, significant overall spalling occurred, with the deepest reaching 1.2 m and support resistance exceeding the safety valve threshold, indicating the initial weighting of the main roof. The initial weighting interval was estimated to be around 40–44 m.

2. Periodic weighting of the panel

Monitoring data indicated that the panel pressure exhibited periodic characteristics. The weighting interval ranged from 6.4 to 16.8 m, averaging 11.6 m, with a dynamic load factor of approximately 1.4. Weighting was generally evident and followed a daily occurrence pattern, occasionally extending over 8–10 cutting cycles.

3. Analysis of pressure influencing factors

The 12,511 panel roof comprises two key strata layers. The first load-bearing layer is approximately 7.7 m thick; the sub-key stratum is about 19.8 m thick, and the main key stratum is around 35 m thick, all consisting of sandy mudstone, forming a typical double key strata roof structure. The sub-key stratum fractures, leading to small periodic weightings, and after multiple occurrences, a large periodic weighting due to the main key stratum fracture is observed, exhibiting strong and distinct pressure manifestations. The fracture intervals of the sub-key and main key strata can be inferred, with structural or internal rock displacement potentially altering the key strata fracture patterns [3].

8.2.4 8.8 m Large Mining Height Technology

The 12,401 panel of Shangwan Coal Mine is the world's first fully mechanized panel with an extra-large mining height of 8.8 m. This panel features innovations and technological advancements in equipment integration, surrounding rock control, mining techniques, and information technology. Practical experience has demonstrated that the selection and matching of equipment for the 8.8-m extra-large mining height panel are rational and capable of meeting the production, support, and transportation requirements of the fully mechanized panel. Additionally, the production capacity has significantly improved. Compared to the 7.0-m large mining height panel in the same coal seam, this extra-large mining height panel can recover an additional 4.05 million tons of coal, increasing the recovery ratio by 20.2% [4, 5].

8.2.4.1 Geological Conditions

The 12th coal seems fourth panel area at Shangwan Coal Mine is located in the central part of the mine. The panel area extends approximately 4.05 km in the east–west direction and 5.7 km in the north–south direction, covering an area of 23.1 km^2, with an elevation ranging from + 1030 to + 1086 m. The coal thickness in the panel area varies between 7.32 and 10.79 m, with an average minable thickness of 9.02 m. The coal seam dip angle is between 1–3°, indicating a relatively stable coal seam. The geological reserves in the panel area total 250 million tons, with a designed minable reserve of 182 million tons, and 12 extra-large mining height panels are planned, providing a service life of 13 years.

The 12,401 panel is the initial panel in the fourth panel area of the 12th coal seam, with an inclined length of 289.2 m and a strike length of 5254.8 m. The coal seam dip angle is 1–3°, with a thickness ranging from 7.56 to 10.79 m and an average thickness of 9.16 m. The depth ranges from 124 to 244 m, covering a mining area of 1.572 km^2, with geological reserves of 20.594 million tons and minable reserves of 19.3 million tons. The comprehensive mechanized mining method is employed, using the inclined longwall retreating method for single-pass full-height extraction with full caving for goaf management.

The immediate roof comprises mudstone, with a thickness of 5.68–20.34 m, a compressive strength of 11.3–13.2 MPa, and a Protodyakonov coefficient of approximately 1.32, indicating low strength and unstable properties. The main roof consists of gray-white fine-grained sandstone with a fine-grained sandy structure, good sorting, and porous argillaceous cementation, containing plant fossils. Its thickness ranges from 2.10 to 8.07 m, with a compressive strength of 18.7–36.9 MPa and a Protodyakonov coefficient of about 2.67, indicating strong and unstable properties. The main roof is gray-white siltstone, with gray-black layers, siltstone structure, argillaceous cementation, and various bedding features, ranging in thickness from 0.20 to 0.52 m, with a compressive strength of 14.5–36.6 MPa and a Protodyakonov coefficient of about 2.32. The immediate floor is black-gray mudstone with a mudstone structure, flat fractures, dense blocky structure, containing plant fossils, ranging in thickness from 0.96 to 1.29 m, with a compressive strength of 13.2–354 MPa and a Protodyakonov coefficient of about 2.15. The comprehensive stratigraphic column of the coal seam is shown in Fig. 8.5.

Fig. 8.5 Stratigraphic column of borehole 12,401 at Shangwan Coal Mine

8.2.4.2 Equipment Configuration of the Panel

1. Shearer on the 8.8 m extra-large mining height panel

Shendong Company led the development of a lightweight, large-dimension, and highly reliable cutting unit with a maximum load capacity of 1250 kW. They developed a high-speed, highly reliable heavy-duty shearer travel system, optimizing the sprocket teeth shape and structure to ensure the sprocket's load capacity and reliability. The MG1100/2925-WD double-drum shearer (Fig. 8.6) has the following main technical features: cutting height range of 4.3–8.6 m, production capacity of 6000 t/h, installed power of 2925 kW, maximum traction of 982 kN, machine height of 4145 mm, drum diameter of 4300 mm, drum center distance of 17,780 mm, drum cutting depth of 865 mm, maximum undercutting amount of 477 mm, and a machine height of 4828 mm with the cover raised.

2. Hydraulic support on the 8.8 m extra-large mining height panel

The ZY26000/40/88D double-column shield type support has a working resistance of 26,000 kN and a cylinder diameter of 600 mm, with a center distance of 2.4 m. The support uses a three-stage shielding structure with a maximum shielding height of 4.38 m. Enhancements were made in support strength selection, structural impact resistance, and active support capability. The large-flow hydraulic system compatibility technology ensures a descent stroke of 10 mm and a bottom lifting time of 4.3 s per support, with a moving support time of 3.49 s per support. The rising stroke of 100 mm to the initial support force takes 2.467 s per support, keeping one work cycle time within 10 s, achieving a moving speed of 14.03 m/min, meeting the fast-moving requirements for the 8.8 m hydraulic support.

3. High-strength, large-capacity AFC

Due to the incompatibility of traditional AFCs with the ZY26000/40/88D hydraulic support, the SGZ1388/3 × 1600 high-strength large-capacity AFC was developed. It features frequency conversion automatic speed regulation, online intelligent monitoring, online fault diagnosis, automatic chain break detection, and real-time coal flow monitoring, achieving international advanced levels. Innovations include:

(1) A 2.4 m extra-heavy AFC middle trough was development. The previous maximum length of middle troughs was 2.05 m, which was insufficient. The new design features an assembled welding structure, with the middle plate thickness of 60 mm and bottom plate thickness of 40 mm.
(2) Heavy-duty AFC tail automatic tensioning system was developed. This system detects the internal pressure of the tail telescopic cylinder, compares it with preset control program values, and adjusts the cylinder using an electromagnetic valve to maintain chain tension within a reasonable range, prolonging the life of major components.
(3) Intelligent speed regulation system for AFC was developed. This system integrates three 1600 kW frequency conversion motors with frequency converters

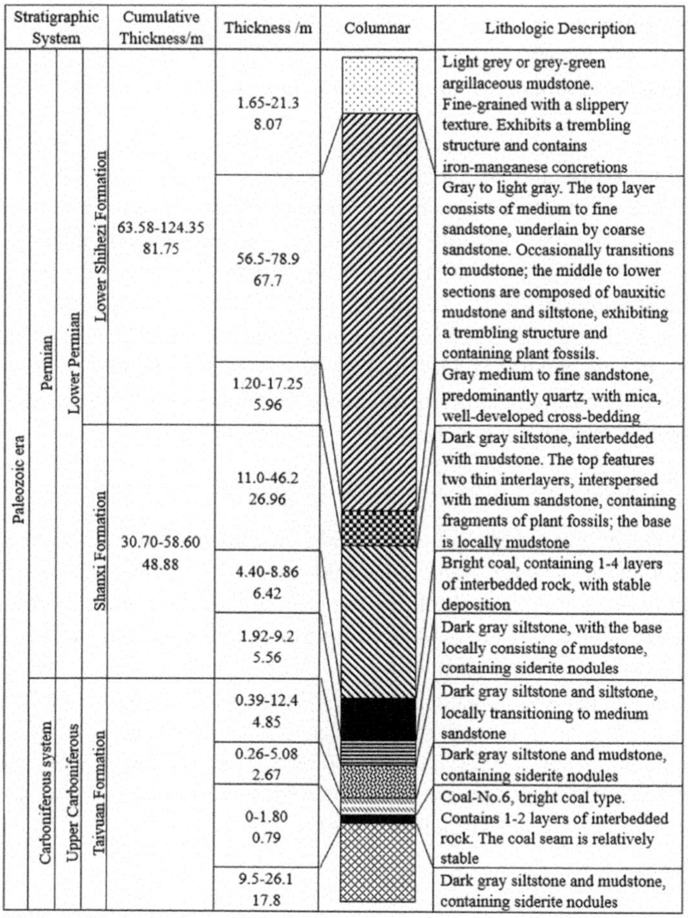

Fig. 8.6 MG1100/2925-WD double-drum shearer

and drive motors. By detecting the load on the AFC and relevant Parameters from the shearer, it adjusts speed for optimal energy efficiency, achieving significant energy savings.

The SGZ1388/3 × 1600 AFC can transport up to 60.0 t/h, providing a high-yield, cost-effective domestic alternative to imported equipment, significantly reducing production costs.

1. 6 km extra-long transport distance intelligent single-point drive belt conveyor

The successful development of a belt conveyor with a bandwidth of 1.8 m, an extra-long transportation distance of 6000 m, a head-end drive, and a conveying capacity of 5000 tons per hour has redefined the national standards for belt conveyors in the coal mining industry. An innovative whole-machine drag reduction technology for belt conveyors was invented, featuring lightweight, low-resistance, long-life idler rollers with a resistance as low as 1.1 N. Compared to the industry's other standard idlers, these new rollers have reduced rotational resistance by more than 60%. Considering the resistance from indentation, PVC belt surfaces were selected, and the belt speed was increased from 4 to 4.5 m/s, reducing the overall machine resistance by more than 20%.

8.2.4.3 Innovative Technologies of the Panel

1. Dedicated support transportation roadway design

To address the installation and withdrawal of large equipment, and to avoid the drawbacks of large-section transport and return air roadways, such as maintenance difficulty and increased costs, a dedicated support roadway was innovatively designed in the 12th coal panel district. This roadway facilitates direct access to the panel from the 2nd auxiliary transportation inclined shaft, improving transportation efficiency and providing multiple uses, including as a return air roadway during the final panel extraction.

2. Extra-large cross-section coal roadway excavation technology

The 12,401 panel cut roadway has a width × height of 11.4 m × 6.3 m, with a maximum linear span of 15.6 m. An innovative stepped excavation, timely support, and secondary bottom cutting process ensured safe and efficient excavation.

3. Extra-long single roadway ventilation technology

The dedicated support roadway in the 12th coal panel district has a cross-section of 6.0 m × 5.8 m and a length of 5340 m. To address ventilation needs, a FBDYNO7.5 (2 × 55 kW) variable frequency local ventilator with a 1200 mm flexible duct was used, along with multiple anti-leakage measures.

4. Large cross-section roadway support technology

The designed cross-section of the fully mechanized panel set-up room is 1.4 m wide by 6.3 m high (width × height) with the width increasing up to 14.4 m at the end bracket area. This roadway, excavated in an extra-thick coal seam, is classified as an extra-large section coal drift. At the time, there were no global precedents for the construction techniques or support parameters for such roadways. Shangwan Coal Mine determined reasonable roof and rib support parameters through a combination of theoretical calculations, numerical simulations, and engineering analogies. For roof support, a combination of anchor rods, "W' type steel strips, lead wire mesh, and anchor cables was used, while the ribs were supported by fiberglass anchor rods, plastic netting or steel mesh. The anchor cable model used was T21.6 mm × 800 mm, arranged in a 2 m × 2 m grid. The roof rods were 22 mm × 220 mm threaded steel, installed in sets of 11 with a spacing of 1 m. The main rib utilized 27 mm × 2400 mm fiberglass anchor rods, arranged at intervals of 1.2 m × 1.0 m. The auxiliary rib was supported with 18 mm × 2100 mm round steel anchor rods, installed in sets of 5 with a spacing of 1.2 m × 1.0 m.

5. Extra-large mining height fully mechanized mining technology

The longwall panel employs a high-cut fully mechanized mining technology that extracts the full height in one pass, with the equipment arranged along the roof. The panel inclination length is 289.2 m, with an advance length of 5254.8 m. The designed average coal cutting speed of the shearer is 7 m/min, with a cycle operation time of 57.7 min. A total of 128 hydraulic supports are deployed. To monitor the working resistance of the supports and the amount of roof subsidence in real-time, sputtered thin-film type column pressure sensors with automatic monitoring capabilities and displacement sensors based on the magnetostrictive principle are installed on the support columns, top beams, and bases, respectively. The measurement accuracy is controlled within $\pm$ 1 MPa and $\pm$ 1 mm. This setup provides comprehensive and systematic baseline data for analyzing the mining pressure manifestation characteristics and the interaction between the supports and the surrounding rock in the 8.8 m extra-large mining height panel. Bottom coal is left along the roof during recovery. Core sampling tests on the roof of the auxiliary transport roadway and the return airway of the longwall panel show that the compressive strength of the roof generally exceeds that of the coal. The mining process on the panel has been optimized to leave bottom coal along the roof, enhancing the control of the longwall panel roof.

6. Intelligent control and information technology

By integrating equipment information and intelligent control data, an intelligent control platform and collaborative control center were established. These included fault diagnosis, expert decision centers, big data analysis, video visualization, high-reliability power supply systems, and mine pressure monitoring, achieving the world's first intelligent mining operation for extra-large mining height panels.

8.3 Large Mining Height Technology in High Gas Thick Coal Seam

8.3.1 Large Mining Height with High Gas Coal Seams in Sihe Mine

8.3.1.1 Geological Conditions

Sihe Mine is a primary production mine of the Shanxi Jincheng Anthracite Mining Group Co., Ltd., mainly exploiting the No. 3 coal seam. The coal is black, predominantly bright coal, characterized by a metallic-glass luster, hardness, and brittleness, with some areas of the coal being soft and friable. It generally contains two layers of interbedded rock, with an overburden thickness ranging from 199 to 349 m. The average thickness of the coal seam is 6.6 m, with a dip angle ranging from 1 to 10°. In 2005, the approved production capacity of the Sìhe Mine was 10.8 million tons per annum (Mt/a). The Sìhe Mine comprises two mining areas: the east and the west. The eastern mining area is classified as a high-gas mine, while the western area is known for coal and gas outbursts. The mine's absolute methane emission ratio is 1128.44 m^3/min, with a gas extraction efficiency of 80.67%. Specifically, the eastern mining area has an absolute gas emission of 497.8 m^3/min (with 177.8 m^3/min being vented and 320 m^3/min extracted), achieving a gas extraction efficiency of 64.28% and a relative gas emission of 20.66 m^3/min. The western mining area, on the other hand, has an absolute gas emission of 630.64 m^3/min (with 40.32 m^3/min being vented and 590.32 m^3/min extracted), and a gas extraction efficiency of 93.6% [6]. A comprehensive geological columnar diagram of the Sìhe Mine is shown in Fig. 8.7.

Thickness/m	Column	Lithology	Description
18.40~5.50 14.16		Fine-grained Sandstone	Light gray, fine-grained sandy texture, sub-angular, poorly sorted, pore-filling clay cement, locally basal silica cement, contains some coal lenses, with indistinct cross-bedding.
20.34~5.68 8.89		Siltstone	Light gray, layers appear dark gray, silt-sized grains, clay cement, exhibits horizontal, wavy, and cross-bedding.
3.81~1.99 2.68		Sandy Mudstone	Light gray, sandy-mud texture, flat fracture surfaces.
8.07~2.10 6.54		Fine-grained Sandstone	Light gray, fine-grained sandy texture, well sorted, pore-filling clay cement, contains plant fossils.
1.75~0.52 1.10		Mudstone	Light gray, muddy texture, flat fracture surfaces, contains plant fossils, massive structure.
10.79~7.56 9.26		1-2 Coal Seam	Black, shiny and dull, locally pitchy luster with coarse banded structure, trapezoidal fractures, contains pyrite films and clay inclusions, layered structure.
1.29~0.96 1.10		Mudstone	Light gray, muddy texture, flat fracture surfaces, contains plant fossils, massive structure.
3.80~0.98 1.99		Siltstone	Light gray, sandy-mud texture, clay cement.

Fig. 8.7 Comprehensive Geological Columnar Diagram of the Sihe Coal Mine

8.3.1.2 Layout of the Panel

The 4301 panel at the Sihe Mine is 300 m long, equipped with hydraulic supports from Zhengzhou Coal Mining Machinery Group, Beijing Coal Machinery Company, and Jinding mining machinery and mining Co., Ltd. These supports have a maximum height of 6.0 m and a rated working resistance of 12,000 kN, totaling 173 units. The equipment selection includes a SZZ1200/315 scraper transfer machine and a PCM250 crusher from Zhangjiakou Coal Machinery Factory. A MG400/920-GWD shearer from the Shanghai Coal Science Institute was initially used, later replaced with an SL-500 shearer after the completion of mining. A JOY 3 × 855 kW triple-drive AFC is employed, with a total installed power of 5070 kW.

The layout of the 4301 panel remains the traditional setup of three entries and two exits, spanning five roadways. Mining commenced on October 1, 2009, and concluded on July 30, 2010, with a total advancement of 2353 m and coal production of 6.85 Mt. The panel underwent multiple challenges, including initial mining, passing over subsidence columns, air gaps, steep inclines, faults, thin coal seams, normal recovery, and final mining stages, achieving all expected indicators. main technical specifications for the panel with large mining height of 6.2 m in height and 300 m in length include a record monthly production of 915,000 tons in June 2010, an average daily output of 235.4 Mt, and a peak daily output of 35,232 t. The mining efficiency reached 218 tons per worker shift, with a 23.4-m advance per 10,000 t mined, and an extraction ratio of over 95% on the panel.

8.3.1.3 Methane Management

(1) Successively added gas drainage pump stations at the bottom of Eastern Air Shaft and Xiaodongshan Shaft to expand the drainage area and enhance the gas drainage capacity.
(2) Evolved from single-seam close-range drainage to regional progressive drainage.
(3) The gas drainage pipeline system adopts a dual system combining goaf drainage and pre-drainage at the panel, ensuring the safe enclosure of the goaf.
(4) Employed kilometer directional drilling rigs and domestic series of standard drilling rigs for planned and targeted drilling of gas pre-drainage holes in the coal seam. Up to now, the gas drainage advances at Sihe Mine reaches over 8000 m per day, with a drainage volume of approximately 1.2×10^6 m^3/min.
(5) Increased mining height, reduced coal loss in the goaf, and decreased gas emissions from the goaf.
(6) Effectively implemented multiple measures, ensured exhaustive drainage, and achieved standard effects. The air supply in the large mining height panel area was reduced from about 10,000 m^3/min to currently 6000–8000 m^3/min, significantly improving the working environment at the mining panel. The 300-m panel employs kilometer drills and MK series drills for combined drilling and

drainage, resolving the gas drainage issues of extended large mining height panels.

8.3.1.4 Initial Mining Technique

During the initial mining of the large mining height panel, the method of allowing the roof to collapse on its own was first adopted at the 2301 first panel. Due to the roof supports at the set-up room (start of the panel), such as anchor rods and cables, not being able to collapse in time with the advancement of the supports, a gap was left between the rear of the supports and the protective beams. This gap did not fill in promptly, leading to substantial accumulations of gas. The immediate roof typically collapsed in large areas after being pressured, posing a constant dual threat of roof falls and gas during the initial mining process.

Subsequently, the initial mining technique at the 4301 panel was optimized to the way as follows: During the installation of supports, the cable trays that supported the roof at the set-up room were removed one by one, and after the installation of supports, controlled blasting was employed using blast holes to fracture the coal in the roof at the set-up room. This method involved drilling two blast holes per support set, one at the front and one at the back, with each hole having a depth of 2.0 m. Each blast hole was loaded with seven cartridges. This approach effectively damaged the coal body in the roof at the set-up room, facilitating its timely collapse. The conditions of the roof and gas significantly improved and ensured the continuous advancement of the panel during the initial mining period. The initial mining duration was also reduced from the original 7–10 to 5–7 days.

8.3.1.5 Roof Control

Roof control at the 4301 large mining height panel is a crucial component of onsite management, which is strengthened in the following ways:

(1) Intensify the management of initial support force of the supports, ensuring it is no less than 24 MPa. Fully utilize the capabilities of the telescopic top beams and the first and second level guard board on the hydraulic supports, strictly control the end distance of the beams, execute advance and follow-up to enhance the timeliness of the support provided by the hydraulic supports.
(2) Strengthen the pressure management of the emulsion pump station at the panel, ensuring it is not less than 30 MPa.
(3) Specifically for the challenging aspect of roof control in the coal wall's triangular area, strengthen the coal mass in advance with Ruimi material to enhance the intrinsic support strength of the coal body.
(4) Increase the range and intensity of advanced support; for roadways affected by dynamic pressure, shed framing support is implemented to maintain structures 100 m ahead with supports, ensuring the support strength is no less than that of one beam with four pillars. By implementing the above measures, no roof

fall accidents occurred during the retreat mining process at the panel, achieving outstanding results in roof control at this mine's large mining height panel.

8.3.2 Large Mining Height with High Gas Seam in Jiaoping Mining Area

8.3.2.1 Panel Overview

The 2301 panel of the Xiashijie Coal Mine is the first mining panel for the 3^{-2} coal seam. The 3^{-2} coal seam is positioned in the middle part of the Yan'an Formation, above the 4^{-2} coal seam, with a separation distance of approximately 0–30 m. This seam typically appears as part of a coal group, with coal being visible in most areas of the minefield except for a few isolated points. The overall structure of the panel is an anticline, with the axis located towards the northwest of the center along the panel. The general dip angle of the coal seam ranges from 3° to 8°, averaging 5°, with parts nearly horizontal. The coal seam contains 2 or 3 layers of silty mudstone and mudstone with gangue, and can include up to 7 layers, making the structure of the coal seam quite complex. The strike length of the panel is 1720 m, with a dip length of 210 m, and the coal seam thickness ranges from 4.75 to 6.29 m, averaging 5.53 m. Due to factors such as gas, the designed production capacity of the panel is 1.7 million tons per year.

8.3.2.2 Equipment Suite for the 2301 Large Mining Height Panel

The equipment of 2301 panel is as follows:

ZY10500/27/58 type shield hydraulic support with support height 2700–5800 mm, center distance 1750 mm, working resistance 10,500 kN (P = 41.8 MPa), support strength 1.18–1.22 MPa, movement interval 800 mm; each hydraulic support is manually operated.

MG750/1860-GWD type electric traction shearer with production capacity 4200 t/h, mining height range 3–5.5 m, total installed power 1860 kW, supply voltage 3300 V, drum diameter 2800 mm, cutting depth 800 mm, features onboard AC variable frequency drive for traction, wireless remote control, control station wired control, onboard control, AC variable speed drive, constant power control for the cutting motor, temperature protection, and short circuit, overcurrent, and overload protection for the traction system, dust suppression internal and external spraying.

SGZ1000/1400 type AFC with conveying capacity 2500 t/h, installed power 2 × 700 kW, supply voltage 3300 V, scraper chain middle double chain (compact chain) Φ42 mm × 146 mm, unloading method end unloading.

SZZ1000/400 type transfer machine with conveying capacity 2500 t/h, motor power 400 kW, voltage 3300 V, scraper chain middle double chain, ring specification Φ34 mm × 126 mm. PCM200 type crusher with throughput capacity 2200 t/h, motor power 250 kW, voltage 3300 V. BRW400/31.5 emulsion pump station with flow ratio 400 L/min, pressure 31.5 MPa.

8.3.2.3 Industrial Test of the First Mining Panel with Large Mining Height

In the Xiashijie Coal Mine, the diagonal cutting in section of the 2301 panel is 55 m long, including a 16.27-m-long shearer body and an average curved section length of 22.46 m for the AFC. The normal coal-cutting segment for the shearer spans 155 m. The shearer maintains similar speeds in both upward and downward cutting directions, with cycles approximately the same duration, averaging 94 min per cut, resulting in about 188 min to complete a bi-directional cutting cycle. The periodic weighting length ranges from 10 to 26 m, averaging 17.4 m. The dynamic load factor during the periodic weighting ranges from 1.03 to 1.93, averaging 1.37, indicating significant pressure. The average working resistance of the supports is 4824 kN, which is 45.9% of the rated resistance of 10,500 kN, indicating a substantial margin in support resistance.

The 2301 panel utilizes a "U + I" ventilation system with a strategy of one roadway and two exits, where air is supplied through the transport roadway and returns through the return and high-level gas drainage roadway. The airflow to the panel is stable between 1500 and 1700 m^3/min, with methane concentrations in the return air ranging from 0.25 to 0.35%, averaging 0.3%. There have been no methane over-limit alarms at the upper corner or panel, thus not impacting the equipment utilization ratio at the panel. The methane concentration in the high-level gas drainage roadway averages 5.5%. When the advance distance of the panel equals twice its length, the total methane drainage amounts to 29.2 m^3/min, with the high-level gas drainage roadway contributing as much as 17.4 m^3/min, accounting for 60% of the total drainage.

In the 2301 panel, three sets of temperature probes and four sets of gas sampling bundles are arranged. Based on the actual measurement data and numerical simulation analysis results, the distribution characteristics of the "three zones" in the goaf area are depicted in Fig. 8.8. On the side of the return roadway, the zones are distributed as follows: 0–25 m is the cooling zone, 25–80 m is the oxidation zone, and beyond 80 m lies the suffocation zone. On the transportation roadway side, 0–35 m is the cooling zone, 35–95 m is the oxidation zone, and beyond 95 m is the suffocation zone.

From the division of the "three zones", it is known that the sum of the maximum widths of the measured cooling and oxidation zones on both sides of the panel is 95 m. The shortest natural ignition period for the 3^{-2} coal seam is 28 d. After calculations, the lowest safe advancement speed for the panel is determined to be 3.39 m/d. The monthly advancement of the 2301 panel ranges from 100 to 135 m, with an average of 119 m/month. The average advancement speed of panel, 4 m/d, is greater than the lowest safe advancement speed, therefore the residual coal in the oxidation zone remains in a low-temperature oxidation stage and has not entered the accelerated

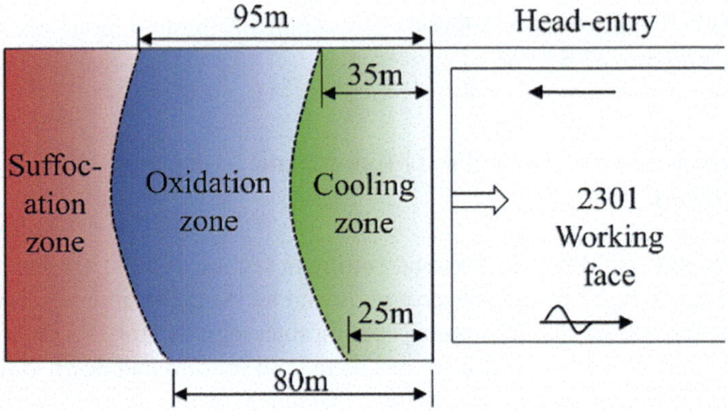

Fig. 8.8 Division of the "three zones" in the goaf area of the panel

oxidation stage before moving into the goaf area. During the observation period, no acetylene (C_2H_2) was detected, and the concentration of carbon monoxide (CO) was also low, indicating that no spontaneous combustion occurred in the goaf area.

8.3.3 Large Mining Height with High Gas Seams in Dongqu Coal Mine

Shanxi Coking Coal Group Co., Ltd. Dongqu Mine has a designed production capacity of 4 million tons per year, employing flat-entry development and inclined longwall mining methods. The primary coal seams mined are the #2, #4, #8, and #9 seams from the Taiyuan Group and Shanxi Formation, consisting of coking coal, lean coal, and basin coal, with a total mineable coal seam thickness of 13.4 m. With the rapid development of the mine, rational planning of the mining field, and significant increases in production, Dongqu Mine has evolved from the original single-level mining at + 973 m to dual-level mining, concurrently mining at the + 860 m level.

The 28,202 panel at the + 860 m level is Dongqu Mine's first large mining height panel with high gas coal seam. Due to its proximity to the Tunlan Mine and the high amount of gas present, the absolute gas emission was estimated to be 45 m^3/min before mining began. Although borehole gas extraction was employed, solely relying on ventilation and borehole extraction from this coal seam was not sufficient to achieve thorough gas control due to factors such as drilling and sealing technique. To address these challenges, Dongqu Mine, in line with the actual conditions of the mine, implemented comprehensive methane control measures to manage gas effectively on the panel, solve issues of gas exceedance, eliminate safety hazards, and achieve high productivity and efficiency. These measures included a comprehensive gas control technology that primarily utilizes high and low-level gas drainage roadways, pipelines buried in the coal seam, upper corners, and the goaf, supplemented by ventilation. This method successfully enables panels with high gas levels to operate in a low-gas state. This approach marked a shift from preventing gas accidents to preventing gas exceedance, ensuring the safe production of the coal mine.

8.3.3.1 Panel Overview

The coal seam structure at panel 28,202 is complex but stable. The thickness of coal seam #8 ranges from 2.2 to 3.8 m, with an average thickness of 3.3 m and a bulk density of 1.37 t/m^3. The seam includes one to three layers of partings; the upper coal of coal seam #8 has a thickness ranging from 0.3 to 0.8 m, with an average of 0.6 m. At some locations, coal seam #8 bifurcates from its upper coal, with parting thicknesses ranging from 0.0 to 1.5 m, averaging 0.42 m. The entire thickness of coal seam #8 and its upper coal, which ranges from 3.3 to 5.3 m with an average of 4.3 m, is mined in one operation at this panel using a fully mechanized inclined longwall retreat mining method, and the goaf is handled using the full caving method. The panel consists of a belt roadway, a track roadway, and set-up room for the panel. The belt roadway is 1300 m long and the track roadway is 1200 m long; both have a rectangular cross-section, with a clear width of 4.5 m and a clear height of 3.5 m. The set-up room is 206 m long, with a clear width of 9.0 m and a clear height of 3.5 m, supported by anchor rods, anchor cables, and steel rib arches. A "U" type ventilation system is used during the recovery process, with the belt roadway serving as the return airway and the track roadway as the intake airway.

8.3.3.2 Comprehensive Methane Control Technology for the Panel

For the panel 28,202, the absolute gas emission is 40 m^3/min. A combination of ventilation and gas extraction methods is employed for control. Specifically, the amount of gas removed through ventilation is 16 m^3/min, and the amount extracted is 24 m^3/min. The techniques utilized include high-level and low-level gas drainage roadway as well as gas extraction from the upper corners and the goaf areas in this coal seam.

High-level gas drainage roadway position: For panel 28,202, the high drainage gallery is constructed along the roof of the No. 7 coal seam. The design places the first 570 m parallel to the belt entry, offset by 25 m, followed by a 0.5° counterclockwise deviation for the subsequent 750 m. The total length of the project is 1320 m, and the cross-section of the roadway is semi-arch shaped. The layout is divided into three levels. At 23–35 m from the panel, 450 m are arranged; at 35–40 m from the panel, 475 m are arranged; and at 40–45 m from the panel, 165 m are arranged. The final segment extends from 45 m down to 30 m from the panel, covering 230 m.

Low-level gas drainage roadway position: Generally, mining adopts a top-down approach. When advancing the upper coal seam return airway, gas emission phenomena often occur at the floor of these roadways. To address the issue of gas emissions from the floor at the panel, Dongqu Mine conducts gas extraction from the depressurized No. 9 coal seam below. A low-level gas drainage roadway is constructed within the No. 9 coal seam 8.9 m below panel 28,202, extending along the length of the panel. It is situated between the belt roadway of panel 28,202 and the track roadway of panel 28,204, protecting the coal pillars (offset by 12 m from the belt roadway of 28,202). The roadway features a trapezoidal cross-section with an area of 12.09 m^2 and is designed to be 1225 m long. Additionally, to enhance the effectiveness of the gas extraction in the low-level gas drainage roadway, extraction boreholes are drilled along both sides of the roadway, spacing each parallel borehole 5 m apart, with a bore diameter of 113 mm. A total of 372 boreholes are drilled, with an average depth of 190 m.

As the panel progresses, the gas extraction volume from the high-level gas drainage roadway increases gradually. When the panel reaches 529 m, the extraction volume peaks at 25.91 m^3/min, and by 679 m, it stabilizes around 16 m^3/min. Throughout the recovery period, the average extraction volume is 15.62 m^3/min, making up 56.1% of the total average extraction rate. Initially, the gas concentration during extraction is high, however, after a week, it decreases to 3.5%. As the panel advances, the concentration gradually increases again, reaching a high of 13% and averaging 8.04%. After the panel has advanced 309 m, the concentration stabilizes around 10%. The high-level gas drainage roadway is noted for its stable and reliable performance, providing substantial extraction volumes and effectively managing return air and upper corner gas. This prevents excess gas concentrations at the panel, ensuring smooth recovery operations.

The low-level gas drainage roadway sees concentrations ranging from 20 to 55%, with an average of 34.22%. The extraction volume there varies between 4.38 and 23.26 m^3/min, averaging 8.76 m^3/min, which represents 31.4% of the total average extraction volume. This roadway significantly enhances the extraction effectiveness for gas in the adjacent lower layers.

Practice has shown that at Dongqu Mine's panel 28,202, the adoption of a comprehensive gas control strategy, which primarily utilizes high- level and low-level gas drainage roadways and is supplemented by ventilation, has eliminated gas as a potential safety hazard. This approach has fully leveraged the potential of large mining height panels, allowing for rapid recovery under conditions of large mining heights. It ensures high productivity and efficiency at the panel and prevents incidents related to gas exceedance.

8.4 Large Mining Height in Large Dip Angle Thick Coal Seam

8.4.1 Equipment Selection for Panels

Yangchangwan Coal Mine is the first large-scale mine with a capacity of 10 million tons per annum to be established and operational in the Ningdong mining area. The minefield extends 6.4 km in strike length and 12.8 km in dip width, covering an area of 51.6 km^2. The total geological reserves of the minefield amount to 911.79 million tons, with designed utilizable reserves of 628.93 million tons and designed extractable reserves of 455.38 million tons. The No. 2 coal seam is the principal seam mined at Yangchangwan, holding geological reserves of 370.21 million tons and extractable reserves of 171.25 million tons, accounting for 40.6% and 37.6% of the total geological and extractable reserves of the mine, respectively. The thickness of the No. 2 seam ranges from 4.42 to 10.15 m, with an average of 7.2 m. Specifically, the large mining height panel Y110206 has an average coal thickness of 6.2 m.

The first large mining height panel with a height of 6.2 m at Yangchangwan Coal Mine is arranged at the panel Y110206 located on the southern wing of the Yangchangwan minefield. The upper set-up room of the panel is located 30 m north of the 56# exploration line, and the lower set-up room is situated 64 m north of the same line. The false roof of the Y110206 panel consists of approximately 0.3 m of mudstone, which is laminated, soft, and prone to spalling. The immediate roof is composed of two layers of grayish-white to dark gray fine sandstone and siltstone, with a thickness of about 4.6 m, incorporating two coal seams approximately 0.15 m thick in the upper-middle part, which are well-layered and easily fractured.

The complex production geological conditions at the large mining height panel of Yangchangwan Coal Mine impose higher requirements on the design, manufacturing, and normal use of supports. Based on the mining conditions of the No. 2 coal seam, after comprehensive demonstration and research, and considering factors such as the stability of the hydraulic supports with large heights, the telescopic ratio of the columns, and the actual underground transport conditions, it was ultimately decided to use a two-column shield-type large mining height hydraulic support. The maximum support height for the these supports should be no less than 6.2 m. As for the reasonable working resistance of the supports, based on the research results of rock layer control theory, the panel roof pressure intensity is approximately calculated as 4–8 times the weight of the overlying stratas above the mining height, giving a theoretical support strength of 0.96 MPa. The effective support width of the supports is calculated at 1750 mm, and the effective roof spacing is calculated at 5000 mm, hence the working resistance of the supports is calculated as 8400 kN. The selection and parameter determination of hydraulic supports for large mining height panels under complex conditions also need to consider the need to improve coal wall stability. Taking all factors into account, the final determined working resistance of the supports is 10,000 kN. To effectively prevent instability of the coal wall and leakage at the roof, the supports are specially designed with a five-drawer structure

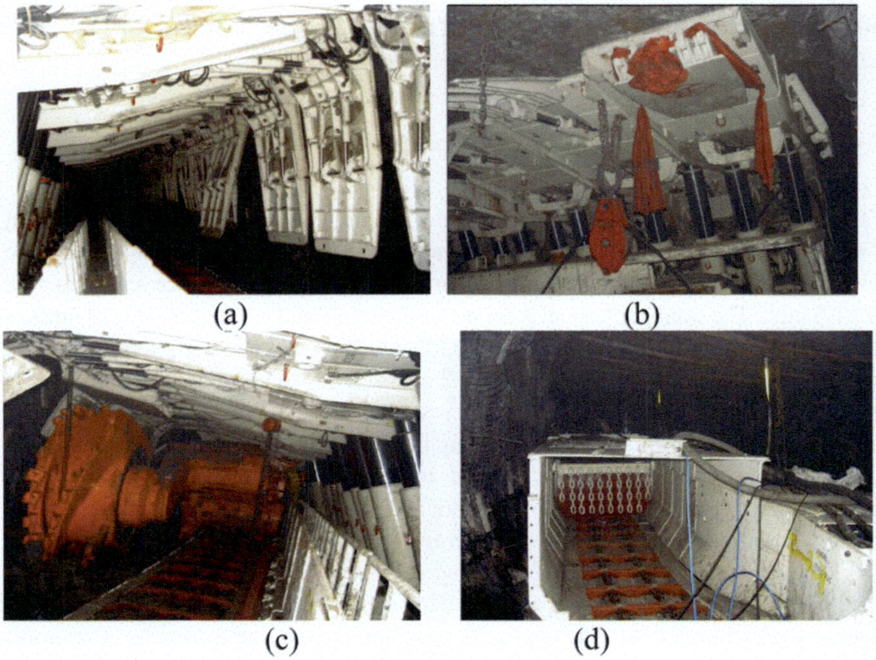

Fig. 8.9 Main mining equipment in the Y110206 large mining height panel: **a** Hydraulic support; **b** end support; **c** shearer and AFC; **d** transfer machine and crusher

for the telescopic beams, enhancing the strength of the telescopic and top beams, increasing the travel of the telescopic beams, and achieving timely roof protection. The effective application of rib protection device can significantly reduce the likelihood of instability in the large mining height panel. In the design of the support selection, two-level rib protection device technology is employed, with rib protection device 1.6 m wide and lengthened to 2.7 m, covering an area of 4.32 m^2.

For the large mining height panel Y110206 at Yangchangwan Coal Mine, it was ultimately decided to use equipment including the ZY10000/28/62D hydraulic support, the MG900/2245-GWD shearer, and the SGZ1250/2565 AFC [7]. This equipment setup is illustrated in Fig. 8.9, with the complete set of equipment and detailed parameters in Tables 8.12, 8.13, 8.14 and 8.15.

8.4.2 Safety Assurance Technology for Large Mining Height Panels

In the application of large mining height technology at the Yangchangwan Coal Mine's Y110206 panel, several major issues have been encountered, including rib spalling, spontaneous combustion and fire risks in the coal seam, support collapses,

Table 8.12 Complete equipment set for large mining height panel Y110206

Equipment name	Manufacturer	Model number
Electric traction double drum shearer	Tiandi Science & Technology Co. Ltd. Shanghai Branch	MG900/2245-GWD
Two-post double telescopic shield intermediate support	Zhengzhou Coal Mining Machinery Co., Ltd., Pingdingshan Coal Mine Machinery Co., Ltd. and Beijing Coal Mining Machinery Company	ZY10000/28/62D
Two-post double telescopic shield transition support	Beijing Coal Mining Machinery Company	ZYG10000/26/55D
Two-post double telescopic shield end support	Zhengzhou Coal Mining Machinery Co., Ltd	ZYT10000/26/55D
AFC	Ningxia Tiandi Benniu Industrial Group Co., Ltd	SGZ1250/2565
Transfer machine or reclaimer	Ningxia Tiandi Benniu Industrial Group Co., Ltd	SZZ1350/525
Crusher	Ningxia Tiandi Benniu Industrial Group Co., Ltd	PLM4500
Self-advancing tail device	Ningxia Tiandi Benniu Industrial Group Co., Ltd	DY1600

Table 8.13 Main technical specifications of the ZY10000/28/62D hydraulic support

Item	Parameter	Item	Parameter
Support width (m)	1.65–1.85	Support strength (MPa)	1.03–1.06
Working resistance (kN)	10,000	Top beam length (mm)	4430
Initial support force (kN)	7912	Push force (kN)	504
Working height (mm)	2800–6200	Pull force (kN)	989
Support area (m^2)	9.5	Floor pressure ratio (MPa)	1.90–2.18
Support structure weight (t)	41.0	Support center distance (m)	1.75
Transport dimensions (mm)	8230 × 1650 × 2800	Moving frame step distance (mm)	865

and the AFC's tendency to creep upwards or slip downwards. These issues have significantly impacted the cutting speed of the shearer and have adversely affected the performance of the integrated mining technology.

8.4.2.1 Coal Wall Stabilization and Control Technology

Changes in the dip angle of the panel significantly impact the stability of the coal wall. Given the challenging conditions of large undulations along the strike at Yangchangwan Coal Mine's Y110206 panel, a zonal coal wall instability prevention and control technology under complex conditions is proposed. When the panel

Table 8.14 Main technical specifications of the MG900/2245-GWD Shearer

Item	Parameter	Item	Parameter
Weight (t)	144	Cutting depth (m)	0.865
Adaptable mining angle (°)	≤ 25	Service life (MT)	≥ 20
Installed power (kW)	2245	Major overhaul cycle (MT)	6
Mining height range (m)	2.7 ~ 6.1	Rated voltage (V)	330
Maximum production capacity (t/h)	4500	Maximum traction speed (m/min)	24.4
Maximum traction force (kN)	1104	Cooling method	Internal and external spray
Cooling water pressure (MPa)	10	Cooling water flow ratio (L/min)	400

Table 8.15 Main technical specifications of the SGZ1250/2565 scraper conveyer

Item	Parameter	Parameter	Parameter
Transport capacity (t/h)	5000	Laying length (m)	320
Total installed power (kW)	2565	Service life (MT)	14
Unloading height (mm)	1600	Rated voltage (V)	3300

is mined horizontally or on a downward incline, the coal wall stability is relatively better, and conventional wall instability prevention techniques are employed. However, when mining on an upward incline, where coal wall stability is comparatively worse, it is necessary to employ advanced grouting to reinforce the coal body in addition to conventional techniques, to effectively control the stability of the coal wall.

8.4.2.2 Anti-tipping and Anti-sliding Technology for Panel Supports

For the mining of large dip angel thick coal seams with large mining heights, Y110206 implements the following five measures to prevent the collapse of the support on the panel:

(1) Strict control over the mining height of the panel is enforced, and the use of supports beyond their specified supporting height is strictly prohibited.
(2) Supports must be raised evenly and tightly, ensuring solid contact with both the roof and the floor.
(3) Side guard boards and hydraulic jacks should be used to timely adjust the spacing between supports to prevent them from being too far apart, which could lead to collapses.

(4) In cases where a soft floor causes a support to collapse, plate girders or sleepers may be laid in front of the support base during setup, with the length depending on the specific conditions.
(5) Misalignments in the supports must be promptly corrected to prevent the height difference between adjacent supports from exceeding two-thirds of the side guard board's height.

8.4.2.3 Control for the Upward Movement and Downward Sliding of AFC

Due to the fast advancing speed of the panel, the phenomenon of the AFC upward movement and downward sliding will persist throughout the production process of the entire panel. If not controlled properly, the AFC may move upward and downward, leading to blockages or insufficient clearance of the safety exits. In response to this situation, the following measures have been implemented at the Y110206 panel:

(1) Equip hydraulic supports with lateral shifting and lifting devices. In inclined or steeply inclined panels, relying solely on the movement of side guard boards to adjust the spacing between supports is inadequate for controlling the supports' upward movement and downward sliding, especially in scenarios with significant incline angles and heavy supports. In these instances, the force exerted by the support side shields is inadequate for adjusting the support spacing. Consequently, it is essential to incorporate lateral shifting and lifting devices to regulate the supports' upward movement and downward sliding effectively, thereby managing the AFC's upward and downward motions.
(2) Implement pseudo-tilted mining. Extend the pseudo-tilted length of the panel to reduce the slope angle, thereby controlling the upward or downward movement of the AFC. Typically, in fully mechanized panels, the pseudo-tilt angle is adjusted based on experience within the range of 2–6° (10–15 m). However, with variations in roadway width, the empirical pseudo-tilted length may not be applicable to fully mechanized panels, necessitating real-time adjustments of the pseudo-tilted length with changes in roadway width. This adjustment presents challenges for managing the lower end of the panel, emphasizing the need for proactive prediction and proactive control measures to address variations in roadway width.
(3) Adjust the direction of shifting for the AFC. During the shifting process of the AFC on the panel, changes in the shifting direction can lead to a tendency for the conveyor to move upward and downward in the conveyor's curved sections. Therefore, based on the actual circumstances, ascending or descending shifting is employed to control the conveyor's sliding or jumping. When the AFC's upward movements and downward sliding are within reasonable limits, minor adjustments to control the upward movements and downward sliding are made using an alternating upward and downward shifting method.

8.4.3 Mining Situation at the Panel

Field mining practices demonstrate that the overall performance and technological level of the large mining height equipment at the Y110206 panel are well suited to the complex mining conditions at the Yangchangwan Coal Mine. They meet the safety and rapid advancement requirements of the fully mechanized panel, achieving safe, high-yield, and efficient production. At the Y110206 large mining height panel of the Yangchangwan Coal Mine, the shearer startup ratio ranges from 74 to 92.6%, with a coal recovery ratio of 93.75%. The panel typically mines coal with 12–15 cuts per day, reaching up to 18 cuts, yielding 25,000–30,000 tons of coal per day, with a peak daily production of 37,000 tons. The monthly average coal output at the panel exceeds 800,000 tons, peaking at 860,000 tons, and the annual production target of 10 million tons at the panel has been achieved.

Field mine pressure observations indicate that the overall load distribution of the supports at the Y110206 panel follows a normal distribution. The average initial support force of the supports is 6656 kN, which is 84% of the rated initial support force, and the average working resistance is 6983 kN per set, accounting for 69.9% of the rated working resistance. During the first weighting of immediate roof, the average working resistance of the supports is 7058 kN, with a maximum of 9092 kN. During the periodic weighting of the main roof, the average working resistance of the supports is 7786 kN, reaching a maximum of 9326 kN. In the periodic weighting phase of the main roof, the average working resistance of the supports is 8590 kN, with a maximum of 9497 kN. The ZY10000/28/62D hydraulic support system has performed effectively, ensuring the stability of the panel roof and creating favorable conditions for enhancing the stability of the coal wall at the panel.

The Yangchangwan Coal Mine pioneered the domestication of a complete set of large mining height equipment for complexly buried conditions with a 6.2 m thick coal seam with a large dip angle. It successfully addressed prominent issues such as stability control of supports, instability of the coal panel, roof leakage, instability of the shearer, and sliding of the AFC. This achievement led to a successful trial of the 6.2 m-mining-height fully mechanized mining technique in large dip angle thick coal seams under complex conditions. It established key domestication technologies for a complete set of equipment for 6.2 m-mining-heightfully mechanized mining in large dip angle thick coal seams. Equipped with the first set of domestically produced equipment for the mining of 6.2 m-mining-height flammable coal seams in large dip angle extra-thick coal seams globally, this endeavor exemplifies successful experiences in the domestication and heavy-duty exploration of fully mechanized equipment for large dip angle extra-thick coal seams with large mining heights.

Based on previous mining experiences at various mines, in 2020, the intelligent fully mechanized mining at the 111,801 panel of the Meihuajing Coal Mine in the

Fig. 8.10 Layout of the large mining height 111,801 Panel

Ningdong Mining Area proceeded smoothly, as shown in Fig. 8.10. In the Ningxia Coal Industry, the first thick coal seam with a large dip angle (average angle of 26°) showcased successful automation of the fully mechanized panel under complex conditions, achieving a 100% accuracy ratio in shifting hydraulic support along with the shearer, leading to a 20% reduction in manpower through automation.

References

1. Luo W (2023) Major scientific and technological innovation achievements and practices of CHN Shendong coal group. Coal Sci Technol 51(2):1–43
2. Yang JZ (2017) Study on overlying strata breakage and strata behaviors law of 7.0 m mining height working face. Coal Sci Technol 45(8):1–7
3. Yang JZ (2017) Research on key mining technology of fully-mechanized working face with 8 m large mining height. Coal Sci Technol 45(11):9–14
4. Yang JZ, Liu QJ (2020) Analysis and measured of strata behavior law and mechanism of 8.8 m ultra-high mining height working face. Coal Sci Technol 48(1):69–74
5. Xu G, Zhang Z, Yang JZ et al (2022) Interaction between support and surrounding rock in 8.8 m super mining height working face. J China Coal Soc 47(4):1462–1472
6. An Z (2016) Study on characteristics of coal wall spalling in high cutting fully mechanized working face in Sihe coal mine. Coal Eng 48(8):54–56+60
7. Liu XM (2012) Determination of fully mechanized equipments and application to large mining height face of Yangchangwan Coal Mine. Xi'an University of Science and Technology

The manufacturer's authorised representative in the EU is Springer Nature Customer Service Centre GmbH, Europaplatz 3, 69115 Heidelberg, Germany. If you have any concerns regarding our products, please contact ProductSafety@springernature.com

Printed and bound by CPI Group (UK) Ltd, Croydon, CR0 4YY
01/10/2026
02225926-0018